595369

614. 8 JEN

WITHDRAWN FROM LIBRARY STOCK

D1342084

Risk-Reduction Methods for Occupational Safety and Health

Risk-Reduction Methods for Occupational Safety and Health

Roger C. Jensen
Montana Tech of the University of Montana

A JOHN WILEY & SONS, INC., PUBLICATION

Copyright © 2012 by John Wiley & Sons, Inc. All rights reserved

Published by John Wiley & Sons, Inc., Hoboken, New Jersey

Published simultaneously in Canada

No part of this publication may be reproduced, stored in a retrieval system, or transmitted in any form or by any means, electronic, mechanical, photocopying, recording, scanning, or otherwise, except as permitted under Section 107 or 108 of the 1976 United States Copyright Act, without either the prior written permission of the Publisher, or authorization through payment of the appropriate per-copy fee to the Copyright Clearance Center, Inc., 222 Rosewood Drive, Danvers, MA 01923, (978) 750-8400, fax (978) 750-4470, or on the web at www.copyright.com. Requests to the Publisher for permission should be addressed to the Permissions Department, John Wiley & Sons, Inc., 111 River Street, Hoboken, NJ 07030, (201) 748-6011, fax (201) 748-6008, or online at http://www.wiley.com/go/permission.

Limit of Liability/Disclaimer of Warranty: While the publisher and author have used their best efforts in preparing this book, they make no representations or warranties with respect to the accuracy or completeness of the contents of this book and specifically disclaim any implied warranties of merchantability or fitness for a particular purpose. No warranty may be created or extended by sales representatives or written sales materials. The advice and strategies contained herein may not be suitable for your situation. You should consult with a professional where appropriate. Neither the publisher nor author shall be liable for any loss of profit or any other commercial damages, including but not limited to special, incidental, consequential, or other damages.

For general information on our other products and services or for technical support, please contact our Customer Care Department within the United States at (800) 762-2974, outside the United States at (317) 572-3993 or fax (317) 572-4002.

Wiley also publishes its books in a variety of electronic formats. Some content that appears in print may not be available in electronic formats. For more information about Wiley products, visit our web site at www.wiley.com.

Library of Congress Cataloging-in-Publication Data:

Jensen, Roger C., 1945-
   Risk-reduction methods for occupational safety and health / Roger C. Jensen.
       p. cm.
   Includes bibliographical references and index.
   ISBN 978-0-470-88141-5 (cloth)
   1. Industrial safety. 2. System safety. 3. Risk management. 4. Industrial hygiene. I. Title.
   T55.J46 2012
   363.11′6–dc23   2011039278

Printed in the United States of America

ISBN: 9780470881415

10 9 8 7 6 5 4 3 2 1

Contents

27.   Violent Actions of People

Preface

My initial motive for developing this book was to provide a textbook for the next generation of Occupational Safety and Health (OSH) professionals to learn systematic methods that will help them throughout their career. As work progressed, I came to realize the book's content could also be useful to anyone with OSH responsibilities in companies currently building or upgrading their OSH programs, including many companies located in later developing countries such as India and China.

Most of the material in the book presents well-established methods and practices familiar to professionals with broad backgrounds that include industrial hygiene, occupational safety, and occupational ergonomics—three specialties I collectively refer to as OSH. I included a few innovations that I think OSH professional and students will find useful, particularly in the rigor used to define fundamental terms, the modeling method for analyzing incidents, and the consistent use of nine risk-reduction strategies.

I wrote this book for three types of readers. For students preparing for a career in OSH, the book can help them learn to approach OSH in a systematic manner. For people with OSH responsibilities in companies going through the process of upgrading their OSH programs, this book can serve as a resource to help pull together the program components needed to meet the basic safety and health needs of their employees. For OSH professionals who know everything about safety, health, and environment in the industrial sector where they work, reading this book may help them see how the OSH practices used in their industry are instances of the practices used in many sectors.

I organized the 27 chapters into five parts: (I) background, (II) analysis methods, (III) programmatic methods for managing risk, (IV) risk reduction for energy sources, and (V) risk reduction for other than energy sources. Part I provides general background for appreciating the later chapters and clarifies the fundamental terms hazard, risk, and risk reduction. Part II describes some system safety tools OSH processionals should know. Part III describes some common components of OSH programs and synthesizes all risk-reduction tactics into nine risk-reduction strategies used extensively in subsequent parts of the book. Part IV contains chapters on the hazard sources involving energy exchange—kinetic energy, electrical energy, acoustic energy, thermal energy, fires, explosions, pressure, electromagnetic energy, and severe weather and geological events. Part V addresses hazard sources other than energy sources— hazardous conditions found in workplaces, chemical substances, biological agents, musculoskeletal stressors, and the violent actions of people. Each chapter points out applications of how known practices fit within the nine risk-reduction strategies.

My thinking is that this book differs from other books on OSH and system safety in three primary ways. First, unlike other books, this one uses a deductive approach—starting with fundamental definitions and nine risk-reduction strategies, the book demonstrates that thousands of the hazard control measures familiar to OSH professionals are instances of these strategies. Second, the book takes an international approach by not treating any particular set of regulations, directives, or standards as authoritative.

A third unique feature is treating as one field the presently distinguished specialties of occupational safety, industrial hygiene, and ergonomics. During my long career working in OSH, I have witnessed a trend in which each specialty develops its own identity, holds its own conferences, has its own journals, and operates its own professional credentialing program. I would like to see this silo-building trend replaced by a shift toward unifying into one overarching field. Although I do not use this book to expressly advocate for this position, I hope that reading the entire book will convince some readers to share my viewpoint.

It is my hope that professors who teach a course in system safety will adopt this book. It is appropriate for undergraduate seniors or graduate students who have previously completed introductory courses on OSH topics. Studying this book should provide students enough system safety expertise for an OSH career and help them appreciate how the material learned in prior courses fits into a cohesive package. Students will find the book more relevant to OSH than books written for system safety professionals, and doing the end-of-chapter exercises (as explained in chapter 1) will help them improve their cognitive abilities for application, analysis, synthesis, and evaluation.

ROGER C. JENSEN

Acknowledgments

I am grateful for the very helpful editing provided by Terri Porter. I greatly appreciate the technical and editorial feedback on an early draft used as the text for an online course. Individual students who contributed most were Thomas Brown, Tyler Fortunati, Britani Laughery, Adam House, and Allan Torng.

I am fortunate to have a wonderful wife, Marian, who put up with my extended hours working on this book project and too few hours spent sharing the fun things in life. Our daughter Lea Jensen, an industrial hygienist, provided inspiration for this extended undertaking. I owe much to another relative, my cousin Jeffrey Lee, for making me aware of the career opportunities in occupational safety and health.

During my career, I have been fortunate to associate with some high-quality people who enabled me to grow professionally. While working 22 years with the National Institute for Occupational Safety and Health, I had some supervisors and mentors I want to acknowledge. Francis Dukes-Dobos and Austin Henschel introduced me to ergonomics. James Oppold, as Director of the Division of Safety Research, allowed me to develop new lines of research on the topics of safety training effectiveness, symbols on warning signs, stairway falls, machine safeguarding, and effects of climatic factors on safety-related behavior. Some of the associates I worked most closely with were John Etherton, Timothy Pizzatella, James McGlothlin, Patrick Coleman, and James Collins. After my NIOSH experiences, I joined a technical services company where I had the privilege of working with John Howard, Julian Christensen, Bill Askren, and Gary Williamson doing diverse projects on product safety, occupational safety, and ergonomics. Since 1999 I have been on the faculty of Montana Tech of the University of Montana, where I have enjoyed working with some wonderful and supportive associates in the Safety, Health, and Industrial Hygiene Department.

The author was partially supported by Grant No. OH008630 from CDC/NIOSH. The contents are solely the responsibility of the author and do not necessarily represent the official views on NIOSH.

Part I

Background

Part I lays the foundation for the entire book. Chapter 1 explains the multidisciplinary perspective used throughout—a perspective built on traditional occupational safety and health (OSH), enhanced by contributions from system safety, public health, and educational psychology. Chapter 2 delves into definitions of three terms used extensively in this book—hazard, risk, and risk reduction. Chapter 3 provides examples of common types of conceptual models and charting methods used in the book and the safety and health professions.

These background topics are fundamental building blocks for the four subsequent parts of the book that provide the content applicable to the practice of occupational safety and health. Part II explains several practical systematic methods for anticipating hazards, assessing risks, and analyzing systems encountered in occupational settings. Part III discusses programmatic and managerial methods for reducing risks. Part IV gets into the technical aspects of reducing risks associated with various forms of energy. Finally, part V addresses risk reduction for occupational hazards not directly linked to energy.

Chapter 1

Multidisciplinary Perspective

Throughout this book, the field of OSH is viewed broadly to include traditional occupational safety, industrial hygiene, occupational ergonomics, and, to a lesser extent, environmental pollution. To make the book internationally applicable, governmental regulations of the United States and other countries are rarely mentioned. All mathematics uses international units. In this and other chapters, italic font is used for titles of books and journals, and for the first use of technical terms defined at the end of the chapter.

Much of part I is based on information covered in traditional OSH books and journal articles. Concepts and methods from three other fields—system safety, public health, and education—are used to enrich and expand the basic OSH concepts and methods described in this book. Contributions from these three fields are provided in the following three sections.

1.1 SYSTEM SAFETY CONTRIBUTIONS

The specialty known as system safety developed in response to needs of the defense and aerospace industries to reduce the enormous costs from failed missile launches and crashed aircraft. After World War II, the United States and the Soviet Union engaged in a race to gain a military advantage. During this period of rapid technological advances, safety took a back seat, and numerous failures occurred during the testing and operational phases of these new systems.[1] Safety remained in the background during the 1950s and 1960s when a common practice was to design and build missiles and aircraft, fly them, investigate crashes, identify the apparent problems, fix those problems, and continue operations. This "fly–fix–fly" approach killed many pilots and destroyed many expensive missiles and aircraft.

The U.S. Air Force took the lead in changing the fly–fix–fly approach to one involving increased safety input during the design and testing phases of missiles, aircraft, and other major acquisitions. In particular, the Air Force published two sets of requirements: (1) *System Safety Engineering for the Development of Air Force*

Risk-Reduction Methods for Occupational Safety and Health, First Edition. Roger C. Jensen.
© 2012 John Wiley & Sons, Inc. Published 2012 by John Wiley & Sons, Inc.

Ballistic Missiles, 1962; and (2) *General Requirements for Safety Engineering of Systems and Associated Subsystems and Equipment*, 1963.

The other branches of the U.S. Department of Defense (DoD) followed suit in 1966 with a broadly applicable standard for military acquisitions. A revised edition titled *System Safety Program Requirement (MIL-STD-882B)* came out in 1969 that has since been modified several times. These developments created a need for specialists to perform the required safety analyses. System safety career positions were available primarily in the DoD, the many defense contractors, and the National Aeronautics and Space Administration (NASA).

In 1973, some of those who pioneered the field formed an international professional society to support the new specialty known as system safety engineering. Now named the System Safety Society, it publishes the *Journal of System Safety* and annually conducts an international conference. More can be learned by visiting the organization's website (www.system-safety.org).

The annual International System Safety Conference provides opportunities to learn about diverse applications of safety analyses. Although many of the presentations focus on safety issues in the military and aerospace industries, applications in other domains continue to grow. One major area of growth is in the transportation domain, where the focus is on improving the safety of passenger trains, buses, ferryboats, harbor traffic, and commercial aviation. Another growth area has been consumer products, where risk assessment has become commonplace.

A diverse set of safety analysis tools has been developed since the early days of system safety.[1,2] This book addresses a few of the tools considered most appropriate for use by OSH professionals. But before jumping into the tools, readers need to learn what system safety is today. The following definition of *system safety* comes from a book by Roger Brauer: "System safety is the application of technical and managerial skills to the systematic, forward-looking identification, and control of hazards throughout the life cycle of a system, project, program, or activity."[3]

This definition contains several significant words and phrases deserving comment. System safety indicates a concern for a *system*, a word referring to a mix of equipment, property, and people interacting in an environment for some purpose. Table 1.1 may help clarify this vague description by pointing out different options for defining system levels, from the narrow to the very broad.[4] At the narrowest level, a system can consist of equipment functioning without humans. The next level adds an individual interacting with equipment. At a somewhat higher level, a system can be a group of employees interacting to accomplish the employer's objectives. At an even broader level, a system can be employees from multiple employers performing their respective functions to achieve broader objectives. The broadest level listed in Table 1.1 adds consideration of influences from applicable governmental regulators and societal values.

In the definition of system safety, the phrase "application of technical and managerial skills" indicates the practical orientation of the field. System safety developed as a technical field, but expanded to address the critical role of using managerial systems to implement safety-related practices and procedures.

Table 1.1 Examples of Systems at Different Complexity Levels

System level[a]	Occupational example
Equipment without human	A building heating system with thermostats, furnace, and air circulation ducts
Individual and equipment	A plumber repairing a leaking faucet. An OSH manager composing a memo on her personal computer
Workgroup level	An assembly line with interactions among employees and their workstations, supervisors, equipment, and materials
Multiple workgroups	A construction site with work being performed by employees of a general contractor and several subcontractors
Highest	All employers in a region or country operating under the same laws and regulatory processes

[a]These levels are adaptations of those described by Erik Hollnagel in Ref. 4.

The "forward-looking" phrase in the definition indicates attention on the future—necessarily involving anticipating problems that might occur. In contrast, a backward-looking focus attends more to investigating past incidents with the intent of assigning blame. A backward-looking focus is driven by the needs of politicians and parties to personal injury litigation, with system safety professionals seeing incident investigations as an opportunity to learn things potentially useful for the future. The core of the system safety community embraces the forward-looking focus by making use of systematic analyses, lessons learned from past incidents, and applicable standards. Another part of the forward-looking focus involves integrating controls into systems to mitigate damage during an incident. Familiar examples are occupant protection features of modern cars like seat belts, air bags, and safety glass in windows. Other examples are engineering devices and software used for monitoring and controlling the complex processes found in industrial systems such as nuclear power plants and chemical processing facilities.

The phrase "identification, and control of hazards" refers to the logical, interrelated steps of first identifying hazards within the system and then determining appropriate means to control those hazards. These steps are almost identical to those used in the practice of occupational safety, industrial hygiene, ergonomics, and pollution prevention. History has shown that hazards can easily be overlooked if systematic processes are not used.

"Throughout the life cycle" reflects the importance of thinking about the full life of a system during the development stage in order to head off future problems. For example, if a project involves hazardous materials, how will the materials be disposed of at the end of the project? How will ship bodies be dismantled and the materials recycled? What will become of outdated weapon systems? What will become of old respirators?

The phrase "system project, program, or activity" indicates that system safety tools and expertise apply to various projects, programs, and activities involving a broad range of systems. Examples of these references to systems are a new fleet of aircraft, a project to develop a prototype, a program for an ongoing organizational function, or an activity such as performing maintenance on equipment.

The OSH community has historically underutilized system safety tools. Those who practice system safety as professionals tend to advocate for greater use of their analysis tools by the OSH community. Two advantages of using system safety tools deserve mention. First, the forward-looking focus of these methods can help reduce the risk of harm to people and property. Second, professionals who develop skills using these methods will find that these tools are portable—they travel with the individual throughout the twists and turns of a career and can be easily adapted to OSH practice in different companies, different industries, and even different countries. This book emphasizes the system safety tools most practical for OSH practice: job hazard analysis, risk assessment, failure modes and effects analysis, and fault trees.

1.2 PUBLIC HEALTH CONTRIBUTIONS

The public health community took an interest in injury prevention during the same time the field of system safety was defining itself. Some of the concepts and tools developed in the early days of public health injury prevention remain viable today, and can be useful for risk reduction in the OSH field.

Although the public health community recognizes the burden of traumatic injuries as being a public health concern, the governmental bodies that fund public injury prevention have been reluctant to commit a lot of resources to these programs based on the seemingly persistent yet mistaken belief among the general public and legislatures that injuries are inevitable. That belief was the topic of a classic paper by Dr. William Haddon Jr. in the 1968 volume of the *American Journal of Public Health*.[5] Haddon advocated approaching roadway injury prevention with the perspective of public health and preventive medicine. He especially rejected the prevailing public opinion at the time that roadway "accidents" could be prevented by focusing funds on improving driver performance to the exclusion of any other preventive measures. His effective advocacy led to increased funding for measures addressing prevention of roadway incidents, better protection of vehicles and occupants during a crash event, and more effective post-crash response capabilities. All these types of measures reduce the risks of roadway transportation.

To sell his message, Haddon developed a tabular format for sorting out opportunities to reduce risks from roadway crashes.[6] Figure 1.1 is an example of the sort of table now known as a *Haddon Matrix*. The example has three rows for the phases of a crash and three columns for the factors involved, yielding nine cells for identifying phase-specific countermeasures. In other papers, Dr. Haddon showed how this basic matrix format can be adapted by adding more columns for other factors. It may also be applied in domains other than roadway transportation.

Today, the Haddon Matrix, in several forms, is highly regarded as a fundamental tool for guiding injury risk-reduction programs in many domains. It serves as one of the threads used to weave this book into a cohesive manuscript.

PHASE	FACTORS		
	Human	Vehicle and equipment	Environment
Pre-crash			
Crash			
Post-crash			

Figure 1.1   An example of a Haddon Matrix. Adapted from Ref. 6, Figure 13.

1.3 EDUCATIONAL THEORY CONTRIBUTIONS

In addition to incorporating contributions from system safety and public health, a third field contributed in subtle ways to this book. Known as *learning theory* in education circles, it provides a framework for structuring curriculum for young children through a university education. The reason for explaining this topic is to make the author's intentions transparent to readers. The OSH profession is in the midst of transitioning from rule-following field to a profession more dependent on effectively using higher level cognitive skills. Many of the Learning Exercises at the end of chapters were written to encourage students to use such skills. These experiences should help the next generation of OSH professionals become more skilled at analysis, adept at conceptual thinking, capable at evaluation, familiar with the science behind the practice, and appreciative of theory.

What is meant by higher level cognitive skills? In their often-referenced handbook, Professor Benjamin Bloom and his colleagues at the University of Chicago classified learning into three broad *learning domains*: cognitive, affective, and psychomotor. Within the cognitive domain, Bloom proposed the following six levels of development.[7]

1. Knowledge acquisition.
2. Comprehension.
3. Application.
4. Analysis.
5. Synthesis.
6. Evaluation.

These classifications remain highly respected by educational theorists in spite of various scholarly proposals for modifications and additions.[8,9] For purposes of writing Learning Exercises, the original Bloom levels are quite appropriate and satisfactory. The levels and their relationships are discussed in greater detail below.

Learning starts with basic knowledge acquisition. Preschool and elementary school learning experiences are structured to help the students gradually build a core knowledge, starting with the alphabets, numbers, and telling time. This knowledge provides a foundation for developing abilities for comprehending written words and arithmetic operations. Fostering the transition from the knowledge acquisition level to the comprehension level is integrated into the entire secondary education curriculum.

The third Bloom level, application, involves making a connection between classroom material and the world outside the classroom, especially with regard to connecting ideas and principles learned in books to everyday decisions and actions. For example, a student taking an introductory psychology course who learned the signs of depression in a book and subsequently recognizes those signs in a friend or relative has successfully applied in the real world what he or she has learned in the classroom. In OSH education, internship experiences after taking some OSH courses are extremely valuable for helping students connect what they learn in textbooks to everyday workplaces.

The original Bloom levels were presented as six progressive steps, like rungs on a ladder. Thus, the Bloom concept was that a person needs to develop, for example, levels 1 through 4 in order to develop level 5. Today, the Bloom list may be conceived as having three ordered lower levels (knowledge acquisition, comprehension, and application) with the higher three learning levels at the same level. Figure 1.2 depicts the relationship among these six levels as being shaped like the letter T.

The fourth level, analysis, involves the capability for examining a complex set of ideas to reach an end point. Often, the process of analysis involves breaking down the input information into components more suitable for analysis. For example, in a construction safety class, students may be assigned to write a short essay comparing and contrasting two different policies on employee drug testing. They may approach the assignment by creating a list of pros and cons for each alternative policy. This approach helps to organize the comparison and provide a basis for contrasting the policies.

The fifth level, synthesis, involves taking extensive input information and developing a model to explain how all the inputs form a logical whole. Some examples of models are provided throughout this book. This entire book is an attempt by the author to present a synthesized model of the OSH field.

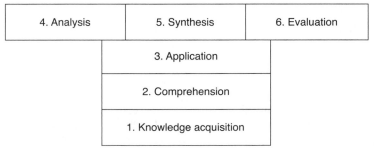

Figure 1.2   Relationship among Bloom's six levels of cognitive development.

Table 1.2   Bloom Level Skills for Topics in Later Chapters

Topic	Bloom level skills
Job hazard analysis	3: Application; 4: analysis
Risk assessment	3: Application; 4: analysis; 6: evaluation
Failure modes and effects	4: Analysis
Fault tree construction	3: Application; 5: synthesis
Fault tree analysis	4: Analysis
Incident investigation	1: Knowledge; 2: comprehension; 4: analysis
Human error	3: Application; 4: analysis

The sixth level, evaluation, involves comparing a specific something against a list of criteria. For example, a governmental agency seeking a contractor for a particular project will make public a description of the project and invite proposals. When proposals are in, agency personnel will review and rate each proposal using the applicable criteria. This skill is used extensively in OSH for periodic evaluations of progress on achieving program objectives.

The Learning Exercises at the end of each chapter contain items calling on a mix of lower and higher level skills. Table 1.2 provides a short list of topics included in parts II and III of this book and the primary types of cognitive skills used for each topic. Parts IV and V call for using the application level to understand how principles developed in earlier chapters apply to very diverse types of hazards.

LEARNING EXERCISES

1. Career paths vary. A person could, for example, be an industrial hygienist and spend an entire career in the mining industry. Or the person could work in various industries for a few years each. Which career path appears most fitting for you? Why?

2. Consider a student named Jane. Her father owned and operated a small roofing company, and Jane worked for him during the summers when she was 18 and 19 years old. As an undergraduate in OSH, Jane did two summer internships, one in building construction and the other in roadway construction. Upon graduating, she took a job in the safety department of a bridge construction company. Every year of her 20-year career, she attended a week-long professional development conference filled with seminars on all topics of safety, industrial hygiene, and environmental protection. She attended only the construction-specific seminars. When the construction industry slumped, she found herself in need of employment in a different industry. She knew her safety-related skills were effective in the construction industry, but all her applications for safety positions in other industries were unsuccessful. What lessons can be learned from Jane's story?

3. Consider another young OSH graduate named Robert. As an undergraduate, he did an internship in OSH with a petroleum company in the pipeline operations. After graduating, he worked for a chemical plant doing process safety analyses. After three years, he changed to a job with an aircraft manufacturer doing system safety analyses. When the aircraft contract ended, he interviewed for a product safety position with a manufacturer of washing machines, dryers, and refrigerators. During the interview, he was asked how his prior jobs prepared him for product safety work in the appliance industry. Imagine you are Robert. How could you use information from this chapter to shape an effective answer?

4. Compare and contrast the career paths of Jane and Robert.

5. Obtain the original article by Dr. Haddon in the *American Journal of Public Health* by following the steps below. After obtaining, read the Background section and write a summary of the main points he makes about (1) terms used when discussing trauma and (2) the etiologic approach used for diseases. The article may be obtained by visiting www.ajph.org, clicking Issues Past and Present, selecting from the grid 1968 and August.

TECHNICAL TERMS

Haddon Matrix	A two-dimensional table for identifying possible countermeasures for public injury problems. It has three rows for the incident phases and three or more columns for system components.
Learning domains	Broad categories for the diverse mental and physical skills humans learn. Bloom defined three categories: cognitive, affective, and psychomotor.
System	An integrated mix of equipment, property, and people interacting in an environment for some purpose.
System safety	A forward-looking and systematic approach to designing safety into a system, project, program, or activity.[3]

REFERENCES

1. Stephans RA. System safety for the 21st century. Hoboken, NJ: Wiley; 2004.
2. Ericson CA, II. Hazard analysis techniques for system safety. Hoboken, NJ: Wiley; 2005.
3. Brauer RL. Safety and health for engineers, 2nd ed. Hoboken, NJ: Wiley; 2006. p. 665.
4. Hollnagel E. Learning from failures: A joint cognitive systems perspective. In: Wilson JR, Corlett N, editors. Evaluation of human work, 3rd ed. London: Taylor & Francis; 2005. p. 901–918.
5. Haddon W, Jr., The changing approach to the epidemiology, prevention, and amelioration of trauma: The transition to approaches etiologically rather than descriptively based. American J Public Health. 1968;58(8):1431–1438.
6. Haddon W, Jr., A logical framework for categorizing highway safety phenomena and activity. *J Trauma.* 1972;12(3):193–207.

7. Bloom BS, Engelhart MD, Furst EJ, Hill WH, Krathwohl DR.In: Bloom BS, Krathwohl DR, editors. Taxonomy of educational objectives: The classification of educational goals. Handbook I: Cognitive domain. New York: David McKay; 1956.
8. Anderson LW, Krathwohl DR, Airasian P, Cruikshank K, Mayer R, Pintrich P, et al. A taxonomy for learning, teaching, and assessing: A revision of Bloom's taxonomy for educational objectives, complete edition. New York: Longman; 2001.
9. Krathwohl DR. A revision of Bloom's taxonomy: An overview. Theory Practice. 2002;41(4):212–218.

Chapter 2

Key Terms and Concepts

The literature on safety and health makes extensive use of three terms—*hazard*, *risk*, and *risk reduction*. This chapter discusses these terms and clarifies their usage in this book.

2.1 HAZARD

2.1.1 Sample of Definitions

Of the many attempts to define hazard, several representative attempts are listed in Table 2.1.[1–8] The first two entries are from dictionaries, the next four from books by respected authors, and the last two from committees. Each definition is separated into three elements: (1) a brief description of a source, (2) words expressing the mechanism of transfer to cause harm, and (3) a description of the harmful consequences. The definitions appear to agree on the following order:

Source → Mechanism of transfer → Harmful consequence.

Other than agreeing on order, the definitions differ markedly when compared element to element. For the source element, some definitions appear to include everything and every activity imaginable. These include the phrases "a source of," "all aspects of technology or activities," and "something." The other definitions provide examples of attempts to be more specific. For the mechanism of transfer element, all eight definitions contain bridging words that differ somewhat but convey the concept that a source requires a means to cause some sort of harm. For the harmful consequences element, all eight definitions contain words about the sort of harm or what will be harmed, but the words differ substantially. Three concise phrases are (1) injury, pain, or loss; (2) harmful effects; and (3) significant harm. The most specific one, found in MIL-STD-882D, is "injury, illness, or death to personnel; damage to or loss of system, equipment, or property; or damage to the environment."

Risk-Reduction Methods for Occupational Safety and Health, First Edition. Roger C. Jensen.
© 2012 John Wiley & Sons, Inc. Published 2012 by John Wiley & Sons, Inc.

Table 2.1   Representative Definitions of "Hazard"

Source of hazard	Mechanism of transfer	Harmful consequences	Reference
All aspects of technology or activities	that produce	risk	1
A source of	exposure or liability to	injury, pain, or loss	2
A condition	that can cause	injury or death, damage to or loss of equipment or property, or environmental harm	3
The potential for an activity, condition, circumstance, or changing conditions or circumstances	to produce	harmful effects	4
Something	that can cause	significant harm	5
An unsafe personal act and/or the unsafe physical or mechanical condition	without which no accident can occur	(accident implied)	6
Any real or potential condition	that can cause	injury, illness, or death to personnel; damage to or loss of a system, equipment, or property; or damage to the environment	7
A condition, set of circumstances, or inherent property	that can cause	injury, illness, or death	8

Because many hazard analysis methods begin with identification of hazards, a definition is needed that sets some parameters for distinguishing what is and is not a hazard.

2.1.2 Proposed Definition of Hazard

The approach to defining hazard is to start with a simple, easily quotable primary definition as the foundation, supplemented by additional definitions of key words in the primary definition. This approach is used extensively in scientific literature when an equation with several variables is presented, followed by specific definitions of each variable. Thus, the primary definition used in this book is

A *hazard* is a source with potential for causing harmful consequences, where

Source is a form of energy, weather or geological event, condition, chemical substance, biological agent, musculoskeletal stressor, or the violent actions of people;

Potential for causing means the source is sufficient to bring about at least one harmful consequence; and

Harmful consequences are outcomes an organization wants to avoid.

Parts IV and V of this book contain extensive discussion of each item in the source list. The harmful consequences element is open ended, so each organization and/or industry group can enumerate whatever outcomes it values and wants to protect.

2.1.3   Additional Rationale and Clarification

This subsection is for readers interested in more in-depth discussion of the foregoing definitions. The source phrases in Table 2.1 include several ways to describe the sources of hazards. The definition from Merriam Webster's Collegiate Dictionary uses the inclusive but vague word "source."[2] The other definitions in Table 2.1 demonstrate the challenge of trying to be more specific or more general. In this discussion, the limitations of these attempts are noted, and the rationale for listing particular sources is provided.

For the source part of the definition, energy is a major component. Energy exists in both potential and transitional states. The potential state of energy involves stored energy such as a compressed or stretched spring, gravitational potential energy, thermal energy within materials, compressed gases, and magnetic fields. The transitional state takes several forms. Kinetic energy consists of materials moving from one place to another, as well as objects rotating. Electrical energy hazards include current traveling in a transmission line and electrons moving in lightning, static discharges, and arcs. Electromagnetic radiation contains harmful forms of ionizing radiation and nonionizing radiation. Chemical energy hazards in the transitional state include active chemical reactions (e.g., fire and explosions) that produce heat, gases, and high pressure. Pressure being transferred from a location of high pressure to a location of low pressure is a form of transitional energy.

Heat energy hazards often arise from other forms of energy such as chemical reactions and electricity passing through a resistor. But the manifestation of heat energy in workplaces deserves specific recognition in a list of transitional energy states. Heat transfers from a warmer to a cooler body through conduction, convection, or radiation. These transfer processes can involve the potential to harm people, property, or the environment. For example, hot objects can burn skin through contact or ignite some flammable vapors, and work in hot environments can transfer enough heat to a person to cause a disorder such as heat exhaustion or heat stroke.

Severe weather and geological events are recognized as hazards involving multiple forms of transitional energy. Hazards of nature include storms, floods, landslides, tornados, hurricanes, drought, wildfires, earthquakes, tsunamis, and volcanic eruptions.

For the source part of the definition, the word "condition" is common. It clearly includes static situations such as a slippery spot on a floor, a stairway with riser heights that vary substantially, and a room containing flammable vapors. It may also include forms of potential energy such as a stretched spring, materials stored overhead, and compressed gas cylinders in a laboratory. A human–machine interface so poorly designed that it invites mistakes could also be considered a condition. A work area with airborne particulates is a condition that threatens the health of those working in the area. Despite its broad scope, "condition" does not encompass everything we recognize as hazard sources; it omits, for example, active/transitional forms of energy, biological agents, flammable materials, and volcanic eruptions. Thus, the word "condition" belongs in a definition of hazard, but is not sufficiently comprehensive to describe everything commonly recognized as a hazard.

Chemical substances, although highly useful to society, are recognized as hazards because of their potentially harmful effects on humans and other living entities. Some chemicals can kill by asphyxiation. Numerous chemicals, notably fuels, are recognized as hazards because of their flammability. Many chemicals are considered hazards because of their inherent explosive, corrosive, or reactive properties. Some increase risk of cancer, genetic mutations, or birth defects in the offspring of those exposed.

Numerous biological agents are sources of occupational hazards. Infectious diseases like flu and colds are threats in all workplaces where people interact. Infectious agents like hepatitis and HIV are especially a concern to healthcare personnel. Plants like poison ivy and poison oak are recognized as sources of allergic reactions. The research and development community has created numerous biological agents capable of harming people. Wild animals, pets, and farm animals are sources of injuries and several infectious diseases.

Musculoskeletal stressors are another important source. Although the vast majority of muscular work is healthy, musculoskeletal stressors become a hazard when the level of stress approaches or exceeds the tolerance of the person's body. The most frequent workers' compensation claims are musculoskeletal injuries and disorders.[9] Events directly causing most of the musculoskeletal injuries are overexertion from excessive lifting, pushing, pulling, holding, carrying, or throwing. Many other musculoskeletal injuries are from bodily reaction to slipping or tripping without falling. Highly repetitive motion accounts for another significant proportion of the musculoskeletal claims.

In addition to hazard sources mentioned above, the violent actions of people are the source of some hazardous situations. Among these are the highly dangerous situations created when armed robbers hold up a bank, terrorists hijack an airplane, or a recently discharged employee shows up at a worksite with a gun intent on shooting a supervisor. Once initiated, these situations can turn in many directions and end with outcomes ranging from no one being hurt to multiple deaths.

The second part of the definitions found in Table 2.1 is the bridge phrase. Although expressed in several ways, the substance of this part is quite similar across all eight definitions. Read as a group, the definitions indicate the bridge phrase should indicate that the hazard exists prior to the harm, that existence of a hazard can cause harm while not going so far as to say the source will cause harm, and that the level of intensity of the hazard source needs to be sufficient to cause the harm.

The rightmost column of Table 2.1 shows different ways to describe harmful consequences. The ANSI/AIHA management systems standard sticks closely to people outcomes—injuries, illnesses, and death.[8] Some of the other definitions go beyond people, including damage to equipment, property, systems, and structures. The differing items identified as being harmed by a hazard reflect differences in the values and backgrounds of the various authors and committees.

To further highlight the difficulty of listing all items we wish to protect from harm, consider this contrast in perspective. A farmer with income from selling eggs would consider the chickens as valued property, and anything that could harm the chickens, like a fox, would be a hazard. In contrast, a wildlife ecologist would view the same fox as a valued part of the ecosystem deserving protection from the farmer's shotgun. Thus, if a definition of hazard incorporates a list of items that could be harmed, then no single definition will suit all authors and all organizations because of their differing values.

Fine-tuning the definition of hazard is more than a point for philosophical discussion. A workable definition is useful to an organization when trying to anticipate, recognize, and control hazards. One processing plant may wish to define a hazard as any threat to harm employees. A second processing plant in the same business as the first may choose to define hazard as any threat to harm employees and visitors, or damage equipment, raw materials, in-process materials, finished product, and the ground on which the plant sits. These examples illustrate some of the endless variations in what organizations might wish to include in a definition of hazard.

In conclusion, any definition of hazard that attempts to list each item of concern will be verbose and may not accurately reflect the values of all organizations. The preferred approach is to have a definition that is both concise enough to be easily remembered and quoted and flexible enough to allow each organization to define whatever it is they value and wish to protect.

2.2 RISK

The word risk is used in numerous ways by the public and in professional circles. Articles in the safety-related journals use risk in three basic ways.

Definition 1 says risk is a probability. Mathematicians agree that all probabilities are numerical quantities with a value in the range of zero to one, and these pure probability values have no units. Sometimes it is convenient to multiply the pure probability value by 100 in order to report it as a percentage. Books on probability and statistics use several notations for probabilities. Common notations to indicate probability of event B are P_B and $P(B)$. Using the second notation, the first definition of risk is

$$\text{Risk } 1 = P(B). \tag{2.1}$$

Definition 2 says risk is the product of probability and severity. This definition is used in the insurance industry and business community to forecast expected

	SEVERITY			
Probability	Catastrophic	Serious	Slight	Minimal
Probable	*High*	*High*	*Moderate*	*Moderate*
Possible	*High*	*High*	*Moderate*	*Low*
Unlikely	*Moderate*	*Moderate*	*Low*	*Low*
Negligible	*Moderate*	*Low*	*Low*	*Low*

Figure 2.1   Example of a two-dimensional risk-assessment matrix.

monetary loss for a particular set of inputs. Expected loss may be expressed in equation form as

$$\text{Risk 2} = E(\text{loss}_B) = P(B)L_B, \tag{2.2}$$

where

$E(\text{loss}_B)$ is the expected value of financial loss from event B,

$P(B)$ is the probability of event B occurring, and

L_B is the estimated financial amount of loss if event B occurs.

This definition may be applied to forecast losses from multiple events. For example, if three events (A, B, and C) have the potential to produce losses of L_A, L_B, and L_C, respectively, then the expected loss may be calculated as

$$\text{Risk 2} = E(\text{loss}_{A,B,C}) = P(A)L_A + P(B)L_B + P(C)L_C. \tag{2.3}$$

Definition 3 says risk is the combination of probability and severity. Some authors call this the "doublet" of probability and severity. When this definition is used, most authors are visualizing a two-dimensional risk-assessment matrix such as that shown in Figure 2.1.

Some organizations add a third dimension when using risk assessment—typically frequency of exposure. If there are three levels of frequency, for example, the risk assessment will use a risk matrix for each frequency category. This could be considered a fourth definition of risk, but for this book it is considered a variation of the third definition.

Which of these three concepts of risk is preferred? Actually, all three concepts are valid and useful in various situations.

Definition 1 is supported by the public health community, particularly the epidemiologists. For example, risk has been defined as "A probability that an event will occur (e.g., that an individual will become ill or die within a stated period of time or by a certain age)."[10]

Definition 2 is a monetary definition, supported by the business community and the related specialty known as risk management. The entire basis for underwriting relies on the ability to predict the expected value of claims for each policy. Underwriters do this by using past experience to estimate the probability and monetary value of claims for the specific policy. The expected loss, or risk, is a summation of all the foreseeable losses using an extension of Equation 2.3. An insurance policy transfers the monetary risk from the insured to the insurer, and the insurer spreads the risk among the many insured clients. This definition is also used in the system safety community. Although Definition 2 is useful when the loss can be expressed in monetary units, it presents problems if the monetary value of an event is controversial. For example, what value should an employer put on an incident that causes an employee's death, brain damage, or paralysis?

The risk-assessment matrices such as the one in Figure 2.1 illustrate Definition 3. The U.S. military has long used such a framework for risk assessments of major procurements. Variations and features of these tables are discussed in more detail in chapter 5.

2.3 RISK REDUCTION

Several terms for efforts to make systems safer are found in the OSH literature. This section explains the rationale for choosing *risk reduction* as the best term for describing the diverse efforts to improve the safety of systems.

Choosing an appropriate term involved trying to satisfy two criteria. The first came from the pioneering concepts of Dr. Haddon, concepts still highly regarded in the public health injury control community. His simple idea was that vehicle crash events involve three phases, and each phase affords opportunities to reduce losses. The original phases were called pre-crash, crash, and post-crash.[11] Haddon's later papers expanded these three phases to include all forms of trauma from energy exchange.[12] Today, these phases are called pre-event, during event, and post-event. Thus, a term was sought that would capture the fundamental concept that harmful events consist of three phases.

The second criterion came from the various definitions of risk discussed in the preceding section. Recall that the first definition involves specifying a medical outcome, and then defining risk as the probability of that outcome. This definition is valuable for medical research, but less useful for OSH because it implies there is only one way to reduce risk—reducing the probability of the event occurring. More useful are the second and third definitions of risk because they include both the probability of the event and the severity of the harmful consequences. A term was

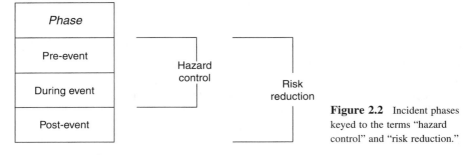

Figure 2.2 Incident phases keyed to the terms "hazard control" and "risk reduction."

sought that would apply to all these definitions. In the process of looking for the best term, some competing terms were considered.

The term *hazard control* was considered because it is used extensively in the safety literature. It implies existence of a hazard and one or more ways to control that hazard. However, it does not encompass any efforts involving post-event response, rehabilitation, or restoration. Figure 2.2 depicts the distinction between the terms hazard control and risk reduction. Using the Haddon phases, the graphic indicates that hazard control is appropriate for efforts (or countermeasures) aimed at the first and second phase. However, it is an inappropriate term if the third phase is included.

Another possible term is the one used by Dr. Haddon—loss reduction. Although this term encompasses the efforts in all three phases, it remains problematic due to its association with monetary losses. Thus, an alternative word was sought. The word risk was chosen because it embodies the concept of harmful outcomes, including but not limited to monetary losses.

Based on the above rationale, risk reduction means to lessen risk, and risk means any of the three definitions discussed in the previous section. Thus, we can reduce risk by

- Reducing the probability of a specified undesired outcome,
- Reducing the severity of the harm,
- Reducing the exposure to a harmful agent, or
- Combining two or more of the above.

The three terms discussed in this chapter figure heavily in the concepts addressed in the five major parts of this book. The chapters in part II address methods for analyzing hazards and assessing risks. Part III contains chapters on program management approaches for reducing risks; and parts IV and V contain chapters on risk-reduction strategies and tactics applicable to each of the hazard sources introduced in this chapter.

LEARNING EXERCISES

1. The definitions of hazard in Table 2.1 are only a sample of many definitions. Review other OSH books or standards to find another definition. Respond to items a, b, and c below.

(a) Provide the definition.

(b) Give the reference to it.

(c) Explain how it might be broken down into three parts as was done with the definitions in Table 2.1.

2. In the *2001 ASSE Dictionary of Terms* (see Ref. 1), the definition of hazard contains the word risk. This same book defines risk as "a measure of the combined probability and severity of potential harm to one or more resources as a consequence of exposure to one or more hazards." In the first definition of hazard in Table 2.1, substitute the foregoing definition for the word "risk."

(a) With the substitution, how would hazard be defined?

(b) What are your thoughts about this?

3. Ericson advocates a three-part hazard description.[13] The three parts are (1) a hazardous element, (2) an initiating mechanism, and (3) the target and threat. Compare these three elements with the three elements used in Table 2.1 (i.e., source, mechanism of transfer, and harmful consequence).

(a) What are their similarities?

(b) What are their differences?

4. This chapter discusses three definitions of risk. Review other OSH books or standards to find another definition. Respond to items a, b, and c below.

(a) Provide the definition.

(b) Give the reference to it.

(c) Indicate which of the three definitions of risk aligns best with the definition you found. If there seems to be no good fit, explain.

5. The psychology literature contains many papers reporting surveys about the public's perceptions of risk and risk-acceptance decisions. These surveys examine how the public views risks such as living near a chemical plant, traveling by commercial air, driving a car, and smoking cigarettes. Results of these surveys often disclose discrepancies between public perception and objective measures of risk.

(a) Do you think risk perception should be included as a fourth type of risk definition?

(b) What is your rationale?

6. Measures of risk are not uniform across industries. Compare the occupational injury statistics, which generally measure risk as a ratio of number of cases to number of person-hours worked during a year, to transportation industries, which traditionally report a ratio of number of deaths (or other events) to distance traveled in miles or kilometers. In both domains, the raw ratios are multiplied by a constant to make the ratio values more convenient for reporting. Ignoring the use of a constant multiplier, comment on what you think is the reason for reporting differently in the occupational domain and the transportation domain.

7. In the occupational domain, some older reports about safety performance used a ratio of number of injuries to number of items produced. Explain why you think labor unions strongly opposed this approach.

8. Two very different indicators of safety find use of the specialties focused on transportation safety. One uses fatalities per million kilometers or miles traveled. A second uses fatalities per 200 000 person-hours worked. Which ratio would be most relevant to each of the following? State your reasons.

    (a) Someone planning a trip from Paris to Munich with options for traveling by plane, bus, train, or personal car.

    (b) Someone contemplating a career as a flight attendant versus another occupation.

TECHNICAL TERMS

Harmful consequences	An organization-specific enumeration of whatever is to be avoided.
Hazard	A source with potential for causing harmful consequences.
Hazard controls	For a specified hazard, various approaches intended to prevent the hazard from causing harm or to reduce the severity; similar to risk reduction but not including post-incident efforts to reduce consequences. See Figure 2.2.
Potential for causing	Phrase meaning the source is sufficient to bring about at least one harmful consequence. Sufficient in this sense considers both the source (e.g., concentration, level, energy level, etc.) and tolerance of whatever might be harmed.
Risk	A general term acknowledging the possibility of an undesirable event. Three definitions are used in the OSH community: (1) probability of a specified event, (2) expected loss, and (3) combined consideration of probability and severity using a risk-assessment matrix.
Risk reduction	Term for any effective means of lessening risk.
Source	When used in the definition of hazard, a source may be a form of energy, weather or geological event, condition, chemical, biological agent, musculoskeletal stressor, or the violent actions of people.

REFERENCES

1. Lack RW, editor. The dictionary of terms used in the safety profession, 4th ed. Des Plaines, IL: American Society of Safety Engineers; 2001.
2. Merriam Webster's Collegiate Dictionary, 10th ed. Springfield, MA: Merriam-Webster; 1995.
3. Roland HE, Moriarty B. System safety engineering and management, 2nd ed. New York: Wiley; 1990.
4. Brauer RL. Safety and health for engineers, 2nd ed. Hoboken, NJ: Wiley; 2006. p. 25.
5. Stephans RA. Systems safety for the 21st century. Hoboken, NJ: Wiley; 2004. Chapter 17. p. 11.
6. Heinrich HW. Industrial accident prevention: A scientific approach, 4th ed. New York: McGraw-Hill; 1959.
7. U.S. Department of Defense. 2000. MIL-STD-882D, Definitions section. Available at http://www.wbdg.org/ccb/FEDMIL/std882d.pdf.

8. ANSI/AIHA Z10 Committee. Occupational health and safety management systems: ANSI/AIHA Z10-2005. Fairfax, VA: American Industrial Hygiene Association; 2005.

9. Klein BP, Jensen RC, Sanderson LM. Assessment of workers' compensation claims for back strains/sprains. J Occupational Medicine. 1984;26(6):443–448.

10. Bailey LA, Gordis L, Green M. Reference guide to epidemiology. In: Reference manual on scientific evidence, 2nd ed. Washington, DC: Federal Judicial Center; 2000. Available at http://www.fjc.gov/library/fjc_catalog.nsf.

11. Haddon W, Jr. A logical framework for categorizing highway safety phenomena and activity. J Trauma. 1972;12(3):193–207.

12. Haddon W, Jr. The basic strategies for reducing damage from hazards of all kinds. Hazard Prevention. 1980;16(5):8–12.

13. Ericson CA, 2nd. What's in a hazard? J System Safety. 2005;41(3):6–9.

Chapter 3

Tools for Analysis and Synthesis

This chapter introduces two types of tools used to aid communication, analysis, and synthesis. Section 3.1 introduces various types of models and section 3.2 discusses charts of various kinds. In both sections, the focus is on the tools rather than the examples.

3.1 USING MODELS FOR SAFETY ANALYSES

Models are representations of systems. This section introduces the idea of modeling systems and provides examples of some types of models useful for safety analyses. The types of models introduced here are physical models, graphic relationship models, physics models, and mathematical models.

3.1.1 Physical Models

Physical models are three-dimensional physical constructions made similar to the actual system. For example, a model train is a scaled-down version of a real train. Flight simulators are made to provide a safe environment for pilot training and skill enhancement. Driver simulators are used for highway safety research. Mock-ups are commonly made during product design, and sculptors often make small models before starting their full-scale work. Physical models are also used for usability testing of products and assembly workstations.

3.1.2 Graphic Relationship Models

System safety projects start with a description of the system that includes a conceptual representation of the system. *Graphic relationship models* are practical for modeling a specific system or a generic system.

Risk-Reduction Methods for Occupational Safety and Health, First Edition. Roger C. Jensen.
© 2012 John Wiley & Sons, Inc. Published 2012 by John Wiley & Sons, Inc.

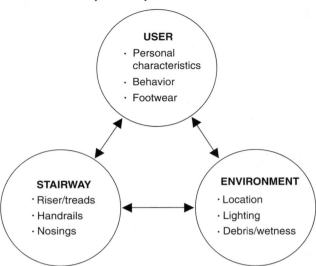

Figure 3.1   Example of graphic relationship model for user–stairway–environment system. Adapted with permission from Ref. 1.

    The example of a graphic model representing a generic system comes from a study by Cohen et al. (Figure 3.1).[1] They reported findings from in-depth investigations of 80 stairway falls. Their information about the cases is organized into a graphic model depicting the interactive nature of three system elements: user, stairway, and environment. The graphic model is an effective way to show that a change in one element may, and often does, affect one or both the other elements.

3.1.3 Physics Models

Models based on the laws of physics are called, in this book, *physics models*. They typically consist of a diagram paired with one or more equations. Readers may review their college physics textbook to see numerous examples. In one popular physics textbook, the authors provide a general problem-solving strategy in which a diagram is the second of eight steps: read problem, draw diagram, label physical quantities, identify principles, choose equations, solve equations, substitute known values, and check answer.[2] Thus, even physics textbook authors recognize the utility of developing a diagram before choosing equations.

    Electrical circuit diagrams are one type of physics model. These display components of a circuit arranged as the actual circuit. The equations for describing current, voltage, and resistance are the same for the actual circuit as for the diagram. Some people call such models analog models because the diagrams are analogous to the actual circuit.

    Another kind of physics model uses a free-body diagram as the analog for actual objects and forces. These diagrams depict a solid object (free body) with vectors

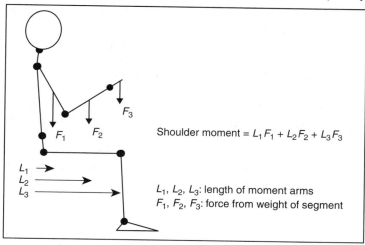

Figure 3.2   Example of force diagram for a biomechanical model.

indicating forces, directions, and lines of action. The equations paired with the diagram indicate how the attributes of the vectors relate. For example, suppose the system of concern is an industrial workstation in which a seated person works with arms forward in a fixed position. Because the position is held fixed, the whole body is treated as a free body. The diagram in Figure 3.2 may be used to show gravitational forces acting on the upper arm, forearm, and hand of the seated person. These forces create a clockwise moment on the shoulder. The shoulder muscles can hold the position by generating a counterclockwise moment of the same magnitude. This sort of model provides the foundation for various biomechanical models used in ergonomics.

Biomechanical models are now available as software. Depending on preferences, these may be called computer-assisted mathematical models or computer models. Also related are mathematical models for human biological systems (see Ref. 3 for an introduction to biomathematical modeling).[3]

Another type of computer model is used for reconstructing vehicle collisions. The analyst inputs known data about roadway attributes, skid marks, weather, damage to vehicles and roadway elements, and witness testimony. The models provide outputs based on engineering computations to indicate the probable sequence of events and speeds of the vehicles. A common use is for litigation to determine fault and liability.

3.1.4 Mathematical Models

Mathematical models are used to define relationships among variables and constants. A model of the exponential growth of a population provides an example. Given the number of people in the present population (N_0) and a proportionality constant (r), the following model allows prediction of the number of people in the population (N) after a period of time (t) has elapsed: $N = N_0\, e^{rt}$.

This type of mathematical model describes a quantitative relationship between a dependent variable and a set of input variables. In this case, if the constant r is known, and values are specified for N_0 and t, the specific value of N will be determined from the calculations. However, because the model validity is based on several assumptions, the calculated N will be only be a prediction, not the precise number of people at that future time.

Statistical models are another type of mathematical model. A common use is to describe experimental results involving the effects of experimental variables on a dependent variable. These models include an error term to account for the inexact nature of the relationship. For example, an investigative team might test 10 human subjects in a laboratory to learn about the relationship between heart rate and clothing while pedaling on a stationary bicycle. The test data may be used to construct an analysis of variance model:

$$HR_{ij} = \mu_{ij} + S_{i\cdot} + C_{\cdot j} + \epsilon_{ij},$$

where HR_{ij} is the dependent variable (heart rate) of subject i in clothing condition j, μ_{ij} is the grand mean heart rate for all subjects in all clothing conditions, $S_{i\cdot}$ is the effect of subject i averaged over all clothing conditions, $C_{\cdot j}$ is the effect of clothing condition j averaged over all subjects, and ϵ_{ij} is the term to account for the variations unexplained by the other terms in the model.

The models presented above illustrate the diverse types of models used for safety analyses. Several more are described later in chapters. Equally useful are various charting methods.

3.2   USING CHARTING METHODS

Various charting methods are useful for sorting out complex systems and processes. The charting methods introduced here are process flow diagrams, organization charts, event trees, and analytical trees.

3.2.1   Process Flow Diagrams

Process flow diagrams provide a flexible methodology for modeling interrelationships among processes and order. For one common use—planning projects—project team members identify each work process they will need to complete the project. Then they organize the processes in the form of a process flow diagram. For another common use—diagramming industrial processes—these charts depict the order in which a material or product flows through the various processes from initial input to final product.

Process flow diagrams are relatively easy to develop once the processes are broken down and understood. Software is readily available and often packaged with software suites, word processing programs, and presentation software.

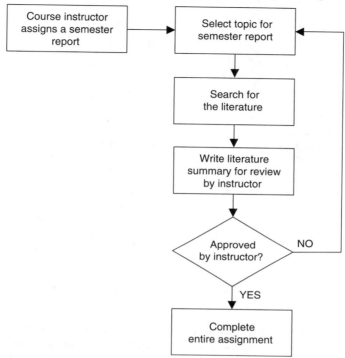

Figure 3.3 Example of a process flow diagram for a semester project report.

The most conventional arrangement for process flow diagrams is to depict flow from the top downward. Sometimes it is useful to have the flow proceed from left to right or in a mixture of directions. Occasionally, the flow is presented in a circular arrangement. The conventional symbols use rectangles to depict processes, diamonds to depict decisions, and lines with arrowheads to show the ordered flow. An example of how a process flow diagram can be used to model processes is shown in Figure 3.3. All but the first and last rectangles in the example have one input arrow and one output arrow. Other options for developers of process flow diagrams include having multiple arrows on the input side to show multiple inputs to the process and connecting multiple arrows on the output side to show multiple outputs.

A common misuse of multiple outputs from a rectangle is to show alternative paths. The correct way to show alternative paths is to use a diamond. Outputs from diamonds often involve alternative paths based on a decision indicated by text inside the diamond. The example presents the decision as a simple question calling for a "yes" or "no" answer. Diamond shapes may also be used to indicate a choice between three options by starting each option path on a corner of the diamond not occupied by an input arrowhead.

The model depicts a process in a college course. The instructor assigns students to select a topic for a semester project. The students are to pick a topic, search for relevant literature, write a description of what they found, and submit their proposal to the

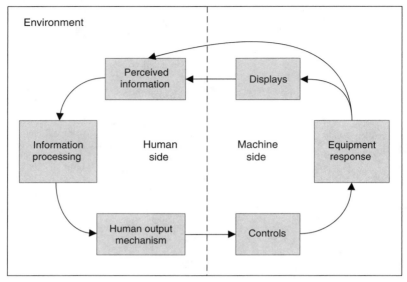

Figure 3.4 Example of a circular process flow diagram.

instructor for approval. If the proposal is not approved, the student goes back to the topic selection process, and using feedback from the instructor, develops a more suitable topic. Once the instructor approves the topic and literature review, the student proceeds to finish the project.

Many books on human factors and ergonomics contain a graphic human–machine interaction model. The authors differ on the complexity and the fine points of presentation, but agree on the overall concepts.[4] These models depict the two major components—the human and the machine—operating within an overall environment. Figure 3.4 depicts the human component on the left and the machine component on the right. Within these major components are process boxes and arrows arranged in a circular manner to depict a counterclockwise flow of processes within each major component. While a person is interacting with a machine, such as driving a car in city traffic, the person frequently repeats the cycle depicted in order to meet the ever-changing requirements for collision-free travel.

3.2.2 Organization Charts

Organization charts are top-down charts used extensively to help structure organizational management and concisely communicate the organizational hierarchy. Software for creating such charts is readily available and easy to use.

This form of chart finds use in system safety for modeling complex systems. A basic step is breaking down a complex system into units more amenable to manufacturing and safety analyses. Organizations involved in large system

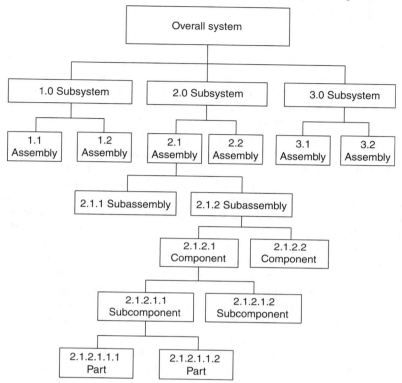

Figure 3.5   Example of an organization chart for breaking down a complex system.

development projects can make good use of organization charts when parceling out safety analysis projects and allocating resources to the subsystems and assemblies most critical to system safety and reliability.

Figure 3.5 shows a general model for breaking down a complex hardware system. It also illustrates a top-down system for numbering various units of the overall system. In the illustration, the overall system is broken down into three subsystems numbered 1.0, 2.0, and 3.0. Beneath each subsystem are assemblies numbered to indicate their parent subsystem. This systematic approach is followed in each branch as it divides into smaller units identified as subassemblies, components, subcomponents, and parts.

In system safety, the organization chart provides the format for a functional tree. A functional tree has at the top a statement of the overall system goal. Below that, the first tier breaks down the overall goal into more than one major function. The lower tiers in each branch may show the systems involved in achieving the respective function. Lower breakdowns may resemble those in Figure 3.5. Examples of function trees are provided in books on system safety.[5,6]

3.2.3 Event Trees

Event trees are tools to help analyze possibilities over time. In these charts, time progresses from left to right. These may take different forms, depending on the wishes of the analyst. Commonly, the left column indicates an undesired/initiating event. Branching moves left to right through various intermediate events, with the branches splitting at each intermediate event. Typical splits are based on answering yes or no to a simple question or indicating the success or failure of a safety mechanism. Each branch ends in the outcome described in the last column on the right.

An example event tree is shown in Figure 3.6. It applies to the foreseeable event of a roofer falling. From the initiating event on the left, the path divides at each intermediate event resulting in multiple paths. At the right of the event chart, each path ends with a likely outcome. In the example, the roofer may have been wearing a harness connected by a lanyard to a secure cable on the roof. The first intermediate event asks whether the fall was arrested by a lanyard. In that case, the first path is the top one marked "yes." The most likely outcome would be a painful experience when the fall is rapidly arrested, but no substantial injury. For the roofer who was not tied off, the second intermediate event accounts for the possibility of landing on a bush or other object that would cushion the impact. In addition, the third intermediate event allows the possibility of the roofer landing on his head/neck versus avoiding this worst-case impact position. Thus, this example event tree identifies five foreseeable paths between the initiating event and the outcome.

Event trees are also useful for following the probabilities of each path at each juncture. Examples of these are found in systems safety literature, including books by Modarres and by Ericson.[6,7]

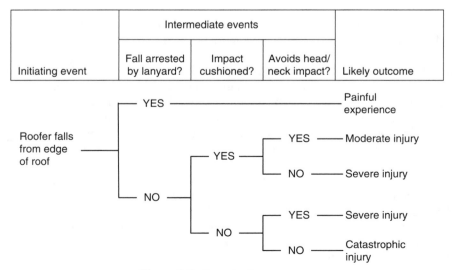

Figure 3.6 Example of an event tree.

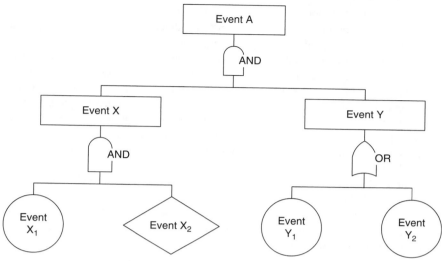

Figure 3.7 Example of an analytical tree.

3.2.4 Analytical Trees

Analytical trees are charts that show logical relationships among events and conditions. Usually, the top event is the one of greatest interest, while the lower events may be thought of as causes of the events higher in the chart. These are called trees because they resemble the shape of an evergreen tree, with the widest spread at the bottom and the narrowest at the top. Figure 3.7 shows a simple arrangement.

The box shapes represent events or conditions. Text inside the boxes describes the event or condition. Other events and conditions are indicated by ovals and diamonds. Between these boxes, ovals, and diamonds are logic gates. The most common gates indicate logical conditions that, if met, allow upward flow. For example, in the tree shown, the left-hand side branch indicates that event X will occur only if event X_1 and event X_2 occur. The right-side branch indicates that event Y will occur if either event Y_1 or event Y_2 occurs. The top event will occur if both event X and event Y occur. Chapters 7 and 8 are devoted to the most common type of analytical tree used in safety analyses—the fault tree.

3.3 SUMMARY OF PART I

The three chapters in part I provide background for the entire book. Chapter 1 introduces the multidisciplinary foundations for this book, in which traditional OSH knowledge is supplemented by concepts from system safety, public health, and educational theory. Specialists in system safety use numerous systematic methods for examining systems to anticipate hazards and find feasible means of reducing risks. Pioneers in public health injury prevention programs contributed logical ways of looking broadly at nonoccupational injuries to find risk-reduction opportunities acceptable to the public. Educators provide the theoretical structure for the end-of-chapter exercises aimed at strengthening higher level cognitive abilities.

Three important terms are discussed in Chapter 2. The first section compares and contrasts several definitions of hazard. The section ends with a proposed definition that includes a list of seven hazard sources—forms of energy, weather or geological events, conditions, chemical substances, biological agents, musculoskeletal stressors, and violent actions of people. The second section explains the three definitions of risk. It concludes that each definition of risk has utility. The third section provides the rationale for using the term risk reduction in the title of this book and in many places throughout the book.

Chapter 3 introduces several useful tools for analyzing and synthesizing systems—some of these tools are models of processes and relationships, others are charting methods used extensively in system descriptions and analyses. Many authors use these tools to enhance communication in their papers, reports, and books. Applications of these tools are illustrated in several chapters in parts II–V.

LEARNING EXERCISES

1. Find a textbook for a college course. Search the book to find a model. Models are common in economics, biology, toxicology, and engineering. If you look in a physics or engineering textbook, do not use equations found at the beginning of a chapter for your example (these are not representations of any particular system). Instead, use an example problem in which the author has described a particular situation and reduced it to a free-body diagram or analogous mechanical system.
   (a) Write a brief description of the model you found.
   (b) Indicate your view as to what type of model it is.

2. Consider the graphic relationship model for the generic user–stairway–environment interaction shown in Figure 3.1. Assume that you need to investigate an incident of a man seriously injured from falling while descending a flight of stairs. The man admitted to being distracted by a pretty woman who was ascending the stairs. Explain how you might use the model to help you think about the distraction factors in relation to other factors.

3. Make a process flow diagram to depict what a college student goes through during an academic year to secure a job when school lets out. It could be an internship position or a career position. Allow various possibilities including (1) the first interview is botched so no offer is made, (2) the second interview lands a so-so offer, and (3) the third interview leads to a fabulous offer. Include at least one process, one decision, and one feedback loop (an arrow going from a lower decision diamond back to an earlier process indicating the need to reprocess something).

4. Suppose you are asked by a power tool manufacturer to help with a comprehensive safety analysis of a new model handheld rotary power saw.

You decide to start by making a functional tree using an organization chart format.

(a) Make a list of major functions suitable for the first tier of the tree.
(b) Using software available to you, make the chart with the top box indicating what you regard as the goal of a person-tool system and putting the functions in the first tier. If charting software is unavailable, you can (1) use spreadsheet software, with cells serving as the boxes in an organization chart, or (2) use the table function in your word processing software. Both alternatives can be made to look similar to an organization chart if you format the cell borders appropriately.

5. Part I of this book has three chapters. What was the rationale for grouping these topics into a unit called part I?

6. For supplemental reading on the use of modeling for safety analyses, an article by Pat Clemens is recommended.[8] He explains how models are used in system safety and provides a critical commentary on sloppy practices.
(a) When he talks about "fidelity" of a model, what does he mean?
(b) Explain his analytical progression of models based on the degree of fidelity.
(c) What sloppy practice does the article author consider most pervasive?
(d) Near the end of the article, Clemens provides advice for minimizing the effects of the errors and assumptions from modeling. Briefly list the items of his advice.

TECHNICAL TERMS

Analytical trees	Charts for analyzing systems using probability and other mathematics.
Event trees	Charts for showing how a specified event can result in multiple paths leading to different end points.
Graphic relationship models	Two-dimensional or three-dimensional depictions of the elements in a system and their interrelationships.
Mathematical models	Generic equations relating variables and constants under various assumptions.
Models	Representations of systems.
Organization charts	Two-dimensional arrangements for depicting the hierarchical structure of units within an organization.
Physical models	Three-dimensional physical constructions made similar to the actual system represented.
Physics models	Generally, simplified drawings representing the variables in an equation paired with the drawing. Some common variables are objects, vectors, time, and distance.

Process flow diagrams   2D graphics showing order of flow among processes and decisions.

Statistical models   Equations to explain relationships among specified variables, including unexplained variability known as error.

REFERENCES

1. Cohen J, LaRue CA, Cohen HH. Stairway falls: An ergonomics analysis of 80 cases. Professional Safety. 2009;54(1):27–32.
2. Serway RA, Faughn JS, Vuille C, Bennett CA. College physics, 7th ed. Belmont, CA: Thompson Brooks/Cole; 2006. p. 16–17.
3. Chandler AP. Human factors engineering. New York: Wiley; 2000.
4. Wilson JR, Methods in the understanding of human factors. In: Wilson JR, Corlett N, editors. Evaluation of human work, 3rd ed. London: Taylor & Francis; 2005. p. 6–10.
5. Bahr NJ. System safety engineering and risk assessment: A practical approach. New York: Taylor & Francis; 1996. p. 191–199.
6. Modarres M. Risk analysis in engineering: Techniques, tools, and trends. Boca Raton, FL: CRC; 2006. p. 30–37.
7. Ericson CA, II. Hazard analysis techniques for system safety. Hoboken, NJ: Wiley; 2005. p. 223–233.
8. Clemens P. Modeling in system safety: A significant source of error. J System Safety. 2009;45(1):13–17.

Part II

Analysis Methods

Part II contains five chapters about analysis methods applicable to workplace risk reduction. Chapter 4 describes job hazard analysis—a common method for analyzing jobs and tasks by listing steps involved, identifying for each step the potentially hazardous events and exposures, and recording appropriate precautions. Extending job hazard analysis to a higher level of sophistication leads to the topic of chapter 5, risk assessment—a process widely used by businesses involved in making and distributing products and equipment to consumers, businesses, and governments. Risk assessments are useful for assessing risk before implementing efforts to control, recording plans to minimize risks, and documenting the rationale for making decisions on residual risks.

Chapter 6 explains the basics of failure modes and the effects analysis—a method used extensively for analyzing equipment to clarify the foreseeable consequences of failures involving parts, components, assemblies, and systems. Chapter 7 presents fault tree diagrams—a type of graphic model useful for understanding the events and conditions that could lead to an undesired outcome specified at the top of the tree, typically a major disaster. A developed fault tree contains extensive information discernible from visual examination, but even more information can be obtained using the analytical methods described in chapter 8. The three methods described are mathematically estimating the probability of specified harmful events, identifying sets of basic events that can cause the undesired event at the top of the tree, and finding common cause failures—single failures with potential to cause a disaster. All methods presented in part II can be applied to the diverse sorts of hazards encountered in the practices of occupational safety, industrial hygiene, and occupational ergonomics.

Chapter 4

Analyzing Jobs and Tasks

4.1 BASICS OF JOB HAZARD ANALYSIS

Job hazard analysis (JHA) is a systematic technique for analyzing the hazards involved in a defined job or task. Many companies and governmental organizations use JHA, but not everyone calls it by the same name. It is also called *job safety analysis*. In this chapter, JHA methodology is explained and discussed in sections 4.1–4.3. Some related types of analyses are introduced in section 4.4.

A conventional format for a JHA form is shown in Figure 4.1. The form consists of an upper section and a lower section, usually on a single sheet of paper. The upper section is for recording information about the task, job title, company, department, analyst, and signature dates. Many forms also have a space for recording applicable work tools and safety equipment. The lower part is a table for recording information about the hazards and risk-reduction tactics.

An initial point of discussion concerns the term *job*. Usually, a job refers to an employee's position (e.g., carpenter, engineer, and receptionist). Such jobs typically consist of multiple functions, and each function can include one or more tasks. JHA is used to analyze a *task*, not an entire job. So if we were more precise on terminology, we would call the technique a *task hazard analysis*. This term, however, has not found acceptance in the safety community, so this book uses the conventional name JHA.

This chapter provides a brief overview of JHAs and some comments on alternative approaches for standardizing JHAs within a company. For a more in-depth treatment of JHAs, consult a book by George Swartz.[1] He provides numerous examples of completed JHAs. Free instructional material on JHAs can be found on Internet sites, including that of the Canadian Centre for Occupational Health and Safety.[2] Both sources provide advice on how to select tasks for analysis and how to use the information to improve workplace safety. For a thorough review of the history of JHAs, with recommendations for getting the most value from the effort, a paper by Glenn is recommended.[3]

Risk-Reduction Methods for Occupational Safety and Health, First Edition. Roger C. Jensen.
© 2012 John Wiley & Sons, Inc. Published 2012 by John Wiley & Sons, Inc.

JOB HAZARD ANALYSIS		
Task/Job Described:		Safety-Related Items:
Job Title/Classification:		
Company:		
Department:		
Analyzed By:		Analyzed Date:
Approved By:		Approval Date:
Step	Potential Hazards	Hazard Controls

Figure 4.1 Typical JHA format.

The JHA table provides three columns for recording the components of a typical JHA. The left column is for listing the steps involved in completing the task. For each step, there may be zero to several potential hazards noted in the middle column. The right column is for recording one or more tactics for reducing the risk associated with the hazard in that row of the table. Some thoughts on how to record information in each column are provided in the following sections.

4.1.1 Step Column

Before embarking on a project to develop JHAs, ask the company human resources department for job or position descriptions for those positions you plan to analyze. Job descriptions generally list the job functions and abilities needed for performing each function. Some job descriptions will list common tasks for the position. The OSH department will want to match the information used for JHAs with corresponding information used in job descriptions.

The job functions listed in job descriptions may be among the elements recorded in the upper part of the JHA form. Figure 4.2 illustrates the hierarchical relationships among a job and its functions, tasks, and steps. This graphic relationship model indicates that a job exists to accomplish certain functions. Within each *job function*, there may be multiple tasks, and most tasks may be broken down into a series of steps.

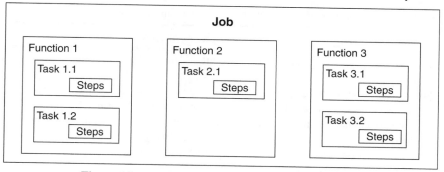

Figure 4.2 Relationships among job, function, task, and step.

Wisely selecting which tasks to analyze can optimize the value added in relation to the resources expended. Some reasons for selecting a task for a JHA may be that the task is new or revised, has a history of injuries, has the potential to cause severe injury, or involves a highly hazardous process or substance.[1–3]

After choosing a task, the analyst needs to break down the task into ordered steps (or other breakdown of activities). The preferred approach is to collaborate with an experienced worker who can explain the steps, demonstrate the procedures, and point out the potential hazards in each step. If a step has zero or one potential hazard, a single row of the table is used. When a step has two or more hazards, using a row for each hazard provides room for applicable information in the middle and right columns.

4.1.2 Potential Hazards Column

The middle column of a JHA form is for recording potentially hazardous events or exposures that might occur in the applicable step. There is no best or standard way of listing hazards in a JHA, but a practical approach is to identify the event or exposure that could directly cause the harm of concern. For occupational injuries and diseases, an organization may choose to adopt the categories discussed below as standard hazards for JHAs. Swartz uses the following abbreviations in the hazard column.[1]

SB **S**truck **B**y

SA **S**truck **A**gainst

CW **C**ontact **W**ith

CBy **C**ontacted **By**

CB **C**aught **B**etween

CI **C**aught **I**n

CO **C**aught **O**n

O **O**verexertion and repetitive motion

FS **F**all to **S**ame Level

Table 4.1   Useful Categories for Use in the Middle Column of a JHA

Event categories	Exposure categories
00: Contact with objects and equipment, unspecified	31: Exposure to electric current
01: Struck against object or equipment	32: Contact with temperature extremes
02: Struck by object or equipment	33: Exposure to air pressure changes
03: Caught in or compressed by equipment or objects	34: Exposure to caustic, noxious, or allergenic substances
04: Caught in or crushed in collapsing materials	35: Exposure to noise
10: Fall, unspecified	36: Exposure to radiation
11: Fall to lower level	37: Exposure to traumatic or stressful event, not classified elsewhere

FB   **F**all to **B**elow

E   **E**xposure to chemical, noise, and so on

Using the above abbreviations, entries in a JHA table might read something like the following:

Worker SB tool dropped from above.

Worker CB collapsed trench walls.

O when lifting portable generator.

CW caustic material.

Worker trips on uneven floor, injured from FS.

Companies might want to use a more comprehensive set of categories with abbreviations. An extensive list of coding categories is available in the U.S. Bureau of Labor Statistics' *Occupational Injury and Illness Classification Manual*.[4] Among these is the attribute known as "type of event or exposure." It contains numerous event categories, similar to those used by Swartz, for injuries involving energy exchange. A list of the energy exchange categories is provided in the left column of Table 4.1. In addition to the energy exchange group, numerous exposure categories are listed in the right column of the table. More detailed subcategories of events and exposures may be found in the above referenced manual.[4]

4.1.3   Hazard Control Column

The right column of a JHA contains text indicating what should be done to prevent an injury or other harm. The column may also be used to record tactics for moderating the hazard or mitigating the extent of harm once an identified incident occurs. For example, workers doing some work at elevation may use personal fall protection equipment as their primary hazard control. If one of them should fall and end up suspended by his lanyard, a JHA for what the other crew members should do could be useful for mitigating further harm to the fallen person and for protecting other crew members from attempting a rescue with no advance planning. The steps in JHAs do not extend into the postinjury activities of first aid to victim, transportation to a medical facility, medical care, or rehabilitation. In the context of the Haddon Matrix, which considers the pre-event, during-event, and post-event phases, JHAs typically focus on ways to reduce risk in the pre-event and during-event phases. Consequently, the third column of a JHA is labeled "Hazard Controls" rather than "Risk Reduction" (see Figure 2.2).

Eight general risk-reduction strategies are listed below and explained in more detail later in this book.[5] Most of these strategies reduce the probability of an undesired event, and some aim to reduce the severity of harm. As such, these strategies afford opportunities for reducing risk.

- Eliminate the hazard.
- Moderate the hazard.
- Avoid releasing the hazard.
- Modify release of the hazard.
- Separate the hazard from that needing protection.
- Improve the resistance of that needing protection.
- Help people perform safely.
- Use personal protective equipment.

These strategies are too broad to be entered "as is" into the third column of a JHA. Rather, the strategy list is intended to help the analyst think broadly about what to put in the third column. Without a list, analysts may resort to the unimaginative approach of putting all the responsibility onto employees by stating they need to be careful, follow procedures, or get refresher training. Using the strategy list may open minds to engineering options deserving consideration.[3]

The actual entries in the right column should be specific enough to identify what should be done and who is responsible for getting it done.

4.2 IMPLEMENTATION

Employers differ on how they develop and use JHAs, starting with identifying who will develop the JHA. A JHA analyst may be someone within the OSH department such as a safety technician or safety intern. Some companies take a different approach. They train the personnel who perform the task to do their own JHA prior to starting the task. For example, maintenance personnel at a mine can do a JHA while preparing to travel to the site. This helps ensure they will remember the hazards, the tools and safety equipment needed, and the precautionary practices involved. The third approach is to train supervisors to perform the JHAs. This widely used approach has the benefits of getting supervisors to think about safety more than they would otherwise do. It may also deepen their understanding of what their employees do.

Procedures are needed for reviewing, approving, and preserving a record. If the analyst is from outside the work group, getting review and feedback from the workers can help improve the quality of the JHA. Similarly, if the supervisor prepares the draft, going through it with an experienced employee may identify misunderstandings and lead to better accord on how tasks should be performed. If the JHA is prepared by the workgroup, some sort of review and approval process at a supervisory level is important. Without a review, the workgroup will begin to feel as though the JHAs they develop are unimportant to the mid-level management, and this feeling will erode the JHA quality and reduce the potential benefits of the process.

The effort involved in creating JHAs warrants follow-up to get the most value. Some uses are mentioned here. When preparing safety training for employees, the JHAs for tasks they perform can provide the basis for instruction. Talking with trainees about the work they do is far more interesting to them than talking generally about rules and regulations that may or may not apply to their jobs. While reviewing a JHA, employees may be invited to share comments and experiences on various steps, hazards, and safe practices. This process engages trainees in the topic—a practice recognized as an important contributor to adult learning.

JHAs also may complement standard operating procedures (SOPs) for employees. The basic steps involved in a task should be the same or similar. Some SOPs list many detailed steps in a process, whereas a JHA may consolidate several small steps into a single larger step. Another difference is that SOPs typically include instructions on how to perform each step, while the JHA usually identifies a step as a process to complete. In addition, SOPs identify all tools and personal protective equipment needed, whereas a JHA may or may not list these items.

For tasks involving the use of tools and equipment supplied by outside vendors, the JHA must account for the procedures specified in the instruction manual (or user guide) accompanying the tool or equipment. Unfortunately, these instructions vary substantially in quality. Some appear to have been developed by engineers as a last step in their design project, with no testing of a representative sample of users to determine comprehension. Some provide only minimal safety information and hazard warnings in the instructions. Some go overboard by including a long list of prohibition statements regarding absurd behaviors. These extended prohibition lists do not belong to a JHA. As a general rule, the JHA should indicate proper actions rather than improper ones. The appropriate place to bring up detailed instructions about tools and equipment is during training. Thus, judgment is required to prepare a JHA that includes the steps to accomplish without becoming overloaded with specific instructions on how to perform each step.

4.3 EXAMPLE JHA

A utility company has a fleet of cars and vans for employee use. The company adopts a policy of encouraging employees to offer assistance to motorists in need. The OSH department decides to develop a JHA for each foreseeable type of assistance. The JHA will be used for training and developing a public assistance manual to keep in each vehicle. The lower section of the JHA they developed for jump starting a vehicle is in Figure 4.3.

This JHA is based on instructions in the owner's manual of the most common car in the fleet. The first step in the JHA is to defer the instructions for the vehicle models. If such instructions are absent, the steps in this JHA would be followed.

In the example JHA, Steps 1–10 are listed in the step column. This particular breakdown is one of many possible ways. Notice that Steps 3, 8, and 10 are listed as a broad step consisting of three substeps. Another analyst could put each of the substeps in a separate row, resulting in total 16 steps instead of 10.

Three suggestions for creating useful JHAs are offered. First, avoid getting overly specific on every little part of each step. For example, detailed instructions on how to position the booster vehicle near the stalled vehicle are unnecessary, as one might reasonably assume the driver knows how to drive. Similarly, detailed instructions on how the spotter will help guide the driver of the booster vehicle into position would be superfluous. A second suggestion is to write the steps with short statements beginning with an action verb. In the example JHA, notice the action words: find, position, prepare, connect, perform, remove, and finish. The action verb sentence format is also used for writing instructions. The third suggestion is to put the actions needed to complete the task in the step column and the safety-related precautions in the third column. It can be tempting to put both in the step column, but that is an inappropriate use of a JHA.

The basic approach used for JHAs has been adapted for numerous other uses. The next section introduces some hazard analysis methods similar to JHA.

4.4 HAZARD ANALYSES SIMILAR TO JHA

When civil engineers and architects plan construction projects, they routinely break down the project into phases. This helps in estimating costs, making schedules, and preparing a proposal to compete for the contract. Construction companies with the better safety and health programs involve the safety manager in the planning process in order to integrate safety into each phase. A hazard analysis tool similar to a JHA, called *phase hazard analysis*, provides a convenient format for the safety manager.

Think of a three-column form such as the one in Figure 4.1, but in the left column record the phases identified by the civil engineers and architects. Within each phase, there will be smaller units that may be called activities. Modify the middle column to record foreseeable major hazards for each activity. At this stage, it is too early in the project to identify every job and task involved in every activity. Instead, this is the time to record major hazards that may be encountered in the phases and activities. In the right column, record the general approach to reducing risk from the hazard. For example, if fall protection railing will be needed in a certain phase, this process will help identify that need and enable inclusion in the budget.

Step	Potential hazards	Hazard controls
1. Find Owner's Guide in one of the vehicles. Find section on jump-starting and follow. If not found, follow the steps below	Gases around batteries can explode if contacted by sparks, flames, arc, or lit cigarette. Contact with sulfuric acid in batteries can burn	Follow instructions for all steps
2. Position booster vehicle near stalled vehicle. Front to front is easiest. Keep some air between the vehicle bodies	Booster vehicle impacts another vehicle, causing minor damage	Operator of stalled vehicle acts as spotter
3. Prepare each vehicle: (a) Open hoods (b) Check all battery terminals for corrosion (c) Check vent caps (should be level and tight)	(a) Vehicle could start rolling downhill (b) Contact with corrosive could be harmful to skin (c) Loose vent caps allow sulfuric acid to escape (d) Electrical surge could damage vehicle electric system	(a) Set parking brake on both vehicles. Chocking wheels is encouraged (b) Use dry cloth or paper to rub corrosion of terminals (c) Tighten battery vent caps using hand (rag or glove recommended) (d) In both vehicles, turn on the heater and turn off all other accessories
4. Connect red cable to positive terminal (+) of stalled vehicle	No hazard	
5. Connect other end of red cable to booster battery positive terminal	Connecting red cable to negative terminal could make jump-start fail or damage battery	Use rag to rub grease and dirt from terminal (+ −) marking. Make sure to connect to +. Use flashlight if needed
6. Connect black cable to booster battery negative terminal (−)	Connecting black to positive terminal could make jump-start fail or damage battery	See above
7. Connect other end of black cable to engine block of stalled vehicle	Incorrectly connecting second end of black cable to stalled vehicle may create a spark capable of igniting flammable vapors	Make connection to a metal part of engine away from battery and carburetor. *Do not connect to negative terminal of stalled vehicle*
8. Perform jump-start: (a) Start the booster engine; (b) start the engine of the stalled vehicle; (c) after success, run both vehicle engines for 3 min before disconnecting cables	Standard transmission vehicles may lunge forward or backward if started in forward or reverse gear	Make sure standard transmission vehicles are in neutral gear before starting engine
9. Remove cables in reverse order: black from engine block of stalled vehicle, black from booster, red from booster, red from stalled vehicle	Incorrectly disconnecting cables may create a spark capable of igniting flammable vapors	Remove cables in specified order
10. Finish task: (a) Remove jumper cables; (b) close both hoods; (c) put cable away	Fingers could get caught between hood and vehicle body	Use both hands to push hood down

Figure 4.3 Example JHA for jump-starting a vehicle.

The phase hazard analysis is not expected to be in final form before the project begins. Once a project proposal is accepted, more time working out details will be justified. The preproject analysis is intended for modification as the project progresses. Planning for the first phase will be completed during the time between contract award and the start of phase 1. Similarly, during phase 1, the phase 2 analysis will be completed, and so on.

In order to limit redundant entries in each phase, the broadly applicable hazards may be specified outside the phase hazard analysis form. For example, if the company considers falling objects as a hazard during all phases of the project and has a general rule that all personnel wear a hard hat while in the construction site, it would be redundant to include that in every phase analysis. It is more convenient to refer to the company's safety manual or make a separate list of safety-related practices applicable to the entire project. This way, a phase hazard analysis will avoid redundancy and focus on phase-specific hazards.

The U.S. Army Corps of Engineers requires contractors to use another adaptation of a JHA called a *position hazard analysis*.[6] One is developed for each employment category (e.g., laborer, carpenter, and pipe fitter) scheduled to work on the project. Two forms are used. The first contains an upper section and a lower section. The upper section asks for the name of the employee, job title, organization, and primary duty location. The lower section has three columns as the basic JHA. The three columns list the tasks performed, hazards, and controls. The second form has three columns for listing equipment needed, inspection requirements, and training requirements for a specified task. Both supervisor and the affected employee must sign the form. Proper use of the two forms should ensure that all personnel working at the site receive information on the safety-related aspects of their job.

LEARNING EXERCISES

1. To practice breaking down a task into steps, consider the following scenario. You are driving a car and have a flat tire. You pull off the main roadway to a location you think will be suitable for removing the flat and replacing it with a spare tire. What steps will be involved in changing the tire?

2. Create a JHA form similar to the one in Figure 4.1.

3. Find a JHA form on the Internet or at your workplace.

4. Compare and contrast the JHA form you created in exercise 2 with the JHA form you find in exercise 3.

5. Fill out the lower part of one of the JHA forms with the steps you identified for changing a tire.
   (a) For each step, fill out the column for potential hazards. Each hazard should be in a separate row.
   (b) For each hazard, record appropriate risk-reduction tactics in the right column.

TECHNICAL TERMS

Job	An employment position that includes at least one function. Sometimes the word is used to identify a particular task.
Job function	In the context of employment, one of the achievements expected from the employee holding a job. Each job has one or more functions. An employment position with no functions would be unnecessary to the employer.
Job hazard analysis	A systematic technique for analyzing each step in a job or task by identifying potentially hazardous events/ exposures and applicable controls for preventing or minimizing harm.
Job safety analysis	Same as job hazard analysis.
Phase hazard analysis	A systematic technique for analyzing each phase of a construction project by identifying foreseeable hazards and determining applicable controls for preventing or minimizing harm.
Position hazard analysis	A document for a specified employment position listing OSH information such as hazards encountered, precautionary practices, tools used, required inspections, and required safety training.
Steps	The ordered elements involved in completing a task.
Task	An assignment for one or more individuals to complete by performing a series of steps or a set of separate activities.
Task hazard analysis	Same as a job hazard analysis.

REFERENCES

1. Swartz G. Job hazard analysis: A guide to identifying risks in the workplace. Rockville, MD: Government Institutes; 2001.
2. Canadian Centre for Occupational Health and Safety. Job safety analysis. Available at http://www.ccohs.ca/oshanswers/hsprograms/job-haz.html.
3. Glenn DD. Job safety analysis: Its role today. Professional Safety. 2011;56(3):48–57.
4. U.S. Bureau of Labor, Statistics., 2007. Occupational injury and illness classification manual. Available at http://www.bls.gov/iif/.
5. Jensen RC. Risk reduction strategies: Past, present, and future. Professional Safety. 2007; 52(1): 24–30.
6. U.S. Army Corps of, Engineers. 2003. Safety and health requirements manual, section 1, EM 385-1-1. Available at http://www.elcosh.org/en/document/92/d000100/Army-Corps-of-engineers-safety-manual.html.

Chapter 5

Using Risk-Assessment Methods

The previous chapter on job hazard analysis showed how a table format helps organize safety analyses. It involved breaking down a task into discrete steps, identifying potential hazards in each step, and specifying appropriate tactics for reducing the risks. This basic approach provides a suitable background for the next topic, *risk assessment* (*RA*). This chapter presents RA as an ordered series of processes usually undertaken as part of a system design project. In practice, however, as the design progresses, analysts need to revisit and update earlier steps to reflect the most current design concepts.

5.1 RISK-ASSESSMENT PROCESSES

Risk-assessment processes have been embraced for many applications where concerns involve safety and health. RA has become a way of doing business in numerous domains.

In the aerospace and military fields, for example, private firms are often retained to develop the next generation of weapons systems, upgraded planes, ships, and space exploration equipment. All the contracts awarded require the firms to conduct RA and other safety analyses using methods recognized in the system safety field. Military organizations wishing to purchase a piece of industrial machinery require an RA conducted by the supplier.

In the domain of consumer products, the laws of product liability have evolved into more clearly defined legal duties of product manufacturers to address hazards. These laws vary from country to country, but international trade has led to considerable harmony in expectations. Product manufacturers find that conducting a sound RA before releasing a product helps make the product safer and goes a long way toward documenting proper attention to risk reduction.

In the OSH field, RA has been gaining respect as an effective method for proactively considering hazards. Some companies ask their workgroups to use a

Risk-Reduction Methods for Occupational Safety and Health, First Edition. Roger C. Jensen.
© 2012 John Wiley & Sons, Inc. Published 2012 by John Wiley & Sons, Inc.

simplified, field-level RA approach to assess risks of a task they are about to undertake. This process helps the group members pause to think about and discuss the steps in the task, methods they will use, tools needed, hazards, and proper ways to minimize risks. Other methods comparable to those used by system safety engineers are suitable for numerous OSH applications such as designing a new assembly line, designing an upgrade to a chemical processing operation, and examining existing operations to find ways to reduce risk.

Basic RA projects follow a common series of processes similar to the following:

1. Define the system and scope of the project.
2. Develop a *preliminary hazard list (PHL)*.
3. Establish risk before risk reduction.
4. Determine hazard-specific *risk-reduction tactics*.
5. Reassess risk after risk reduction.
6. Make decisions about residual risks.
7. Prepare documentation of entire process.
8. Implement tactics and verify implementation.

Numerous authors use a process flow chart for depicting processes similar to those mentioned above.[1–3] The following sections describe these processes.

5.1.1   Define the System and Scope

Before putting much work into an RA project, it pays to define the system boundaries and specify the scope of the project. This approach is similar to that used for a construction project in that the owners and contractors need to agree on the work to be performed, a completion schedule, and a payment schedule. For substantial RA projects, staff time and other resources will be needed. Those who need to approve the commitment of resources will expect a proposal describing the scope of the project, costs to complete, completion schedule, and a projection of value added. Such a proposal starts with a clearly defined system and scope of the project.

5.1.1.1   System Boundaries

How might system boundaries be specified? There is no one-size-fits-all approach. One might consider the system levels outlined in chapter 1 (Table 1.1). OSH professionals will generally work on projects at the second level of complexity, involving equipment and the people who use it. A practical way to start defining the system for an RA project is to consider each element in Figure 5.1 and write a description of what aspects will be included and not included for each element. Not all the elements apply to all RA projects, but each warrants consideration. Each element included in a system description contributes to the overall clarification of what is expected from the RA analysts. Some basic questions needing answers are listed below.

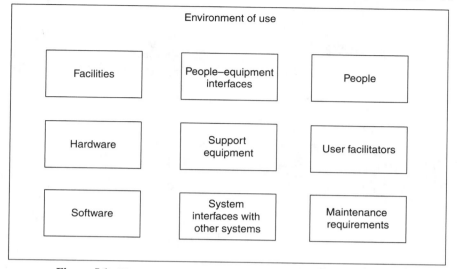

Figure 5.1 Elements to consider specifying in an RA system description.

- In what general environment will the system operate?
- In what facility or facilities will the system be used?
- What hardware will be part of the system?
- What software will be part of the system?
- What people–equipment interfaces will be part of the system?
- What support equipment will be considered part of the system?
- What interfaces with other systems will be part of the system?
- What will be the role of people within the system? What will be their skill levels?
- What manuals, training, procedures, and other human performance facilitators will be considered part of the system?
- What maintenance specifications and equipment will be considered part of the system?

After developing a draft system description, considerable discussions and several iterations may be needed to reach a description of the system boundaries satisfactory to all concerned. A system description provides half the foundation for an RA project. The other half is a description of the RA project scope.

5.1.1.2 Project Scope

The scope of the RA analysis needs definition. Both the organization funding the RA and the RA analysts benefit from a clear understanding of what will be assessed and

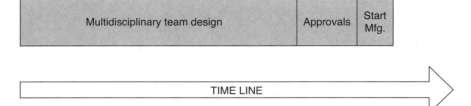

Figure 5.2 Contrast between series and concurrent design processes.

when and how this will occur. In design projects, it is especially important to clarify the role of the RA analysts in the overall design process, which can vary considerably depending on whether the process will follow a series model or a concurrent engineering model. The distinction is depicted graphically in Figure 5.2.

In a *series design process*, a group of engineers works on the design. Once they consider their design acceptable, they send their plans to other groups for review. The reviews may be performed independently by specialists for safety, manufacturability, human usability, marketability, legal matters, and others. After some time, these reviewers may approve the design or send it back to the engineers for changes. The slowest of these reviewers becomes a bottleneck for completing the final design. In Figure 5.2a, the series process shows the reengineering taking place in a relatively short time. Obviously, the time frame can vary tremendously given that the design and review steps might involve multiple iterations. Eventually, the design will satisfy all reviewers, and the process of manufacturing or construction may begin.

Many companies now use a *concurrent design process* similar to that shown in Figure 5.2b. In the concurrent design model, a multidisciplinary team is assembled to come up with a design ready for manufacturing or construction. The team might include representatives from engineering, safety, manufacturing, legal, human factors, and marketing. By integrating multiple inputs into the early design, the team hopes to complete a final design suitable for organizational approval in less time than would be required using a series model.

Safety and health specialists prefer the concurrent design process because it allows integrating safety into the design. As part of the design team, the safety expert gets a much better understanding of the design options, roadblocks, and trade-offs required to complete the project, and being on the team is a more comfortable role that being relegated to the role of reviewer. Thus, as part of the RA project scope, it pays to clarify the role of the safety personnel in relation to others involved in the design process. Some other points to clarify are listed below.

- What system *life cycle phases* are to be included in the RA?
- Will the RA analysts have office space near the engineers on the design team?
- Does the budget for the engineers include appropriate time for them to converse with the RA analysts?
- Will the RA analysts have any responsibility for monitoring implementation of the risk-reduction tactics described in the final RA report?
- Will the RA analysts have a clear end point when their work is considered finished?

5.1.2 Develop Preliminary Hazard List

The RA process may begin much like the JHA methods studied in the previous chapter. Recall that in a JHA a task is broken down into a series of steps and for each step potential hazardous event scenarios are identified. This basic approach is one of several options for starting an RA. Other ways to break down a system into discrete units include using the system elements depicted in Figure 5.1. Once the units are sorted, they are recorded in the first column of a spreadsheet and allowed an open-ended number of rows to be used in subsequent processes. For each unit, potential hazardous events and exposures are listed in another column. This initial list of potential hazards is referred to as a PHL.

A team approach may be used to develop both the PHL and the follow-up processes. Team members with diverse backgrounds should be able to develop a comprehensive list of hazards. Some approaches they might use are described in a paper by Bruce Main.[2]

Be aware that making a PHL can generate resistance among people unfamiliar with RA who fear the implications of creating a PHL. Imagine, for example, how a manager of a power tool company might worry about such a list being used in litigation to show the company knew about the hazards associated with the use of its products. Often the lawyers for these companies have the same fears. It takes some education to convince them that using the RA approach can actually help during product litigation because when the RA process is completed, and written into a proper report, the company will have documentation of all its efforts to reduce the risks associated with each hazard. The company can then offer the report as evidence of its diligent efforts to make the product suitable for safe use. Getting to this point requires starting with a complete PHL and revisiting it at various times during the design process to add or remove items based on new information and insights. With the PHL as a starting point, the RA process continues with the assessment of risk before risk-reduction tactics are implemented.

5.1.3 Establish Risk Before Risk Reduction

Before determining how to reduce the risks of each hazard in the PHL, the RA team should assess those risks based on the assumption that no prior risk-reduction efforts have been made. Doing so provides a *baseline RA* suitable for documenting safety gains from various efforts to reduce the risks.

For each hazard in the PHL, the team discusses the foreseeable severity of the harm and the probability of that harm occurring. The team will decide on typically three to five categories of severity and establish a similar number of ordered categories for probability of the harmful event (e.g., remote, unlikely, likely, and very likely).[2]

Spreadsheets are convenient for organizing the information. Figure 5.3 illustrates a basic column arrangement. The first five columns (C1–C5) are shown in Figure 5.3a, and five more columns (C6–C10) are shown in Figure 5.3b. The first column (C1), as is typical for RA formats, is for listing the task or other logical unit of analysis. The second column (C2) is for describing the potential hazards or exposures associated with the step or unit being analyzed.

To the right of the hazard columns are three columns for risk-related entries. The first two (C3 and C4) are for recording the categories of severity and probability, respectively. The third is for entering an overall *risk level*. The risk level (also called risk index or risk score) is determined from a *risk matrix* shown in Figure 5.4. Others may be found in various system safety books, standards, and papers.[2,3,4]

The example risk matrix in Figure 5.4 illustrates a 4×4 matrix with 16 cells. The cells are labeled to indicate categories of risk level. In this example, only three categories are used: low, moderate, and high. At this point in the RA process, the analysts have a baseline rating of risk level for each hazard in the PHL. In the next process, various tactics for reducing the risk level for each hazard will be added to the RA table.

5.1.4 Select Risk-Reduction Tactics

The process of identifying risk-reduction tactics also benefits from the team approach. Multiple perspectives and different kinds of expertise should contribute to a more effective result. A *risk-reduction priority* scheme (often called a *hierarchy of controls*) is selected and used. These differ in the details, but are consistent in having a preferred order starting with the possibility of eliminating the hazard, followed by considering engineering controls, and subsequently looking for helpful administrative/behavioral controls.

The choice of control hierarchy often depends on the sort of system being analyzed. If, for example, the system involves a machine tool, a hierarchy applicable to machine tools would be appropriate. The lists in the Main and Manuele papers are from a voluntary standard applicable to machine tools.[2,3] Similarly, if the project involves a military system, a particular hierarchy may be suggested or required by the contract. For OSH applications, the choice may be wide open. An option for OSH applications is presented in Table 5.1. It provides in the left column a three-level priority list and in the right column applicable strategies for reducing risk.

(a) Columns 1–5 of a risk-assessment worksheet

C1	C2	C3	C4	C5
			Without Risk-Reduction Tactics	
Analysis Unit	Potential Hazards	Severity	Probability	Risk Level
1				
2				
3				
4				
5				
6				
7				
8				

(b) Columns 6–10 of a risk-assessment worksheet

C6	C7	C8	C9	C10
		With Tactics		
Tactics	Severity	Probability	Risk Level	Status
1				
2				
3				
4				
5				
6				
7				
8				

Figure 5.3   RA worksheet format: (a) columns C1–C5 and (b) columns C6–C10.

The priority categories in Table 5.1 are intended to encourage the design team to consider the three priorities in order. Within each priority category, the strategies are in no particular order. For example, if a particular hazard cannot be eliminated, the second priority comes up for consideration. Within the second priority, the engineering strategies listed are equal in terms of preference. Any strategy that will help reduce risk is open for consideration.

		Severity of Harm			
		Catastrophic	Serious	Slight	Minimal
	Probable	High	High	Moderate	Low
	Possible	High	High	Moderate	Low
Probability	Unlikely	Moderate	Moderate	Low	Low
	Negligible	Low	Low	Low	Low

Figure 5.4   Example of a 4 × 4 risk matrix.

Another important point about hierarchies warrants comment. Selection of one of the engineering or administrative strategies does not preclude selecting others. For example, a design team looking at the in-running nip point of a conveyor belt may choose a guard to separate the hazard from that needing protection (Strategy 5). The team may also decide to place a warning on or near the guard about following lockout and tagout procedures whenever maintenance is needed (Strategy 7). A well-marked release switch might be installed to enable release of the conveyor belt tension in case someone gets caught in the nip point (Strategy 9).

After appropriate risk-reduction tactics are selected, they are added to the overall RA spreadsheet in the column to the right of the risk-level column. When multiple

Table 5.1   Three-Level Priority Order for Considering Risk-Reduction Strategies

Priority	Strategy[a]
I. Control by eliminating the hazard	1. Eliminate the hazard
II. Control through engineering methods	2. Moderate the hazard
	3. Avoid releasing the hazard
	4. Modify release of the hazard
	5. Separate the hazard from that needing protection
	6. Improve resistance of that needing protection[b]
III. Control through administrative methods	7. Help people perform safely
	8. Use PPE
	9. Expedite recovery

[a] These strategies are a modification of those originally proposed by Jensen.[5]

[b] Attempts to change employees through means such as stretching and exercise programs belong in priority III.

tactics are specified, it is helpful to identify them with separate lines so that the columns to the right can be used to document how each tactic affects risk.

5.1.5 Reassess Risk After Risk Reduction

After risk-reduction tactics have been identified, each is recorded in the RA spreadsheet. Figure 5.3 shows applicable columns. The design team needs to reassess the risks for each hazard by assigning ratings of probability, severity, and risk level under the assumption that each risk-reduction tactic will be implemented. The *reassessed risks* are documented in columns C7, C8, and C9. This process provides documentation of the anticipated safety benefits of each tactic.

When multiple tactics apply to a particular hazard, the analysts need to decide how many rows to use. If, for example, a hazard has four tactics identified, the analysts could put all four in a single row, or they could use more than one row. The example later in this chapter illustrates an approach that, for a specific hazard, uses one row for all tactics that will reduce the severity and a second row for all tactics that will reduce the probability. This approach may prove helpful in three ways. First, it helps the design team think through its rationale for specifying each tactic. Second, it provides documentation of the design team's thinking for future reference. Third, it may help in the next process of assessing the acceptability of risk remaining after each tactic is implemented.

5.1.6 Make Residual Risk Decisions

A key part of the RA process is clarifying risks remaining after tactics are implemented. The organization will use this *residual risks* information to make decisions about going forward with the project. The decision is made assuming the risks will be controlled as indicated in the RA. The decision typically will take one of the two forms: (1) the organization agrees to accept the residual risk, or (2) the organization does not actually accept but is willing to tolerate the residual risks. Each of these has advocates and opponents. Learning Exercise at the end of the chapter asks readers to think about the pros and cons of each way of stating the decision.

5.1.7 Document RA Project

A thoughtful final report is essential. It may become the critical piece of evidence in a legal action years after the RA is completed and in the consumer product business this is quite likely. Someday, somehow, someone may get injured using the product. The product manufacturer, the common defendant in a product liability suit, can use the RA report as evidence of having paid proper attention to safety prior to releasing the product for sale. To satisfy legal standards, the RA must have been performed in good faith and with due diligence. Thus, the report should include factual information about the level of effort put into the RA and related safety activities, qualifications of the design team members, and summaries of discussions reflecting sincere attempts to make the product as safe as practicable.

5.1.8 Implement the Tactics

The project is not completed when the report is finished. Management processes are needed to ensure proper implementation of the risk-reduction tactics. For example, engineering controls will need quality control processes. For workplace projects, behavioral controls will need administrative systems to support implementation. A process for verifying implementation of these risk-reduction tactics will help avoid overlooking something and provide documentation that may be valuable for establishing regulatory compliance and due diligence.

With products and equipment sold to consumers, businesses, or governments, a mechanism for obtaining feedback is important. Think of how many products have a toll-free number on the package for reporting problems. These numbers are specifically for getting customer feedback. The feedback needs to be recorded and periodically reviewed to identify problems with safety or quality. For high-cost products, such as motor vehicles and machine tools, a system for tracking ownership is helpful in case there is a need to contact owners about product recalls or part replacement.

5.2 EXAMPLE RISK ASSESSMENT

An example of an RA for a simple product is provided in this section. Readers are apprised that different analysts performing an RA on the same system will differ in their judgments about probability and severity as well as the selection of tactics applicable to particular hazards. The following example of a laser pointer design project is intended to illustrate the eight RA processes, not to make judgments about the technical aspects of the product.

Start by defining the system and scope of the project. The system is a laser beam pointer used by lecturers and instructors. It is both a consumer product and a tool for use in business and educational establishments. The company name is not mentioned in this example. The design team considered the system to be the laser beam electronics, housing, batteries, packaging, and a paper insert containing instructional and precautionary information. Thus, the initial design concept specified a product having the following attributes:

- The battery-powered beam will begin by considering continued use of the same diode laser as used on the company's prior product (a wavelength of 670 nm and power output <5 mW).

- The beam and batteries will be enclosed in housing similar to a fat pen, with appearance clearly different from the company's previous models. The housing will have an on/off switch and a clip to hold the product in a shirt pocket. One end of the housing will have a small hole for the beam. The housing will be <150 mm in length.

(a) Columns 1–5 of risk-assessment worksheet

C1	C2	C3	C4	C5
		Without Risk-Reduction Tactics		
Analysis Unit	Potential Hazards	Severity	Probability	Risk Level
1. Laser beam	Exposure of someone's eye to laser beam	Serious	Unlikely	Moderate
2. Housing edges	Hand contact with sharp edge of housing cuts skin	Slight	Possible	Moderate
3. Housing structure	Product housing damaged from falling to floor	Slight	Possible	Moderate

(b) Columns 6–10 of risk-assessment worksheet

C6	C7	C8	C9	C10
	With Tactics			
Tactics	Severity	Probability	Risk Level	Status
1.1. Limit laser power to class 2	Slight	Unlikely	Low	Acceptable
1.2. Reduce chance of shining in someone's eye by providing warnings, using fail-to-inert-mode design, and discouraging holding-down button	Slight	Negligible	Low	Acceptable
2. Bevel all sharp edges	Minimal	Negligible	Low	Acceptable
3. Make housing sturdy	Minimal	Unlikely	Low	Acceptable

Figure 5.5   Example RA worksheet for laser pointer example.

- The laser pointer will be packaged in an attractive box.
- Instructions will be provided on a piece of paper inserted in the box.

Second, the RA team—consisting of an electrical engineer, a mechanical engineer, an expert in ergonomics/human factors, and a marketing specialist—developed a PHL of foreseeable hazards. This PHL contained three hazards: exposure of someone's eye to laser beam, contact with a sharp edge of the product housing cuts the user's skin, and the product housing is damaged from falling to the floor. These are recorded in column C2 of the upper RA spreadsheet in Figure 5.5.

The third task was to establish a risk level before implementation of the risk-reduction tactics. The matrix in Figure 5.4 was used for assessing risks. It uses four ordered categories for severity and four for probability. The 16 cells in the matrix are labeled for risk levels. The RA team judged the possibility of the lecturer pointing the laser light directly into the eye of an attendee as having a severity of serious with a probability of unlikely. That put it into the risk-level category of moderate. The RA team judged the possibility of a sharp edge of the tool cutting the lecturer's skin as having a severity of slight with a probability of unlikely. Remember, this rating is based on the assumption of no prior efforts to address this hazard (i.e., the designers of the laser housing had made no prior effort to round off the sharp aspects). The RA team assigned the third potential hazard, damage to the tool housing from impact with the floor, a severity rating of slight and a probability of possible. That put it in the risk level of moderate. These ratings are shown in C3, C4, and C5 of Figure 5.5a.

Fourth, the team identified feasible risk-reduction tactics for each hazard. These were entered into column C6 of the spreadsheet in Figure 5.5b. The actual text entered in C6 may be kept brief by writing the full explanation in the report. Notice that a second row was added for the laser beam to distinguish between the risk-reduction tactic to limit severity and the three tactics to reduce probability.

Row 1 applies to restricting the severity of an eye injury by limiting the power of the laser beam to the minimum needed for effective use of the pointer. The beam used in the company's previous laser pointer projected at 670 nm and was classified as a class IIIA laser under the U.S. Food and Drug Administration classification system (class 3R in the *IEC* 60825 system). Technological developments subsequent to the company's earlier design produced a safer beam that projects light in an optical range easier for human visual detection.[6] By adopting the new technology, the design team was able to make a laser pointer that projects a beam closer to the optimal visual range using less power. The new product projects at 635 nm and will be classified as a class II laser (class 2M in the IEC 60825 system). Although this lower classification means fewer required precautions (e.g., laser symbol and text warnings), the company wants to exceed the minimum requirements because it aspires to be world class in product safety.

Row 2 is used for the three tactics identified for reducing the likelihood of shining the laser beam into someone's eye long enough to cause damage. Tactic 2 is to place a warning message on the product housing and a second one on the instructional insert stating that the user should avoid shining the light at anyone's eyes. Such a tactic is not required for class 2 lasers, but the company asked the RA team to include these extra precautions. Tactic 3 is to design the actuator button using fail-to-inert mode, which means that when the user is not pressing the button, beam is not projected. Tactic 4 is to make the actuator button in such a way as to cause the user some discomfort after pushing on it for 30 s. The thought behind this tactic is to discourage speakers from getting into the habit of constantly pressing the button.

In the fifth process, the revised risk ratings are recorded in columns C7, C8, and C9 of the RA spreadsheet in Figure 5.5b. Column C10 is for notes on the status of tactics used in each row. It may be used for recording whether or not the residual risk has been accepted or tolerated, what remains to be done, who is responsible for getting it done, and whether implementation has been verified. Many organizations use multiple columns for status data in order to keep track of specific matters such as status of hazard control implementation and verification.

Figure 5.6 graphically shows how the tactics will change the risk level of the laser beam from moderate to low. The same basic process is used for the other hazards. Learning Exercises are provided at the end of this chapter to encourage a more hands-on experience.

(a)

No controls	Severity of harm			
Probability	Catastrophic	Serious	Slight	Minimal
Probable	High	High	Moderate	Low
Possible	High	High	Moderate	Low
Unlikely	Moderate	Moderate	Low	Low
Negligible	Low	Low	Low	Low

(b)

Tactic 1	Severity of harm			
Probability	Catastrophic	Serious	Slight	Minimal
Probable	High	High	Moderate	Low
Possible	High	High	Moderate	Low
Unlikely	Moderate	Moderate	Low	Low
Negligible	Low	Low	Low	Low

(c)

Tactics 2, 3, 4	Severity of harm			
Probability	Catastrophic	Serious	Slight	Minimal
Probable	High	High	Moderate	Low
Possible	High	High	Moderate	Low
Unlikely	Moderate	Moderate	Low	Low
Negligible	Low	Low	Low	Low

Figure 5.6 Graphic depiction of how RA tactics reduce risk of laser beam.

The sixth process, making decisions about residual risks, is not carried out by the design team. Their role is to provide the technical information to a high-level manager for the decision. The company considering the new laser pointer needs to consider every risk in the final risk assessment matrix. Some guidelines are helpful to sort out the easy approvals from those needing more thought. For example, the hazards with a low-risk classification may be considered approvable without discussion. Any hazard with a high-risk classification will be disapproved. Those in the moderate-risk classification would require discussions prior to deciding to approve or disapprove. In the example, all the hazards were in the low-risk classification and easily approvable.

In the seventh process, the individuals who conducted the RA are responsible for writing a full report on their project, including the residual risk decisions resulting from the project. A suggested organization is the conventional scientific format with sections for methods, findings, discussion, conclusions, references, and appendices. Appendices are useful for including documentary material few people will want to read. For example, the initial PHL may be put in an appendix, even though it was modified several times during the project. Minutes of meetings may be included in an appendix. Appendices are also useful for including copies of standards and guidelines referred to in the report. An important consideration in preparing a report is to imagine it being read by lawyers after someone has been injured.

The last process in an RA is implementing the tactics. An organization that makes the effort to conduct an RA will certainly want to make sure the tactics are implemented. As obvious as this may sound, investigations of major incidents have revealed failures to implement known tactics. In aerospace and some other fields, a formal process to verify the implementation of each tactic is used. Verification can be considered the last step in the implementation process discussed in this chapter.

LEARNING EXERCISES

1. Suppose you are concerned about a particular task performed quarterly by a pair of electricians. You already have a JHA for the task. It has 12 steps, beginning with assembling the equipment they will need and obtaining a permit for a lockout/tagout procedure and ending with putting away the equipment they used. Answer the following questions.
   (a) Refer back to Table 1.1. What system levels would you include in your project?
   (b) For your system description, consider the components in Figure 5.1. List the components and indicate (briefly) what, if anything, involving each component would be useful to include.
   (c) How would you describe the scope of your RA project?
   (d) What would be the obvious breakdown for column C1 of an RA table?

2. The example in the text shows how to enter risk ratings for each hazard before implementing any risk-reduction tactics. After the controls are implemented, revised risk ratings are needed. In the example of a laser pointer, the change is

shown in Figure 5.6 for the laser beam hazard. Show the change for the sharp edge hazard in a similar way. In other words, create an RA matrix, circle the risk level before the tactic, and draw an arrow pointing to the risk level after the tactic.

3. These two items concern the connection between tactics and the risk-reduction priorities shown in Figure 5.1.
   (a) For the tactic of limiting the laser beam power, determine which priority it fits into and which strategy or strategies apply.
   (b) For the tactic of putting warnings on the product and the insert, determine which strategy or strategies apply.

4. After completing a risk-assessment project, some residual risks are normal. A person high in the organization is asked to approve or disapprove going forward. The form of this decision has generated controversy. Comment on strengths and weakness of the two ways of stating the decision.
   (a) To accept the residual risk.
   (b) To tolerate the residual risk.

TECHNICAL TERMS

Baseline RA	The assessment of risk for each hazard in a PHL based on the assumption that no prior risk-reduction efforts have been made. A risk matrix is used to establish the baseline risk level.
Concurrent design process	An approach to a design project that uses a team of people from multiple disciplines and departments working side by side to complete the design.
Hierarchy of controls	A preferred order of options for reducing the risks associated with a hazard, also known as a *risk-reduction priorities*.
IEC	International Electrotechnical Commission.
Life cycle phases	A term encompassing the various phases a system may go through, including initial design, user and prototype testing, final design, manufacturing, deployment, main use phase, and disposal.
Preliminary hazard list	A list of hazards identified early in the RA process, with the intent of assessing the risks associated with each hazard on the list.
Reassessed risks	The result of assessing risks for each hazard under the assumption that all noted efforts to reduce the risk have been implemented. It uses the same risk matrix as used for the baseline RA.
Risk assessment (RA)	An ordered series of processes usually undertaken as part of a system design project.
Risk level	Categories of risk used for risk assessment.

Risk matrix	A tabular array of cells that are assigned words or numbers as labels for ordered categories. The common tables for risk assessment use rows and columns with ordered categories of severity and probability.
Risk-reduction priority	Same as hierarchy of controls.
Risk-reduction tactics	For a specified hazard, various approaches to lessen the probability of a harmful event or reduce the severity of the harm.
Series design process	An approach to a design project that involves completing the engineering design processes before involving the expertise of other departments.

REFERENCES

1. Bahr NJ. System safety engineering and risk assessment: A practical approach. New York: Taylor & Francis; 1996. p. 206.
2. Main BW. Risk assessment: A review of the fundamental principles. Professional Safety. 2004;49 (12):37–47.
3. Manuele FA. Prevention through design: Addressing occupational risks in the design and redesign processes. Professional Safety. 2008;53(10):28–40.
4. Piampiano JM, Rizzo SM. Measuring risk & its variables objectively. Professional Safety. 2012;57 (1):36–43.
5. Jensen RC. Risk reduction strategies: Past, present and future. Professional Safety. 2007;52(1):24–30.
6. Miller G, Yost M. Nonionizing radiation. In: Plog BA, Quinlan PJ,editors. Fundamentals of industrial hygiene, 5th ed. Chicago: National Safety Council; 2002. p. 315.

Chapter 6

Analyzing Failure Modes

When systems consist of multiple parts, there is a justifiable interest in understanding the possible effects of a part failing. This interest led to development of a methodology known as *failure mode and effects analysis (FMEA)*. Originally developed by reliability engineers, the system safety community recognized the usefulness of FMEA and incorporated it into its arsenal of methods.

6.1 RATIONALE FOR FMEA

An FMEA is a bottom-up analysis technique because the analysis starts with a list of numerous items in the system and examines what might happen if any of them should fail. The starting item is often a specific part, but it may be a subcomponent, component, subassembly, assembly, or subsystem. This *inductive approach* of reasoning contrasts with the *deductive approach* used in the next two chapters on fault trees.

An FMEA helps clarify which items are most critical for the safety of the system. Sometimes, the failure of a single part can result in total system failure, and some system failures include death or serious injury. By identifying any *safety-critical items*, engineers can focus on solutions to avoid such failures. They may be able to procure or manufacture an item with greater reliability, specify a larger safety factor, or design in redundancy so it takes multiple part failures to cause system failure. Numerous books on system safety are available for those seeking more in-depth information.[1–3]

6.2 BASIC FMEA METHODOLOGY

Systems often consist of many elements. Figure 6.1 depicts a common convention for organizing all the items in a system. It uses an indenture format, similar to the outline of a report, to show the naming convention. The indenture farthest to the right contains the smallest units, usually specific parts. The reasons for this unusual list format is to

Risk-Reduction Methods for Occupational Safety and Health, First Edition. Roger C. Jensen.
© 2012 John Wiley & Sons, Inc. Published 2012 by John Wiley & Sons, Inc.

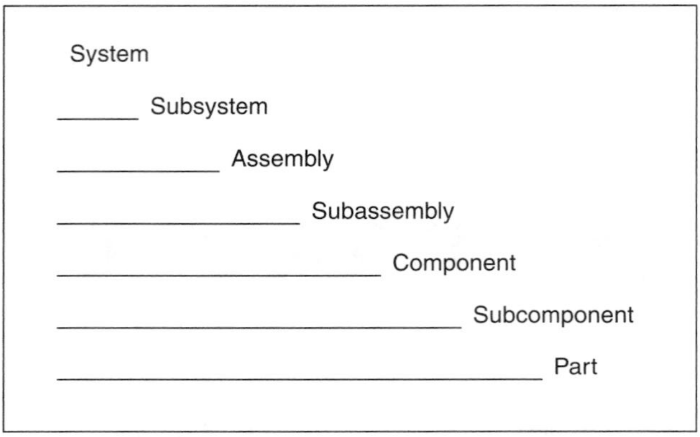

System

_____ Subsystem

_____ Assembly

_____ Subassembly

_____ Component

_____ Subcomponent

_____ Part

Figure 6.1 Concept of system indenture levels.

introduce the concept known as *indenture level*, a concept useful when choosing the level to begin an FMEA.[1]

A complex system such as an aircraft or a ship has thousands of parts. It would be an incredible task to analyze every single part to determine how it might fail and what effect each failure might have at higher levels. Therefore, an initial consideration in conducting an FMEA involves selecting the subsystems that are important to overall safety. In a ship, for example, items affecting the structural integrity of the hull are critical for the safety of the entire crew, whereas items in the dining areas are far less important for system safety. Thus, little would be gained by performing a comprehensive safety analysis for every fastener used in the dining tables, but analyzing fasteners in the hull structure is obviously important for ship safety. Consequently, choosing the initial starting place, or indenture level, is a very pragmatic decision.

Once the starting place is chosen, a trained analyst is assigned a unit for analysis. A form something like the worksheet shown in Figure 6.2 is used. The upper section is similar to that used for a JHA or RA. It has spaces for indicating what is being analyzed, who performed the analyses, who reviewed and approved it, and when the analyses were completed. The lower section of the worksheet is for the technical information.

Like numerous safety analyses, columns are used for delineating the various types of information, and rows are used for the items analyzed. The information sought in each column in Figure 6.2 is explained below. Examples of other forms, with additional columns for other types of information, are provided in other books.[1,2]

Column 1 is for identifying specific items. If the analysis starts at the lowest indenture level, a specific part will be identified. If the analysis starts at a higher level, the analyst records the applicable subcomponent, component, subassembly, assembly, or subsystem.

Column 2 is for noting the mode or modes of failure for each item. For example, a valve may fail in an open mode or a closed mode. A bolt may fail in shear, tension, or bend. Some items may fail to operate at the correct time, while others may fail by not stopping at the proper time. Each *failure mode* can have different consequences. It is

Failure Modes and Effects Worksheet				
System/subsystem:				
Analyst:		Date:		Page
Approved by:		Date:		Total pages:
1 Item	2 Failure mode	3 Effects on other components	4 Effects on higher in system	5 Explanations

Figure 6.2   Example of an FMEA worksheet.

not uncommon for a specific item to have multiple failure modes. Stephans provides an example of a fire sprinkler system analyzed with an FMEA.[2] He lists three failure modes for sprinkler heads: fails to open, open prematurely, and deliver an inadequate spray pattern. Using a separate row for each failure mode facilitates both analysis and recordkeeping for each foreseeable failure.

Column 3 is for stating the foreseeable effects of the failure on other items such as those located nearby or in the same group as the failed part. These effects may be at the same or next higher indenture level. For example, if a part fails, we want to know how that will affect other parts and the subcomponent it belongs to. If a subcomponent fails, we want to know how that will affect the component it belongs to. This inductive approach of moving to successively higher indenture levels helps clarify the safety significance of each item and failure mode.

Column 4 is used for noting the effect at a higher indenture level than the one in column 3. This involves working upward from the effects noted in column 3 to the entire system or to an intermediate level considered appropriate. When considering effects, the analyst should be pessimistic, noting the worst-case effect rather than the hoped-for effect. Being pessimistic expands the consideration given to avoid such outcomes. Many FMEA forms have another column for noting the effects of each failure mode on the overall system. Although examples are not shown in Figure 6.2, readers can refer them in the books mentioned in the reference list.

Column 5 in Figure 6.2 may be simply for recording observations and comments. For example, while performing the FMEA, an analyst may wish to record an idea for addressing an identified concern so that the idea will not be forgotten. Other relevant comments might include detailed attributes of parts mentioned, assumptions, relevant standards, sources of information, and meanings of abbreviations.

6.3 BEYOND THE BASICS

This section introduces some extensions of traditional FMEA that system safety engineers developed and use. Some organizations extend the FMEA by incorporating parameters of risk. As in risk assessment, columns are used for the probability of each failure, the severity of the effects, and some sort of index for risk level. This approach is called failure modes, effects, and criticality analysis (FMECA) to distinguish it from an ordinary FMEA. Organizations may prefer an FMECA when they need to assess how critical the effects of different failure modes would be on the whole system. An FMECA can provide insights useful for allocating engineering efforts to the most important failure modes.

A second variation in FMEA methodology, known as a functional FMEA, begins with a top-down approach for breaking the system into branches.[3] The first tier of branching may be based on system function or subsystems. Lower tiers extend the branching downward. Then, an FMEA applied to each branch helps to distinguish the functions most important to system safety. With that resolved, resources for safety analyses can be directed only to the branches important to system safety. Ericson provides an excellent explanation of the theory behind functional FEMA approaches involving functions, hardware, or hybrids.[3] These three approaches to FMEA use a basic FMEA format, but provide different means to the same end—clarifying what warrants the greatest commitment of engineering resources for improving system safety.

Another use of the FMEA method is to identify safety-critical items. For example, the U.S. National Aeronautics and Space Administration (NASA) uses an extension of the FMEA method to assign items to categories based on how critical their failure would be to mission success.

FMEAs can be useful during the design phase of systems by identifying critical failure items. These may then be used by the design team to consider alternative designs such as adopting a larger safety factor, specifying highly reliable components, or increasing redundancy.

LEARNING EXERCISES

1. Explain why the FMEA approach is considered an inductive approach, rather than a deductive one.
2. Use an FMEA chart such as the one in Figure 6.2 to analyze sprinkler heads in a wet pipe fire protection system. Wet pipe systems have water throughout the

overhead pipes. The water is kept from spraying by the sprinkler heads. When the temperature of a sprinkler head exceeds a trigger point, it opens and lets the water discharge in a spray pattern similar to an umbrella shape. Each sprinkler is connected to a water pipe, and the pipes and sprinklers are distributed throughout the room in rows. The pipe for each row is connected to a feed line shared by all the other pipes in the room. For this exercise, format a spreadsheet with the five columns shown in the lower section of Figure 6.2, with a row for the column header, and rows 2 and 3 for listing the following two sprinkler head failures: (1) opens when there is no fire and (2) remains closed when there is a fire. Fill out the remaining cells in each row.

3. For the example of sprinkler heads, imagine at least one possible "common cause" that would make multiple sprinklers fail to function properly.

TECHNICAL TERMS

Deductive approach	An approach to logical reasoning that uses a general principle or theory to reach a solution or conclusion about a specific set of facts.
Failure mode	Any of the ways an item could fail to function as intended.
Failure mode and effects analysis (FMEA)	A systematic method for analyzing a system for reliability or safety by starting with items, such as parts and components, determining how these items could fail (failure modes), and what would be the effects of each of these failure modes at a higher level of indenture.
Indenture level	The levels of elements that constitute a system. Indenture levels from higher to lower are systems, subsystems, assemblies, subassemblies, components, subcomponents, and parts.
Inductive approach	A way of logically analyzing by starting with details or facts and working toward a coherent conclusion. Detectives use this approach when they first gather evidence, and then try to develop an evidence-based explanation for the crime.
Safety-critical items	Parts, components, or assemblies that are essential and necessary for safe system operation and support.

REFERENCES

1. Bahr NJ. System safety engineering and risk assessment: A practical approach. New York: Taylor & Francis; 1996. p. 145–151.
2. Stephans RA. System safety for the 21st century. Hoboken, NJ: Wiley; 2004. p. 156–167.
3. Ericson CA, II. Hazard analysis techniques for system safety. Hoboken, NJ: Wiley; 2005. p. 235–259.

Chapter 7

Constructing Fault Trees

7.1 INTRODUCTION TO FAULT TREES

Fault trees are useful tools for understanding how undesired events may occur. Their main purpose is to help engineers design a system so that one or more specific events can be avoided. Thus, fault trees are tools for integrating safety into a system design process.

The need for a fault tree may become apparent during a risk assessment. Say the risk assessment identifies two hazardous events the organization is unwilling to tolerate or accept. For each of these, a fault tree could prove helpful for understanding both causal events and options for reducing risk.

Fault trees are deductive trees in the shape of evergreen trees, with a point at the top and widening at lower levels. This chapter addresses the processes and methods for constructing fault trees—starting with the basics and progressing to some more advanced methods. The next chapter discusses mathematical processes and other analytical methods for use with fault trees.

7.1.1 Common Symbols and Arrangements

A fault tree is a type of graphic analytical tree used to model a specific undesired event. This *top event* is represented by a box containing a concise description of the undesired event. Under the top event box are various symbols; the most common are shown in Figure 7.1.

Numerous variations in these symbols are used in commercial software. Those in Figure 7.1 are traditional with one exception. In this chapter, *logic gates* have text to help readers learn to recognize the two most common logic gates—*AND gates* and *OR gates*. System safety engineers do not need labels.

In fault trees, relationships between events and gates are shown as lines without arrows. Under the top event are tiers of other events organized into logical branches. As tiers descend toward the bottom, the events become more specific. At the bottom of

Risk-Reduction Methods for Occupational Safety and Health, First Edition. Roger C. Jensen.
© 2012 John Wiley & Sons, Inc. Published 2012 by John Wiley & Sons, Inc.

Traditional symbols	Explanation

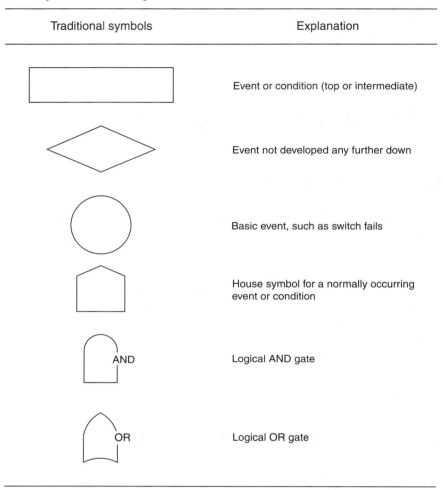

	Event or condition (top or intermediate)
	Event not developed any further down
	Basic event, such as switch fails
	House symbol for a normally occurring event or condition
AND	Logical AND gate
OR	Logical OR gate

Figure 7.1   Key for most common fault tree symbols.

each branch, a specific fault or failure serves as an end point. The generic example in Figure 7.2 illustrates terminology.

The top event of a fault tree is usually an undesired event or condition. Examples of undesired events are "person contacts high-voltage current source" and "fire starts in warehouse." Examples of undesired conditions are "slippery spot in grocery store aisle" and "carbon monoxide level in room exceeds 35 ppm." Typically, a system could have many undesired events and conditions. From these, a very small number may be chosen for modeling as fault trees. Rather than spending time on less significant outcomes, more value comes from time spent on high-severity events and very undesirable conditions.

After the top event is identified, a logic gate is placed under it. The AND gate and the OR gate are the two most common types. The AND gate indicates that the top

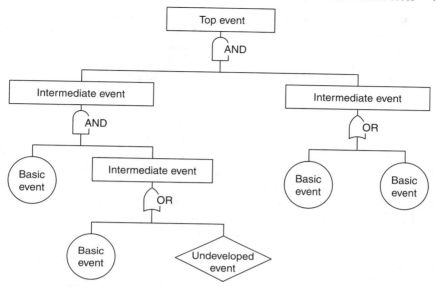

Figure 7.2 General locations of symbols on a typical fault tree.

event will occur only if all events immediately under it occur. The OR gate indicates that the top event will occur if any of the events under it occurs. One can test the logic of a tree with either of these gates by asking if the events under it are "necessary and sufficient" to cause the event above the gate.

The next tier down the fault tree is referred to as the first tier or tier 1. Tier 1 commonly consists of rectangles containing brief text to indicate an *intermediate event* or *intermediate condition*. Sometimes, tier 1 will have symbol of a house, a diamond, or a circle. The house is used for a normally occurring condition. For example, if the fault tree is for a fire in a warehouse, this tier could have a house for the existence of oxygen. When a house symbol is used, the branch ends because there is no reason to explain how a normally occurring condition exists.

Another symbol that may be used in tier 1 is a circle. It indicates a basic event that does not warrant further development, such as a specific part failing. A circle is inappropriate if there is a need to examine how the event could occur. To illustrate the difference, consider a crane rope snapping. An analyst may choose to show the crane rope snapping as a basic event in a circle. But doing so terminates further inquiry into the events leading to the rope snapping.

A rectangle is used when the event will have another gate under it. A diamond is used when a branch is extended to a point where the analyst cannot justify further time or effort to explain it. Thus, a diamond ends a branch with the message that further development is not considered worthwhile at this time.

The symbols shown in Figure 7.2 may vary somewhat in appearance. The main reason is that different software programs provide somewhat different graphics for the same symbol. This is particularly the case with the AND gate and the OR gate. Some software provides gate labels using a format such as G1, G2, and G3. Some show the probability of the event above the occurrence of the gate, and some will have no text of any kind. Also, some software facilitates entering text directly inside the symbol, while other software uses textboxes located atop a symbol.[1] Because of these variations, reports on fault tree analyses should provide a *symbol key* to indicate the meaning of each symbol. Examples of software-generated fault trees can be found in many articles in the Journal of System Safety and in the books on system safety cited in the reference list.

Two terms, *fault* and *failure*, often encountered in discussions of fault trees need explanation. The word fault applies when the desired state did not occur. A particular kind of fault, a failure, applies when a specific item in the system is unsuccessful. Thus, a fault may occur due to one or more component failures or due to some other cause such as an incorrect sequence of actions. In fault trees, the top box is for a fault. The circles representing failure events are found at the lowest end of tree branches.

The developed fault tree sorts out the various ways the top event can occur. A system that allows the top event to occur with only one or two basic failures may be referred to as a *failure-intolerant system*. In contrast, a well designed system can continue functioning even though several components have failed. That sort of system may be called a *failure-tolerant system*. A section in the next chapter explains how to use a fault tree to identify the smallest set of failures that can cause the top event.

The words fault and failure are used in the naming of two common system safety analysis tools introduced in this book. A fault tree begins with a fault (undesired event or condition) at the top and works downward through one or more tiers to the level of basic failures. In contrast, an FMEA begins with basic item failures and investigates what effects each failure might have on the tiers above it.

7.1.2 Example Fault Trees

Numerous books on system safety provide examples of fault trees.[1-7] Two common examples are a flashlight and a simple electrical circuit. These two examples, as well as a wildfire example, are provided in this chapter.

A fault tree for the flashlight example is shown in Figure 7.3. Note that the diamonds in the first tier indicate a decision by the analyst not to develop the branch further. If the left diamonds were developed, it would be similar to the middle branch with the batteries. The on/off switch branch was not developed further because the tree is to illustrate only tree construction.

Fault trees express logic graphically. Engineers tend to like graphic models, while many people prefer to express logic verbally. A verbal equivalent to the top event and first tier of Figure 7.3 is as follows. The flashlight will fail to provide light if

- The light bulb fails,
- There is a lack of battery power, or
- The on/off switch fails.

A verbal approach as this may be used as an aid in sorting out the logic of a tier prior to starting the drawing. Several Learning Exercises at the end of this chapter are intended

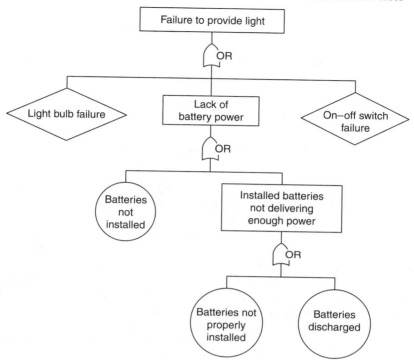

Figure 7.3 Example of a flashlight fault tree.

to help people appreciate how the verbal and graphic approaches provide equivalent expressions of the logic.

Look at the second tier of the middle branch. The two branches involve mutually exclusive conditions—the flashlight either has no batteries or has at least one battery. This approach is often a useful concept for developing fault trees. It forces logical development under an OR gate. In the event box for installed batteries, instead of simply saying the batteries are installed, we want to state some kind of failure event. In the above tree, it says "Installed batteries not delivering enough power."

For the second example of a fault tree, consider the simple electrical system modeled by the circuit diagram in Figure 7.4. The source of power is a battery. There are three switches in the hot side of the circuit (A, B, and C). These are normally in the open position. There is a light in the circuit. If someone wants the light on, they should flip toggle switches A, B, and C to the closed position. How might the pilot light fail to illuminate? First, draw a tree with the top event being "Fail to illuminate."

Under the top event, an OR gate is appropriate because any one of several faults will cause the top event. Any one of the three switches left in the open position will make the entire circuit open and ineffective. Also, if the bulb fails for some reason, or battery power is lacking for some reason, the bulb will fail to illuminate.

The overall arrangement of the fault tree shows the three switch faults grouped under a single branch. Another analyst could elect to place all five fault events in tier

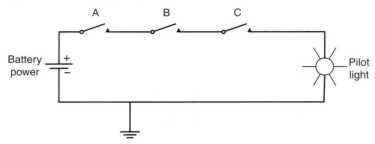

Figure 7.4   Circuit diagram with three normally open switches.

1—a perfectly logical option. Most fault trees can be organized in multiple ways. In the case of Figure 7.5, the analyst felt that the tree would be understood more easily by grouping the switches together in a branch.

The symbols in Figure 7.5 were chosen to illustrate some points about fault tree construction. In the left branch (open circuit path), an intermediate event box is used. The OR gate under it can be opened for movement upward if, and only if, at least one of the lower events occurs. The gate may be thought of as a one-way opening that, when conditions below are met, allows influence from the lower tier to the upper tier.

At the lowest tier in this tree, circles are used for the switch failures to indicate that these are basic events. Thus, the left branch ends properly, with basic events at the bottom of each branch. The middle branch (bulb failure) and the right branch (battery failure) are in diamonds. The diamond symbol indicates an event the analyst chose not to develop further. A different analyst may have elected to extend one or both of these branches to explain why these failures might occur.

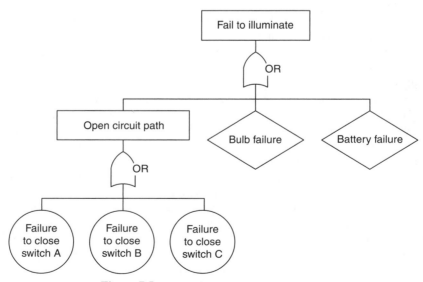

Figure 7.5   Fault tree for circuit in Figure 7.4.

Once a fault tree is completed, each branch should end with an appropriate symbol. Appropriate symbols to terminate a branch are the circle (basic event), the diamond (undeveloped event), and the house (normally occurring condition). A transfer triangle (shown later in this chapter) may also appear at the bottom of a branch on a page, but it does not mean the end of the branch. It means the branch is continued at another place in the report.

The electrical circuit example also serves to illustrate the importance of paying attention to how a failure might occur—the failure mode. A simple electrical switch can fail open or fail closed. In the preceding circuit, the top event will not occur if there is a failure in closing any of the switches. That means the switch stayed open when the system operators wanted it closed.

Our third example applies to an undesired fire event. The first point is the importance of making the top event specific. Simply putting "Fire" in the top event may lead to misunderstanding. For example, on a camping trip, you want to start a campfire, a fire is a success rather than a fault. Similarly, you want burners in your furnace to burn. Many other fires, however, are undesirable. Examples of undesired fires are kitchen grease fires, house fires in general, automotive engine fires, and fires in wildlands (e.g., forests, grasslands, and agricultural fields). A fault tree for a wildland fire is shown in Figure 7.6.

Under the top event, the AND gate indicates the top event will occur only if all four events in the first tier occur. Notice the use of diamonds for the two left branches. This indicates the analyst chose not to develop these branches further. It also simplifies the tree for the purpose of illustrating the use of an AND gate. It could, however, be productive to extend each of these branches. For example, the first tier provides nothing useful about the sources of heat, amount of heat energy needed, or that the fire needs heat to ignite as well as to continue. The second diamond tells us next to nothing about the vegetation sources encountered in wildland fires. Moving across the tree, the circle is used for chemical chain reaction. This is considered, for this fault tree, to be a basic event. It is actually a somewhat complicated interaction of heat level, fuel availability, fuel condition, weather, and supply of oxygen. The fourth branch of the tree indicates that oxygen occurs normally in outdoor locations on planet Earth.

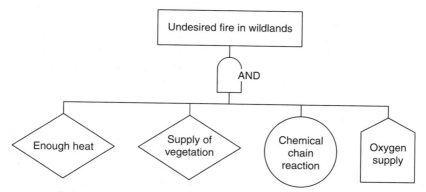

Figure 7.6   First tier of a fault tree for undesired fire in wildlands.

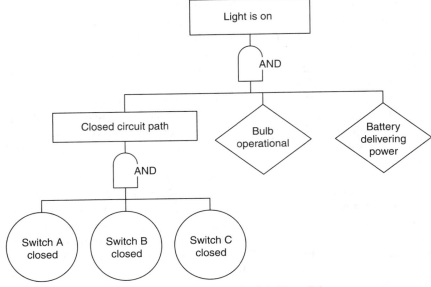

Figure 7.7   Success tree for circuit in Figure 7.4.

The three fault trees in Figures 7.3, 7.5, and 7.6 illustrate modeling an undesired event using symbols and logic applicable to fault trees. What if we want to model a desired event? For example, when camping we want to start a campfire, and we want our flashlight to illuminate when we switch it to the "on" position. For these applications, we can use the same symbols and similar logic, but we put the successful event at the top and call it a positive tree or a *success tree*.

7.1.3 Example Success Tree

The electric circuit depicted in Figure 7.4 was used to explain a fault tree, but it may also be used to explain a success tree. To achieve success, all three switches need to be closed, the bulb must be operational, and the battery must be delivering enough power. A success tree is shown in Figure 7.7.

Table 7.1 summarizes changes made in the fault tree (Figure 7.5) to convert it into a success tree (Figure 7.7). Notice that gates were flipped from being OR gates to being AND gates, and event descriptions inside the symbols were rephrased to change from stating undesired events to stating desired events.

7.1.4 Common Mistakes

When learning to construct fault trees, four mistakes are common. The first is using lines with arrowheads. Proper fault trees use simple lines to connect the various symbols. The second is ending a branch with an event box. Each branch should end with a proper *terminal event* symbol, a circle, a diamond, or a house.

Table 7.1   Converting a Fault Tree to a Success Tree

Element	Fault tree	Success tree
Gates	OR gate	AND gate
Top event	Fail to illuminate	Light is on
Left branch, tier 1	Open circuit path	Closed circuit path
Left Branch, tier 2	Failure to close switch A*	Switch A closed[a]
Middle branch, tier 1	Bulb failure	Bulb operational
Right branch, tier 1	Battery failure	Battery delivering power

[a] Same for switches B and C.

The third common mistake when constructing fault trees is to connect a gate directly to a gate under it. This construction makes the logic difficult for readers to follow and may even confuse the analyst. Thus, when constructing fault trees, avoid having a gate under another gate without an event or condition between them. Figure 7.8a illustrates a gate stacking error. A corrected tree is shown in Figure 7.8b. The correction involved inserting an event between the gates. For example, the new event will be whatever happens when both intermediate events C and D occur.

The fourth common mistake is skipping a tier. This may occur because the analyst writes in an event box the word "and" or the word "or." Consider this a red flag deserving careful review. When the word "and" is needed to describe an event, it may be more appropriate to have another tier under to box with an AND gate. To illustrate, an event box reading "Person slips and falls" is actually two events. A person can slip without falling, and a person can fall without slipping. Similarly, when the word "or" is needed to describe an event, it may be more logical to create another tier under the

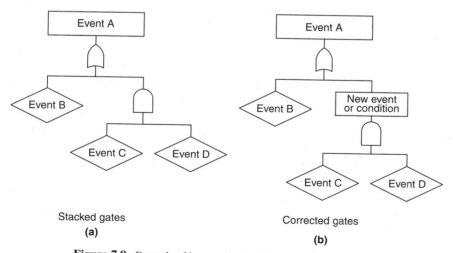

Stacked gates
(a)

Corrected gates
(b)

Figure 7.8   Example of improperly stacked gates and correction.

event box with an OR gate. An event box reading "Person slips or trips" is actually describing alternative events, not one event.

This concludes the introduction to constructing fault trees and success trees. Professionals who work with fault trees regularly find some additional symbols and techniques useful. Some of these are introduced next.

7.2 ADDITIONAL FAULT TREE TOOLS

Systems safety engineers who regularly work with fault trees find several additional symbols useful. Because standardization tends to be industry specific, it is essential to provide in reports a key to all symbols used. Figure 7.9 shows the symbols key for this discussion.

The first two symbols in Figure 7.9 show transfer triangles. These triangles are useful when a fault trees grows too large for a single page in a report. They may also be used to avoid repeating identical branches for particular failure events. As a reader examines the tree downward from the top event, a branch will seem to end with a triangle attached under an event box. There will be some sort of note to tell the reader where to look for further development of the branch. The note format in Figure 7.9 is from the book by Stephans.[7] For example, on page 1 of a tree, a transfer triangle may send the reader to page 4 to see the branch developed further down. To assist readers in following the logic of the branch, it is useful to duplicate the event and triangle at the top of the continuation page. The branch continues downward until it ends with an appropriate terminal failure or transfers to other pages. Whatever format convention is used, focus on clearly communicating your logic to readers who may attempt to understand your tree.

The DELAY gate symbol is used to show that once the event below occurs, some time passes before the event above occurs. For example, consider a mixing vessel in a chemical plant. The process was designed for a reaction involving two input chemicals and one output chemical. The chemical engineers designed it, so the inflow will equal the outflow. If something goes awry, such as material input exceeds output, there will be a buildup of material in the vessel. A pressure relief valve should open to release pressure. If it fails to open, the pressure inside the vessel will increase. The fault tree could make use of a DELAY gate to inform about how long it will take for the pressure buildup to exceed the vessel capability.

The *exclusive OR gate* symbol is used when the upper event or condition will occur if exactly one of the lower events occurs. If two or more of the lower events occur, the gate will block upward movement through it to the upper event.

The ellipses depicted in the lower rows of Figure 7.9 are used by professionals to clarify or supplement an attribute of the associated gate or event. For most fault trees, the developer can limit the use of ellipses by using AND gates and OR gates. In other words, if you develop a fault tree with numerous ellipses, you may want to revisit the tree to determine if your logic is clear.

The INHIBIT gate is above a single-input fault event. It may be used to indicate a factor that inhibits or facilitates movement upward through the gate.

Symbols	Description

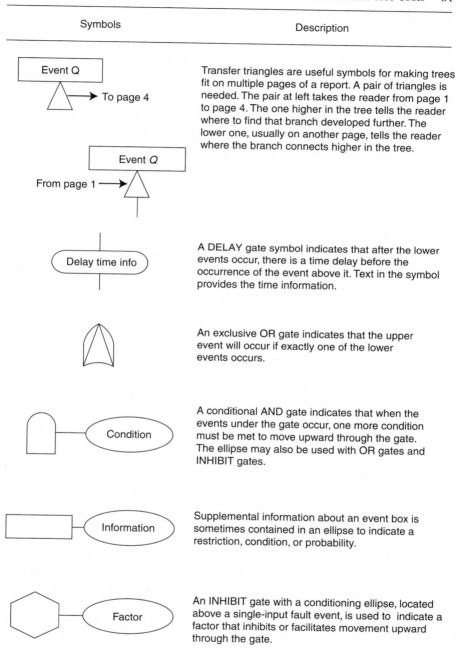

Transfer triangles are useful symbols for making trees fit on multiple pages of a report. A pair of triangles is needed. The pair at left takes the reader from page 1 to page 4. The one higher in the tree tells the reader where to find that branch developed further. The lower one, usually on another page, tells the reader where the branch connects higher in the tree.

A DELAY gate symbol indicates that after the lower events occur, there is a time delay before the occurrence of the event above it. Text in the symbol provides the time information.

An exclusive OR gate indicates that the upper event will occur if exactly one of the lower events occurs.

A conditional AND gate indicates that when the events under the gate occur, one more condition must be met to move upward through the gate. The ellipse may also be used with OR gates and INHIBIT gates.

Supplemental information about an event box is sometimes contained in an ellipse to indicate a restriction, condition, or probability.

An INHIBIT gate with a conditioning ellipse, located above a single-input fault event, is used to indicate a factor that inhibits or facilitates movement upward through the gate.

Figure 7.9 Key for additional fault tree symbols.

LEARNING EXERCISES

The outline form of expressing logic (see section 7.1.2) can be useful for developing logical fault trees and success trees. Several of these Learning Exercises are for developing abilities to translate between the verbal and the graphical approaches for expressing logic.

1. For the fault tree in Figure 7.3, write out the logic for the left branch, second tier, using an outline format. You should try three ways the bulb could fail.

2. Study the fault tree in Figure 7.6 depicting elements for an undesired wildland fire. Write out the logic for the first tier using the outline format.

3. What does the fault tree in Figure 7.6 tell you about strategies for extinguishing a fire?

4. Examine the success tree in Figure 7.7. Write in outline format the logic for the first tier.

5. This chapter contains several mentions of software for creating professional looking fault trees. A useful first step is to examine your software resources. You might try the integrated software suites to find graphic software. Try using whatever software you have to create a fault tree similar to that in each of the following.
   (a) Figure 7.3
   (b) Figure 7.6

6. A professor scheduled a test at 11 a.m. on Wednesday. A student might fail to show up for the test.
   (a) If you want to make a fault tree for this, what text would you put in the top event box?
   (b) What type of gate would you put under the top event?
   (c) List reasons for this "no-show."
   (d) Group the reasons for no-show into crisp categories suitable for placing under the gate in a fault tree. If you have more than five reasons, consider consolidating some of the reasons into categories suited to a fault tree.
   (e) Use the outline format to write the final version of the logic you would use to make a fault tree for the student failing to show up for the test.

7. Suppose you are safety manager for a package shipping company such as UPS. Many packages are shipped by air during the night. After arrival at local airports, the packages are loaded into delivery vans or trucks. Past experience indicates that occasionally one of your drivers is cited by police for driving a vehicle without two headlights shining. You have not had any expensive losses, but you want to be proactive to avoid a large loss in future. Consider using a fault tree to better understand how this could happen and how it might be prevented in the future. You would need to decide whether to make the top event an undesired event or an undesired condition.
   (a) What would you put as the top event for an undesired event?
   (b) What would you put as the top event for an undesired condition?

8. In the fault tree for a flashlight failing to provide light (Figure 7.3), there are three branches under the OR gate in tier 1. Under the middle branch, there is another OR gate, and under it are two branches (tier 2).
   (a) Would it be logically correct to simply raise the pair of branches in tier 2 up to tier 1?
   (b) What reason can you offer for using the branching system shown in this chapter, rather than adding more events to the first tier?
   (c) What if we were to replace the box for "lack of battery power" with the three basic (circle) events in the middle branch. That would make the entire fault tree have only one tier. Would it be logically correct?

9. Gain experience constructing logic trees by making a success tree for making a campfire. Try including tiers 1 and 2.

TECHNICAL TERMS

Basic event	Failure of a specific item, depicted in a fault tree as a circular symbol with text describing the failure.
Failure	Term used for the event of an item not functioning as intended. Failures are a subset of faults.
Failure-intolerant systems	Term for systems that can be substantially harmed by the failure of one or very few parts.
Failure-tolerant systems	Term for systems that were designed and built to withstand occasional failures of one or more parts.
Fault	Word indicating that the desired state did not occur.
Fault tree	A top-down arrangement of symbols connected with logic gates and lines to indicate how failures at lower levels can cause undesired events and conditions higher in the tree.
Intermediate condition	A condition within an analytical tree located below the top event and above the terminal event in a branch.
Intermediate event	An event within an analytical tree located below the top event and above the terminal event in a branch.
Logic gates	Symbols in analytical tree, located between tiers, serving the function of allowing or disallowing passage from the lower tier to the next higher tier. An AND gate allows passage if all requirements under it are met. An OR gate allows passage if any one of the requirements under it is met.
Success tree	A top-down arrangement of symbols connected with logic gates and lines to indicate how success of the top event depends on the success of numerous lower events.
Symbol key	An explanation of the symbols used in a fault tree or other graphics.
Terminal event	An event at the bottom of a branch of an analytical tree.

Top event   The event located at the top of an analytical tree, shown
                                               as a rectangle. In fault trees, it is for an undesired event
                                               or condition. In success trees, it is for a desired event or
                                               condition.

REFERENCES

1. Ericson CA, II. Hazard analysis techniques for system safety. Hoboken, NJ: Wiley; 2005.
2. Bahr NJ. System safety engineering and risk assessment: A practical approach. New York: Taylor & Francis; 1997.
3. Brown DB. Systems analysis and design for safety. Englewood Cliffs, NJ: Prentice Hall; 1977.
4. Hammer W. Handbook of system and product safety. Englewood Cliffs, NJ: Prentice Hall; 1972.
5. Roberts NH, Vesely WE, Haasl DF, Goldberg FF. Fault tree handbook. Report No. NUREG-0492. Washington, DC: U.S. Nuclear Regulatory Commission; 1981.
6. Roland HE, Moriarty B. System safety engineering and management, 2nd ed. New York: Wiley-Interscience; 1990.
7. Stephans RA. Systems safety for the 21st century. Hoboken, NJ: Wiley; 2004.

Chapter 8

Analyzing Fault Trees

The previous chapter dealt with constructing fault trees. This chapter discusses methods for analyzing fault trees by computing probabilities, finding cut sets, and considering common-cause failures.

8.1 QUANTITATIVE ANALYSIS BASED ON FAULT TREES

The *quantitative analysis* of risk helps numerous individuals involved or affected by the systems designed and built to serve a function. The engineers working on a system design project consider risk data to recognize safety-related weaknesses so they can make improvements before completing the design. Managers of the organizations involved want risk data to help them decide whether to accept particular residual risks. If the system might affect a local community, the risk data may help residents feel more confident that the company has taken risk seriously and incorporated appropriate risk-reduction tactics into the facility design. An effective approach is to provide risk information in a quantitative format, and fault trees provide a useful foundation for choosing proper equations to compute risk probabilities. Because the computations are based on probability, this chapter starts with a brief review of applicable probability.[1]

8.1.1 Applicable Probability

Probability is a branch of mathematics dealing with chance occurrences. Basic probability is a value in the range of zero to one that indicates the likelihood of a defined event resulting from a trial, experiment, or occurrence. For example, a crane moves an I-beam from the ground to position on a structural frame of a building. Outcomes of that occurrence can be categorized as no damage or damage. Another example, using a time-based approach, is an industrial operation that is performed for 100 h. The outcomes events can be categorized as having functioned successfully or unsuccessfully.

In this book, the probability of occurrence of event X is represented by $P(X)$. Other equivalent notations are P_X and $Pr(X)$.

Risk-Reduction Methods for Occupational Safety and Health, First Edition. Roger C. Jensen.
© 2012 John Wiley & Sons, Inc. Published 2012 by John Wiley & Sons, Inc.

Outcome of one employee hour   Outcome of drive to work
            (a)   (b)

Figure 8.1 Two examples of Venn diagrams.

A useful tool for understanding probability is the Venn diagram—a visual model of the possible outcomes of a single trial, experiment, or occurrence. A rectangle represents the entire sample space of possible outcomes. Outcomes of special interest, called events, are shown as circles within the rectangular sample space. For our purposes, the circles represent the unwanted events. For example, consider the specific instance of an employee working for 1 h. The outcomes of that working hour could be classified as not injured or injured. Figure 8.1a depicts these outcomes in the Venn diagram. Another example illustrates the application of probability to transportation. Consider the specific instance of a person driving a car from home to work. Outcomes of the journey may be classified into three categories: (1) arrived on time, (2) arrived late, or (3) failed to arrive that day. In the Venn diagram in Figure 8.1b, circles represent the two undesired events.

For fault tree analyses, probability values for the failure events at the bottom of the tree are needed. These values may come from assumptions, empirical data, or estimates. Probability values from assumptions are often used as examples in text-books on probability and statistics. Examples commonly used are coin tosses, which assume $P(\text{heads}) = P(\text{tails}) = 0.5$, and the roll of a single die, which assumes each die is balanced so $P(\text{each side}) = 1/6$. Probability values from empirical data are obtained from multiple trials conducted as formal experiments or from past records of similar events. Probabilities from estimates may be developed from discussion among members of a system design team.

Using the probability values for lower events in a fault tree, the probabilities of events at the next higher level can be computed, and from those values, probabilities for progressively higher levels can be computed until the top event is reached. The formulas for these computations are explained next.

The equations presented in this chapter use algebra. In many of the system safety books, and probability textbooks, the preferred computations use Boolean algebra. However, Boolean algebra is not essential for typical applications encountered in occupational safety and industrial hygiene. Readers interested in learning about the Boolean approach may consult other books.[2–4]

8.1.2 AND Gates

Figure 8.2 includes four ways to think about fault tree analysis. Two graphic ways are the Venn diagram and the corresponding fault tree. Two other ways of

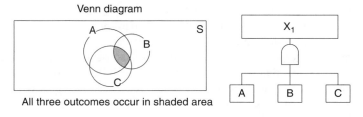

$P(X_1)$ = Probability of A,B, and C occurring

$P(X_1) = P(A) \times P(B) \times P(C)$

Figure 8.2   Understanding the AND gate.

expressing the probability of X_1 are the sentence format expression and the algebraic formula.

The three circles in the Venn diagram correspond to the three failure events in the first tier of the fault tree. The fault tree indicates that the top event (X_1) will occur if, and only if, events A, B, and C occur.

A small space in the middle of the Venn diagram shows where all three circles intersect and that space represents failure X_1. Thus, within the large sample space (S), all events are successes except the highlighted space in the middle. The probability of X_1 is the algebraic product of the probabilities of the events A, B, and C. The equation $P(X) = P(A)P(B)P(C)$ may be used for most AND gates located above three events in a fault tree. However, this formula assumes that the probabilities of the events are independent.

When the assumption of *independence* does not apply, the equation needs modification to account for the dependence. An example of dependence would be if $P(C)$ is initially 0.005, but when event B occurs, the value of $P(C)$ changes to 0.01. Thus, the probability of C, given that B had occurred, is $P(C|B) = 0.01$. In such case, the latter value goes into the formula instead of $P(C)$.

An example may help to illustrate the value of having a situation in which multiple failure events must occur before the top event occurs. Suppose that A, B, and C are independent events with occurrence probabilities of 0.01, 0.005, and 0.002, respectively. The probability of the top event, X, is computed using the multiplication formula given previously:

$$P(X) = P(A)P(B)P(C),$$

$$P(X) = (0.01)(0.005)(0.002),$$

$$P(X) = 0.000\,000\,1 = 1 \times 10^{-7}.$$

Notice that event X is very unlikely to occur (one chance in 10 million). This is why we like to have an AND gate under an undesired event—because it takes multiple failure events to make the top event occur.

If there are more than three events under the AND gate, the probability of the upper event can be calculated by multiplying the probabilities of all the lower events. It is just a simple extension of the previous equation. Thus, if a fault tree has event X above an AND gate, and events A, B, C, D, . . ., N below it, $P(X)$ is calculated using Equation 8.1:

$$P(X) = P(A)P(B)P(C)P(D) \cdots P(N). \tag{8.1}$$

This is the general equation for an AND gate with any number of independent events under it.

8.1.3 OR Gates

An OR gate means the top event will occur if any of the lower events occurs. Two methods for computing the probability of the top event are presented here. The first method yields an accurate value when the events under the gate are mutually exclusive and yields an approximate value otherwise. Figure 8.3 depicts the mutually exclusive case with a Venn diagram, a fault tree, a sentence format, and an equation.

The approximation formula shown in Figure 8.3 applies to three events under an OR gate. The more general formula in Equation 8.2 allows adding the probabilities of any number (n) of events (E_i) under the OR gate.

$$P(X) \approx \sum P(E_i). \tag{8.2}$$

Figure 8.4 depicts the situation where the events are not mutually exclusive (i.e., the event spaces in a Venn diagram overlap). The shaded area in the Venn diagram is the union of the three circle events.

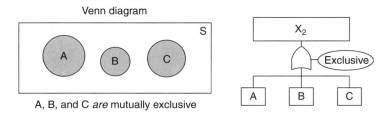

Venn diagram

A, B, and C *are* mutually exclusive

$P(X_2)$ is the probability of A or B or C occurring when the events are mutually exclusive.

$P(X_2) = P(A) + P(B) + P(C)$

Figure 8.3 The OR gate for mutually exclusive events.

Venn diagram

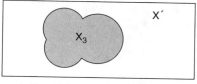
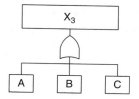

A, B, and C *not* mutually exclusive

Events not mutually exclusive under *OR gate*.
From full sample space, $P(S) = 1.0$, subtract probability of
the white space: $P(A') P(B') P(C')$

$P(X_3) = 1.0 - P(A') P(B') P(C')$
where
$P(A') = 1.0 - P(A)$
$P(B') = 1.0 - P(B)$
$P(C') = 1.0 - P(C)$

Figure 8.4   The precise formula for calculating the probability of three events under an OR gate.

The formulas in Figure 8.4 use the notation that the probability of a fault event is $P(X)$ and the probability of the background event is $P(X')$. Together these represent all possible outcomes of an event. They are called complementary because together they make up the whole. Thus, we can say

$$P(X) + P(X') = 1.0 \tag{8.3}$$

or for fault events

$$P(\text{fault}) + P(\text{not fault}) = 1.0.$$

When an expression for $P(X)$ or $P(X')$ is needed, simply rearrange Equation 8.3 to obtain Equations 8.4 and 8.5.

$$P(X') = 1.0 - P(X), \tag{8.4}$$

$$P(X) = 1.0 - P(X'). \tag{8.5}$$

Equation 8.5 is the one for computing the probability of the event above an OR gate. The key to using it is getting the correct values for $P(X')$. A three-step algebraic approach is explained below.

The first step in OR gate computations is to determine probability values for the complements of each failure event under the gate. This is achieved for each circled event by subtracting the failure probability from 1.0 according to Equation 8.4. If the events are denoted A, B, and C, we would compute $P(A')$, $P(B')$, and $P(C')$. The second step is to apply an adaptation of Equation 8.1 to compute $P(X')$ by multiplying the complements of every circled event in the sample space:

$$P(X') = P(A')P(B')P(C')$$

The third step is to use Equation 8.5 to compute $P(X)$.

An example should make this clear. Suppose the sample space represented by the Venn diagram has three circled fault events. The probabilities of events A, B, and C are 0.01, 0.02, and 0.005, respectively. To compute the probability of X, first use Equation 8.4 to determine probabilities of the complements of each failure event:

$$P(A') = 1 - P(\text{A}) = 1 - 0.01 = 0.99,$$

$$P(B') = 1 - P(\text{B}) = 1 - 0.02 = 0.98,$$

$$P(C') = 1 - P(\text{C}) = 1 - 0.005 = 0.995.$$

Second, insert these values into Equation 8.1 to compute $P(X')$:

$$P(X') = P(A')P(B')P(C'),$$

$$P(X') = (0.99)(0.98)(0.995),$$

$$P(X') = 0.9653.$$

The third step is to use Equation 8.5 to compute $P(X)$:

$$P(\text{X}) = 1.0 - P(X'),$$

$$P(\text{X}) = 1.0 - 0.9653 = 0.0347.$$

The results may be compared with the value obtained with the approximation method (Equation 8.2). Using the same event probability values, the approximation formula yields

$$P(\text{X}) = P(\text{A}) + P(\text{B}) + P(\text{C}) = 0.01 + 0.02 + 0.005 = 0.0350.$$

This computed value is very close to that obtained using the first method. This will be the case when the probabilities of all the events being multiplied are low, as in the example.

This concludes the discussion of basic building blocks for quantitative fault tree analysis. The next two sections introduce qualitative methods for identifying cut sets and finding common-cause failures.

8.2 IDENTIFYING CUT SETS

A very practical aspect of fault tree analysis involves identifying sets of events that could cause the top event. Consider a design team working on a system development project. They would like to know what basic failures and faults are most critical to safety. With that information, the team can adjust its design to avoid those events.

The term *cut set* refers to a collection of basic failures that will cause the top failure event. Finding all the cut sets in a small fault tree may be achieved by examining the branches and thinking through combinations needed. But as fault trees expand downward and laterally, identifying cut sets becomes more complex. In addition, system designers are particularly concerned about which cut sets involve the fewest number of basic failures. Basic events in these *minimum cut sets* provide logical targets for design team emphasis.

Nearly every book on system safety describes at least one systematic method for locating cut sets and minimum cut sets. For this book, a method suitable for OSH projects is presented.[5] It involves systematically working from the top of the tree down through each tier to find the various cut sets. A basic step in identifying the most critical failures involves identifying sets of basic events that could cause the top event. The analyst first examines each set for redundant events, dropping all but one, and then reviews the remaining sets to find those with the minimum number of basic event. The idea is to design the system so the top event will occur only if numerous basic failures occur. An example is the best way to explain how the method is used.

Figure 8.5 shows a fault tree with two branches in the first tier. Begin by assigning labels to the various events. A convention for identifying events is to use numbers for basic events and letters for intermediate events. Each basic event gets a unique number and that number is used even though the event is part of multiple branches. In this illustration, basic events 1, 2, and 3 happen to be in both branches.

The general approach is to work downward from the top event. Sets of events are recorded in a tabular format. When working downward through an AND gate, record each event under the gate in a separate column. When working downward through an OR gate, record each set in a new row of a table. To illustrate with this example, the AND gate requires two events (A and B) to produce the top event.

The example uses a spreadsheet setup as shown in Table 8.1. The leftmost column is for the number or each step in the process. The second column from the left is for a set identifier of the form "*n.m*" where *n* is the number of the step and *m* is a number assigned to each set created in that step. The next columns are the worksheet columns; in this case, only four columns are needed. Further to the right is a column for

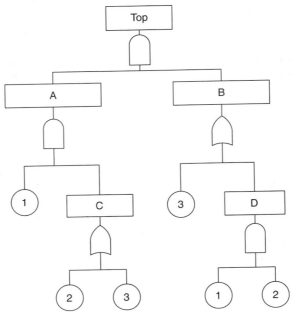

Figure 8.5 Fault tree to illustrate cut sets.

recording how the analyst created that set. The far right column asks if the set is a cut set. The answer will be "yes" when the set contains only numbers.

The first step is to record events A and B in separate columns of the worksheet. After this step, the goal is to replace all intermediate events (the letters) with basic events (the numbers). In Step 2, the left branch of the tree is taken to the second tier while keeping the right branch unchanged. This involves working downward from A through an AND gate. At the second tier, the A branch divides into two more branches: basic

Table 8.1   Filled Worksheet for Identifying Cut Sets

Step	Set ID	C1	C2	C3	C4	Procedure	Cut set?
			Worksheet columns				
1	1	A	B			Record first tier events in top row	No
2	2	1	B	C		Replace A with 1 and C. Keep B as is	No
3	3.1	1	3	C		From set 2, replace B with 3	No
3	3.2	1	D	C		From set 2, replace B with D	No
4	4.1	1	3	2		From set 3.1, replace C with 2	Yes
4	4.2	1	3	3		From set 3.1, replace C with 3	Yes
4	4.3	1	D	2		From set 3.2, replace C with 2	No
4	4.4	1	D	3		From set 3.2, replace C with 3	No
5	5.1	1	1	2	2	From set 4.3, replace D with 1 and 2	Yes
5	5.2	1	1	3	2	From set 4.4, replace D with 1 and 2	Yes

event 1 and intermediate event C. These two events plus event B make a set that satisfies the top event, and this set is recorded in the second row of the table. Only one event is allowed in a cell, so it takes three columns to record the three events in the set. One of those events is recorded in column C1. Column C2 retains the B event, and column C3 is used for the second event under A. Thus, the second row contains the set {1, B, C}.

Step 3 takes on the second tier of the right branch. It has an OR gate indicating that event B will take place if either basic event 3 or intermediate event D occurs. These are entered into the worksheet as shown. Since this is an OR gate with two events under it, two rows are needed. These rows are for set 3.1 and set 3.2. The two sets are {1, 3, C} and {1, D, C}. Neither of these sets is a cut set because each contains one or more letters. So, "No" is entered in the far right column of each of the rows.

Step 4 continues the effort to replace intermediate events with basic events. This time intermediate event C in sets 3.1 and 3.2 is replaced by the basic events under C (events 2 and 3). Since these are under an OR gate, two rows are needed to replace set 3.1, and two more rows are needed to replace set 3.2. Thus, set 3.1 will result in two new sets: {1, 3, 2} and {1, 3, 3}. These cut sets are entered in the table as set 4.1 and set 4.2, respectively. Similarly, to replace C in set 3.2, two more rows are needed. These are entered as sets 4.3 and 4.4. Neither of these new sets, {1, D, 2} and {1, D, 3}, is a cut set. Step 5 involves replacing the only remaining intermediate event in set 4.3 and set 4.4. Replacing event D in each set with basic events 1 and 2 leaves cut sets in sets 5.1 and 5.2.

The Table 8.1 worksheet now shows four rows with numbers and no letters. These cut sets are examined to identify redundant basic events. In set 4.1, there are no redundant events. Set 4.2 has one redundant element (event 3). The extra 3 should be dropped to yield the set {1, 3}. Set 5.1 repeats both event 1 and event 2, so one of each should be dropped. That leaves the following cut sets:

$$\text{Set } 4.1 = \{1, 3, 2\},$$

$$\text{Set } 4.2 = \{1, 3\} \quad \text{a minimum cut set,}$$

$$\text{Set } 5.1 = \{1, 2\} \quad \text{a minimum cut set,}$$

$$\text{Set } 5.2 = \{1, 3, 2\}.$$

With this analysis method, errors by the analyst are foreseeable. It is therefore suggested that each cut set be verified by shading a printed version of the tree as illustrated in Figure 8.6. Parts (a) and (b) of the figure show the minimal cut sets {1, 3} and {1, 2}, respectively. The verification procedure starts at the bottom of the tree and works upward. In the upper tree, basic events 1 and 3 are shaded first, and the paths upward are traced by shading the gates and events that provide an upward pathway to the top event. In the lower tree, events 1 and 2 are shaded, and the path upward is shaded in the same manner.

Finding minimal cut sets provides useful information to system designers. However, caution is needed to avoid focusing all attention on the minimum cut set

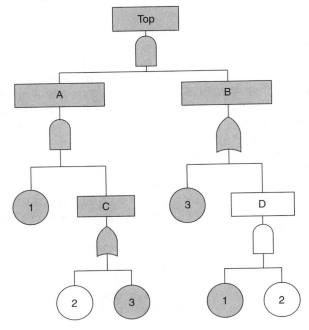

(a) Minimum cut set {1,3}

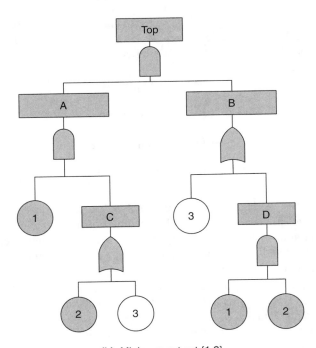

(b) Minimum cut set {1,2}

Figure 8.6   Two minimum cut sets for example fault tree.

paths. There are typically many cuts sets. Although the minimal cut sets warrant the most attention, others should not be ignored. This is illustrated in a study about a bridge collapse analyzed with a fault tree.[6] There were cut sets containing a single element, such as a plane crashing into the bridge, and a ship colliding with a span due to being too tall for the clearance. But the failure path that actually caused the collapse consisted of multiple elements. The concrete footing of one of the four support piers deteriorated after 32 years of water erosion. This led to a failure in the footing and collapse of the pier. Soon after the initial pier collapsed, adjacent piers were overloaded and failed. Ten people were killed.

8.3 FINDING COMMON-CAUSE FAILURES

Redundancy has made possible very respectable safety records for large and complex systems with high potential for disasters such as those found in aviation, nuclear power generation, and chemical processing plants. However, disasters involving these complex systems have occurred, and lessons have been learned from the investigations. Among these lessons is one about not allowing a safety-critical function to fail due to a single cause that affects multiple items in the system—a failure known as a *common-cause failure*.

Consider the fault tree in Figure 8.7. The top event is the failure of a safety-critical subsystem. Under it is an AND gate. In tier 1, assemblies A, B, and C are considered redundant. Each assembly has a different design, but some parts are the same. A design team may look at this and, using Equation 8.1, calculate that the probability of all three redundant assemblies failing is very small. This would make the design team quite confident about avoiding the top event. However, further examination of the system will reveal a common-cause failure.

Tier 2 has subassemblies D, E, F, G, and H. Tier 3 has components J and K. The basic failure events are parts. Look first at the left branch, starting at the bottom tier. If part 3 fails, component J will fail, subassembly D will fail, and assembly A will fail. Thus, the failure of part 3 is a single cause for the failure of assembly A. Now look at the middle branch. If part 3 fails, subassembly E will fail. The failure of E will cause assembly B to fail. Now look at the right branch. If part 3 fails, component K will fail. That failure will pass upward to make subassembly H fail, and this will make assembly C fail. From all this, we can see that a failure of a single item (part 3) will cause failure of each of the three assemblies—assemblies initially considered redundant.

The fault tree in Figure 8.7 illustrates how a safety-critical function can fail due to a single part failure even though it has three assemblies under an AND gate. The source of this problem is the interdependence of the three assemblies. This lack of independence means that simply multiplying their failure probabilities yields an unrealistic value for the upper event probability.

How might the vulnerabilities to safety-critical functions originate? Ericson provides several examples, some of which are mentioned here.[7] In one example, an airliner crashed after multiple hydraulic lines were damaged by a single event in the rudder area. This led to loss of hydraulic fluid in multiple lines, loss of control of the

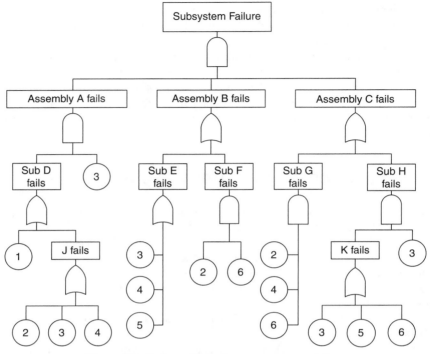

Figure 8.7   Fault tree illustrating a common-cause failure.

aircraft, and a disastrous crash. Common causes may also stem from two redundant assemblies designed and manufactured by the same manufacturer. In that situation, a defect in design, a part, or the manufacturing could make both assemblies vulnerable to failing at the same time for a common reason. In other words, the two assemblies are not independent. Similarly, a common cause may stem from having the same firm perform maintenance on two or more redundant assemblies. A single maintenance specialist might make the same error on all similar assemblies. Other origins involving errors may include errors in the specifications, design, production, and installation of the components and assemblies. Another potential source is unexpected outputs from software shared by elements of multiple safety-critical subsystems. Environmental influences may also adversely affect performance of multiple elements of components that support redundant assemblies. These are just a few of the possible origins of common causes. Quite often, common cause origins are easy to overlook and challenging to find.

The systems for which common-cause failure analysis may be used range in complexity. For less complex systems such as that in Figure 8.7, a good fault tree and some attention may be enough to identify common-cause failures. For complex systems, a systematic approach is needed. Typically, a project is funded for a team of engineers and system safety professionals to perform the common cause analysis. For OSH professionals who need to learn more about the process, the Ericson book has an excellent chapter on common-cause failure analysis.[7]

8.4 SUMMARY OF PART II

Part II consists of five chapters on methods OSH professionals can use for risk identification and analysis. The first (chapter 4) introduces the tool known as job hazard analysis (JHA). The chapter provides a typical format and discusses each of the three columns for recording the analysis. The JHA development process begins by breaking down a task into steps. The analyst then identifies any potential hazardous events and exposures in each step and records in the third column appropriate hazard controls for each potential hazard. Two variations on the basic JHA are mentioned. The phase hazard analysis serves as a practical tool for integrating safety and health into construction projects. The position hazard analysis serves as a tool for documenting that the employer thought through the safety-related requirements of each position and shared that information with each employee.

The second methodology, risk assessment (RA), is described in chapter 5 as a widely used process for businesses involved in making and distributing products and equipment to consumers, businesses, and governments. RA also provides a structured approach for OSH professionals to assess occupational systems to improve understanding of potential hazards and identify suitable risk-reduction tactics. The ability to apply RA processes is becoming an essential tool for careers in occupational safety, risk management, and industrial hygiene.

Chapter 6 introduces the third methodology—failure mode and effects analysis (FMEA). It is useful for identifying items in a system most critical for system performance and safety. The direction of analysis is opposite to that used for constructing fault trees. To construct a fault tree, the analyst starts at the top of an imaginary evergreen tree and works downward. To perform an FMEA, the analyst starts low in the tree and works upward. The methodology of FMEA uses a tabular form with rows and columns to structure analysis of specific items. For each item, possible failure modes are entered into the form. Each mode is characterized in terms of effects farther up the tree.

The fourth methodology, described in chapter 7, is constructing fault tree diagrams. A well-constructed fault tree provides insight into the causes of a specific undesired event or condition. The tree can help system engineers identify vulnerabilities and safer design options. Constructing these diagrams requires knowledge and comprehension of the system, as well as an ability to synthesize the information into a logical graphic model.

The fifth methodology in part II (chapter 8) extends the fault tree construction topic by explaining three methods for analyzing fault trees. First, fault trees provide a basis for computing the probability of the top event by starting at the bottom of the tree and using failure probabilities of basic events to compute the probability of events at progressively higher tiers. Second, fault trees provide a foundation for identifying the sets of basic events that can cause the top event, including identifying the minimum cut sets. Third, fault trees provide a structure for seeking out possible common-cause failures—obscure situations in which a single cause can have detrimental effects on multiple items initially thought to be independent and redundant. All five chapters in section II concern methods currently used by OSH professionals and their associates to reduce the risks of harmful events and exposures.

LEARNING EXERCISES

1. Consider the case of an organization with 40 full-time employees that during the course of a year had two recordable injuries involving two different individuals.
    (a) Sketch a Venn diagram to depict the outcomes (injured or not injured) for 1 year of work by one individual employee.
    (b) If no changes are anticipated in the work or the OSH program, we can project the same results next year. What is the projected probability of an individual employee being injured on the job next year?
    (c) What is the probability of an individual employee completing next year without an injury?
    (d) Venn diagrams often use the size of circles to represent the proportion of total space in the diagram applicable to the specified outcome. If that were done in this case, what portion of the space in the rectangle should be allocated to the injury circle?

2. Consider a fault tree with an AND gate under the top event. The first tier contains four failure events with respective failure probabilities: 0.005, 0.002, 0.01, and 0.0001.
    (a) What equation should you use to compute the probability of the top event?
    (b) Using the equation, what numerical value do you get for the top event probability?
    (c) For the top event, what is the probability value of the complementary event space (i.e., success probability)?

3. This is an exercise in probability computations involving a two-tier fault tree. It is the same example fault tree as used in chapter 7, Figure 7.5. For the example, use the following probability values for each event.

Failure to close switch A	0.001
Failure to close switch B	0.001
Failure to close switch C	0.001
Bulb failure after 2000 h of operation	0.03
Battery not working at 2000 h	0.005

   (a) Given the probabilities above, compute the probability of an open circuit path using the most accurate method.
   (b) Say you have used the flashlight for 2000 h with the same bulb. The batteries have been changed occasionally, but the overall probability of failing remains 0.005. What is the probability of the top event occurring when you turn the flashlight on? Use the most accurate method and your answer to (a) for calculating your answer.
   (c) What if the tree had been constructed by putting all five failure events in the first tier? Show the probability calculations for that arrangement

and compare the computed value with your answer to (b). Use Equation 8.5.

(d) Compute (c) using Equation 8.2. Compare the probability values you obtained using the two equations.

4. For the two qualitative methods described in this chapter, concisely distinguish between their purposes.

5. Regarding minimum cut sets, the discussion mentioned a bridge collapse. What point was made?

6. Part II of this book has five chapters. What was the rationale for grouping these topics into this part?

TECHNICAL TERMS

Common cause	A single origin for the failure of multiple system components, assemblies, or subsystems.
Common-cause failure	Phrase for multiple failures of system components, assemblies, or subsystems originating from a common cause, usually referring to failure of safety-critical items.
Cut set	Term used in fault tree analyses to describe a collection of basic failures that, if all occur, will cause the top event.
Independence	In probability theory, N events in a sample space are independent if the probability of any event is unchanged by the occurrence of another event.
Minimum cut sets	Term used in fault tree analysis for the cut set, or cut sets, containing the least number of events. Also called minimal cut sets.
Quantitative analysis	Term referring to a numerical-based analysis.

REFERENCES

1. Walpole RE, Myers RH. Probability and statistics for engineers and scientists, 4th ed. New York: Macmillan; 1990. p. 1–33.
2. Ericson CA, II. Hazard analysis techniques for system safety. Hoboken, NJ: Wiley; 2005. p. 198–204.
3. Roberts NH, Vesely WE, Haasl DF, Goldberg FF. Fault tree handbook. Chapter XII. Report No. NUREG–0492. Washington, DC: U.S. Nuclear Regulatory Commission; 1981. Available at www.nrc. gov/reading-rm/doc-collections/nuregs/staff/sr0492 (accessed July 6, 2011).
4. Andrews JD, Moss TR. Reliability and risk assessment, 2nd ed. New York: ASME; 2002. p. 201–267.
5. Stephans RA. System safety for the 21st century. Hoboken, NJ: Wiley; 2004. p. 178–180.
6. LeBeau KH, Wadia-Fascetti SJ. Fault tree analysis of Schoharie Creek Bridge Collapse. J Performance Constructed Facilities. 2007;23(4):320–326.
7. Ericson CA, II. Hazard analysis techniques for system safety. Hoboken, NJ: Wiley; 2005. p. 397–421.

Part III

Programmatic Methods for Managing Risk

Part III contains five chapters about risk-reduction methods implemented through OSH programs. Chapter 9 addresses incident investigation programs starting with a description of a closed-loop process to ensure that lessons learned are used to strengthen operational weaknesses. It presents policy issues to address an organization-specific incident investigation policy. It explains tools to help incident investigators organize evidence into a series of events and conditions leading to and following the harmful event. It concludes with a graphic model depicting the interrelationships among (1) the regular functioning system operating normally, (2) an initial deviation from normal going uncorrected, (3) the end of control, (4) subsequent events leading to a harmful outcome or near miss, and (5) post-incident events affecting the ultimate outcome.

Chapter 10 addresses human errors, explaining the concept of errors in relation to system tolerance, summarizing a multidisciplinary classification system, and offering two analysis methods for addressing safety-related errors and rule violations. Chapter 11 proposes consolidating all risk-?reduction strategies into nine categories and explains how these fit into the traditional three-level risk-reduction priority scheme: (1) eliminate the hazard, (2) use engineering controls, and (3) use administrative controls.

Chapter 12 discusses some common components of most OSH programs, starting with the importance of defining the organization's OSH program aspirations, followed by essays on training, warnings, safety devices, emergency preparedness, and sanitation/housekeeping. Chapter 13 provides an overview of OSH program management, specifically addressing the overarching topics of safety culture, management systems, and ethical policies.

Chapter 9

Incident Investigation Programs

People investigate harmful incidents for various reasons.[1] For safety professionals, the primary reason for investigating incidents is to learn about system weaknesses and vulnerabilities so that corrective actions can be identified and implemented, and a secondary reason is to comply with recordkeeping requirements. For legal advisors, the main interest is to properly obtain evidence in order to prepare for possible litigation. For the news media, a single "cause" is sought for a news story. For regulatory agencies, the main reason is to determine what regulations may have been violated. In addition to these common reasons, an incident investigation may be conducted for at least three other reasons. One is to look for evidence of criminal behavior, such as a fire started by an arsonist. A second is to determine the economic value of losses in order to obtain some financial indemnity from an insurer. A third is to learn about ways to make the organization more effective by strengthening management processes (not just safety-related processes). To accommodate these varying needs and perspectives, it helps to have a written policy and standard practices in place before an incident occurs.

This chapter consists of five main sections. The first three sections address the concept of a closed-loop process, common issues to resolve in an *incident investigation policy*, and basic investigative processes. The fourth one explains some useful tools to help investigators with their analysis, and the fifth section provides a graphic model useful for visualizing the various phases and events leading to and following the incident.

Incident investigations fit somewhere on a continuum from shallow to deep. Shallow investigations typically look only at the conditions and events close to the incident, and the report consists of descriptive information called for by an incident investigation form. Deeper investigations extend to root causes—the weaknesses and vulnerabilities that allowed or led to the events and conditions most directly associated with the incident. Clearly, deeper investigations have the potential to yield more insightful conclusions and useful recommendations.

Risk-Reduction Methods for Occupational Safety and Health, First Edition. Roger C. Jensen.
© 2012 John Wiley & Sons, Inc. Published 2012 by John Wiley & Sons, Inc.

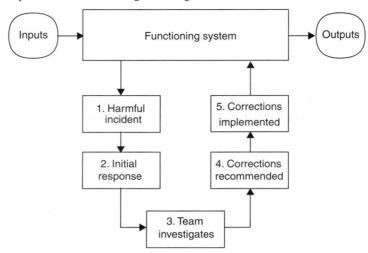

Figure 9.1 Graphic model of a functioning system with a closed-loop process for addressing harmful incidents.

9.1 CLOSED-LOOP PROCESS

A foundation for understanding organizational options for incident investigations begins with understanding the concept of a closed-loop process. The graphic model in Figure 9.1 is used to explain the relationship between a normally functioning system and a closed-loop incident investigation process. The upper part of the model depicts the normal activities of the system simplified into (1) the inputs to the system, (2) the system performing in a normal, predictable manner, and (3) the outputs. The lower part of the model shows the flow of processes for responding to the occasional *harmful incident*.

System inputs include materials, utilities, people, and other resources entering the system. System outputs are goods and services valued by customers and usually some waste. The box depicting a functioning workplace system signifies an operating entity. Some workplace systems are defined by location, such as manufacturing plants, chemical processing facilities, a group of people working together in an office building, construction sites, hospitals, mines, ships, airports, and farming operations. Other workplace systems are defined by an area or region in which they operate, such as local governments, entities operating railroads, courier service providers, and businesses engaged in transportation. Within workplace systems, there are daily variations that are normally tolerable. Successful systems have capabilities to monitor and detect when things start going awry and to make corrections in a routine, relatively easy manner. The correctable deviations are regarded as being within the tolerance of the workplace system. Some deviations go uncorrected, and a portion of these start a chain of uncontrolled events that leads to a harmful outcome.

Figure 9.1 uses rectangles to represent the five processes in a closed-loop investigation. It applies to investigations conducted by a team seeking to learn how to make the workplace safer. It may apply to *form-driven investigations*, provided the investigators are well trained, they come up with useful recommendations, and the organization follows up on those recommendations. When the decision is to conduct a team investigation, all five processes in Figure 9.1 are performed—effectively closing the loop.

The remainder of this chapter emphasizes the importance of each organization tailoring policies to fit its unique needs. No attempt is made to describe a best practice or a preferred "how to" approach for post-incident investigation. Background material for this chapter comes from the author's experiences integrated with advice from numerous authors, especially Bahr, Hughes, and Oakley.[2–4]

9.2 POLICY CONSIDERATIONS

A written policy provides the organizational guidance about what to do after a harmful incident. The organization may conveniently make the policy a section of its official policy for administering an OSH program. Table 9.1 provides an outline of the

Table 9.1 Overview of Some Policy Issues for Responses to Harmful Incidents

Policy application	Issues to resolve
Defining scope of incident investigation policy	What types of incidents will be investigated?
Deciding on depth of investigation	What criteria will be used for deciding which incidents to have a team perform an in-depth investigation?
	Will decision on depth of investigation be based on actual outcome or on what could have been the outcome had the response been ineffective?
Assigning and preparing investigators	How does organization choose investigators? What training will investigators get?
Collecting and preserving evidence	What procedures are expected for (1) collecting evidence and (2) preserving evidence?
	Will policy explicitly call for getting evidence on (1) the normal process and (2) the history of equipment maintenance?
Analyzing evidence	Should all teams use a particular analysis tool?
Reaching useful recommendations	Does organization want to have guidelines for the recommendations contained in investigation reports?
	How will draft recommendations be reviewed prior to finalizing the investigation report?
Communicating findings, recommendations, and lessons learned	Does organization want a policy on (1) assigning corrections to appropriate individuals, (2) a formal verification procedure, (3) distributing the report, or (4) sharing lessons learned outside the organization?

processes and associated policy issues the organization may choose to include in an incident investigation policy. The remainder of this section discusses these issues without attempting to specify a one-size-fits-all approach.

The beginning of the closed-loop process depicted in Figure 9.1 is an incident of some kind. Organizations need to define what incidents to include and exclude from their incident investigation policy. Will the policy apply to employees, the environment, product users, equipment, facilities, or others? What are the legal requirements for investigations? What level of severity will trigger an investigation?

Another question to resolve is what sorts of events will be investigated with a full team doing an *in-depth investigation*. A policy could specify criteria for equipment damage in terms of estimated financial loss. A policy could specify criteria for human injuries based on severity. If the organization uses a risk-assessment matrix, it may also be useful to mirror the severity categories used in their risk matrix. For example, say the company uses four categories of severity in its risk assessments (e.g., minor, moderate, serious, and catastrophic). It may be advantageous to use the same severity categories in the policy on depth of investigations. Advantages of this include being consistent, limiting complexity, and facilitating comparison of actual harm to the level of harm shown in the applicable risk-assessment document.

The severity of harm can be affected by the post-incident response. A rapid response by emergency medical service personnel can make the difference in life or death. The response by an environmental response team to a spill (e.g., an overturned tanker truck leaking a chemical) can greatly limit the amount and spread of the chemical and thereby reduce the cost of restoring the soil. The response of a fire department to a structural fire can make the difference in a total loss or a repairable structure. Because post-incident response often affects severity, a question for consideration is: Should the policy be based on the outcome severity after the response or on the potential worst case had there not been an effective response? For example, in a petroleum refinery, a small fire started, and an employee noticed it and used a portable fire extinguisher to extinguish it. If based on actual outcome, it would be considered a minor incident warranting an ordinary investigation, but viewed from the perspective of the potential worst case, it would be considered serious incident warranting a full-scale, in-depth investigation.

Some organizations encourage personnel to report near-miss incidents based on the idea that for each actual injury there are many instances of someone coming close to being injured. By obtaining reports and investigating many near-miss incidents, OSH professionals can uncover hazardous conditions, events, or exposures suitable for correcting before any harm occurs. Organizations that ask employees to report near-miss incidents will want a policy or process for investigating or otherwise following up on the reports.

Another policy decision involves terminology. Some people passionately debate the pros and cons of terms used to describe incident outcomes. The U.S. military community uses the word *mishap*. The word "accident" is commonly used in general industry, construction, in reference to major transportation disasters, and by coroners to classify cause of death. The public health community uses words specific to human health such as injury, death, and disease, while avoiding the word "accident" because a substantial portion of the public associates the word with an unpreventable or random event. No attempt is made in this chapter to either criticize these other terms or advocate for the term "harmful incident."

One final policy issue warrants comment. Organizations vary on how much time and money they want to put into incident investigations. Differences result in part from how OSH leaders view the benefits from investigating past events versus time spent on proactive OSH projects. Some safety professionals liken investigating past incidents to a dog chasing its own tail—a never-ending circle of investigating and fixing problems. Some organizations choose to allocate more OSH resources to proactive programs. There is no one-size-fits-all allocation formula.

9.3 INVESTIGATIVE PROCESSES

Critical parts of an incident investigation include (1) collecting and preserving evidence, (2) reaching evidence-based conclusions, (3) developing useful recommendations, and (4) communicating findings, recommendations, and lessons learned.

9.3.1 Collecting and Preserving Evidence

Common types of evidence are photographs, scale drawings of the location, statements by witnesses, and documents in paper or electronic form. Before allowing changes in the scene, take numerous photographs, keep a log of the date, time, and location, and preserve all of it. Obtain measurements for making a scale drawing of the area. Mark or tag tools, equipment, and other items for clear identification in case of litigation. As soon as possible, interview *eyewitnesses*—people who were in the area of the incident and who possibly saw, heard, or smelled something relevant to the incident. Other witnesses to interview are people who have knowledge of how the operations are supposed to work and how things actually work. Interviews with these people may provide information for identifying deviations that may have contributed to the events preceding the harmful incident.

Readers who occasionally investigate workplace incidents can easily identify published material containing advice on investigation methods and procedures.[3,4] For readers who find that their jobs involve regular incident investigations, training courses are typically available through professional societies or agencies in their business domain. Some common domains with need for frequent investigations are found in government agencies that investigate chemical plant fires and explosions, workplace fatalities, and crashes involving aircraft, trains, and buses. Special courses are available for fire and police department investigators who need to look for evidence of arson (deliberate fires started with criminal intent). Property insurance companies also provide training for investigating fires to determine if the fire was covered by the policy, if a third party might have some liability, and the value of losses.

Easily overlooked in incident investigations is the need for evidence with regard to the normal processes, practices, and controls in the affected system. This information will facilitate comparing what normally occurs with the actual events leading to the incident being investigated. That comparison may indicate *deviations from normal* that created a hazardous condition or resulted in a normally effective control being ineffective. An example would be learning that a particular machine frequently malfunctions or that an employee routinely followed a procedure that differs from the standard operating procedures.

9.3.2 Reaching Evidence-Based Conclusions

The various sources of evidence may be viewed as pieces of a puzzle. Once the pieces are assembled, the investigative team can begin putting the puzzle together. Some analytical tools for helping an investigative team complete the puzzle are discussed in this section.

Incident investigators should strive for evidence-based conclusions. While that may sound obvious, it can be tempting to do the opposite. The opposite is reaching a conclusion and then sifting through the evidence to find support for that conclusion, while discounting or ignoring evidence that conflicts with the conclusion. To avoid that pitfall, organizations may choose to specify methods for keeping investigators on track. Three logical steps for taking the path from evidence to conclusion are

1. Develop the accident sequence,
2. Analyze it, and
3. Determine causal factors.[4]

In Step 1, the investigation team uses the evidence to explain the sequence of events. It is often helpful to start by describing the scene where the incident occurred. This can include such information as the place and conditions existing before, during, and after the incident. Typical types of evidence for this step are photographs and a scale drawing of the location.

In Step 2, the investigative team seeks to provide a plausible explanation for all the reliable evidence. Some techniques the team may use to analyze the evidence are described in books by Oakley and Stephans.[4,5] One such tool—the events and causal factors chart—is explained later.

In Step 3, the investigation team reviews their results from Steps 1 and 2. From this foundation, the team extracts the factors that contributed, enabled, or directly caused the events in the sequential chain. Following the three-step process should lead to evidence-based conclusions as well as identifying factors in need of change.

Investigators should recognize their investigation is what statisticians call a sample of one. Therefore, investigators should be cautious about trying to generalize their findings beyond the scope of their investigation. It is best to limit the findings and conclusion to your case. Later, others may develop generalization after reading your report and many other reports. Also keep in mind that the ordinary occupational incident investigation team lacks the resources, and perhaps the expertise, to adequately assess organization-level issues. This contrasts with highly funded multidisciplinary investigations of major disasters involving ships, passenger planes, nuclear power plants, petroleum refineries, and space flights.

9.3.3 Developing Useful Recommendations

As a practical matter, the investigation team should attempt to focus on making recommendations that are feasible to implement economically, technically, and legally. Consider the alternative. If the team makes a recommendation that never gets implemented, the organization may find its nonresponsiveness used against it in future litigation. To illustrate how such a scenario might unfold, consider a general contractor in the building construction business. They had a subcontractor employee seriously hurt when he fell from a roof. The investigation team recommended that the general contractor should closely monitor and enforce fall protection for elevated work performed by all subcontractor employees. Now suppose three years later an employee of a subcontractor falls from a roof and gets seriously hurt. The employee could sue the general contractor claiming failure to exercise due diligence by not making sure all subcontractor employees always use fall protection. As evidence, the injured employee's attorney can introduce the report from three years ago that recommended the general contractor should enforce fall protection by its subcontractors. The attorney will argue that the general contractor failed to exercise due diligence by not fully implementing the recommendation in the prior report. In essence, the recommendation may have imposed a legal duty on the general contractor that did not previously exist. Thus, before finalizing its report, it will be wise for the investigation team to consider each recommendation for legal implications as well as for economic and technical feasibility.

Another matter to consider when developing recommendations involves prior risk assessments. If the organization already has an RA applicable to the investigated events, it makes sense to review it. The RA can provide information useful for the incident investigators, and the findings from the incident investigation can be useful for reassessing the RA.

If not all the risk-reduction tactics in the RA were in place when the incident occurred, the investigation team would most likely recommend actions to reimplement the tactics. If all the tactics were in place, the team needs to reassess the RA and consider revisions to correct deficiencies. Another option would be to conclude that the risk was acceptable before the incident and continues to be acceptable into the future. This second conclusions could be quite appropriate for an incident of equipment damage. But if an employee was harmed, this conclusion would leave the impression that the employer is coldhearted and lacks genuine concern for employees. In reality, people expect the investigation team to make some sort of recommendations for change, either to reimplement risk-reduction tactics or modify the RA.

9.3.4 Distributing Findings, Recommendations, and Lessons Learned

Before wrapping up the investigation, the team needs to distribute their findings, recommendations, and lessons learned. Clearly, the final investigation report should be distributed to the individuals most affected by the incident. If not given a copy, they might suspect something important is being hidden from them. A distribution list should include everyone with a possible role in implementing recommendations.

Although an incident investigation team is responsible for making recommendations, the team has no authority to assign managers to implement the recommendations. Therefore, an organization intending to have a closed-loop process needs to transfer recommendations from the incident report to managers with authority to implement. Since busy managers may not find the time to read the entire report searching for findings and recommendations applicable for their units, some mechanism is needed to highlight the specific responses applicable to each manager. One approach could be to distribute the report to managers with an accompanying memo highlighting items applicable to the manager. Another approach could be an oral presentation on the investigation at a management meeting.

There is also a need to follow up on implementation. A process to verify proper implementation of corrective measures contributes to overall safety by providing (a) impetus for managers to implement recommendations applicable to their operations, (b) assurance for the organization that corrections were made, and (c) documentation of closing the loop.

The investigation team may want to share important lessons learned with others. Some organizations with multiple sites have mechanisms for sharing lessons learned at one site with counterparts at other sites. Similarly, some industry groups have shared reporting of lessons learned. The nuclear power industry, for example, developed an information sharing system because of a mutually recognized need to operate all nuclear plants in the world safely. Another example of sharing is found in the fire protection community. Journals and magazines routinely publish reports about major fires and explosions, and these typically contain lessons learned.

9.4 PRACTICAL TOOLS FOR INCIDENT INVESTIGATORS

In this section, practical tools for incident investigation teams are explained. The first tool helps the team use the evidence to develop a sequential chart of events and associated contributors to those events. The second tool can help the team think broadly about potential corrective actions to reduce future risks. The third helpful tool is the use of prior investigation reports found in published literature and reputable websites. The fourth section briefly explains three other analysis tools.

9.4.1 Events and Causal Factors Chart

A useful tool for sorting out the sequence of events is depicted in Figure 9.2. Called an *events and causal factors* chart, it displays the events in boxes arranged in order from left to right. The text in the event boxes indicates the actor (person or thing) and the action. The action description should simply state the action and avoid judgmental words such as unsafe, improper, or erroneous. Ovals connected to events are for noting conditions and factors associated with the event. For the example chart, there was one

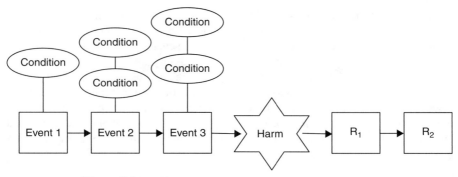

Figure 9.2 Basic format of an events and causal factors chart.

condition relevant to event 1 and two conditions relevant to events 2 and 3. The point where the harm occurred is depicted by a star-shaped figure. Although some organizations use a diamond shape, this may not be ideal because it gives too many meanings to the diamond—in the conventional flowchart a diamond shape is used for decisions, and in analytical trees a diamond shape is used for an undeveloped event or condition. In the figure, two response events (R_1 and R_2) are depicted to the right of the star. These are useful for documenting the performance of the first response and subsequent actions to mitigate the harm.

An events and causal factor chart can help not only the investigation team, but also those who read the final report. It depicts the main events and associated conditions, and it provides an organizational structure for writing the incident scenario. Although the chart shown makes this whole effort seem very simple, possibly trivial, it can be invaluable when sorting through scenarios that are more complex. For one thing, the "conditions" described in the chart can be used for various types of information relevant to the event, four of which are noted below.

1. Enabling conditions that had to occur in order for the event to occur (e.g., the machine was connected to the building's electrical system). This kind of condition is recognized in legal circles as being one element of a negligence case. It is the element involving proof that the harm would not have occurred "but for" the condition existing.

2. Omitted events or nonevents such as noting the victim omitted a step in a procedures (e.g., entering a confined space without first checking the concentration of flammable vapors).

3. Contributing factors such as a warning sign was not visible, lighting was poor, or the employee had been working extra hours for three straight days.

4. Basic facts relevant to the event such as noting the distance of a fall, the weight of an item involved, or the fact that the company required a particular item of PPE.

The text in ovals should be concise and avoid judgmental adjectives. Some organizations and authors will use an additional symbol to show why a particular condition existed (Oakley uses a hexagon).[4] Thus, events and causal factors charts provide considerable flexibility for sorting out the sequence of events, conditions associated with the events, and the causes of the conditions and events. All this information will directly help the investigators understand how and why the incident occurred.

An example of an occupational fatal fall is offered to illustrate how the events and causal factors chart can be used to clarify the sequential events. Figure 9.3 uses three rows to organize events into three phases. The top row contains the preliminary events that set up and allowed the man to fall. The second row contains the sequential events from loss of control to injury. The third row contains the relevant response events subsequent to the injury.

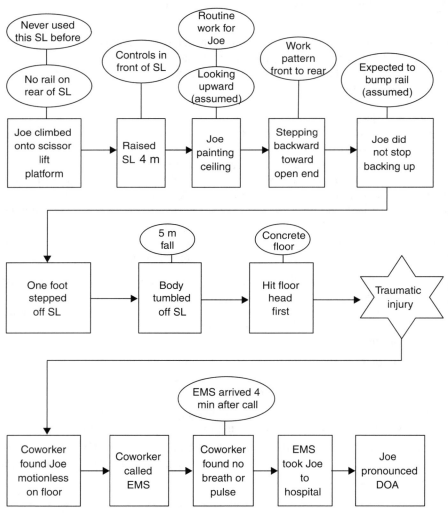

Figure 9.3   An event and causal factors chart for fall from a scissor lift.

A painter named Joe was painting a high-bay ceiling with a roller brush. He was working from a scissor lift (SL) elevated to about 4 m, making his center of gravity about 5 m above the floor. The SL was borrowed from another subcontractor with a crew at the site. The railing on the lift platform was intact on the front and both sides. Elevation controls were attached to the front rail. When the SL was new, the rear of it had a swinging, self-closing rail at approximately hip height and a chain for use as a midrail. Workers could enter the platform by either crawling under the rail or using the rail as a door. However, the old borrowed SL no longer had the rail. It also did not have a chain.

Information obtained from other painters after the incident enabled the investigators to determine the normal method they use to paint a high ceiling from a SL. Painting a ceiling involves constantly looking upward. Workers fill their brush, and then start in the front of the platform and work backward until they feel their hip bump the back rail. They even call this a "bump bar." Once they feel it, they walk back to the front to begin another cycle. After obtaining photographs of the SL, statements from a coworker, and interviews with other painters, the event sequence was pieced together as shown in Figure 9.3.

The chart begins with Joe climbing onto the SL platform. Other investigators may elect to start further back in time in order to elucidate why Joe borrowed a SL instead of using one owned by his employer. They might also ask why Joe would choose to use a SL he could easily see was lacking a rail. Thus, one of the issues facing incident investigators is where to begin.

Benner proposed a place to begin an investigation.[6] He recommended looking for an event that started an uncontrolled chain of events based on a general model of a functioning system. A brief explanation of this model will help readers appreciate his rationale. Think of the daily activities of a functioning system operating in a normally controlled manner or in homeostasis. In that state, many events go as expected, and the occasional deviation from normal (Benner called it a perturbation) can be easily dealt with by human response or engineered adjustments. When something unusual occurs and fails to be detected and adequately corrected, homeostasis is lost—initiating of a potentially uncontrolled chain of events. The chain of events may lead to various outcomes ranging from no harm to very serious harm. Once the investigators know what events led to the harm, Benner advises that they take their investigation back in time to the original deviation. At that starting point, the investigators try to determine why and how the deviation occurred and what happened afterward.

In the illustration, the deviation was identified as Joe climbing onto the platform of the unfamiliar SL with a missing gate. That was the deviation event—when he deviated from his normally successful method of painting a ceiling using his own company's SL. Thus, the event sequence in the top row of Figure 9.3 starts with the deviation event and ends with the failure of the last hazard control—Joe keeping mental focus on his body position relative to the opening. The ergonomics principle of single-channel processing tells us that humans are poor at concurrently performing two or more mental processing tasks. Joe was trying to concurrently process both body position and the roller brush applying paint to the ceiling. His normally successful stop in the correct position was the only fall prevention tactic. The failure of that tactic ended his postural control.

The second row of the chart begins with Joe no longer having control of his body—he was falling and could do nothing about it. Other *intermediate events* in this row led to his head hitting the concrete floor, causing traumatic injury.

It is the nature of intermediate events that multiple outcomes are possible. Outcomes can range from catastrophe to close but unharmed. In the example case, outcomes would have been different if Joe had been wearing a harness connected by a lanyard to a fixture on the platform, if he had tumbled differently and landed on a different body part, or if he had landed on a half-empty cardboard box instead of a hard floor. All these variations in possible outcomes illustrate the importance of including the details of the intermediate events in the chart.

To appreciate a chart such as this, it helps to initially read the descriptions in the event boxes, and then go back and look at the ovals. The two ovals above the first event box illustrate different types of conditions. The first (never used this SL before) was a contributing condition. The second (no rail on rear of SL) was an enabling condition. The condition above the second event box (controls in front of SL) is a fact that merely makes clear that Joe must have gone to the front of the SL to raise the lift. The two ovals above the third event are contributing factors for the painting work Joe was doing. The first one (routine work for Joe) was determined from witness statements obtained from other painters who had worked with Joe. The second (looking upward) was assumed as being necessary to properly paint a ceiling. The oval concerning work pattern being from front to rear is to make clear that an assumption was associated with this event. The last oval in the top row (expected to bump rail at rear) was another assumption based on interviews with other painters.

Investigators can include in their events and causal factors chart any events or conditions that helped mitigate the severity of harm. These are generally located just to the left of the star symbol in the chart. For example, if Joe had landed on a half-empty cardboard box, that would be identified as an event that reduced the deceleration at the bottom of his fall. As another example, if a dropped object impacts the hard hat of a worker below, the mitigating effect of the hard hat would be shown as an event. If the driver of a car loses control and the car hits a tree, mitigating events might be "seat belt holds driver in seat" and "air bag cushions driver's head." On the other hand, if a normally effective safety device was not in use or unavailable, the chart should indicate the absence in an oval.

Figure 9.4 contains two charts for car wrecks. Figure 9.4a shows how to include the role of safety devices. Since the contributions of the seat belt and air bag occur at the same time, they are shown as parallel events. Figure 9.4b shows how to indicate the lack of safety devices. The two ovals are placed above the event they influence— driver slams into steering wheel. Thus, events and causal factors charts can show safety devices contributing (in event boxes) or failing to contribute (in ovals).

The events and causal factors chart is a useful tool for understanding and presenting a harmful incident. It may also help the investigators identify potential recommendations for system improvements. Another useful tool to help with recommendations is the Haddon Matrix.

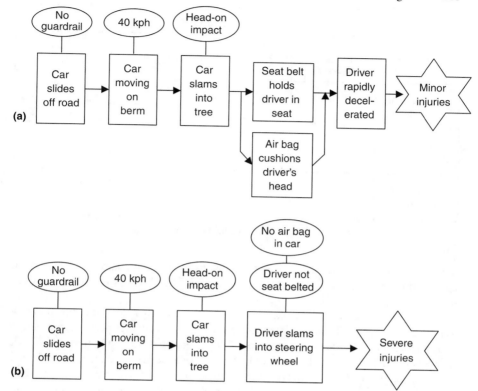

Figure 9.4 Event charts for severity-mitigating factors that helped (a) and failed to help (b).

9.4.2 Haddon Matrix for Countermeasures

The investigation team may use a Haddon Matrix to generate and organize possible recommendations for what Haddon called "countermeasures" applicable to highway injuries and fatalities.[7] He presented a matrix with three rows to categorize the phases of a highway crash: pre-crash, crash, and post-crash. Columns were for elements of highway systems that could be modified. The cells of the matrix were used to record possibilities for making the system safer. Obviously, the cells are intended only for short descriptions of each potential countermeasure.

Figure 9.5 shows a form of Haddon Matrix with three rows and four columns suitable for most workplace systems. The three rows are like Haddon's except terms more suited for occupational incidents are used. The "pre-incident operations" row is for noting ways the system can better control deviations from normal operations. Success with items in this row will prevent an uncontrolled chain of events. This row is also useful for noting preparations for the second and third phases. The second row is for noting ways to alter the event sequence between the loss of control and the

PHASE	System Element			
	Human	Equipment	Environment	Management systems
Pre-incident operations				
During incident				
Post-incident response				

Figure 9.5 Matrix for risk-reduction recommendations.

harmful event. This is the row for entering ways to stop the chain of events, avoid aggravating factors, and use mitigating factors. The third row is for noting ways to make the response as effective as possible.

When using a Haddon Matrix, confusion can occur when deciding in which row to put safety devices. To appreciate this confusion, first consider personal protective equipment such as hard hats and personal fall protection devices. On the one hand, it makes sense to put them in the pre-incident phase because that is when personnel need to start using the devices. On the other hand, these devices provide mitigating value in the during-incident phase. Using another example, seat belts in cars, the driver and passengers need to latch their seat belts in the pre-incident phase, but the injury mitigating value would be realized in the during-incident phase. Using both rows is an effective way to get the most benefit from a Haddon Matrix.

Similar confusion can arise with response activities. One could argue that planning and preparation for effective responses occurs in the pre-incident phase, so these activities belong to that row. One could also make the argument that the value from first responders occurs in the post-incident phase, so that row would be more appropriate. Different organizations may prefer one or the other row. Again, using both rows is the most effective way to benefit from a Haddon Matrix.

9.4.3 Learn from Prior Incident Investigations

OSH professionals can learn innumerable lessons by studying incident reports found in journals, magazines, and websites of governmental agencies. Unfortunately, many in the OSH field fail to believe that lessons learned in one industry may be useful to other industries. Lessons learned about weaknesses in management system are often applicable across industries, as the following examples illustrate.

A large team of highly-regarded professionals investigated the 1986 Space Shuttle Challenger explosion. They dug deeply into NASA's management system and practices. One of the most notable chapters in their report was entitled the "The Silent Safety Program."[8] That title came from findings that major decisions about launching were made by a team of engineers and managers with no input from a safety representative. Moreover, the placement of safety offices within the organizations was "under the supervision of the very organizations and activities whose efforts they were supposed to check." These findings supported a recommendation to change the organization by placing the safety, quality, and reliability functions at a higher position within NASA, where it would have more independent authority. This lesson about making the safety function independent of production management applies to many organizations, not just NASA.

The more recent Space Shuttle Columbia disaster investigation also delved into management systems.[9] Both NASA investigations provide lessons for other organizations regarding the need for a closed-loop process something like that described earlier in this chapter.

In the nuclear power industry, much has been learned by examining multiple investigation reports. The in-depth investigations of specific incidents at nuclear power plants have provided insights into specific cases, but to maximize lessons learned for the benefit of the worldwide nuclear power industry, it helps to examine these cases to find common underlying lessons. Matthews examined multiple incidents and arrived at two important conclusions.[10] The first was that a history of no accidents at a facility tends to make people complacent. The second was that safety performance requires both safety-oriented managers and safety-oriented personnel. One without the other is inadequate. These lessons from nuclear industry incidents also apply to other industries.

The trend for investigations of seriously harmful incidents, particularly in the chemical industries, has been to examine the role of management systems and organizational culture. An investigation of an explosion in a French dynamite factory serves as an example.[11] The investigators demonstrated that information about management systems can be obtained with modest increase in investigation resources by using an evidence collection approach focused on a list of desirable organizational dimensions. This finding might be useful to many other organizations, but unfortunately many people in the safety field will look at a report like this as being applicable to only one French dynamite factory. This narrow perspective discourages the transfer of useful information from one industry to another.

9.4.4 Other Analysis Tools

System safety specialists have developed numerous techniques for analyzing past incidents. The summary descriptions in this section are for three practical techniques.

The change analysis technique involves comparing two situations or event scenarios. It was originally developed for comparing an existing system with a proposal to change the system. It can also be used after a harmful incident to compare the events that occurred with the events in an ideal scenario (e.g., the standard operating procedures). As an example, the change analysis might reveal an employee was using work practices different from the standard practices used in the past. To perform this analysis, the investigators develop two sequential event charts—one for a no harm scenario and the other for the harmful incident scenario. The events are then compared directly to find differences.

The barrier analysis technique helps the investigation team identify various means by which the system prevents a hazard from harming a "target" (that which we want to protect). Known as barriers, these means include physical barriers and administrative approaches. Some are no brain devices such as barriers used for machine guards and radiation shields. Others are engineered systems capable of monitoring processes, identifying when safe tolerances are exceeded, and responding in an appropriate manner, such as initiating corrective responses or communicating with employees by causing an alarm or warning light to activate. Employees perform barrier functions when they monitor processes to detect and correct deviations. Administrative barriers include work practices, checklists, and procedures. Some software programs function as barriers by not allowing operators to select options inconsistent with safe procedures. The team performing a barrier analysis looks for barriers that were in place but inadequate, barriers that should have been in place but were not, and barriers that might be helpful if adopted in the future.

The analytical tree technique helps the investigators use deductive logic to find failures and faults that contributed to the unwanted events. For example, if a fault tree has the top event matching the event being investigated, the team can work downward to examine each lower fault event. Evidence can be examined for each lower event to determine if it occurred or not. This analysis is particularly useful for eliminating some hypothesized failures from being the cause.

9.5 METHOD FOR MODELING HARMFUL INCIDENTS

Numerous respected experts have proposed models of the accident process, but none of these models has achieved broad acceptance. The model presented here makes use of the three Haddon phases, the concept of deviations and *deviation-control mechanisms*, and the flexibility of the events and causal factors charting tool.

Figure 9.6 depicts a model for integrating the concepts presented in this chapter. The three rows are those of Haddon, slightly modified to model occupational incidents. These phases are labeled pre-incident operations, during incident, and post-incident responses. In the core of the graphic, activities and events are depicted according to phase. During pre-incident operations, the system operates as it should. This does not mean it operates perfectly. At various times, equipment malfunctions, people make errors, and other deviations from normal occur. Most workplace systems have informal or formal mechanisms for avoiding, detecting, and responding to deviations. These deviation-control mechanisms include various means for addressing deviations to restore normal operations. Foreseeable deviations that have the potential for causing serious harm warrant multiple control mechanisms, each capable of stopping a particular deviation from developing into a full-blown problem. Each

Phase	Generic model for a harmful incident
Pre-incident operations	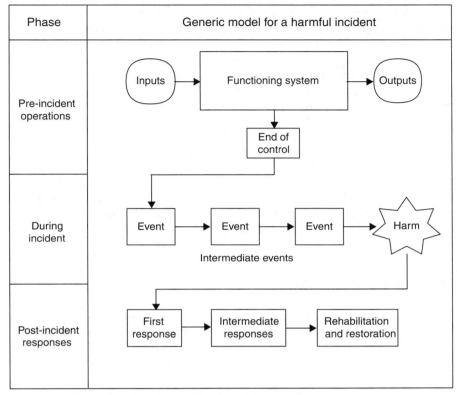
During incident	
Post-incident responses	

Figure 9.6   Generic model for a harmful incident organized by phase.

additional independent deviation-control mechanism makes the system more resilient to the effects of that particular deviation.

The graphic model depicts the functioning system as usually working with various deviations being controlled. This is essentially the same starting point as the "deviation model of accidents" attributed to Kjellén.[12,13] The concept is that deviations from normal operations, if uncorrected, can start an uncontrolled chain of events, possibly ending in a harmful outcome. The concepts of deviations and deviation-control mechanisms should lead incident investigators to seek an understanding of (1) the manner in which the system normally operates, (2) the initial deviations, (3) the deviation-control mechanisms that were supposed to control deviations, and (4) the reasons the mechanisms were ineffective in the incident under investigation.

The next phase of the model begins once a deviation manages to get through or around all deviation-control mechanisms. Events in this during-incident phase, such as those shown in Figures 9.2, 9.3, and 9.4, are charted in the second row of the model. Starting with an out-of-control event, subsequent intermediate events open up various possibilities for the outcome. Intermediate events can alter the end result by aggravating the harm, mitigating the harm, or avoiding harm. Some of the no harm cases will be recognized as near misses. The model in Figure 9.6 depicts a case ending as a harmful event because that is usually the reason for doing the investigation, but the charting model is equally suited for analyzing near-miss incidents.

The third phase is the same as Haddon's third phase. It contains the first response events, secondary responses, and longer term events that involve rehabilitating an injured employee, returning an injured employee to work through a modified duty program, restoring the environment, repairing damaged equipment, and similar efforts to reduce the final extent of harm.

Figure 9.6 may fit some incidents such as a glove, while others may not fit the model very well. Some examples of good fits are single workplace incidents resulting in traumatic injury or damage to property, a short-term contact with a chemical or blood-borne pathogen, cases of asphyxiation, and onset of heat or cold disorders. It is not suited for modeling harm that occurs over a period within a functioning system, such as cumulative trauma disorders from highly repetitive hand activities, lung diseases that develop gradually over long periods of exposure to airborne dust, and diseases attributed to long-term exposure to low concentrations of chemicals.

LEARNING EXERCISES

1. Incident investigators have different reasons for investigating. Which reason serves as the basis for following the closed-loop process described in Section 9.1?

2. Explain the role of barriers for supporting normal system operations.

3. What is the use of a written incident investigation policy?

4. Why should an incident investigation policy define precisely what incidents will be investigated?

5. Why should the post-harm response be included in an incident investigation policy?

6. An incident investigation policy can provide guidance for deciding the level of an investigation. What three factors are mentioned for inclusion in that guidance?

7. What are the three essential topics to include in the incident investigation policy regarding evidence?

8. Suppose a person was an eyewitness to an event that injured a coworker. An hour after the incident, the witness wrote a description of what she heard and saw, and then signed and dated the document. Is what she described evidence or fact?

9. Explain what an incident investigation team can learn about barriers from a prior RA of the process involved in an incident.

10. Suppose you chair an incident investigation team. After completing a report, you present it to the organization's highest managers. You get asked several questions. One manager asked if the report distribution could be limited to

those with a need to know. Other managers used the expressions "sweep it under the rug" and "don't air you dirty laundry." What reasons can you give for taking the opposite tack and widely distributing the report?

11. Regarding the example of the painter falling from a scissor lift platform (section 9.4.1), answer these questions.
    (a) What hazard control tactic was used to keep Joe from falling off the back of the platform?
    (b) Explain why this hazard control had a low chance of being effective.
    (c) What ergonomics principle explains why the hazard control tactic had low reliability?

12. Give two examples of lessons learned by investigators of nuclear power incidents that may be broadly applicable to other technical industries.

13. The speaker at a seminar suggested two techniques for helping supervisors improve their injury investigation reports.[14] Provide a brief response for each of the two techniques.
    (a) Four reports from another plant of the company were obtained—two good ones and two poor ones. These were shared with the supervisors during a training session and used for discussion. How do you think that using these reports as a basis for discussion among the supervisors could help improve the quality of supervisors' investigations and reports?
    (b) Adopt a standard practice of having supervisors present their reports to managers or the safety committee. How do you think that process might help improve the quality of the reports?

14. Which risk factor, severity or probability, is reduced by an effective barrier?

15. Which risk factor, severity or probability, is reduced by mitigating factors?

TECHNICAL TERMS

Deviation-control mechanisms   Various means for addressing deviations to restore normal operations. Engineered systems and personnel function as deviation controllers by monitoring processes, detecting deviations, and taking corrective actions. Effective deviation-control mechanisms make a system more resilient.

Deviations from normal   The difference between how an employee or business process actually performs and the performance considered normal. Generally, small deviations are tolerated or easily corrected and large deviations are more challenging.

Eyewitnesses   Individuals who were in the area of the incident and who possibly saw, heard, smelled, or otherwise sensed something relevant to the incident.

Form-driven investigation	The process of collecting data on incidents as required for a report form. These are usually performed by one person who gathers the information required by the form.
Harmful incident	A sequence of events resulting in harm to a valued entity—most commonly injury to persons, damage to property, roadway crashes, fires, and contaminant releases to the environment. A near-miss incident is similar to the initial events but ends without harm.
Incident investigation policy	An organization-specific guide for investigating incidents.
In-depth investigation	The process of collecting evidence, developing the incidence sequence, analyzing it, determining causal factors, reaching conclusions, and making recommendations. In-depth investigations are usually performed by a team of people with different expertise and generally involve incidents that ended in a high-severity outcome.
Intermediate events	Events occurring during the period starting with the moment system control is lost and ending with a harmful outcome or near miss.
Mishap	An unplanned event or a series of events resulting in death, injury, occupational illness, damage to or loss of equipment or property, or damage to the environment.[15]

REFERENCES

1. Benner L, Jr. Five accident perceptions: Their implications for accident investigators. J. System Safety. 2009;45(5):17–21, 44.
2. Bahr NJ. System safety engineering and risk assessment: A practical approach. New York: Taylor & Francis; 1996. p. 191–199.
3. Hughes B. Incident investigation: Evidence preservation. Professional Safety. 2009;54(9):55–57.
4. Oakley JS. Accident investigation techniques: Basic theories, analytical methods, applications. Des Plaines, IL: American Society of Safety Engineers; 2003.
5. Stephans RA. Systems safety for the 21st century. Hoboken, NJ: Wiley; 2004. p. 253–259.
6. Benner L, Jr. Accident investigations: Multilinear events sequencing methods. J Safety Research. 1975;7(2):67–73.
7. Haddon W, Jr. A logical framework for categorizing highway safety phenomena and activity. J Trauma. 1972;12(3):193–207.
8. Presidential Commission on the Space Shuttle Challenger Accident (Rogers, Commission). Report of the Presidential Commission on the Space Shuttle Challenger Accident. Report Number NTIS-64. Springfield, VA: National Technical Information Service; 1986. p. 152–162.
9. Manuele FA. Is a major accident about to occur in your operations: Lessons to learn from space shuttle challenger explosion. Professional Safety. 2004;29(5):22–28.
10. Matthews RB. Nuclear safety: Expect the unexpected. Professional Safety. 2005;30(12):20–27.

11. Le Coze J. Accident in a French dynamite factory: An example of an organisational investigation. Safety Science. 2009;48(1):80–90.
12. Kjellén U. The deviation concept in occupational accident control—I: Definition and classification. Accident Analysis and Prevention. 1984;16(4):289–306.
13. Kjellén U. Prevention of accidents through experience feedback. London: CRC; 2000. p. 36–58.
14. Moriarty B. How to develop and sustain an effective accident investigation program. Presented at the AIHce2011 Conference. May 19 2011, Portland, OR.
15. U.S. Department of Defense. February 2000. MIL-STD-882D, Definitions section. Available at www.wbdg.org/ccb/FEDMIL/std882d.pdf.

Chapter 10

Human Error Reduction

Learning from past experience is not limited to investigating harmful incidents. Another learning source comes from analyzing past human errors and violations of safety rules. This chapter presents some concepts of human error, classifications of safety-related errors and rule violations, and two approaches for finding countermeasures.

10.1 CONCEPTS OF ERRORS

There seems to be many concepts of human error. This section introduces some of those, starting with a concept from the law of negligence. It comes up when one person sues another for allegedly acting negligently or failing to act when under a duty to act. When translated into the concept of human error, there are two categories.

1. Errors of commission—when someone makes an error while trying to perform a task, and

2. Errors of omission—when someone with a duty to act does not attempt the action.

Common errors of commission in workplaces occur when an employee tries to perform some action and fails to execute it in the proper way. Common errors of omission in workplaces occur when an employee tries to perform a multistep task and somehow manages to skip a step. These two broad types of errors may be viewed as a starting place for developing subcategories that are more useful for understanding causes of errors and developing effective countermeasures.

The system safety community approaches human errors much like it approaches part failures—they want to know the probability of human errors in particular situations, and they want to clarify factors influencing human errors. One of the underlying concepts of this approach is the recognition that humans make many errors. Most are easily identified and corrected, and others have relatively trivial consequences. One might say such common errors are within the *system tolerance*.

Risk-Reduction Methods for Occupational Safety and Health, First Edition. Roger C. Jensen.
© 2012 John Wiley & Sons, Inc. Published 2012 by John Wiley & Sons, Inc.

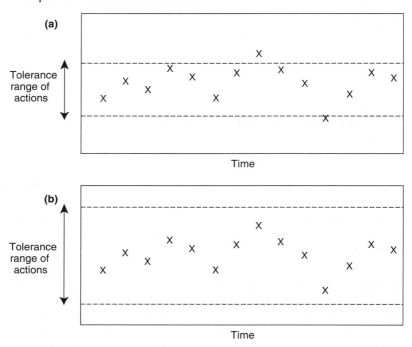

Figure 10.1 Control charts contrasting systems with (a) narrower and (b) wider tolerance for human actions.

Even in safety-critical domains such as aviation, modern systems are designed to tolerate most ordinary human errors. Figure 10.1 shows the concept of system tolerance in the form of two control charts, each with dashed lines depicting limits of the range of human actions the system can tolerate. Figure 10.1a shows a narrower range of acceptable conduct, with two of the data points being outside the system tolerance range. Figure 10.1b shows a system more tolerant of human variability. The same behaviors are plotted on both charts, but the system shown in Figure 10.1b tolerates all the actions.

One definition of human error is "an out-of-tolerance action within the human—machine system."[1] Consider this definition in light of the two control charts in Figure 10.1. For the same series of human actions, the system in Figure 10.1a, with the narrower tolerance range, has two errors, while the system in Figure 10.1b, with the wider tolerance range, has zero errors. Thus from this perspective, a human error is not solely defined by human behavior; it is defined by both the behavior and the system in which that behavior takes place.

Literature from the system safety community provides some fundamental concepts for quantifying error probabilities. The basic variable for error rates is the *human error probability (HEP)*, defined as the ratio of number of errors that occurred (N_{errors}) to the number of opportunities for errors to occur (N_{opp}).[1] When a human error is an intermediate event in a fault tree, HEP may be used when computing the probability of an item failure. However, HEP values are imprecise because of limitations in how they are obtained and factors that can affect them. Consider the following examples of how HEP data might be obtained.

First, think about a power press setup for actuation using a foot pedal. The press operator loads the part into the point of operation, pulls his hands away, and actuates the stroke using the foot pedal. One of the known errors occurs when reaching forward. The operator's center of gravity gets too far forward, causing the operator to apply pressure downward with his toes.[2] If this occurs while the toes are over the pedal, the machine will stroke. In order to obtain a HEP value, one could take extended videos of a few operators performing the cycle many times. After a lengthy time reviewing the videos looking for instances of the operators starting to lose balance, an analyst could compute the HEP and use it for fault tree computations.

It should be clear from the example that getting accurate HEP values can be tedious as well as difficult. A more practical approach to obtaining a HEP value for a task is to search the literature or the Internet. Two examples of the sort of HEP values one might find are (1) general error of omission for items embedded in procedure (0.001) and (2) technician "seeing" an out-of-calibration instrument as "in tolerance" (0.01).[3] When searching literature, be sure to include search terms that have meanings interrelated with human error. A common term in the literature, *human reliability*, refers to the probability of a human correctly performing a specified task under stated conditions for a specified period. For example, to find the reliability of a roofer complying with fall protection procedures, you could ask: What is the probability that a roofer will comply with fall protection rules for a complete shift of roofing work?

Efforts to find data on human error rates and reliability are further complicated if various situational circumstances are considered. The numerous factors influencing error rates and reliability are often referred to as performance influencing factors or *performance-shaping factors*.[1,4] Those connected to the individual—*internal per-formance-shaping factors*—include factors such as experience with the task, training, attitude, motivation, and mental models. Those external to the individual—*external performance-shaping factors*—include factors such as workplace temperature, lighting, noise, pace of production, communications, safety climate, supervision, and fellow workers.

Both types of performance-shaping factors can influence the HEP in a positive or negative way. Thus, if a fault tree analysis requires an HEP value, the analyst may be fortunate enough to find data for the average HEP of similar events. If there appear to be significant performance-shaping factors, an adjustment to the published HEP value would be appropriate.

In addition to the legal and system safety concepts of human error mentioned above, there are several others. A concept proposed in a book by James Reason focused on making systems more tolerant of variations in human actions.[5] He characterized operational systems as having multiple barrier-like defenses for preventing faults and errors from disrupting normal system operation. Each defensive mechanism can effectively stop some, but not all, types of errors from propagating. Defenses include organizational factors, supervisory factors, preconditions for *unsafe acts*, and unsafe acts. The defenses are depicted in a graphic as slices of Swiss cheese aligned in order. The idea is that a fault or error needs to find a straight path through all slices in order to seriously disrupt the system. A fault or human error may lead to events that get through a hole in one slice but get stopped by the next slice. If there are several independent slices, each with a few holes, the likelihood of a hole in the first slice being aligned with a hole in the second slice is small, but not zero. Having a third slice in the array further reduces the probability of holes being aligned in all three slices. In Chapter 8, this concept was explained in terms of probability. If three successive slices have holes constituting 5%, 8%, and 10% of their respective area, and the holes are independently located, the probability of an open path through all three slices is $P(\text{all}) = 0.05 \times 0.08 \times 0.10 = 0.0004$. Investigating errors using the Swiss cheese model involves identifying the defenses and looking for the holes in each. That sort of analysis can lead to a deeper understanding of the pathway to the incident and help identify ways to, in essence, reduce the size and number of holes in each slice. This Swiss cheese model is widely recognized and respected.

A well-published author on human error, Sidney Dekker, distinguished two views of error.[6,7] The older view regards the people who commit error as the problem, while the newer view regards the organization/system as the problem.[6,7] The newer view opens many doors for organizational improvements that can reduce the probability of human error, increase the system tolerance for human error, or both.

Leading authors on the subject of human error share many of the same concepts. One is that a classification system for types of human errors is essential because different types of errors require different error-reduction strategies. Another area of agreement is the importance of understanding the background factors affecting errors. A well-researched, multidisciplinary approach to classification is explained in the next section.

10.2 COMPREHENSIVE CLASSIFICATION SYSTEM

Several attempts to classify human errors were examined by Wiegmann and Shappell.[8,9] After analyzing relevant theories from experts with diverse disciplinary backgrounds, they determined that the classification system needs to include both errors and rule violations. The umbrella term they chose for these, unsafe acts, indicates that the classification system applies to only acts relevant to safety. Their system is presented in Figure 10.2 and summarized in this section.

The left column of Figure 10.2 identifies two major groups of unsafe acts—human errors and violations. Within the first major group—errors—are three types identified in the mid-major category. Decision errors may be thought of as thinking errors. A person may plan a course of action that is inappropriate for the situation, misinterpret available information, make a poor choice based on an incorrect assumption, or choose a less safe option when a safer alternative is available. Skill-based errors involve a person performing a familiar task she is skilled at doing,

Major group	Mid-major group	Specific group
1. Errors	1.1. Decision errors	1.1.1. Rule-based decisions 1.1.2. Choice decisions 1.1.3. Ill-structured decisions
	1.2. Skill-based errors	1.2.1. Attention failures 1.2.2. Memory failures 1.2.3. Technique errors
	1.3. Perceptual errors	1.3.1. Misperception 1.3.2. Misjudgment
2. Violations	2.1. Routine violations	2.1.1. Habitual and willful 2.1.2. Usually system related
	2.2. Exceptional violations	2.2.1. Out of character and not condoned by management

Source: Based on the concepts of Wiegmann and Shappell.[8]

Figure 10.2 Taxonomy of unsafe/inappropriate acts as per Wiegmann and Shappell.

but on a particular occasion, she does it incorrectly. Some examples of skill-based errors are having a lapse of attention, forgetting something, omitting a step in a procedure, or incorrectly executing an intended action. Perceptual errors involve a failure of the senses (i.e., hearing, seeing, smelling) to capture some sort of signal, or sensory inputs being misprocessed by the brain. Many of these errors in the first major group are part of the everyday lives of people.

Within the second major group—violations—are two different types identified in the mid-major category. The first type is the routine violation of safety relevant rules. In workplaces, it is not uncommon to find that personnel routinely perform a task without following company rules or standard procedures. This sort of regular behavior exemplifies a willful or habitual violation. The second mid-major type of violation involves a normally safety compliant employee performing an out-of-character act not condoned by management. A deeper investigation of such a violation may reveal underlying reasons. In the case of a willful violation, the person may have thought his approach was equivalent to or better than the standard procedure; in the case of a routine violation, the person may have been working in his habitual manner without thinking about any applicable rule. In other words, the individuals who committed these errors may have been qualified and conscientious employees who thought they were doing their assigned work safely and effectively.

Classifying an error or violation begins a process aimed at understanding the underlying factors and conditions leading to the unsafe act. Building on the prior work of Reason, Wiegmann and Shappell developed taxonomies for each of three background factors: (1) preconditions for unsafe acts, (2) unsafe leadership, and (3) organizational influences.[8,9] They named the entire set of taxonomies, together

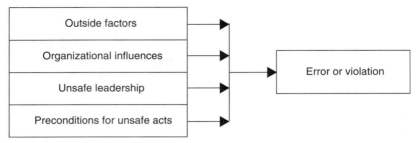

Figure 10.3 Depiction of four sets of background factors that influence errors and violations according to Patterson and Shappell.[10]

with detailed explanations of each, the human factors analysis and classification system (HFACS).

An exemplary analysis of historical cases applied the HFACS to more than 500 mining accident reports from Queensland, Australia.[10] One outgrowth of that project was the addition of a fourth background factor, outside influences, which in the domain of mining is the influence of regulation. The same paper presents the HFACS modified for the mining industry. This modified version appears much better suited for OSH applications than the original version. Figure 10.3 depicts the concept that four types of background factors influence errors or violations.

The process of finding countermeasures starts with classifying a particular error or violation into an appropriate subcategory (Figure 10.2) and then identifying the background factors relevant to the error or violation addressed. The effort put into this leads to a better understanding of the underlying or root causes of the error, and this understanding will help identify appropriate countermeasures.

10.3 METHODS FOR FINDING COUNTERMEASURES

Methods for finding suitable countermeasures for unsafe acts depend on how the problem is approached. The proactive approach involves anticipating future unsafe acts, while the reactive approach involves looking for ways to correct a historical problem with unsafe acts.

The proactive approach involves anticipating unsafe acts employees might make while performing a task. A systematic method makes use of a worksheet similar to the JHA form. Figure 10.4 shows a format suitable for recording the information. Column 1 of the worksheet is for listing steps in the task, just as was done on a JHA form. Columns to the right are for entering foreseeable unsafe acts, the initial/immediate consequence, any recovery opportunities, and tactics for reducing the probability or consequence of the act. Readers should recognize that after identifying a foreseeable unsafe act, the analysis process is a bottom-up, inductive approach similar to that used in an FMEA, except that the starting point is an unsafe act instead of a part failure. The "tactic" column is useful for noting existing methods for mitigating the effects as well as new tactics that might improve the situation.

Task/step	Foreseeable unsafe act	Initial consequence	Recovery opportunities	Tactic

Figure 10.4 Worksheet for proactively analyzing foreseeable unsafe acts.

The reactive approach is used to address a historical problem with unsafe acts—perhaps a problem was revealed because several different individuals performing the same task made the same mistake multiple times. Examples might include roofers failing to tie off, or mine equipment operators occasionally parking their equipment improperly. In such situations, organizations should look for opportunities to change system attributes either to improve system tolerance or to change factors that contribute to unsafe acts.

An approach for finding appropriate countermeasures (used by Wiegmann and Shappell) uses small groups.[11] After classifying the unsafe act according to the HFACS taxonomies shown in Figure 10.2, the group considers each unsafe act in terms of five elements: the human, technology, environment, task, and organization.[8-10] Then the group identifies ideas for countermeasures. Finally, the group examines each potential countermeasure for feasibility, acceptability, cost, effectiveness, and sustainability.

Both proactive and reactive methods provide a structured approach for identifying countermeasures. Any success resulting from the effort depends on the problem-solving skills and creativity of the participants.

LEARNING EXERCISES

1. Briefly explain the difference between errors of commission and omission.
2. Explain the relationship between human error and system tolerance.
3. Distinguish between the labels "performance-shaping factors" and "performance influencing factors." Which do you think best fits the underlying concept?
4. Explain external performance-shaping factors.

5. Explain internal performance-shaping factors.

6. What are the variables in the HEP ratio?

7. What specific group in the right column of Figure 10.2 fits the following scenario? A truck driver drove through an intersection while the light was red. When interviewed later, he said he was driving toward the sun and could not tell the light was red.

8. What specific group in the right column of Figure 10.2 fits the following scenario? An employee tripped on a floor mat and fell. When interviewed, she said her mind was on something other than walking when she suddenly tripped.

9. What specific group in the right column of Figure 10.2 fits the following scenario? A clerical worker had a box full of files that needed moving to another office. She knew it was quite heavy, so she looked around for a suitable cart. She could not find a cart, and she wanted to get the job completed before lunch, so she decided to lift and carry the box herself. In the process, she injured her lower back.

10. Develop an ability to use the worksheet in Figure 10.4 by considering the following scenario from the 1992 Summer Olympics in Barcelona. Based on a description by Steven Casey, relevant background and events were as follows.[12] In the synchronized swimming competition, part of the individual performance events is the compulsory routine in which four elements are prescribed. Competitors perform the four compulsory routines. Judges are chosen from a pool of qualified judges who come from countries with no competitors in the event. Judge Lobo, a Brazilian, was one of the judges. Judges start with a score of 10 and make deductions for various weaknesses in the performance. At the end of each routine, judges entered their score using a keypad similar to a telephone keypad with 0–9 digits plus a 10 key and a 1/2 key. After each routine, their scores were entered, and the computer determined the average score used as the final performance measure. In 1992, the favorite was a Canadian named Sylvie Frechétte. She performed extremely well. Judge Lobo rated the performance a 9.7, but her fingers punched 8.7 on the keypad. She soon saw the score on the large scoreboard and realized her error. She tried unsuccessfully to change it using the keypad at her station. She called a referee for assistance. The referee from Japan and Judge Lobo had difficulty communicating in English. They called the head referee, an American. After a few minutes checking the rulebook, she ruled that a change in the score was not allowed. As a result, Sylvie Frechétte finished second and an American won the gold medal. For good reason, the Canadians were in an uproar about the ruling and were especially mad at the American official. Table 10.1 shows a partial worksheet for analyzing foreseeable errors. The entries for columns 1 and 2 are given. Understand that the worksheet is for future planning and not for documenting the actual events in the 1992 Olympics. Compose the words that you would enter in third, fourth, and fifth columns of each row.

Table 10.1   Columns 1 and 2 of a Worksheet for Analyzing Potential Judging Errors

Task/step	Error
1. Observe entire routine	1.1. Fail to pay attention to all elements of routine
2. Decide on score	2.1. Choose score inconsistent with scoring rules
2. Decide on score	2.2. Misperceive an element of routine
2. Decide on score	2.3. Misjudge an element of routine
2. Decide on score	2.4. Normally fair judge allows nationality to influence score
3. Enter score by keypad	3.1. Punch wrong key

TECHNICAL TERMS

Human error probability (HEP)	The likelihood of a person performing a particular action erroneously.
Human reliability	The probability that a person can correctly perform a specified task under stated conditions for a specified period.
Performance-shaping factors	The various factors that influence the probability of human error and reliability of human performance. Internal factors are influences connected to the individual. External factors are influences outside the individual.
System tolerance	The range of actions a system can accommodate while maintaining normal function. Examples of a system not functioning normally include interruption of production, human injuries, product damage, and property damage.
Unsafe acts	Safety relevant human errors and violations of company rules.

REFERENCES

1. Bahr NJ. System safety engineering and risk assessment: A practical approach. New York: Taylor & Francis; 1996. p. 151–161.
2. Trump TR and Etherton JR., Foreseeable errors in the use of foot controls on industrial machines. Applied Ergonomics. 1985;16(2):103–111.
3. Stephans RA. System safety for the 21st century. Hoboken, NJ: Wiley; 2004. p. 138.
4. Kim JW, Jung W. A taxonomy of performance influencing factors for human reliability analysis of emergency tasks. J Loss Prevention Process Industries. 2003;16:479–495.
5. Reason J. Human error. Cambridge: Cambridge University Press; 1990.
6. Dekker S. The field guide to understanding human error. Aldershot, UK: Ashgate; 2006.
7. Dekker S. Reconstructing human contributions to accidents: The new view on error and performance. J Safety Research. 2002;33:371–385.

8. Shappell SA, Wiegmann DA. The human factors analysis and classification system–HFACS (DOT/FAA/AM-00/7) Washington, DC: U.S. Federal Aviation Administration; 2000. Available through the National Technical Information Service, Springfield, VA.

9. Wiegmann DA, Shappell SA. A human error approach to aviation accident analysis: The human factors analysis and classification system. Burlington, VT: Ashgate; 2003.

10. Patterson JM, Shappell SA. Operator error and system deficiencies: Analysis of 508 mining incidents and accidents from Queensland, Australia using HFACS. Accident Analysis Prevention. 2010;42:1379–1385.

11. Shappell SA, Wiegmann DA.Workshop 12: Managing human error in complex systems. Human Factors and Ergonomics Society 52nd Annual Meeting; September 22, 2008.

12. Casey S. The atomic chef and other true tales of design, technology, and human error. Santa Barbara, CA: Aegean; 2006. p. 54–62.

Chapter 11

Risk-Reduction Strategies

11.1 CONCEPTUALIZING "STRATEGIES"

This chapter addresses strategic thinking about risk reduction. In contrast to thinking about specific hazards, or about rules, regulations, and standards, the idea of thinking strategically involves stepping back and looking for broader principles applicable to all occupational hazards.

11.1.1 Terminology

Organizations typically function by establishing broad goals and measurable objectives. *Strategies* for achieving objectives are general approaches, and *tactics* are more specific means for implementing strategies. Figure 11.1 depicts the relationships among goals, objectives, strategies, and tactics.

For an OSH department, an objective might be to reduce risk in a particular operation, task, or business unit, with measurement of risk based on risk assessments performed before and after making changes. For such an objective, a *risk-reduction strategy* would be a general approach for reducing risk, and a *risk-reduction tactic* would be a more specific means for implementing the risk-reduction strategy.

The terms strategies and tactics also have some usage in the domain of mathematical problem solving. An insightful book by Professor Zeitz at the University of San Francisco teaches problem solving using a top-down approach starting with a strategy, then a tactic, and finally with various tools of mathematics.[1] This approach differs from that found in typical mathematics textbooks that explain the tools and give students problem sets at the end of each chapter. With these traditional textbooks, students are expected to find—through some unspecified process—solutions to the problems. What usually happens in a class is that some students seem to "get it" and successfully solve most problems, while other students become discouraged because they cannot seem to find these unspecified paths for problem solving. Professor Zeitz developed the top-down approach to help all

Risk-Reduction Methods for Occupational Safety and Health, First Edition. Roger C. Jensen.
© 2012 John Wiley & Sons, Inc. Published 2012 by John Wiley & Sons, Inc.

Figure 11.1   Relationships among goals, objectives, strategies, and tactics.

students discover this general path to problem solving. A similar approach is advocated in this book, even though the "problems" differ from those encountered in mathematics.

In OSH, the "problems" are the potential hazards identified by one or more of the proactive hazard analysis methods discussed in previous chapters. After identifying potential hazards, how can we decide what approach will best fit our particular problem within the context of the constraints and policies of our organization? By thinking of these issues as problems that need to be solved, we can trace a path to the solution from strategies to tactics to the tools for implementation.

An example of this way of thinking in the public health domain comes from the pioneering work of Dr. Haddon. While working on the public health problem of roadway injuries and fatalities, he determined that legislators and the general public had very limited vision for possible solutions. In his communications, Dr. Haddon chose the word strategy for his landmark public health papers on controlling hazards. For his 1973 paper in the *Journal of Trauma*, he chose the title "Energy Damage and the 10 Countermeasure Strategies."[2] In a later paper published in the journal *Hazard Prevention* in 1980, he again used strategies in the title "The Basic Strategies for Reducing Damage from Hazards of All Kinds."[3] In these papers, and others, he emphasized the value of finding solutions to injury problems by following a logical path from strategies to tactics to the tools for implementation. Dr. Haddon's pioneering ideas and papers remain the primary source of public health theory for injury control. Thus, the words strategies and tactics are used in this book to sustain the theoretical model started by Dr. Haddon and to encourage professionals in the OSH community to think broadly and logically about the various potential solutions to occupational hazards.

11.1.2   Adapting the Original Strategies

In addition to the 10 strategies proposed by Haddon, other respected authors have proposed alternatives strategy lists. An influential paper by Johnson in 1975 contained a list of 12 strategies for controlling energy hazards.[4] Both Johnson and Haddon founded their lists on the premise that all injuries result from the transfer of energy from an energy source to that which is harmed, be it a person, animal, environment, or property. Subsequent authors proposed lists encompassing more than energy hazards.[5,6]

Building on the work of these prior authors, the best ideas were consolidated into nine risk-reduction strategies.[7] Subsequent modifications led to the following list.

1. Eliminate the hazard.
2. Moderate the hazard
3. Avoid releasing the hazard.
4. Modify release of the hazard.
5. Separate the hazard from that needing protection.
6. Improve the resistance of that needing protection.
7. Help people perform safely.
8. Use personal protective equipment (PPE).
9. Expedite recovery.

The above list applies to energy sources and other hazard sources encountered in occupational settings. Efforts in each phase contribute to reducing overall risk by reducing the probability of the harmful event occurring or the extent of damage resulting from the incident. Table 11.1 illustrates how the strategies align with Haddon phases using common practices for entering a confined space.

Table 11.1 Practices for Confined Space Entries Arranged by Haddon's Phases

Phase	Confined space practice	Strategy
Pre-event	Lock out any potential pathways for hazards to gain entry	Avoid releasing the hazard
	Use checklist to assure the entering person has donned appropriate PPE, rescue equipment is operational, and communications devices are working	Use PPE and help people perform safely
	Station a person outside the space to (1) monitor the person inside in case rescue becomes necessary, (2) ensure rescue equipment remains operational, and (3) assist entrant with egress	Help people perform safely
During-event	Extract entrant due to an unusual threatening occurrence	Separate the person from the hazard in the confined space
Post-event	Provide first aid and transport to suitable medical facility	Expedite recovery

Some readers may think that working toward a classification of risk-reduction strategies is a waste of time. In response, OSH professionals might do well to recognize that classification systems are a foundation of each major branch of science—physics, chemistry, biology, and geology.[8] While each field has a history of scientific classification (also known as taxonomy), the one developed for classifying biological species is most well known to the general public. Carl Linnaeus developed and published a taxonomy for all living things, now known as the Linnaean classification system.[9] At the time of his publication (1735), his taxonomy may have seemed as a waste of time to many people. Even Linnaeus may not have foreseen the long-lasting effects of his pioneering work. Since the original publication, biological scientists have made many additions and other modifications due to discoveries of previously unknown forms of life. The prolonged and focused effort biologists have made toward a comprehensive and updated classification system underscores the value of scientific fields having sound classification systems. Each of the other scientific fields has an analogous history of developing classification systems for the major aspects of their field.

Dr. Haddon wrote about the importance of having scientific classification systems for the emerging field of injury control. He advocated approaching the highway injury and fatality problem by using methods already established in the recognized fields of science. In particular, Haddon advocated for classification systems for both the etiology (causes) of highway crashes and the countermeasure strategies for reducing risk.[10] This second classification system is discussed next.

11.2 THE NINE STRATEGIES

This section explains the nine strategies for risk reduction. Some explanations include a list of common tactical approaches. While this section mentions a few specific applications of the tactics, the Learning Exercises contain questions designed to help readers appreciate the connection between strategies, tactics, and common workplace applications.

11.2.1 Eliminate the Hazard

After identifying a hazard, the first control option to consider is the possibility of eliminating the hazard from the system. This is highly desirable because eliminating the hazard is a 100% reliable solution for avoiding harm from that hazard. Tactics for this strategy are the following:

- Avoid creating the hazard in the first place, and
- Remove an existing hazard.

11.2.2 Moderate the Hazard

Often a hazard can be modified to reduce the harm it might cause. This strategy has no effect on the probability of a harmful event occurring. Instead, it moderates the hazard so less harm is caused when the harmful event occurs. The strategy may involve several tactics for reducing the potential harm such as the following:

- Reduce the energy level to no more than what is needed for functionality,
- Reduce the intensity of energy transfer,
- Reduce the concentration of a hazardous air contaminant, and
- Substitute a less hazardous material for a more hazardous material.

11.2.3 Avoid Releasing the Hazard

A common strategy in occupational settings is to control a known hazard by preventing it from escaping from an enclosure or location. Similarly, we can avoid starting a fire or explosion by preventing the coexistence of a fuel, an oxidizing agent, and an ignition source. Risk-reduction tactics that avoid releasing a hazard are the following:

- Enclose potentially hazardous materials within appropriate containers,
- Contain electrical energy within insulated circuits,
- Enclose sources of radiation within appropriate shields,
- Lockout and tagout potential sources of energy or materials, and
- Avoid the coexistence of fuels, oxygen, and an ignition source.

11.2.4 Modify Release of the Hazard

This strategy applies to hazards that might be harmful if released from a controlled location. Engineering approaches implemented prior to a release can provide mechanisms for controlling the manner of release to minimize any harm that might result. The tactics are the following:

- Control the rate of release,
- Control the location of the release, and
- Stop the released hazard to avoid further harm.

This strategy does not include PPE, such as hard hats, impact resistant glasses, and footwear with toe protection. Their use does not modify the release of the hazard, so this sort of PPE does not fit well into Strategy 4. Impact protective PPE is a last line of defense from a previously released hazard, such as a falling object or a small projectile. In this list of nine strategies, the use of impact protective PPE fits in Strategy 7, which is specifically for the use of PPE.

11.2.5 Separate the Hazard From That Needing Protection

Hazards with known locations may be controlled by designs that ensure separation of the hazard source from the persons, animals, environment, or property needing protection. The tactics of this strategy are the following:

- Separate by distance,
- Separate by locations, and
- Separate by a barrier.

When the focus is people protection, separation by distance involves designing operations so that personnel are located a safe distance from the hazard. It also applies to facility layouts that put hazardous operations a safe distance from buildings occupied by numerous people. Separation by location involves locating a hazard where people are not exposed. Separation by barriers involves placing between a hazard source and people something that will prevent harm, such a fence, barrier, shield, or guard.

11.2.6 Improve the Resistance of That Needing Protection

Several opportunities exist for improving the ability of things to resist hazards. Consider a laptop computer. If you drop an ordinary laptop from a meter onto a hard floor, it has a good chance of being damaged. What if you drop a "ruggedized computer" made for military use in hostile environments? It should easily take a 1 m drop without sustaining damage. The difference is that one computer is more resistant to impact than the other. This same concept is captured in Strategy 6. Some common tactics for improving resistance include improving the ability of

- Equipment and tools to withstand impacts and vibrations,
- Electronic devices to withstand power surges,
- Structures to survive severe weather events,
- Buildings to resist fire,
- Materials to withstand rusting and corrosion,
- Product containers to withstand rough handling during the distribution processes, and
- Humans to resist a disease through vaccination.

11.2.7 Help People Perform Safely

Reducing risks in occupational setting depends heavily on the employees. In the field of human factors, the term *facilitators* refers to the many ways of designing systems to help personnel correctly perform their tasks.[11] The common tactics for Strategy 7 are listed below.

- Design human–machine interfaces to minimize errors and maximize correct performance.[12]
- Provide warnings to notify and remind people of hazards, as well as to communicate appropriate precautions.

- Design work to match task demands with the capabilities and limitations of the workers.
- Provide personnel with excellent task training and safety-related training.
- Design work so that the convenient method is also the safest method.
- Design equipment and work demands to tolerate foreseeable human errors.
- Help employees prepare their bodies for the stresses faced on the job.
- Conduct operations in a manner that minimizes likelihood of workplace violence.

11.2.8 Use Personal Protective Equipment

The use of PPE provides personnel a means of moderating the extent of harm from hazards not otherwise controlled. As such, PPE is regarded as the last line of defense. PPE usage shows up frequently in JHAs and risk-assessment tables. Recall that when using a risk-assessment approach, hazards are listed in a column, usually the left one. To the right are columns for probability of a particular event, the severity of the associated harm, and a risk index of some sort. Further to the right is a column for risk-reduction tactics. A typical application of PPE in an RA would be to first list in a row of the table a risk-reduction tactic aimed at controlling the hazard, and then on a lower row enter the use of PPE to provide a last line of defense. Keep in mind that the PPE strategy is not for changing the probability of an event or exposure; it is for reducing the severity. Common tactics for PPE are the following:

- Protect body parts from impacts from objects,
- Protect body parts from repeated pressure by damping and distributing forces,
- Greatly reduce the concentrations of air contaminants in breathing air, and
- Provide a barrier to protect skin and eyes from contact with hazardous chemicals.

11.2.9 Expedite Recovery

This strategy corresponds to the post-event phase of harmful events and exposures. Although preparations for a response occur during the pre-event phase, the implementation takes place in the post-event phase. There are many ways to prepare for potentially harmful events in the workplace. At the basic level, employers provide first aid kits at the worksite, as well as personnel with at least basic training for administering first aid. Employers with operations having potential to spill hazardous materials on soil or into a body of water know the importance of advance preparation for quick response in case a spill occurs. Advanced preparation is a key elements in all of the following Strategy 9 tactics:

- Administer effective first aid,
- Expedite transit of injured employees to emergency facilities,

- Refer employees with possible occupational diseases to appropriate medical providers,
- Respond promptly and effectively to hazardous material releases, and
- Implement steps in a business continuity plan.

11.3 PRIORITY FOR APPLYING STRATEGIES

This section explains how and in what order to consider the strategies within the context of risk assessment and a design team project. Recall that the team starts by identifying all foreseeable hazards and harmful incidents that might occur without any risk-reduction tactics in place. After listing them in a spreadsheet or table, the team estimates the probability of each incident and the severity expected, and using a risk-assessment matrix assigns a risk level.

Once the risks are classified, the design team begins to explore options for reducing each risk. The list of risk-reduction strategies may be useful in two ways. First, it may help the design team generate control options. Second, it should help structure their search for options by considering the strategies by order of priority. Figure 11.2 shows an order for considering the various strategies. If a team chooses to use this scheme, its first consideration would be eliminating the hazard. If this is not feasible, the second priority options, *engineering controls*, would be considered. These strategies lead to solutions that have high reliability when properly designed, installed, and maintained. In contrast, the third priority strategies, administrative practices, are less reliable because of their heavy dependence on human effort and behavior.

A couple more points regarding priorities and strategies warrant clarification. First, selecting a particular strategy from the second group does not preclude also selecting others from the second or third group. Selecting an engineering control and one or more *administrative controls* is a common approach. The other point of clarification concerns the Strategy 6 approach of making things more resistant to

Priority	Strategy
I. Eliminate the hazard	1. Eliminate the hazard
II. Use engineering controls	2. Moderate the hazard 3. Avoid releasing the hazard 4. Modify release of the hazard 5. Separate the hazard from that needing protection 6. Improve resistance of that needing protection
III. Use administrative practices	7. Help people perform safely 8. Use PPE 9. Expedite recovery

Figure 11.2 Strategies arranged into three priorities.

harm. Most applications of this strategy are permanent modifications to physical things. An application to humans would be an immunization that lasts a lifetime. Attempts to make employees more resistant to musculoskeletal stressors, such as a stretching program at the beginning of a shift, fit better in Strategy 7—help people perform safely. These types of efforts also belong to priority III because of their dependence on regular human effort.

LEARNING EXERCISES

The intent of these Learning Exercises is to provide readers an opportunity to bridge the gap between general strategies and specific applications (see section 11.3). Each numbered item in these Learning Exercises corresponds to the strategy in section 11.2 with the same number. Many of the exercises involve practices widely recognized throughout the OSH community. For each specific scenario, readers are asked to provide only one solution using the applicable strategy. A complete solution for many of the scenarios would require more than one tactic.

1. Perform exercises a–c below using Strategy 1—eliminate the hazard.
   (a) A nursing assistant manually lifts a patient into a bathtub because the patient is unable to support his own weight or assist in the transfer. The nursing assistant has been complaining of frequent backaches. As an ergonomics consultant, what solution do you suggest?
   (b) In a fish canning facility, diesel powered forklifts are used for much of the transport. In the past 5 years, there have been two instances of the building ventilation system shutting down. In both instances, the carbon monoxide levels in some locations rose to twice the limit recommended (e.g., threshold limit value). The company facilities engineers say they cannot achieve 100% reliability with the ventilation system. As the company's industrial hygiene consultant, what solution do you suggest?
   (c) A company involved in the transportation of refined petroleum has a facility where railroad tank cars are filled with product stored in large tanks. To complete the filling operation, a worker must climb on top of the tank car in order to access the infill ports. The company's present approach to fall protection is that it has a cover over the location and a rail with a short lifeline hanging down. The worker is required to wear a fall protection harness and hook up his or her lanyard to the lifeline when atop the tank car. This takes time, and sometimes a worker will fail to hook up, so it is not a perfect system. As the company's safety engineer, what solution can you suggest?

2. Perform exercises a–c below using Strategy 2—moderate the hazard.
   (a) An older industrial building has a fire extinguishing system equipped with Halon extinguishing agent. The company is making an effort to brand itself as an environment-friendly company. Halon is recognized as being detrimental to the earth's ozone layer, but the facility needs a fire protection system. As an environmental engineer, what solution do you suggest?

   (b) In a shop, there is a pathway for visitors marked by yellow lines. Some
        high school students were taking an escorted tour of the shop when one of
        them walked slightly out of the path and bumped the sharp edge of a metal
        table. The student tore his shirt and lacerated his abdomen. There was a lot
        of concern about the wound becoming infected and the possibility of legal
        action. The facility manager wants to continue allowing escorted tours
        through the shop. As a safety engineer, what do you suggest to moderate
        the sharp edge hazard?
   (c) Heat from a furnace in a foundry keeps a room warm during the winter.
        However, during the summer, the combination of furnace heat and warm
        humid external air in the room can become difficult for workers to
        tolerate. There have been instances of heat exhaustion and the occupa-
        tional health physician is concerned about a future case of heat stroke.
        What solution do you suggest?

3. Perform exercises a–c below using Strategy 3—avoid releasing the hazard.
   (a) A room within a petrochemical facility contains various piping for
        flammable materials. You recognize the possibility of leaks that would
        raise the concentration of flammable vapors in the room. If the vapor
        concentration rises to the flammable range, it is vulnerable to ignition.
        Workers in the room use various steel tools and you are concerned about a
        spark being created by these tools because it might ignite the flammable
        vapor and cause considerable harm to the employees, facility, equipment,
        and overall operation. As a safety engineer, what do you suggest for
        avoiding this harmful outcome?
   (b) A construction company has laborers working in a trench 2 m deep. A
        safety engineer sees this and expresses grave concern about the possi-
        bility of the dirt walls collapsing on the laborers. She explains to the
        foreman that ground pressure from the sides of the trench push on the dirt
        laterally as if trying to fill up the opening. If you were the safety engineer,
        what would you recommend to avoid collapse of the dirt wall?
   (c) A graduate student in chemistry wants to study certain mixtures of
        chemicals. One of the chemicals emits hazardous gases when exposed
        to air. The student proposes to conduct the experiments in a general-
        purpose laboratory room in the campus chemistry building. As the
        campus safety and health officers, what do you require to avoid releasing
        the hazardous gas into the room?

4. Perform exercises a–c below using Strategy 4—modify release of the hazard.
   (a) A chemical processing facility is planning an addition. One major
        component is a chemical mixing vessel for an exothermic reaction. The
        chemical engineers recognize what they call a "remote" possibility of a
        system variation that would lead to a dangerous rise in pressure if the
        following were to occur. First, the internal pressure rises about the
        capacity of the vessel. Second, the seams in the vessel walls will be
        challenged. Third, the weakest seam will fail and allow some gas to

escape. Fourth, the escaping gas will rapidly open the seam and external air will be contaminated. Workers in the area would thus be adversely affected. The company will be liable for air pollution fines and the damage caused. What do you suggest for designing the vessel to avoid these outcomes?

(b) A replacement building is being designed for a facility that makes cardboard. This facility will be replacing a similar building that was destroyed by an explosion. Inside the building, one stage in the cardboard making process creates a highly explosive mixture of air and small particles. Although numerous engineering controls are being planned to prevent this from recurring, engineers admit to a remote possibility of a major explosion existing even with their designed controls. When talking to the architects, you explain how the previous building was destroyed when the rapidly expanding gases from the explosion knocked down the walls and blew the roof straight off. What do you suggest to the architects so that the next explosion does not destroy the walls and roof of the building?

(c) A building renovation project is being planned. One foreseeable hazard is electrical energy from using power tools. The project engineer tells you of the intent to use building outlets with three prongs (hot, neutral, and grounding wires) for all power tools, and they will need to use extension cords. A safety engineer reviewed the project and tells the project engineer that having a power tool grounded is not enough to protect the tool user in case of a short in the tool housing. This is because the worker will experience severe current exposure before the circuit breaker opens and stops current flow. What do you recommend for ensuring the current flow will be stopped quickly enough to protect the worker from a lethal shock?

5. Perform exercises a–c below using Strategy 5—separate the hazard from that needing protection.

(a) A positive clutch mechanical power press is used for processing various items, so it needs frequent changes in setup. The principal hazard is the point of operation where the upper die strikes the lower die to create the desired output. Two company employees have had fingers amputated on the machine during the past 10 years. The company safety director, who actually works half-time on safety and half in human resources, seeks the advice of a safety engineer. What can the safety engineer recommend for assuring the operator's hands are at a safe distance from the point of operation?

(b) A conveyor belt is being planned for moving coal dust from a mine to a nearby power plant. The safety engineer on the design team pointed out the hazard of in-running nip points. An employee representative on the design team asked fellow workers about their safety-related concerns. Several were concerned about getting clothing or body parts caught in the nip point on one end of the conveyor. It seems the walkway leading to the restrooms passes dangerously close to the end of the conveyor, and sometimes they are in a hurry. How might the new conveyor be located so that only qualified maintenance personnel will be able to get near the nip point?

(c) A new casino is being planned. Among the various potential hazards identified in the risk assessment is the possibility of an armed robbery. The main concern is that cashiers may be threatened with a gun. What do you suggest for building suitable protection for the cashiers from the hazard of armed robbery?

6. Perform exercises a–c below using Strategy 6—improve the resistance of that needing protection.

(a) A governmental agency wants to build a new office in a location prone to earthquakes. If you were asked for advice, what can you suggest for improving the resistance of the building to damage from an earthquake?

(b) Administrators of a hospital want to protect their staff from possible infection with hepatitis B. A vaccine is available. How is a vaccination program an example of Strategy 6?

(c) A forest not far outside a city is filled with dead tree materials and considerable undergrowth. Experts on forest fires describe the forest as being a tinderbox (i.e., it that will burn rapidly if a fire should start during hot weather). The vast majority of city residents want to avoid having their forest burned. What option can you suggest for improving the resistance of the forest to fires?

7. Perform exercises a–e below using Strategy 7—help people perform safely.

(a) The shop in an industrial facility has four pedestal rotary grinders. Each uses a different grinding wheel. The shop foreman is concerned that occasionally a worker might make a mistake and put a grinding wheel of the wrong size on a grinder. That can results in a rotation rate exceeding the capability of the grinding wheel. History shows numerous instances of a grinding wheel breaking up in an explosive fashion and projecting particles into the face of the operator. What solution do you suggest for helping personnel select and install the correct grinding wheel?

(b) An assembly workstation has a fixed height of 72 cm. Four workers work in that station during the various shifts spread throughout the month. One of the best workers happens to be the tallest. He has been complaining about bothersome pain in his upper back from spending so much time with his head tilted down and his torso leaning forward. The company medical director considers these complaints as early indicators of a developing upper back disorder. As the company ergonomist, what do you suggest to help this worker avoid an upper back disorder?

(c) A large haul truck transports ore within a mine. When backing up, the driver can use the mirrors on each side of the cab, but still has a large blind space behind the truck. What might a safety engineer recommend to help both driver and pedestrians avoid a backing over fatality?

(d) A room in a large government facility has an automatic fire protection system. If activated, it quickly fills the room with an extinguishing agent that will snuff out a fire. However, if a person is in the room, this is a serious threat because the new atmosphere is unfit for human breathing. The installation company recognized this threat and built the system, so a buzzer sounds for 60 s before the extinguishing agent is released. The buzzer was intended to convey the message to promptly exit the room and shut the door. A fault tree used to analyze the top event of an employee being in the room when the extinguishing agent is released revealed a weakness. What if personnel in the room hear the buzzer but fail to understand the intended message? What solution can you suggest for correcting this vulnerability in the system?

(e) A diamond mine deep in the earth requires miners to work long hours in intense heat. Owing to a history of heat strokes, the mine wants to institute a program to prepare the new hires before sending them into the hottest areas of the mine. If you were asked for suggestions, what would you suggest and how does that relate to Strategy 6?

8. Perform exercises a–c below using Strategy 8—use personal protective equipment.

(a) At a building construction site, personnel working on the lower floors might be injured by an object falling from higher floors. What do you recommend for those on the lower floors?

(b) A company cleaning up an old chemical plant is concerned about its employees who initially encounter unknown chemicals. What do you recommend for protecting these individuals from breathing hazardous air contaminants?

(c) A small company specializing in carpet installation received complaints from employees about knee pain. They use a tool known as a "knee kicker" to force the carpet tightly against the walls. These tools have one end with sharp teeth-like protrusions for gripping the carpet edge and the other end has a pad that the carpet installer kicks with his knee to force the carpet against the wall. What do you recommend for these carpet installers?

9. Perform exercises a–c below using Strategy 9—expedite recovery.

(a) A farm located in a hot climate requires many field workers. The owner is aware of the risk of an employee developing a heat stroke that, if not cared for promptly, could cause the employee to die or suffer permanent brain damage. What can you suggest to ensure that an employee suffering from a heat stroke is cared for promptly?

(b) A trucking company transports liquid pesticide from a manufacturing facility to regional distribution points. On the basis of the prior experiences of other trucking companies, the owners are aware that a truck can tip over under some circumstances and spill pesticide onto the roadway, road edges, and water adjacent to the roadway. The company would be fully liable for any environmental damage. What can you suggest the owners do?

(c) A construction company is planning to bid on a bridge construction project in a rural location. The safety director knows from past projects in rural sites that medical care is often delayed by the admission practices at local medical facilities. Injured employees have been known to wait as long as an hour to see a doctor. What can you suggest the safety director do to overcome this slow access to medical care?

TECHNICAL TERMS

Administrative controls	Risk-reduction tactics dependent on human behavior and effort, making administrative controls less reliable than engineering controls.
Engineering controls	Risk-reduction tactics designed and built into a system to control a hazard with minimal dependence on human action. Engineering controls are considered more permanent and reliable than administrative controls.
Facilitators	Engineering and administrative practices aimed at helping personnel perform tasks safely and correctly.
Risk-reduction strategy	A general approach for reducing the risk associated with a hazard.
Risk-reduction tactic	A specific means for implementing a risk-reduction strategy.
Strategies	General approachs for achieving an objective.
Tactics	Means or methods for implementing a strategy.

REFERENCES

1. Zeitz P. The art and craft of problem solving, 2nd ed, Chapter 1. Hoboken, NJ: Wiley; 2007.
2. Haddon W, Jr., Energy damage and the 10 countermeasure strategies. J Trauma. 1973; 13(4): 321–331.
3. Haddon W, Jr., The basic strategies for reducing damage from hazards of all kinds. Hazard Prevention. 1980; 16(5): 8–12.
4. Johnson WG. The management oversight and risk tree. J Safety Research. 1975; 7(1): 4–15.
5. Asfahl CR, Rieske DW. Industrial safety and health management, 6th ed. Upper Saddle River, NJ: Pearson Prentice Hall; 2004. p. 61–62.
6. Manuele FA. On the practice of safety. Hoboken, NJ: Wiley; 2003.
7. Jensen RC. Risk reduction strategies: Past, present, and future. Professional Safety. 2007; 52(1): 24–30.
8. Hazen RM, Trefil J. Science matters: Achieving scientific literacy, 2nd ed. New York: Anchor Books; 2009. p. 19–23.
9. Hazen RM, Trefil J. Science matters: Achieving scientific literacy, 2nd ed. New York: Anchor Books. 2009. p. 266–270.
10. Haddon W, Jr., The changing approach to the epidemiology, prevention, and amelioration of trauma: The transition to approaches etiologically rather than descriptively based. American J Public Health. 1968; 58(8): 1431–1438.
11. Peacock B, Laux L. "WARNING: Do not use while sleeping"—The role of facilitators. Ergonomics Design. 2005; 13(4): 5-6, 28–29.
12. The Eastman Kodak Company. Kodak's ergonomic design for people at work, 2nd ed. Hoboken, NJ: Wiley; 2004. p. 269–371.

Chapter 12

Common Components of OSH Programs

This chapter discusses selected components of OSH programs—OSH *program aspiration*s, training, warnings, safety devices, emergency preparedness, and sanitation and housekeeping. All these topics are referred to multiple times in subsequent chapters, so presenting them here reduces redundancy.

12.1 OSH PROGRAM ASPIRATIONS

Organizations differ in the level of their OSH program aspirations. At the lowest level are employers who pay no attention to OSH. Managers of these organizations may be heard saying things like "we do not need an OSH program because we have experienced employees who know how to work safely," or "we do not need a formal safety program because our work is not dangerous." When employees of these organizations incur an injury, a cursory investigation invariably concludes the injury was caused by the unsafe work behavior of the injured employee or a coworker. That shallow conclusion reinforces the employer's belief that a safety program is unnecessary.

At the next higher level are organizations focused on compliance with applicable rules, standards, and codes. Managers in these organizations typically equate compliance with safety—a belief that is not entirely justified for several reasons. First, countries and other jurisdictions differ on the comprehensiveness of their standards and codes. If a particular hazard is not covered by a regulation or code, it can easily be overlooked. In the United States, for example, there are no national regulations on musculoskeletal stressors associated with heavy manual materials handling or repetitive motion. Second, standards adopted for regulatory purposes are often outdated, no longer reflecting the practices found in reputable voluntary standards. In the United States, for example, the national standards for air contaminant exposures contain many outdated exposure limits the industrial hygiene community considers inadequate. Third, standards typically emphasize workplace conditions while placing minimal requirements on the behavior of employees. This unregulated

Risk-Reduction Methods for Occupational Safety and Health, First Edition. Roger C. Jensen.
© 2012 John Wiley & Sons, Inc. Published 2012 by John Wiley & Sons, Inc.

factor plays an important role in many occupational injuries and diseases. So a compliance-focused OSH program is much better than no program, but less effective than it could be.

The highest levels of OSH program aspirations are for organizations choosing to perform at a "best practice" or "world-class" level. These organizations comply with applicable regulations, but go far beyond that. They often set their own standards at levels stricter than the regulatory standards. They typically subscribe to a continuous improvement methodology for their OSH programs, incorporating cycles of performance measurement, evaluation of findings, and correction of weaknesses. An example of how a continuous improvement process can be incorporated into OSH training programs is provided next.

12.2 TRAINING

Training employees for safe job performance is a core function of OSH programs. Two basic reasons for OSH training are the employees' right-to-know and their on-the-job behavior. Regarding the first reason, it is widely recognized that employees deserve to be informed of the hazards they might encounter on their job, and what practices they should follow to avoid harm. When the hazards are chemicals, the phrase "right to know" is used extensively and adopted into the regulations of many countries. However, the idea of "right to know" should not be limited to chemicals. It should certainly include any type of hidden hazards. Whether the right-to-know principle should apply to hazards that are open and obvious is a question worthy of debate. All employers, especially those aspiring to have world-class OSH programs, should attempt to ensure that every employee potentially exposed to a hazard will be informed enough to recognize the hazard and know the precautions for avoiding harm. The primary vehicles for enabling employees to recognize and avoid hazards are training and warnings. This section describes training processes and the next section discusses warnings.

Regarding the second reason for OSH training, it is generally believed that training influences how employees perform their jobs. So prevalent is this belief that studies to check it are rare. In one study, forklift operators with a large regional warehousing company were assigned to one of three experimental groups.[1] One group received no training, another received training alone, and the third received training plus performance feedback. Performance was measured by extensive behavioral observations before and after the interventions. The proportion of proper operating behaviors for each group provided the performance measure for comparison. Comparing group performance change from pre- to postintervention, the no-training group changed minimally, the training-only group improved considerably, and the training plus feedback group improved somewhat more than the training-only group. Thus, at least this one study found evidence supporting the belief that safety training translates into improved on-the-job safety performance.

The flowchart model depicted in Figure 12.1 provides an overview of processes involved in OSH training.[2] As a generic model (i.e., not for any particular topic), it can

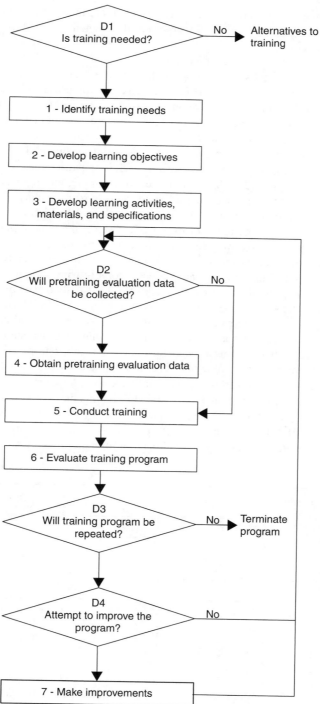

Figure 12.1 Generic training process model. Adapted with permission from Ref. 2.

serve as a framework for all OSH training within an organization, regardless of the particular topic. The generic model incorporates the elements needed for continuous process improvement.

The rectangles in the model depict processes that vary, depending on the subject of the training. Diamond shapes in the model indicate decision points. From these decision points, some arrows indicate feedback to earlier processes. These are the key branches for supporting the continuous improvement process. To facilitate the following explanations, the elements in Figure 12.1 are labeled using a number for processes and the letter D followed by number for decisions.

According to the mode, an organization considering a new training program will first decide if training is the most suitable approach for a particular safety and health problem (D1 in Figure 12.1). There may be, for example, an engineering control sufficiently effective to eliminate the need for training. In many instances encountered in the OSH field, training is required by regulation and is often a requirement of large businesses that contract out much of their work. Even if the underlying motive for establishing a training program stems from a requirement, course developers should write a statement of goals for the training that emphasizes the safety and health benefits rather than simply saying the goal is to comply with a requirement.

Each training program begins with needs (No. 1), usually driven by a history of harmful incidents, by regulatory requirements, by the findings of a JHA or RA, or by a desire to improve the employees' abilities to recognize hazards and participate in hazard control. Several methods for determining training needs are summarized in a paper by Cekada.[3] For training related to a particular job, one practical methods involves (1) defining job requirements, (2) specifying tasks within each function, (3) determining potentially hazardous events and exposures for each task, (4) defining precautions the employee should take, and (5) listing the skills and knowledge required to safely perform each task. A *task-analysis chart* such as that shown in Figure 12.2 facilitates organizing this information in a format the course developers can use to delineate learning objectives for the training.

Function	Task	Knowledge	Skills
Function A	A1		
	A2		
	A3		
Function B	B1		
	B2		
	B3		
	B4		
Function C	C1		

Figure 12.2 Task analysis worksheet relating functions, tasks, knowledge, and skills.

Effective learning objectives focus on the needs of trainees. In contrast, a poor practice is to list course objectives from the perspective of the instructor. For example, "to demonstrate how to properly don and doff personal protective gear" is much more appropriately worded as "at the conclusion of the training, each trainee will demonstrate the skill to properly don and doff the personal protective gear." This format helps course developers and instructors focus the course on what the trainees take away from the training. The learning objectives developed in process 2 should drive the third process—develop learning activities, materials, and specifications.

Each objective needs a plan to help trainees develop the applicable knowledge or skill. Because OSH training involves adults, best-practice course development makes use of adult learning theories.[4-6] A basic difference between working adults and college students is that college students have more tolerance for lectures. Experts on adult training agree that lecture sessions should not be too long. In the morning, when trainees are fresh, their attention spans may be as long as 20 min. After lunch, attention spans will be shorter. A second factor to consider is the mode of learning. Some adult learners absorb a lot from a lecture, while others get little. Some learn from watching a professionally developed training CD about the topic, some learn best from touching and using items, and some learn best from participating in peer discussions. The point is, in order to reach all trainees in a group of adults, the better courses offer several modes for learning the same material or skill.

Part of the course development process is to allocate time for each topic and develop a lesson plan for each time unit in the course outline. The lesson plan will identify multiple learning activities for each unit. Many trainers use topic relevant games or hands-on activities to stimulate interest and involvement. Training on respirators, for example, might use a mixture of lecture, demonstration, and hands-on practice with the equipment.

Generally, classroom training is most suited to achieving objectives involving knowledge, while hands-on training is most suited to objectives involving skill development. However, knowledge gained can be reinforced through hands-on training, and some skills can be developed in a classroom setting.

Many training programs start with a test of some sort to document trainee knowledge and skill before the training (D2). There are three very good reasons for pretesting. First, instructors can compare pretraining results with similar posttraining test results to provide a measure of training effectiveness. It is the only way to find out how much trainees gained from the course. The second reason is to determine to what extent trainees already have the knowledge and skills prior to sitting through the training. Depending on the results of the pretest, the instructor can adjust the instructional level and material to match the existing expertise of trainees. The third reason is that the process of taking the pretest contributes to the overall learning experience of attending the course by providing trainees time to concentrate on the material, recall and reinforce older memories of the material, and recognize weaknesses in their knowledge and skills.

When the course is ready for delivery, some ingredients that strongly influence effectiveness include having an instructor with appropriate expertise, suitable facilities and audiovisual equipment, and access to equipment for hands-on activities.

Table 12.1   Training Assessment Techniques

Assessment technique	What is assessed?	Comments
1. Trainee evaluations at end of course	Trainees' opinions of their experiences, instructor's performance, and facilities	Can help instructor learn what trainees like and dislike
2. Testing trainees at end of course	Trainees' knowledge and skill at end of course	Very valuable if tests assess course learning objectives
3. Pretest to posttest comparison	Trainee learning due to the course	The two tests must be comparable in difficulty and material
4. Instructor's assessment after course	Instructor's impression of course attributes that could be improved	Useful if instructor follows up by making improvements for future offerings
5. Trainee performance on their jobs before and after course	Changes in work behaviors apparently due to the course	Obtaining observational data pre- and post-course is resource intensive

Topic relevant games and small group activities support trainee interest, attention level, and learning.

Table 12.1 summarizes various techniques for assessing a training course—process 6 in the flowchart. Assessment results provide the basis for the instructor to make changes with intention of improving the training. The next diamond in the flowchart (D3) asks if the course will be repeated. If not, the training model ends here. Otherwise, the process continues to D4 that asks if there will be an attempt to improve the program.

If no improvements are needed, the course is ready to be repeated with no changes. If improvements are needed, the flowchart leads to the last process box in the model, which calls for making whatever changes would be beneficial. After making the changes, the flow chart has an arrow from process 7 up to the start of the next course offering.

A record of attendance and successful completion for each trainee provides important documentation. Many programs also create a card or certificate for each successful trainee to keep for his or her personal documentation.

12.3 WARNINGS

Like training, *warnings* and similar safety-related messages provide another approach for both informing employees of hazards and influencing work behavior. These devices communicate through human auditory, tactile, olfactory, and visual senses. Each is reviewed in this section.

An auditory warning serves best to alert people to an uncommon situation. Auditory alarms are excellent for getting attention, and the sound distributes in all directions so that people can detect it regardless of the direction of their head relative to the source. Traditional alarms (sirens, buzzers, horns, and bells) are limited in that they convey a simple message to the effect that something different is happening, which is why these devices are sometimes called auditory alerts. The person hearing the sound must figure out what to do. Various advances in the design of auditory alarms have extended the options. Devices are now available with options for pitch, waveform, rhythm, and intensity.[7] With a bit of training, personnel can learn to distinguish the sounds and recognize the intended meanings. In addition, advances in voice synthesizer technology have created options for adding verbal messages to the traditional auditory warning devices.

A common use of auditory warnings is for backing-up alarms on large vehicles. A key reason why these devices are used, and required, is that when a large vehicle is backing up, large areas behind the vehicle are not visible. These blind spots have been implicated in many run-over fatalities. To compensate, the alarms installed on the back of a large vehicle emit an obnoxious beeping sound while the vehicle is backing up. The hope is that people behind the vehicle will become aware of the approaching vehicle and move aside. No one knows how many lives have been saved by these devices, but they are by no means foolproof—people have been backed over even when the vehicle backing-up alarm was working. Some of these unfortunate incidents occurred at construction sites and mines, where multiple large vehicles routinely operate. A common belief is that the personnel on foot become complacent to the frequent sounds of backing-up alarms and begin to ignore the sounds. When that happens, the safety devices become less effective.

Devices based on human tactile senses typically use vibration to alert people. When the device is next to the skin, the user can easily detect the vibrations. These devices are used extensively in cell phones and in the devices that restaurants use to notify waiting customers that their table is ready. A review of possible devices that could be used to notify an emergency responder when exposed to radiation concluded that the vibratory type is desirable for getting attention, but the combination of vibratory device and a spoken message would be most effective.[7]

For the danger of exposure to odorless and colorless gases, adding an odorous chemical provides a mechanism for warning personnel that the gas is present. Examples are the addition of pyridine to argon and methyl mercaptan to natural gas. Limitations are similar to those of auditory warnings.

Lights are used on police cars, fire trucks, ambulances, wrecker trucks, and utility vehicles to attract the attention of other drivers and pedestrians. An advantage of flashing lights is their noticeability, even when viewed in the periphery of one's visual field. Flashing lights are also used in buildings for fire notification. Although fire alarms are extremely loud auditory devices, for building occupants who are deaf, the flashing lights serve as an alternative warning to the loud alarm. Because a bright flashing light reflects off wall and ceiling surfaces, even an occupant who is looking in the opposite direction of the light source will notice the flashes of reflected light.

A weakness of flashing lights is the very limited information conveyed. A person seeing a flashing light may or may not understand what it means. Someone who misinterprets the warning could choose a course of action that leads to increased danger (a decision error).

Another kind of lower key visual warning is found on displays in the control rooms of advanced facilities, ships, aircraft, motor vehicles, and many other engineered systems. Common forms of these include a light that comes on where none usually exists, a green light that changes to red, or a light that starts blinking. These warnings draw attention initially, but a weakness is that personnel can become complacent or mistrustful. If you visit a control room of a large plant, look at all the lights and gauges. Sometimes you will see a blinking light, ostensibly warning of an abnormal condition. If you ask the operators why they ignore the light, they will likely tell you that experience has convinced them the problem is not the equipment being monitored, but it is a defect in the warning light system. As a visitor, you will not know if they are right or wrong, but you will have to wonder why they are so willing to continue operations when they know something is not working as it should.

Signs are commonly used in workplaces to convey safety-related information. Many of these are warnings about particular hazards, and some signs provide general information. Signs are valuable in four ways. First, for personnel already familiar with the work area, signs located near hazards serve as reminders of the hazard. Second, for personnel unfamiliar with a work area, signs serve to notify them of the hazard. Third, safety signs serve to inform personnel of precautionary behavior. Fourth, a value not widely recognized is that the quality of workplace signage can affect the safety culture of the organization. Quality signage in a workplace conveys the subtle message that management really cares about employee safety. On the other hand, a facility filled with old, rusted, dust covered signs conveys the opposite message.

Worldwide standardization efforts for safety-related signs have nearly achieved harmonization for colors and symbols. This success is largely due to the contributions of numerous researchers and their studies.[8,9] Formats for sign layouts have not achieved international harmonization due to the standard formats used in the United States differing from those adopted internationally.

The two examples of workplace signs in Figure 12.3 illustrate common layouts in the United States. Signs contain spaces (panels) for designated content and standards specify format options for each panel. For the signs in Figure 12.3, the top panel contains a signal word in all capital letters within a specified background color. For safety-related signs used to warn of a hazard, the standard format of signal word panels reflects both theory and considerable empirical research into the colors, words, and symbols. Red conveys the level of greatest hazardousness. Less hazardousness is conveyed by yellow and orange. The signal word DANGER conveys greater hazardousness than CAUTION or WARNING. The sign in Figure 12.3a includes an internationally recognized triangle with an exclamation inside. When used in the signal word panel as shown, the symbol increases the impression of hazardousness.[10] Standards for signs used for communicating messages relevant to safety in general, rather than a particular hazard, use blue or green backgrounds in the signal word panel.

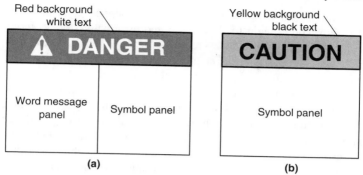

Figure 12.3 Example formats for workplace warning signs in the United States.

Under the signal word panel, the sign may have a text panel, a symbol panel, or both. In the text panel is a succinct message about the hazard and potential harm, followed by a text message about appropriate precautionary behavior. Many workplace signs provide the text messages in two or more languages. The symbols used in symbol panels should not be invented by an artist or a creative attorney. There are numerous symbols with standardized formats that have been tested for comprehension and found suitable for a specific message.[11] Some advantages of safety symbols are that they communicate effectively without the need for workers to be literate, their meaning may be captured by a quick glance, and many symbols are understood internationally.[11] Symbols for indicating a prohibited behavior use a pictogram depicting the behavior behind a diagonal slash running from the lower right to the upper left.

Although the world is slowly moving toward harmonized safety signage, flexibility is needed to allow local traditions and stereotypes. Local testing can be used to check if the international standards need modification for a particular population. An example of local testing used a study population of 50 male employees from industrial enterprises in India.[12] The testing of their perceptions of colors and signal words indicated that their perceptions of hazardousness were similar, although not identical, to those found among participants in studies conducted in other countries. The differences did not appear great enough to warrant development of an India-specific convention for workplace safety signs.

12.4 SAFETY DEVICES

Numerous safety devices are useful for implementing engineering control strategies. The major types are introduced in this section. Devices known as *dead-man controls*, used on many powered tools and equipment, require a human to actively engage a mechanism (e.g., a lever or button) in order to keep the machine operating. The term dead-man control came from the application of the device on train engines. Imagine the early days of locomotives. What if the train engineer died during a run? The train would continue moving as a runaway train. A countermeasure developed for this hazardous situation was a lever in the cab the train engineer was required to hold in position to maintain power to the wheels. If the engineer let go of the lever, for any reason, the train would begin to slow down and eventually stop. A more modern alternative used in many trains today requires the driver/engineer to either press a button or change the throttle position every minute.

The dead-man control design is now used for many power tools such as electric drills, portable power saws, and lawn mowers. The human must actively engage a mechanism to make the tool run. If the human releases the engaging mechanism for any reason, the motor shuts down or the drive mechanism is disengaged, causing the moving parts to come to a stop. One major benefit of this design is that the hazardous movement ceases soon after releasing the engaging mechanism. Thus, if the operator drops the power tool, it does not continue operating. The second benefit comes from placing the engaging mechanism a safe distance from the hazard. Thus, during operation the hand on the engaging mechanism should be located a safe distance from the hazardous energy. There is, however, a concern about possible injury during the brief time between when the control is released and when the hazardous motion fully stops.

What happens if the operator releases the control and then rapidly reaches for the hazardous moving part? This may seem a bit crazy, but consider this actual scenario. A groundskeeper was mowing a lawn with a push-type power lawn mower. There had been 5 days of rain, so the grass was wet and long. Several times, the blades jammed from the wet grass accumulating in the housing around the blades. Each time the groundskeeper released the control, reached under the housing, and cleared the excess grass using a short stick. The more he performed this procedure, the faster he got. Finally, his movement was so fast that he got his hand under the housing before the blades fully stopped.

Power tool manufacturers recognize the need to design tools for shutting down rapidly. To determine how brief this *stop time* must be, they need to know the duration of the operator's movement from the control mechanism to the point of danger (i.e., the *after-reach time*). Determining after-reach time involves experimentation with a sample of people similar to those who will be using the power tool. After collecting several after-reach times from each of the study participant, descriptive statistics are determined for the measured after-reach times (e.g., mean time, standard deviation, and range). From these findings, the manufacturer needs to decide what after-reach time to use for design specifications. After selecting a value for after-reach time, the tool designers need to find a way to make the stop time less than that of after-reach time.

Related to the dead-man control are some system designs widely used in safety engineering.[13,14] The term *fail-safe* applies to a system in which, if a particular failure occurs, the state of the system will not be dangerous. The nondangerous state may be achieved by failing passive or failing active.

The fail-passive design is a direct descendent of the dead-man control. If the operator fails to maintain the engaging mechanism, the power for the equipment drops to its lowest level, often to zero. Thus, the equipment ceases to move or otherwise operate. A fail-passive design is also used for electrical circuits. If the current exceeds the design capacity of the circuit, the circuit breaker should trip, opening the circuit path and stopping the current flow. A ground fault circuit interrupter is another safety device used in many electrical circuits. It monitors the difference in current between the hot wire and the neutral wire. If the difference exceeds a set point (approximately 5 mA), the device causes the circuit to shut down.

The fail-active design is an alternative to the fail-passive design. Safety engineers like fail-active designs for systems that would be dangerous if allowed to shut down at the wrong time. Examples might include an industrial operation that would be severely damaged if shut down suddenly, or a system that serves a critical safety function. For these systems, a failure should not result in shutting down or ceasing operation. The safer alternative is for a fail-active mode to kick in. A common design for workplace applications involves having a normal mode and an alternate fail-active mode. For the basic workplace fail-safe design, once the normal mode fails, the system is left with one means of sustaining the operation or safety function. This mode should be regarded as a temporary state in need of correction as soon as possible. Allowing the safety system to continue without correction would be inviting disaster. Engineers familiar with this issue will design a mechanism to make personnel aware of the temporary operating mode. This might be a blinking light on a control panel, a unique noise, or both. The design should provide a message to make clear to the operator what action is needed.

In addition to fail-active devices, two other designs are useful for sustaining safety-critical functions after a single failure. One is the use of a backup unit. For example, if the power supply to a hospital shuts down for very long, patients would die. Therefore, hospitals have backup generators available to provide power during an electrical outage. The other design, *redundancy*, involves putting two or more components into the system, each one being sufficient to perform the function independently.

Another group of safety devices includes interlocks, lockouts, and lockins.[13] A safety interlock serves the function of allowing, or disallowing, the active energy state of a device such as a microwave oven, a clothes dryer, or an industrial machine. Although there are numerous types, some of the most familiar are incorporated into a door or gate. The interlock acts as a switch that allows the active energy state when closed and disallows operation when open. A lockout keeps a source of energy out of where it should not be. This device is commonly used by maintenance personnel while they work on a machine, clean an empty storage vessel, or make changes in electrical circuits—they first isolate the site from all hazard sources using various switching devices, then lock those setting with a padlock to make sure no one comes along, and finally change the setting. A lockin keeps an energy in the active state (e.g., an electric circuit that must be active) or keeps a hazardous energy from getting out of a normally safe location.

12.5 EMERGENCY PREPAREDNESS

This section discusses organizational planning and preparation for emergencies and catastrophes. The focus is on the role of employers, rather than on the roles of fire departments, police, and emergency medical services; and discussions are limited to fundamental aspects of emergency planning and preparation. Subsequent chapters revisited the topic in the context of emergencies due to fires, explosions, geological events, weather events, workplace violence, and terrorist attacks.

What is meant when people talk about catastrophes and emergencies? An incident may be called a *catastrophe* if the outcomes create exceptional demands on resources needed to respond. Resources include medical personnel, rescue personnel, transport capacity, and hospitals. Incidents considered emergencies include catastrophes and other highly disruptive situations. Brauer defines an *emergency* in terms of three elements.[15] Emergencies are characterized by the following features: (1) arising in a relatively short span of time, (2) having disruptive effects on normal activities, and (3) needing a rapid and effective response to limit the damage. After the onset of typical emergencies, a lack of effective response allows the harm to expand in breadth and depth, whereas an effective response will limit the damage.

In the movies, effective responses to emergencies are shown as the heroic actions of characters played by movie stars. This idealized approach is unreliable and inadequate. Businesses have responsibilities for the consequences of the emergencies they create, and their post-event response will influence their liability. Governmental organizations also have responsibilities to respond to emergencies they create, and some governmental organizations exist primarily to provide response services (e.g., fire department, police departments, and emergency medical services). All these business and governmental organizations benefit by advance planning and preparations for emergency response.

Effective planning requires the time of many individuals, which in turn requires management support. Such support can range from minimal to enthusiastic. A minimal level of support corresponds to a compliance aspiration as discussed in Section 12.1. If management supports only compliance, the applicable laws and regulations will set the bar for the planning effort. For organizations aspiring to a best-practice level, much more is involved. OSH personnel typically advocate for a best-practice program, while other managers may have only compliance aspirations. Getting clear on the level of managerial support early in the process can help avoid endless frustration stemming from different visions by the top managers and the personnel involved in implementation. Because most of the books and journal articles about emergency planning assume a best-practice aspiration, those involved in the planning process would do well to review the advice in such documents and select only those recommendations most suited for meeting organizational needs.

A logical starting point for emergency planning is to assemble a planning team.[15,16] The team will want to start by identifying all the possible emergency incidents that might occur, followed by a shorter list of incidents most likely to affect the organization's operations. Some incidents deserving consideration by business and government organizations are as follows:

- Fires and explosions originating within the organization's property
- Fires and explosions on neighboring property
- Severe weather events

- Geological events such as earthquakes and volcanic eruptions
- Acts of terrorism
- Acts of employee violence
- Aggressive activities by mobs
- Releases of hazardous chemicals from industrial facilities or neighboring facilities
- Disruption of core utility services
- Disruption of electronic systems by lightning, power surges, and cyber attacks

Creating a short list of the most compelling types of emergencies enables a planning team to optimize the value resulting from their efforts. The process involves considering both the probability and the severity of foreseeable events. Similar to the risk-assessment matrix approach discussed in Chapter 5, some potential events will be high in both severity and probability, others low in both, and many in between. Reaching agreement on a risk matrix early in the planning process might prove useful for resolving differences in opinion. For example, an individual on the planning team might advocate for including a particular type of event on the short list because, somewhere in the world, some time ago, an event like that occurred. When presented with such a worst-case argument, the team members may need clarity on the responsibilities of the team. Generally, an emergency planning team is responsible for using its collective judgment to choose events for the short list and to develop emergency plans for those events. The team also has a responsibility to consider the economic realities of the organization or organizations participating in the planning.

Each type of event on the short list needs careful thought so that the team can plan suitable responses. Fires and many other events require plans for evacuating personnel from buildings. Other events call for keeping personnel inside buildings (the lockdown response). Events involving failure of a utility service or means of communication usually involve establishing redundant services or backup systems. Each type of event on the short list needs planning.

The team will want to discuss and decide on a policy on fighting fires. Three basic policy options are (1) evacuating all personnel whenever a fire is detected, (2) authorizing a few selected personnel to use a fire extinguisher on small (incipient) fires while everyone else evacuates, and (3) maintaining a team of well-trained and equipped personnel to respond while all others evacuate. The third option places a rather heavy burden on the organization to provide adequate training and equipment to the response team members.

Another component of planning considers the investigation process. Major incidents tend to involve large losses and raise issues of legal and financial responsibilities. It helps to have a plan for the immediate process of securing the scene, collecting evidence, and preserving evidence. The plan may call for using two teams—one to collect and preserve evidence and the other to analyze the evidence, determine root causes, write a report, and conduct debriefings.[17]

The best emergency plans are comprehensive in their scope. An Internet search identified dozens of templates for emergency action plans that can provide a useful outline and format for developing a plan. A sample template by the U.S. National Institute for Occupational Safety and Health may be a convenient starting place for organizations developing an original plan.[18]

Using a template does not negate the need for an organization to form a team and think through many issues, including answering the following questions. How will the organization create a record of evacuated employees? What method will be used to account for all personnel who evacuated and any who stayed to provide initial response? How will employees interact with outside responders? Will employees assemble in locations out of the way of responders? Will employees participate in directing traffic? Will certain personnel be assigned to assist the responders? Which employer personnel will deal with newspaper and television reporters? Who will contact insurance providers? For more information on issues to consider relative to disaster and emergency planning and management, the National Fire Protection Association has a comprehensive standard[19].

Another emergency-related planning process involves planning for business continuity. The disruption of normal business operations due to a major fire, explosion, or similar disaster can financially destroy an unprepared business. The adverse business impacts that can result from such unexpected events can be mitigated by following what is generally known as a *business continuity plan.*[20] Some elements to address in such a plan are communications with insurance companies, sharing information with the news media, meeting short-term financial obligations, retaining customers, cleanup, salvage, rebuilding structures, and repairing damage to the environment.

12.6 SANITATION AND HOUSEKEEPING

Both facility sanitation and *housekeeping* play fundamental roles in workplace safety and health. Poor sanitation encourages transmission of diseases such as common colds and flu, which means more employees will call in sick. This affects safety in two ways. First, replacement workers often have less expertise than the regular workers and may not fully appreciate the hazards and reasons for safe practices. Second, in many workplaces, when someone calls in sick, the work crews are expected to complete the regular work with a short crew. That puts more workload on each person and may lead to injuries. In nursing homes, for example, the nursing aides work in teams of two. If 12 nursing aides are scheduled for the shift, there will be six teams. If one calls in sick, there will be 11, leaving five two-person teams and one working alone. Since some of the work involves transferring residents (patients) who can provide little assistance, the exposure to musculoskeletal stressors of the nursing aide working alone will be greater than when working with a partner, increasing likelihood of a musculoskeletal injury.

Unsanitary work conditions can contribute to the spreading of hazardous materials. This is especially a concern where hazardous dusts or toxic materials are encountered in limited areas of a site, but due to poor sanitation practices, the contaminants are spread to other parts and to other employees. Another issue in this regard arises when employees wear their contaminated clothing home where the contaminant can be spread to other family members.

Unsanitary facilities provide an attractive harbor for rodents and insects that, along with other pests, can transmit diseases throughout the facility. Once these sorts of pests establish a colony in the facility, they can be difficult to eliminate. Good sanitation is the most effective way to keep these pests from getting that initial foothold.

Basic employee needs for sanitation should not be taken for granted, even in the most developed countries. Some of the most basic needs are a plentiful supply of sanitary drinking water, clean and functioning toilet facilities, hand washing facilities, a sanitary area for eating, and an adequate sewer system. At many industrial sites and mines, employees need locker rooms for changing into and out of work clothes, showering, and storing their gear. In workplaces where employees use personal protective equipment, there is a need for facilities to clean and sanitize equipment such as respirators, hard hats, and nondisposable hearing protectors.

Workplace sanitation also benefits from regular cleaning practices. Sometimes called "housecleaning," these practices include regular removal of trash and other waste materials, dusting, sweeping and mopping floors, cleaning spilled oils and other materials, and cleaning equipment.

Good housekeeping contributes to both risk reduction and a safety culture.[21] Housekeeping involves having places for everything, and actually keeping those things where they belong. Signs of poor housekeeping include pallets, barrels, boxes, and tools placed in all sorts of places; pedestrian pathways cluttered with items people might trip on; and fire exit pathways with items left there by employees who could find no other place to leave the items. In a well-managed facility, work is planned, there are designated places to store equipment, and employees understand that they are expected to perform according to the rules. In poorly managed workplaces, work is loosely planned, there are few rules for storing equipment, and employees perform their tasks according to their judgment.

In general, workplaces incorporating good sanitation, housecleaning, and housekeeping practices convey to employees and visitors the impression of being desirable places to work. These practices, combined with quality safety signage, should positively affect safety culture by conveying the message that management is committed to supporting employee health and safety.

LEARNING EXERCISES

1. Suppose you review some of the OSH training programs already developed and used by an organization. You find each program has a purpose statement, but it appears the various instructors had very different ideas about how to phrase a purpose statement. Develop a generic format for a purpose statement that may be used in all the OSH training courses within your organization. The format will have blank spaces to fill in for each particular course.

2. Develop a plan for a 2 h training program on warning signs. Assume the employees have never had a course like this one. The OSH department recently bought new signs conforming to the most recent standards and replaced all the old signs. Your plan should incorporate adult learning theory, such as diverse ways for learning. You are not being asked to develop the content of the course, but rather to describe how the time will be used. Include statements of the purpose of the training and learning objectives for trainees. Include time for a pretest and a posttest.

3. Each of the types of warnings has strengths and weaknesses. Answer the following questions.
(a) What strengths and weaknesses do auditory beeps and flashing lights share?
(b) What strengths and weaknesses are inherent to warning signs?
(c) If an employee says warning signs are ignored, and therefore useless, how can you respond in support of having warning signs in the workplace?

4. In some management system standards, warnings are among the categories for hierarchy of controls. In these ordered lists, warnings are preferred to administrative controls, and administrative controls are preferred to PPE.
(a) Do you think all types of warnings are more desirable than all types of administrative controls? Explain your rationale.
(b) Do you think all types of warnings are more desirable than all types of PPE? Explain your rationale.

5. For each item below, what type of safety device is in use?
(a) A clothes dryer will not operate unless the door is closed.
(b) A battery-powered smoke detector emits an audible beep when the battery is nearly dead.
(c) A mechanical power press has a curtain of light separating the human operator from the point of operation (the danger point). If the operator's arm or hand is detected by the light curtain, the press will not cycle.

6. A robotic work cell in a manufacturing facility has two safety devices to keep personnel separated from the hazardous movement of the robot. The first is a gate that will prevent the robot from operating while the gate is open. The second is a pressure-sensitive floor mat located between the gate and the robot movement zone. If the floor mat senses pressure, the robot cannot operate. Why are two safety devices needed?

7. Regarding emergency preparedness, concisely respond to the items below.
(a) Explain differences in organizational aspirations.
(b) Explain the two-step process for selecting a short list of emergencies to focus on when planning for emergencies.

8. Consider the following exchange during a meeting of the emergency planning committee. This particular meeting is for selecting a short list of foreseeable emergencies to address in planning. Jim says the chance of a hurricane is so tiny that it should not make the short list of emergency planning. Robert describes an instance where a similar facility was hit by a hurricane resulting in two deaths and extensive damage to the facility. He says he does not want to be held responsible if a hurricane strikes the facility. If you are chairing the committee, what might you say?

9. In planning for responses to some disasters, it becomes apparent that rescue personnel may be exposed to sources of radioactivity. A personal monitoring device is available to track exposure and recognize when the dosage is getting close to the tolerable level. You want to be sure the emergency worker is alerted when his or her exposure level reaches a point where exiting the area is necessary. The person will be wearing full body protective gear. For notifying the person, what type of warning devices would you recommend and what features would be desirable?

10. Engineers are designing a nuclear power station. To maintain the temperature of the reactor core, continuous water around the core is essential. The engineers need to select a valve for controlling the water flow. They have a choice between a normally closed valve and a normally open valve. Either type of valve will be controlled by an electric motor. During a "What if?" session, somewhat asked what if the electric power to the plant should fail? For this scenario, the engineers decided that a fail open valve would be safest. Is this a fail-active design or a fail-passive design? Explain your answer.

11. Continuing with the nuclear power station design, suppose the plan is to have a pair of redundant cooling systems, each with separate valves, electric motors, and pipes surrounding the reactor core. Electric motors control the valves in both cooling systems. If electric power to the plant fails, both cooling systems would be affected in the same way.
   (a) Would a power outage be an example of a common cause failure?
   (b) In computing the probability of both cooling systems failing, would it be correct to multiply their respective failure probabilities?

12. Explain how industrial housekeeping and sanitation can influence the workplace safety culture.

TECHNICAL TERMS

After-reach time	The time between the release of a hand actuator and the instant the hand enters the danger area of the machinery.
Business continuity plan	A proactive plan for a particular business entity describing what will be done after a major fire or disaster to minimize losses and continue the business.
Catastrophe	An incident resulting in harm to people or environment and creating exceptional demands on resources needed to respond.

Dead-man control	A safety device that allows power to a machine only while a lever or other device is engaged by the machine operator.
Emergency	An incident that occurs in a relatively short time, disrupts normal activities, and creates a need for rapid and effective response to limit the damage.
Fail-safe	A term indicating that a particular failure of a system will result in the system being not dangerous. The nondangerous state may be achieved by failing passive or failing active.
Housekeeping	In the context of OSH, various practices supporting decent working conditions, such as regular removal of trash and other waste materials, dusting, cleaning floors, cleaning spilled oils and other materials, cleaning equipment, and keeping everything in its place.
Program aspirations	In the context of OSH, what organizations are trying to achieve with their OSH programs such as complying with requirements or being world class.
Redundancy	A design approach for increasing system reliability by using multiple components to perform the same important function. Each component can perform the function alone. The redundant components may be identical, similar, or different. Component independence is highly desirable for system reliability.
Stop time	The brief time between releasing the control of powered equipment and the hazardous motion fully stopping.
Task-analysis chart	A useful tool for developers of OSH training courses to determine the knowledge and skills trainees need to perform their tasks safely.
Warnings	Various means for notifying people of a hazard so that they have the opportunity to take appropriate precautions.

REFERENCES

1. Cohen HH, Jensen RC. Measuring the effectiveness of an industrial lift truck safety training program. J Safety Research. 1984; 15(3):125–135.
2. Jensen, RC. Safety training: Flowchart model facilitates development of effective courses. Professional Safety. 2005; 50(2):26–32.
3. Cekada TL. Conducting an effective needs assessment. Professional Safety. 2011;56(12): 28–34.
4. Kelly MH. Teach an old dog new tricks: Training techniques for the adult learner. Professional Safety. 2006; 50(6):44–48.
5. Galbraith DD, Fouch SE. Principles of adult learning: Application to safety training. Professional Safety. 2007; 51(9):35–40.
6. Merli CM. Effective training for adult learners. Professional Safety. 2011; 56(7):49–51.
7. Herring SR, Hallbeck MS. Conceptual design of a wearable radiation detector alarm system: A review of the literature. Theoretical Issues Ergonomics Science. 2010; 11(3):197–219.

8. Wogalter MS, Laughery KR. Warnings and hazard communications. In: Salvendy G, editor. Handbook of human factors and ergonomics. 3rd ed. Hoboken, NJ: Wiley; 2006. p. 889–911.

9. Wogalter MS, editor. Handbook of warnings. London: Lawrence Erlbaum; 2006.

10. Jensen RC, McCammack AM. Severity message from hazard alert symbol on caution signs. *Proceedings of the Human Factors and Ergonomics Society 47th Annual Meeting*, Santa Monica, CA: Human Factors and Ergonomics Society; 2003. p. 1767–1771.

11. Deppa SW. U.S. and international standards for safety symbols. In: Wogalter, Dejoy, Laughery, editors. Warnings and risk communication. London: Taylor & Francis; 1999. p. 477–486.

12. Borade AB, Bansod SV, Gandhewar VR. Hazard perception based on safety words and colors: An Indian perspective. International J Occupational Safety Ergonomics. 2008; 14(4);407–416.

13. Brauer RL. Safety and health for engineers, 2nd ed. Hoboken, NJ: Wiley; 2006. p. 98–102.

14. Hammer W, Price D. Occupational safety management and engineering, 5th ed. Upper Saddle River, NJ: Prentice Hall; 2001. p. 192–200.

15. Brauer RL. Safety and health for engineers, 2nd ed. Hoboken, NJ: Wiley; 2006. p. 537–545.

16. Schroll RC. Industrial fire protection handbook, 2nd ed. Boca Raton, FL: CRC; 1999. p. 183–192.

17. Resimius M, Stiller J. Effective preparedness training: Improving response to major incidents. Professional Safety. 2010; 55(3):18–20.

18. National Institute for Occupational Safety and, Health. Emergency action plan (template). Available at http://www.cdc.gov/niosh/docs/2004-101/emrgact/emrgact.doc.

19. National Fire Protection, Association. NFPA 1600, Standard on disaster/emergency management and business continuity programs, 2010 ed. Available at http://www.nfpa.org.

20. Clas E. Business continuity plans: Key to being prepared for disaster. Professional Safety. 2008; 53 (9):45–48.

21. Becker JE. Implementing 5S to promote safety and housekeeping. Professional Safety. 2001; 46 (8):29–31.

Chapter 13

Tools for Managing OSH Programs

A typical career progression within the OSH field begins with an education in a relevant field. The first job involves application of material learned in school. These entry-level positions often involve considerable attention to compliance with rules and regulations. As the person gains experience, managerial aspects of the job become more significant. This chapter provides background on three managerial topics important to the advancing OSH professional—safety culture, management systems, and ethical policy.

13.1 SAFETY CULTURE

Since the mid-1990s, the level of interest in *safety culture* has skyrocketed. There are many speakers on these and related topics at professional development conferences and many articles on safety culture in the professional journals. This section briefly explains fundamental terms and concepts regarding organizational characteristics relevant to safety culture.

Organizations differ considerably in their values, attitudes, and practices. These differences appear in many business areas such as quality, productivity, financial policy, ethical practices, human resources, environmental protection, and employee safety and health, all of which contribute to the organization's culture. Thus, safety and health is but one component of a complex mixture of values, attitudes, and practices that make up organizational culture.

Many OSH managers strive to improve their part of the mixture—the safety culture—with the understanding that doing so is a long-range goal. The safety-related values, attitudes, and practices within an established organization take time to modify. It requires gaining the genuine support of people at all levels of the organization.

The path to achieving long-range goals usually starts with establishing objectives. In most organizations, objectives include measures of performance, a time

Risk-Reduction Methods for Occupational Safety and Health, First Edition. Roger C. Jensen.
© 2012 John Wiley & Sons, Inc. Published 2012 by John Wiley & Sons, Inc.

frame, and a target level of achievement. For objectives involving safety culture, a challenge is finding a suitable measure of performance.

13.1.1 Safety Climate Surveys

A suitable measurement approach has been evolving to meet the need for tracking objectives involving the safety culture. It involves using surveys to learn about *safety climate*. Safety climate is a part of safety culture we can measure using questionnaire surveys of employees at all levels of the organization.[1-3] These surveys seek to learn about employee perceptions of safety practices and attitudes. An example from one study used multiple survey items for each of the following factors.[2]

- Management commitment to safety
- Supervisor safety support
- Coworker safety support
- Employee participation in safety-related decision-making and activities
- Competence of employees with regard to safety

Constructing an employee perception survey requires considerable effort and sophistication to achieve adequate reliability and validity.[4] Valid surveys contain multiple rating items for each of the factor categories. The survey administrator processes the employee responses to provide information about each factor. Results of perception questions may be useful for one or more of three purposes: (1) tracking objectives, (2) continuous improvement programs, and (3) research. The one already mentioned is tracking objectives. The second is identifying the weaker components of safety climate as a basic step in the continuous improvement program. For example, a survey might reveal that managers think they consistently show support for safety and health practices, while the perceptions of employees may be quite different. Such a finding would prompt the organization to institute improvements during the next cycle of the continuous improvement process.

The third purpose of safety climate surveys is research to advance the understanding of safety climate. An example of this is an employee survey for a large retail business with stores throughout the United States and Canada.[1] Results found a significant association between perceptions of safety climate and the organization's safety-related policies and practices. Two other influential factors found in the survey were the general quality of the work environment and the safety-related communications. Similar retail businesses may regard these findings relevant for comparing to their own safety climate.

When reading research reports of safety climate surveys, attention to the survey sample is critical. If the entire sample consists of employees from one business, results apply only to that business. Likewise, within a large industrial facility such as a vehicle assembly plant, one can expect differences in safety climate among the work units. Generalizing findings from one unit to others in the same organization is problematic because the responses are unique to the surveyed population.[5] For research purposes, the most valid use of safety climate surveys is for learning about factors within the surveyed population. Generalizing may be more acceptable if based on results of similar surveys of employees in a similar type of businesses. For example, if similar surveys of four large retail companies consistently found a relationship between safety policies and perceptions of safety climate, one may cautiously conclude that the relationship applies to large retail businesses in general. It would be unwise to assume that those same findings would apply to organizations in another sector, such as health care, energy, or manufacturing.

A smart approach to safety climate surveys is to develop standardized, industry-specific survey instruments. An example of this approach is being pursued by some people in the healthcare industry.[6] Potential uses might include comparing different sectors within the healthcare industries, comparing different healthcare entities with peers, and learning from their collective experiences more about factors that influence safety climate in healthcare settings.

Organizations concerned with measuring safety climate need to decide how often to administer surveys. The use of a yearly frequency, so common in many areas of business, is probably too frequent. Since attitudes and perception change gradually, a frequency in the range of two to five years may be adequate. For organizations with objectives for improving safety climate, the logical survey frequency will correspond to the frequency of the continuous improvement cycle for safety climate.

13.1.2 Getting Value From Safety Climate Surveys

After a survey is completed and analyzed, the organization will identify weaknesses, plan corrections, and implement the corrections. For example, suppose the responses show that employees perceive management's commitment to safety as weak. A plan to correct this weakness may involve efforts to make the safety-related activities of top managers more visible. Top managers can demonstrate their commitment to safety by personally announcing changes in safety-related policies and practices, by accompanying personnel from the OSH department on walk-around surveys and audits, by highlighting safety and health in newsletters and talks, and by meeting personally with injured employees. An approach used at one plant in the automotive industry involved having the plant manager discuss with each injured employee the findings from the investigation of his or her injury.[7] If the manager leaves the impression of sincerely caring, the injured employee will come away feeling the plant manager actually cares about employee safety, and the same employee will share that impression with associates.

As another example, suppose the safety climate survey indicates lackluster employee participation and coworker support for safety and health. Various changes can strengthen those factors. One is to adopt a system for improving safety-related behaviors. Many employers use behavior-focused systems based on established principles from the field of behavioral psychology. A key starting place is getting small groups of employees who do similar work to identify specific behaviors important for working safely. They then screen these to identify those behaviors that are observable and classifiable as being proper or improper. Training provided for the work group members teaches how to observe the behaviors of others in the group, provide face-to-face feedback to their coworkers, accept feedback from others, and file a report. Someone in the OSH department collects observation reports and computes the percentage of observed behaviors considered proper. Some companies post results on an employee bulletin board in the form of group performance. Some key benefits of these programs are the high level of employee involvement, attention to safety, performance feedback from coworkers, and changes in work group behavioral norms. All these activities contribute to safety climate by increasing employee participation and coworker support.

While safety climate surveys are the primary source of information about safety culture, a different perspective on organizational culture warrants mention. Reviews of some major disasters provided interesting findings about attributes of safety culture that open the door for disasters. These were subsequently paired with attributes that discourage disasters.[8] Table 13.1 lists the pairs of attributes. Although the title of the article refers to the nuclear power industry, the attributes contrasted in the table synthesize finding from disasters in space flight, process industries, and nuclear power generation. As such, one may view the safety culture attributes as useful for many industrial sites where major disasters are possible.

Table 13.1   Contrast Between Deficient and Good Attributes of Organizational Cultures

Deficient attributes	Good attributes
Diffuse responsibilities	Clear accountability and openness
Invincible mindset	Respect for limitations of technology
Compliance means safe enough	Keep striving for excellence
"Groupthink" in teamwork	Teamwork practices encourage and respect individual thinking
No systematic processing of relevant experience from elsewhere	Uses process for learning from the bad experiences of others
Lessons learned disregarded	Lessons learned from ourselves and others are communicated and used for improvements
Low priority of safety actions	Safety is paramount
Little preparation for severe events	Prepared and practiced for emergencies
Unnecessary acceptance of hazards in design or operating features	Maximizing safety design/operation before accepting risks
Failure to use of project and risk management techniques	Use proven techniques to aid technical excellence
Safety matters not recognized/integrated into work of organization	Clearly defined responsibilities and authority for safety matters

Based in part on Ref. 8.

13.2 OSH MANAGEMENT SYSTEM APPROACH

Like the importance of safety culture, interest in management systems has sky-rocketed. Standards organizations throughout the world have developed management system standards. A study by Manuele indicates that the various standards are similar in terms of overall outlines of their standards and in the use of *continuous improvement* processes.[9] A standard adopted by the American National Standards Institute (ANSI) is summarized in this section to illustrate the major features of an occupational health and safety management system (OHSMS).[10,11]

Figure 13.1 depicts the major concepts. The large box on top represents the ongoing workplace system in operation with the current OHSMS depicted in the lower left corner of the large box. Using the process in the ANSI standard, the organizations will transform the current OHSMS into an OHSMS modeled after provisions in the ANSI standard.

The processes for initial planning are depicted with two smaller process boxes placed under the large box on the left. The standard specifies that the planning process must involve the leadership of top management and include employee participation. This process should produce a policy statement indicating management's commitment to and vision for the OHSMS. Another important outcome of this initial planning process is identification of the major issues facing the organization, including both hazards and components of the current OSH management systems. The planning team then prioritizes the issues. The plan will describe the methods to be used for regular operation of the new OHSMS. For example, the plan will describe the risk assessment approach to be used, including the risk assessment matrix and guidelines on tolerable risks. Upon completion of the initial planning steps, implementation begins. The implementation will take time, so the diagram shows multiple arrows to indicate a phased implementation process.

The implemented OHSMS will use the methods described in the OHSMS plan. The left column of Table 13.2 lists some of the activities for which written methods will be invaluable. The right column indicates sources in this book for more information about the methods.

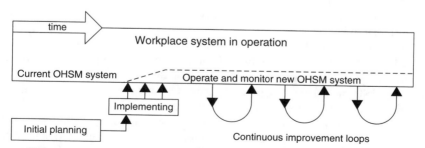

Figure 13.1 Major processes for implementing an occupational health and safety management system.

Table 13.2   OHSMS Activities Needing Written Methods

Activities needing written methods	Source for more information
Identifying hazards	Chapter 4
Conducting risk assessments	Chapter 5
Following a preferred order for hazard controls	Chapter 11, section 11.3
Performing various hazard analyses	Chapters 5, 6, 7 and 8
Investigating injuries and new occupational disease cases	Chapter 9
Training personnel on OSH matters	Chapter 12, section 12.2
Being prepared for emergencies	Chapter 12, section 12.5

This standard does not explicitly include risk reduction involving property damage or environmental pollution, but the organization may choose to include those concerns. Various ongoing monitoring provides assessment data for periodic review. Examples are industrial exposure measurements, injury reporting, behavioral observations, compliance inspections, and audits concerned with the OHSMS implementation.

The plan should also address ways to avoid introducing hazards to the workplaces. Three of these are as follows:

1. A management of change procedure,

2. A plan for incorporating health and safety into the procurement process, and

3. A plan for managing the interactions between the organization's personnel and contractor's personnel working at the same facilities.

On the right-hand side of Figure 13.1 are some *continuous improvement loops*. The ANSI standard specifies that management must review the OHSMS at least annually. The review team is expected to identify weaknesses in the OHSMS and recommend corrections. In addition to the management review, the organization may use other continuous improvement loops to examine the performance of various components of the OHSMS. Each loop includes (1) planning corrective actions, (2) implementing into operations, (3) obtaining and evaluating data collected subsequent to the change, and (4) introducing changes to make the program better. These four processes are often referred to with an easy to memorize string of words such as plan-do-check-act or plan-do-study-act.[12]

The ANSI standard contains numerous required processes as well as recommended practices to give organizations flexibility. Of course, organizations with low aspirations may simply ignore the existence of the ANSI standard or any other voluntary OSH management system standard. Organizations with more ambitious aspirations for their OSH programs may choose to model their OSH management system on the ANSI standard or a standard from another organization such as the International Organization for Standardization (ISO) or the International Labour Organization (ILO). They may apply their chosen standard as is, or take a smorgasbord approach—choosing features that seem to fit the needs of the organization. Most organizations with best-practice aspirations will choose one OHSMS to guide their program.

13.3 ETHICAL POLICIES FOR OSH

OSH departments and staff occasionally need to resolve issues for which there is no clearly right or wrong resolution. These gray area issues arise from the nature of OSH functions. Consider that the OSH manager and staff are employees of the organization. In exchange for their pay and benefits, they owe the organization duties of loyalty and support. At the same time, the main reason the organization has a professionally staffed OSH department is to support all employees through various health and safety programs. When the OSH department is also responsible for addressing environmental concerns, responsibilities extend to supporting the community through pollution prevention programs. The gray area issues typically arise when duties to two or more constituents appear to conflict.

Early writings by philosophers and religious scholars on the subjects of ethics and morals provided a foundation for present thinking. From these roots, two branches relevant to OSH have emerged—business ethics and professional ethics. The business perspective generates extensive media coverage whenever corporate practices contrary to the public interest are revealed. The professional ethics branch consists of profession-based branches, such as medical ethics, legal ethics, and engineering ethics. The professional organizations created by OSH professionals also have codes for ethical practice. This section focuses on the interrelated roles of business ethics and professional ethics applicable to OSH departments and careers.

Many people go through life without much thought about business or professional ethics. They generally assume they have some sort of natural instinct for distinguishing between right and wrong. If they happen to encounter a situation requiring a gray area decision, they will be on shaky ground. They may seek guidance from peers, family members, religious advisers, or the writings of various authors (e.g., see chapter 26 in Ref. 13). However, this approach of waiting until the situation presents itself is not encouraged. The preferred approach—the one advocated in this book—is to be proactive.

A proactive approach involves two parts. One is formulating an ethics policy statement for the OSH department. The other is obtaining top management support for the OSH professionals in the organization who want to practice their profession in accordance with the ethical codes applicable to their profession.

13.3.1 Ethics Policy Statement

The OSH department needs to find a way to formulate an ethics policy that fits the culture of the organization. Large organizations typically have an official statement regarding ethics. The OSH department should align its statements of mission, goals, and ethics with those of the organization. This need to align is not unique to OSH. In organizations, the top-level mission statements of the organization should drive the mission statements for all subordinate units. And the goals of each unit should align with one or more of the organization's goals. Thus, managers are very familiar with the concept of alignment and should have no trouble understanding the need to align the ethics statement of the OSH department with that of the organization.

The OSH program statement of goals should clarify a basic issue facing industrial hygienists: Is it the goal of the organization to comply with the legally required exposure standards or with best-practice exposure standards? A clear statement on this critical matter can head off a common conflict faced by industrial hygienists.

The content of OSH program ethics statements will vary across organizations, but any such statement should address two key issues. One is clarification of the competing duties to support (1) the organization, (2) the employee health and safety, and (3) the environment. Ideally, the statement will be clear enough to provide guidance for the OSH program manager when asked, for example, to support top managers while compromising employee health and safety or environmental protection. The second key issue to address in the OSH program ethics statement concerns the conduct of staff professionals. Ideally, the statement will expressly acknowledge that OSH professionals should conduct their work according to the ethical codes of their profession. If these two issues are addressed in a policy statement that top management has approved, the chances of serious ethical dilemmas for the OSH manager and professional staff are greatly reduced.

13.3.2 Management Approval

Once formulated, the OSH department ethics statement needs approval at the highest feasible organizational level. When seeking management approval, it may prove useful to reference one of the most respected gurus of management, Peter Drucker. In the ethics chapter of his book *Management: Tasks, Responsibilities, Practices*, Drucker asserts that professionals need principles to distinguish themselves from the masses.[14] His first and overarching principle is that professionals "do no harm," which he qualifies by adding the word "intentionally." His second principle is autonomy, which means a person acting in a professional capacity can make decisions within the domain of responsibilities associated with their field. A person is not autonomous if the boss (or client) can control the person's day-to-day activities and dictate decisions or recommendations. Drucker's third characteristic of a professional is respect for privacy. When a professional is entrusted with confidential information, it stays confidential. A professional is expected to know what information is confidential, and if unclear, find out. Some information is normally confidential even without an explicit statement to that effect. Normally confidential information that OSH personnel are likely to encounter include the following:

- Information about company trade secrets or other proprietary rights,
- Employee medical records, and
- Responses from employees to surveys are generally considered confidential.

A professional must be in a position to protect private information without fear of being reprimanded or discharged. Many managers who are familiar with Drucker's writings will find it difficult to oppose an OSH department proposal to formulate a statement of ethics addressing Drucker's three principles of professionalism.

To appreciate why OSH professionals benefit from an ethics code, consider that reputations are built over time. Daily decisions, statements, and behaviors contribute to establishing a reputation. Those in OSH careers can develop a reputation for being ethical by consistently being trustworthy, fair, and honest. All these traits are important for performing OSH functions effectively. Although it can take years to build, a reputation can be destroyed by a single, visible, unethical decision or behavior.

The most effective way to prevent such a damaging career event is to take a proactive approach to ethics. Codes developed by professional peers address the characteristics mentioned by Drucker (do no harm intentionally, work autonomously, and respect privacy) in addition to other issues encountered by those in the profession. The ethical codes and codes of practice specifically for the OSH professions vary, so an OSH professional may chose one that most closely matches their primary specialty. Sources of ethical codes for the OSH professions include those of professional societies and those of organizations that issue professional credentials. The most appropriate choice depends on the country of practice and the professional specialty. For purposes of illustrating the nature of ethics codes, one from the Board of Certification in Professional Ergonomics (BCPE) is in the appendix to this chapter.

13.4 SUMMARY OF PART III

Part III consists of five chapters on programmatic methods for reducing occupational safety and health risks. Chapter 9 summarizes the steps in a closed-loop process for ensuring follow-up to harmful incidents and discusses issues for organizations to consider when developing policies related to each step of that process. It explains tools to help incident investigators organize evidence into a series of events and conditions leading to and following the harmful event. It concludes with an innovative graphic model showing the interrelationships among (1) the regular functioning system operating normally, (2) the initial deviation from normal, (3) a failure to detect and correct the deviation, (4) subsequent conditions and events leading to the harmful event, and (5) post-incident events affecting the ultimate outcome.

A discussion of human error in chapter 10 presents some attempts at classifying errors and an approach for thinking strategically about reducing errors. Lessons may be learned from investigating incidents involving human error by digging deep enough to understand the causes of the human error. The chapter emphasizes a classification system for unsafe acts (errors and rule violations) that OSH professionals can use to identify countermeasures for addressing the underlying cause of the error.

Chapter 11 introduces the concept of "strategies" in the context of public health and occupational safety and health. After a brief review of some notable prior attempts to classify all risk-reduction strategies, a nine-category classification system is provided and briefly described. The nine strategies can be useful for a system design team while stepping through a risk assessment. The strategies fit neatly into a conventional three-level priority scheme—eliminate the hazard, use engineering controls, and use administrative controls. The extensive Learning Exercises provide readers an opportunity to appreciate how the strategies apply to specific OSH scenarios.

Chapter 12 discusses several basic activities and programs that serve as foundations for OSH programs. It begins by pointing out that organizations have differing aspirations for their OSH program, and the aspiration level drives the organization's level of commitment to specific programs. The employee training topic is structured around a flowchart model of the training process that includes continuous improvement. The review of warnings addresses the pros and cons of devices based on sound, vibration, lights, and signs. The overview of safety devices introduces some common engineering approaches for ensuring that equipment failures do not immediately cause serious harm. Section 12.5 summarizes the value of and approaches to emergency planning and preparation. Section 12.6 summarizes the value of sanitation and housekeeping practices to workplace health and suggests that good sanitation and housekeeping practices contribute to employee health and safety culture.

Chapter 13 discusses three managerial topics important to OSH professionals. It begins with the much discussed topics of safety culture and safety climate. Efforts to assess the safety culture of organizations have led to employee survey instruments for determining factors that influence safety climate within the organization surveyed. Because the results of numerous surveys of employees in diverse industries have produced varied conclusions about which factors influence safety climate, the author expresses support for developing and validating industry-specific survey instruments. Occupational health and safety management systems are illustrated by explaining key processes found in an American standard. Section 13.3 discusses ethical policies relevant to professionals in the OSH specialties. The author encourages taking a proactive approach to OSH ethics.

LEARNING EXERCISES

1. Explain the distinction between safety culture and safety climate.

2. A multinational company decides to upgrade its current OHSMS by following the ANSI standard discussed in the chapter. During initial planning, the team members seem confused about the difference between inspections and audits. If you were on the team, how would you explain the distinction?

3. Explain the difference between a proactive and a reactive approach to professional ethics.

4. Assume a recent college graduate is hired for an industrial hygiene position. After completing her first monitoring assignment, she writes a report. The senior industrial hygienist she works for reviews the report and annotates

many changes. Some relate to recommendations in the draft, where the boss changes language stating that certain controls "shall" be implemented to less demanding words such as "could" and "recommended." The new IH is offended by all the changes and considers her professionalism violated because she is not being allowed to work autonomously. Discuss the issues in terms of professional ethics.

TECHNICAL TERMS

Continuous improvement	An organizational practice for improving processes through regular cycles of performance measurement, evaluation of findings, and correction of weaknesses.
Continuous improvement loops	Periodic reviews of processes for evaluating past performance, assessing, planning, and implementing improvements.
Safety climate	The perceptions, attitudes, and beliefs of people at all levels of an organization regarding occupational safety and health practices.
Safety culture	The pervasive values and actions of people at all levels of an organization regarding occupational safety and health.

APPENDIX: EXAMPLE CODE OF PROFESSIONAL CONDUCT

The following is quoted, with permission, from the BCPE. The code is accessible from http://www.bcpe.org/page/codeofethics.

Code of Ethics and Professional Conduct

The BCPE is dedicated to protect the consumer of ergonomists' professional services by (a) establishing, promoting, and revising as necessary standards that reflect the qualifications for the professional practice of ergonomics; (b) establishing procedures for the evaluation of the credentials of those who voluntarily apply for certification by the BCPE, causing the issuance of a certificate to those who have qualified, in the sole judgment of the BCPE, as having met the standards established by the BCPE; (c) maintaining and disseminating a directory of certificate holders on a regular basis; and (d) advancing the field as well as the practice of ergonomics. To promote and sustain the highest levels of professional and scientific performance by its certificate holders, BCPE has adopted this code of ethics. Certificate holders shall, in their professional ergonomics activities, sustain and advance the integrity, honor, and prestige of the ergonomics profession by adherence to the following principles (Adopted May 4, 2002):

Principle 1. BCPE certificate holders shall practice their profession following recognized scientific principles and practices. The lives, health, and well-being of people depend upon their professional judgment. They are obligated to protect the health and well-being of the public.

Principle 2. BCPE certificate holders shall be honest, fair, and impartial. They shall act with responsibility and integrity in all professional actions. They shall adhere to high standards of ethical conduct with balanced care for the interests of the public, employers, clients, employees, colleagues, and the ergonomics profession. They shall avoid all conduct or practice that is likely to discredit the profession or deceive the public.

Principle 3. BCPE certificate holders shall undertake assignments only when qualified by education or experience in the specific technical fields involved. They shall accept responsibility for their continued professional development by acquiring and maintaining competence through continuing education, experience, and professional training.

Principle 4. BCPE certificate holders shall avoid deceptive acts that falsify or misrepresent their academic or professional qualifications. They shall not misrepresent or exaggerate their degree of responsibility in, or for, the subject matter of prior assignments. They shall not misrepresent pertinent facts concerning employers, employees, associates, or past accomplishments.

Principle 5. BCPE certificate holders shall conduct their professional relations by the highest standards of integrity and avoid compromise of their professional judgment by conflicts of interest.

Principle 6. BCPE certificate holders shall act in a manner free of bias with regard to religion, ethnicity, gender, age, national origin, or disability.

Principle 7. BCPE certificate holders shall keep confidential personal and business information obtained during the conduct of their services, except when required by law.

Principle 8. BCPE certificate holders shall seek opportunities to offer constructive service in civic affairs and work for the advancement of the safety, health, and well-being of their community and their profession by sharing their knowledge and skills.

REFERENCES

1. DeJoy DM, Schaffer BS, Wilson MG, Vandenberg RJ, Butts MM. Creating safer workplaces: Assessing the determinants and roles of safety climate. J Safety Research. 2004;35(1):81–90.
2. Seo D-C, Torabi MR, Blair EH, Ellis NT. A cross-validation of safety climate scale using confirmatory factor analytic approach. J Safety Research. 2004;35(4):427–445.
3. Salminen S, Seppälä A. Safety climate in Finnish- and Swedish-speaking companies. International J Occupational Safety Ergonomics. 2005;11(4):389–397.

4. O'Toole M, Nalbone D. Safety perception surveys: What to ask, how to analyze. Professional Safety. 2011;56(6):50–62.

5. Shannon HS, Norman GR. Deriving the factor structure of safety climate scales. Safety Science. 2009;47(3):327–329.

6. Turnberg W, Daniell W. Evaluation of a healthcare safety climate measurement tool. J Safety Research. 2008;39(6):563–568.

7. Boraiko C, Beardsley T, Wright E. Accident investigation: One element of an effective culture. Professional Safety. 2008;53(9):26–29.

8. Long RL, Briant VS. Vigilance required: Lessons for creating a strong nuclear culture. J System Safety. 1999;35(4):31–34.

9. Manuele FA. ANSI/AIHA Z10–2005: The new benchmark for safety management systems. Professional Safety. 2006;51(2):25–33.

10. ANSI Z10, Committee. Occupational health and safety management systems: ANSI/AIHA Z10- 2005. Fairfax, VA: American Industrial Hygiene Association; 2005.

11. Manuele, FA., Advanced safety management: Focusing on Z10 and serious injury prevention. Hoboken, NJ: Wiley; 2008.

12. Adams, EE., Total quality safety management—An introduction. Des Plains, IL: American Society of Safety Engineers; 1995. p. 3–7.

13. Goetsch, DL., Occupational safety and health for technologists, engineers, and managers, 7th ed. Upper Saddle River, NJ: Prentice Hall; 2010. p. 593–605.

14. Drucker, PF., Management: Tasks, responsibilities, practices. London: Harper & Row; 1974. p. 366–369.

Part IV

Risk Reduction for Energy Sources

We come now to a turning point in the book. Parts I–III dealt with some concepts used throughout the book, various systematic analysis methods, and tools for managing OSH programs. Parts IV and V provide chapters on the seven hazard sources introduced in chapter 2—forms of energy, weather or geological events, conditions, chemical substances, biological agents, musculoskeletal stressors, and the violent actions of people. Each chapter provides numerous examples to illustrate how the nine strategies introduced in chapter 11 apply to the hazard sources.

Part IV discusses the first two of the seven categories of hazard sources. Because energy takes many forms, eight chapters are devoted to that hazard source and one is for weather and geological events. The chapters are on kinetic energy, electrical energy, acoustic energy and vibration, thermal energy, fires, explosions, pressure, electromagnetic energy, and severe weather and geological events. Hazards that develop from two or more energy transformations are classified based on the energy category most commonly used by the OSH community.

Each of the energy chapters has a review of the basic physics of the energy source, followed by a discussion or the mechanisms by which the energy can harm people, property, or environment. The third section in each chapter presents numerous examples of accepted practices for reducing the risks associated with the hazardous energy using the nine strategies.

Chapter 14

Kinetic Energy Hazards

Energy comes in many forms. We rely on various forms of energy for our daily living and for making industrial production possible. This chapter begins with a brief review of energy in general before getting into the main topic—*kinetic energy (KE)*.

14.1 ENERGY IN GENERAL

To understand energy, we first need to distinguish among three concepts—energy, work, and power.

Energy refers to the ability to do work.

Work is using a force to move something.[1] To measure work, we need to know the amount of force used and the distance of the movement. If you were to push with all your might on a solid wall, the wall would not move. Although you would accomplish zero work, you would use some of your metabolic energy trying.

Power refers to the rate of work. Expressing power quantitatively involves quantifying work per unit of time.

Energy in an inactive state is *potential energy*. It has potential to perform work, but it is currently dormant. For example, a load held by a crane has the potential to slip from the rigging and fall to a lower level. The gravitational potential energy of the elevated load transforms into an *active state* during a fall. A projectile in motion also illustrates active kinetic energy.

Energy often changes from one form to another. A basic scientific law says that within a closed system, when energy is transformed, the total amount of energy before the transformation equals the total energy after the transformation. Known as the Law of Energy Conservation (or First Law of Thermodynamics), it tells us that energy is neither created nor destroyed, but only changes form.[1]

Forms of energy do not always fit neatly into distinguishable categories. To illustrate how varying perspectives can lead to fuzzy lines between categories, consider two examples. First, the distinction between kinetic energy and electrical

Risk-Reduction Methods for Occupational Safety and Health, First Edition. Roger C. Jensen.
© 2012 John Wiley & Sons, Inc. Published 2012 by John Wiley & Sons, Inc.

energy is generally accepted. However, a physicist may point out that the electrical energy exists because of electrons moving within the copper wires of an electrical circuit; thus, the motion of the electrons may logically be used to classify this as kinetic energy. Of course, this is the perspective of neither the general public nor the OSH community. Second, sound energy is the motion of atoms in the air, so one might logically argue that sound energy is a form of kinetic energy. Again, the OSH community thinks of sound energy as belonging to a separate category.

For this book, the energy category most directly threatening harm is the category used for discussion. Two examples may clarify the point. First, when cooking a meal using an electric stove, the heat of the heating elements is the hazard source most directly threatening to burn the skin if touched. The same hazard could arguably be categorized as an electrical hazard because the heat is created by passing electric current through the heating element that offers resistance, causing the electrical energy to transform to heat energy. Because the most direct threat of injury is a burn, not the threat of electric current passing through the cook's body, the thermal energy category is used.

As a second example, consider that a boulder high on a mountain has *gravitational energy*. While sitting there, it is doing no harm. If it dislodges and rolls downward, it will transform the potential energy into kinetic energy. In the form of kinetic energy, the boulder can become a proximate cause of harm to whatever is in its path, so the source of the hazard is the kinetic energy of the boulder. Using this same rationale, gravitational energy is discussed in this chapter.

Figure 14.1 provides a visual perspective for appreciating energy as part of a sequence of events leading to a harmful event. The graphic, previously introduced in chapter 9, depicts the during-event phase of a harmful event. This phase has an initiating event, leading to intermediate events, leading to the harmful event.

The graphic can be useful both retrospectively for understanding past incidents and prospectively for visualizing sources of energy that pose risk of harm in the future. The ability to envision potential harm facilitates the development of strategies and tactics for reducing the risks associated with the energy source. Throughout this chapter, strategies and tactics are presented for (1) avoiding an initiating event by eliminating the hazard source or keeping the operating system in

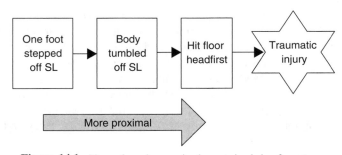

Figure 14.1   Place of varying proximal events in chain of events.

a normal state; (2) if an initiating event occurs, interrupting the sequence of events by blocking or moderating the hazardous energy; and (3) mitigating the severity of harm by using PPE.

14.2 BACKGROUND ON KINETIC ENERGY

The energy of an object in motion is known as kinetic energy. Every discussion of KE begins with the contribution of Isaac Newton. Just after completing his bachelor's degree in 1665 at Trinity College, Cambridge, he was planning to continue with graduate studies. But a plague forced school officials to close the campus. While living at home for two years, Isaac developed the binomial theorem, differential calculus, the law of gravity, and an understanding of objects in motion. On the topic of objects in motion, he synthesized the contributions of other scientists into three interrelated statements. When school reopened, he returned to college and completed his master's degree in 1668. Newton's brilliant developments were so impressive that Isaac Barrow, the Professor of Mathematics at Trinity, resigned his position to make a place for Newton. At age 26, Newton became the Professor of Mathematics at Trinity College, and he did all this without a slide rule, hand calculator, personal computer, or Internet access!

14.2.1 Newton's Laws of Motion

Newton's Laws of Motion are an essential starting point for understanding the processes by which KE causes harm. Newton's three laws are as follows:[1]

First law. A body remains at rest or continues to move in a straight line with uniform velocity if there is no unbalanced force acting on it.

Second Law. An unbalanced force acting on a body will cause that body to accelerate in the direction of the force with an acceleration inversely proportional to the mass of the body. Today, the equation for the Second Law is $a = F/m$, or more commonly $F = ma$, where a is the acceleration vector, F is the vector sum of all forces acting on the object, and m is the mass of the object.

Third Law. For every action there is an equal and opposite reaction. The terms action and reaction are vector forces. The force diagram for floor friction (Figure 14.2) illustrates this law with the downward force of the object weight against the floor being opposed by the reactive force of the floor upward against the object.

Building on Newton's Laws of Motion, other scientists developed additional laws and principles that provide the basis for much of engineering, especially mechanical and civil engineering. First, we will look at the Law of Energy Conservation and Law of Conservation of Momentum in the context of objects moving.

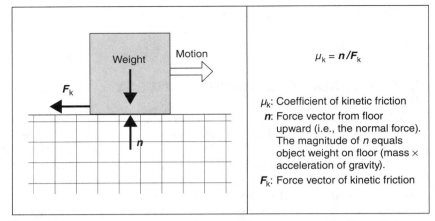

$$\mu_k = n/F_k$$

μ_k: Coefficient of kinetic friction

n: Force vector from floor
upward (i.e., the normal force).
The magnitude of n equals
object weight on floor (mass $\times$
acceleration of gravity).

F_k: Force vector of kinetic friction

Figure 14.2   Free-body diagram of floor friction to illustrate Newton's Third Law of Motion.

14.2.2 Law of Energy Conservation

A moving object possesses kinetic energy. The KE of an object moving in a straight path may be expressed mathematically in terms of its velocity (v) and mass (m) as follows.

$$KE = (1/2)mv^2. \tag{14.1}$$

When a moving object hits a stationary object, the KE of the moving object is not lost—it is preserved in multiple forms. Some of the energy is transformed into heat energy added to each object. Another part is transferred to the stationary object by giving it movement and KE. The object that was moving may retain part of the original KE.

   According to the Law of Energy Conservation, the pre-impact KE will equal the sum of post-impact energies within the closed system. Part of the post-impact energy will go to deforming one or both objects, as well as to heat energy taken up by each of the objects. The impact takes a brief time (Δt) to complete. Afterward, each object has a different energy than before the impact. The post-impact energy within the system consists of the heat energy gained by each object and the post-impact KE of each object.

14.2.3 Law of Conservation of Momentum

To understand more about how impacts cause harm, we turn to the Law of Conservation of Momentum.[2,3] In addition to KE, a moving object has a property known as momentum. The momentum of an object is the product of its mass and velocity:

$$Momentum = mv. \tag{14.2}$$

Thus, for a given object, momentum increases directly with velocity. Think of the difference between tossing a bullet at a target and shooting the bullet with a pistol.

Although the mass of the moving object is the same in both scenarios, the bullet fired from the pistol will have more momentum and therefore much greater effect on the target. Similarly, for two objects moving at the same velocity, the one with greater mass will have more momentum.

According to the Law of Conservation of Momentum, when two objects collide, the momentum before the impact will equal the momentum after the impact. In OSH, the common case involves a moving object hitting a stationary object. In that instance, the pre-impact momentum of the moving object will be distributed between the two objects so that the post-impact momentum of the two objects will equal the pre-impact momentum of the one object. The change in momentum (ΔM) occurs during the brief time period (Δt) from initial contact to completion of the momentum exchange. If we assume for the sake of simplification that the force (F) acting on the impacted body is constant during Δt, we can use the following equation to determine change in momentum:

$$\Delta M = F \Delta t \tag{14.3}$$

Even if the force is not constant, we can obtain a fair approximation of ΔM by using the average force during Δt.

14.2.4 Rotational Motion

Thus far the discussion of kinetic energy has focused on linear motion. Rotary motion is also important in OSH. For example, the rotating blade of a circular saw contains kinetic energy. When it cuts a board, some of that energy is used to break apart the wood fibers. Motors create rotation in a shaft that can be linked with gears, belts, and other power transmission devices to provide useful energy for moving other machine components. The flywheel on a mechanical power press stores rotational energy for use when a press stroke is actuated. When saw blades, shafts, flywheels, and similar devices are rotating, they possess kinetic energy. The equations are more complex than those for linear motion and unnecessary to develop here. Suffice to say that rotational motion also behaves according to Law of Energy Conservation and Law of Conservation of Momentum.

14.2.5 Potential Kinetic Energy

In the practice of OSH, the potential form of kinetic energy is important. An object sitting on the top shelf of a tall storage rack has a kind of potential kinetic energy known as gravitational energy. The energy is normally controlled by the support from the storage rack. If a forklift on the other side of the storage rack happens to collide with the rack, the entire rack may respond by wobbling. The wobbling could cause the object on the top shelf to slide off and fall. During the fall, the original gravitational energy transitions from potential state to active kinetic energy. When the falling object hits the floor, the kinetic energy will again be transferred according to the Law of Energy Conservation.

Gravitational energy plays a key role in many occupational injuries and fatalities. Many injuries result from falling objects hitting people and others result from persons falling. The OSH professional should understand the basic physics model for gravitational potential energy (PE_G). Think of a base level such as a floor and an object positioned at a higher level. The initial potential energy of the object is a function of its mass (m), the downward pull of gravity (g), and the distance it could fall. The potential fall distance (Δd) is the difference between the initial elevation and the base elevation. The equation relating these variables is

$$PE_G = mg\Delta d. \tag{14.4}$$

What this equation tells us is quite intuitive. For a particular object, the mass does not vary. The pull of gravity is very close to constant anywhere on the surface of Earth, with a value of 9.8 m/s. Therefore, the level of energy, PE_G, varies directly with the fall distance.

Another form of potential kinetic energy involves elastic materials. An object with elastic properties may be stretched or compressed a little out of its resting state. The household rubber band exemplifies the concept of storing energy after being stretched. Springs are available commercially for applications involving stretching or compressing. A spring made for stretching is useful for applications where the engineer wants a pulling force between the two ends of the spring. A spring made for compressing is useful for applications where the engineer wants a pushing force between the two ends of the spring. In these cases, the amount of force exerted by the spring is directly related to the length of the deformation (x) and a constant unique to the spring or other elastic material (k). Thus, $F = kx$. The *potential energy of a spring* (PE_{Sprg}) is

$$PE_{Sprg} = (1/2)kx^2. \tag{14.5}$$

This equation applies while the spring is within its elastic range. Once stretched excessively, the spring will not perform according to the equations for springs.

14.3 MECHANISMS OF HARMING

Many occupational injuries and fatalities result from kinetic energy, especially impacts. While impacts transfer kinetic energy during a very brief period, numerous other occupational injuries and fatalities result from kinetic energy transfers that occur more slowly. For example, body parts can get crushed when caught in or between objects. This section explains the relationship between kinetic energy and occupational injuries and fatalities.

Kinetic energy transferred to humans accounts for many occupational injuries and fatalities. The injuries are diverse, but common ones are punctures of the skin or eyes, lacerations, broken bones, spinal injuries, and brain damage such as concussions. Because the body parts harmed may involve muscles and bones, we may be tempted to classify some of these injuries as musculoskeletal injuries. However, classifications based on the source of injury, rather than the result of injury, provide more fruitful opportunities for risk reduction.

Occupational injuries resulting from impacts and slower transfers of kinetic energy account for a large portion of costs. An indication of the relative importance of these various kinetic energy injuries is shown in Table 14.1. It lists the major categories of event types proximal to injuries. The injuries within the data set are based on work-related injuries for which workers' compensation claims were filed and approved. The second column indicates the percentage of costs associated with each event as reported by workers' compensation insurers in 2007. The largest portion of total cost was for falls on the same level. Second on the list were injuries due to an employee falling from a higher to a lower level. Chapter 23 provides in-depth discussion of conditions related to falls on floors and stairways, so the present chapter omits repetition of that material. Other categories listed in Table 14.1 are persons being struck by objects, striking an object, being caught between objects, and highway incidents.

Being caught between objects includes some cases involving KE and KE in combination with other forms of energy. For example, when a worker gets caught between a wall and a piece of mobile equipment, the energy exchanged involves both the kinetic energy of the equipment moving and the energy from the motor driving the movement.

The highway incident category encompasses vehicle collisions, rollovers, and other roadway incidents. The highway incident category differs from the other categories in that it is not based on the most proximal event. To understand this category, think of a model with two proximal event—a vehicle collision followed very quickly by the occupant impacting objects within the interior of the vehicle. The more proximal of these two events is the person impacting an object inside the vehicle, so coding these injuries as struck against injuries would be consistent with other kinetic

Table 14.1   Kinetic Energy Injuries: Percentage of Costs by Type of Event

Event categories	Percentage of costs
Falls on same level	14.6
Falls to lower level	11.7
Struck by object	9.0
Highway incident	4.7
Caught in/compressed by	3.9
Struck against object	3.8
Total	47.7

Data from Liberty Mutual Research Institute for Safety based on costs for workers' compensation cases in the United States in 2007 (www.libertymutual.com/researchinstitute).

energy injuries. But for workers' compensation, such detail is not on the claim form. Therefore, the practical alternative is to simply classify these impact injuries in the broad category for highway incidents.

The categories in Table 14.1 are based on one of several recordkeeping schemes used in workers' compensation systems.[4] The category for "event or exposure" reflects a single event or exposure most directly producing the harm. It may be called the proximate event or proximal exposure. For kinetic energy injuries, human injury results from the proximal event transferring to the humans more energy than can be safely tolerated.

Many of the commonly used protective devices use the technical information discussed in this chapter. Consider two examples. First, what happens when a bricklayer working on a scaffold drops a brick and another worker below is unfortunate enough to be directly in the path of the falling brick? We can compare the effect of wearing a hard hat versus not wearing one using the physics of impacts. In either case, a large portion of the pre-impact momentum will be transferred to the head and a smaller portion will be retained by the brick as it deflects off. Thus, we can assume that ΔM will be the same whether the employee is wearing or not wearing the hard hat.

However, the duration of impact when no hard hat is worn will be very brief, starting when the brick makes initial contact with the scalp and ending when the skull bones stop the brick from further penetrating the head and the brick falls away. For the employee wearing a hard hat, the shell of the hard hat will take the initial impact. A bit of time will pass as the shell descends and transfers some of the energy to the suspension system. The total time from initial impact of the brick on the shell until the brick loses contact with the shell will be longer because the time for the hard hat equipment to deform will be longer than the time for the skull to deform. From Equation 14.3, we see that for the same ΔM, a longer Δt will involve a smaller average force. Thus, one benefit of the hard hat was reducing the average force on the skull from the impact. Another benefit of a hard hat comes into play when the shell reaches its lowest point. When the impact is within the design parameters of the hard hat, there should be some air between the shell and skull due to the suspension system slowing down the shell's descent to zero before it reaches the skull. However, in more powerful impacts, the shell will be driven by the impact into contact with the skull. In either case, the hard hat mitigates the harm from the impact by distributing the forces to a broader surface area on the head. That reduces the force per unit area, which in turn reduces the harm to the skull. In addition, the suspension system consists of stretchable materials that take up some of the initial kinetic energy of the falling brick. This reduces the energy transferred to the head. This, in turn, reduces the load transmitted to the cervical spine, thereby reducing the risk of spinal injury.

The second example of how understanding the physics of mechanical energy can help us understand the mechanisms of kinetic energy injuries is the case of an employee falling. Whenever a person falls from a higher to a lower elevation, the event is classified as a "fall from elevation." Based on the physics involved, the mass of the person has potential gravitational energy as expressed in Equation 14.4

($PE_G = mg\Delta d$). While working at the elevated level, the potential energy is controlled by the work surface under the worker's center of gravity. If for some reason the center of gravity shifts outside the work surface, control is lost and the person begins to fall. During the fall, the gravitational energy transitions into kinetic energy, and both KE and momentum increase as the fall progresses. The maxim "It's not the fall that hurts—it's the landing" accurately conveys the fact that no injury occurs until there is an impact with the lower surface. During the brief time of impact, the KE is distributed between the ground and the human body. If the ground is hard concrete, the impact duration will be extremely brief—almost all the energy goes to the human body in the form of deformation energy and great injury results. If the ground is a more forgiving material, such as loose soil, the impact time will be a bit longer, the deformation of the soil will take some of the energy, and the injury will be less because the force on the human body will be less (see Equation 14.3).

An understanding of the physics of kinetic energy helps OSH professional comprehend how traumatic injuries occur and helps the PPE industry make protective equipment that provides workers with an effective last line of defense from impacts.

14.4 STRATEGIES AND TACTICS FOR KINETIC ENERGY

There are many methods for reducing risks of harm from kinetic energy. Table 14.2 lists the nine risk-reduction strategies in the left column. In the right column are tactics applicable to the corresponding strategy. The following discussion expands the abbreviated text in the table.

Eliminating the hazards is always the first strategy to consider. KE will be eliminated if stationary objects are prevented from moving laterally or falling downward. Some examples are as follows:

When preparing to haul cargo on a flatbed truck trailer, straps and other devices are used to secure it, so it will not get blown off by the wind or a rapid evasive maneuver on the roadway, which would present a KE hazard to other motorists.

When compressed gas cylinders can become torpedo-like projectiles if an upright cylinder falls breaking off the stem and allowing the gas to release through the opening. Such events are eliminated when the cylinders are stored properly (i.e., upright, with the cap in place, and secured to solid objects).

Gravitational energy can be eliminated by performing work at ground level rather than at elevation. For example, some welding tasks on building frames and bridges traditionally performed by structural ironworkers at elevation can be changed so that the work is performed at the ground level. A crane is then used to lift and position the structural assembly into place.

Storing a heavy object on the floor instead of on a shelf eliminates the chance of it falling off.

Table 14.2   Strategies and Tactics for Reducing the Risks of Kinetic Energy

Risk-reduction strategy	Risk-reduction tactics
1. Eliminate the hazard	1a. Eliminate KE by keeping objects stationary by securing in place or locating where movement is prevented
	1b. Eliminate KE by bringing a moving object to zero velocity
	1c. Eliminate gravitational energy by performing work at ground level rather than at elevation or bringing an elevated object to ground
	1d. Eliminate potential energy of a spring by returning it to its neutral length
2. Moderate the hazard	2a. Use no more velocity than needed
	2b. Replace hard object with softer object
	2c. Lengthen duration of impact (Δt) to reduce the average force (F) during impact
	2d. Increase impact area to reduce force per square meter
3. Avoid releasing the hazard	3a. Avoid moving object hitting another object
	3b. Avoid moving object hitting a person
	3c. Avoid letting an object or person with gravitational energy fall
4. Modify release of the hazard	4a. Control rate of release of gravitational energy
	4b. Control direction of the released object or material
5. Separate the hazard from that needing protection	5a. Place a barrier on or around that needing protection from impact
	5b. Design workplaces to separate areas for moving objects from areas used by people
	5c. Design tasks to ensure people are located a safe distance from moving objects
	5d. Design operations to separate by time the co-location of a hazardous moving object and people
6. Improve the resistance of that needing protection	6a. Design and build the housing of instruments and equipment to withstand foreseeable impacts
	6b. Design and build parts and components of instruments and equipment to withstand foreseeable accelerations resulting from impacts
7. Help people perform safely	7a. Use knowledge from ergonomics to design the operator–machine interfaces to minimize errors
	7b. Design highways with clear lane identification to reduce the information processing load on the drivers
	7c. Provide safety signs made to reflect the best available attributes supported by research
	7d. Provide appropriate safety training for personnel who operate the organization's vehicles and heavy equipment

TABLE 14.2 (*Continued*)

Risk-reduction strategy	Risk-reduction tactics
8. Use personal protective equipment	8a. Have employees wear impact-resistant PPE to protect body parts
	8b. Have employees who are at risk of falling from heights use personal fall protection system to limit distance person can fall and avoid impact with surface at lower level
9. Expedite recovery	9a. Be prepared to help employee injured by a fall
	9b. Be prepared to transport injured employee to medical facility

Bringing a moving object to zero velocity is another way to eliminate KE. The fail-passive engineering devices use this tactic for many applications, including rotary power saws and lawn mowers. When the operator releases the actuator, the rotary motion is quickly stopped.

Bringing a stretched or compressed spring to its neutral position is another way to eliminate KE. A standard step taken by maintenance workers before starting a maintenance task is to identify and control all sources of energy that might be threats during the task. When springs are involved, allowing them to return to a neutral length eliminates the hazard. For situations where eliminating the potential energy of a spring is not feasible, the spring may be physically blocked to avoid releasing the hazard (Strategy 3). The workers ensure effectiveness of the block by applying a lockout mechanism supplemented by a tag to inform anyone who might come along that the lock is intentional and must not be removed.

The second strategy is to moderate the hazard. One tactic for moderating the hazards of KE, PE_G, or PE_{Sprg} is to reduce the velocity of the moving object so that the impact force will be less. A second general tactic is to modify the object before it becomes mobile by changing attributes such a hardness, weight, and sharp edges. A third tactic is to modify the impact to either extend the duration of contact or increase the contact area to reduce the force per square meter.

Strategy 3 is to avoid releasing the hazard. When the hazardous energy is KE, the energy is released through deceleration, and the deceleration of primary concern in OSH is a harmful impact. A gradual deceleration is, in general, not harmful. Thus, the method for avoiding release of KE is to avoid a harmful impact. Five examples are as follows:

1. During roadwork, passing vehicles have KE. Traffic flow channeling provides a technique to allow the vehicles to pass by the workers without impacts.

2. Workers using a scaffold for work occasionally drop a tool from a scaffold platform, putting personnel below at high risk. One common method for avoiding this is to tether the tool to something that will arrest the falling tool before it descends very far.[5] A second method is to have a toeboard on the work platform to prevent tools and work materials from falling off when inadvertently kicked by a worker's boot.

3. A fall protection tactic for workers doing work at elevation (e.g., bridge or tall building construction) is to install safety nets below edges and holes where people might fall. Since velocity increases with distance fallen, the shorter fall distance to the net, versus to the ground, involves less velocity. After contacting the net, the person is gradually decelerated to a stop (average deceleration = initial velocity divided by Δt). Thus, the falling person benefits from less velocity and more gradual stopping.

4. On ships, trucks, trains, and planes, many objects and tools have a storage space made specifically to hold a particular item. For example, trucks typically have a specific space for the tools needed to change a tire, which avoids releasing the hazardous projectile during a minor to moderately intense collision (although in a very intense collision the tools might break loose).

5. When someone parks a wheeled vehicle on a slope, it has potential energy. The event of the parked vehicle rolling downhill is eliminated by the combination of setting the parking brake, putting the vehicle in the proper gear for parking, and choking the wheels.

The fourth strategy is to modify release of the hazard. In the early days of elevators, there would occasionally be a rapid descent to the bottom, resulting in fatalities and various serious injuries. Modern elevators have multiple safety devices, one of which is a braking mechanism to limit the velocity of descent.

When an old roof needs replacing, the roofers pull off the old covering. To get the old materials down to a truck, the roofers generally set up a chute with a wide opening at the top and a channel leading down to the truck body. This channeling method keeps the released material under control, so it will not hit something or someone on the street level.

The fifth strategy is to separate the hazard from that needing protection. An example is a long moving conveyor in a location between a work area and the restroom. Workers in a hurry are tempted to climb over or crawl under the conveyor, rather than taking extra time to walk around it. Rather than attempting to change the behavior of all the employees, the employer could construct a pedestrian bridge to allow workers to take the shortest route while maintaining separation from the moving conveyor.

A common method for separating personnel from a hazard is to use barriers, guards, and fences. For example, in a building construction project, the path of crane movement is planned prior to performing lifts. The site authorities will order precautions such as designating the areas under the load path as no personnel locations and setting up portable barriers around the area. Posting signs warning personnel to stay out of the area adds to the reliability of the portable barriers by providing people hazard information so they will choose not to cross the barrier. This example also illustrates increasing the effectiveness of an engineering control (separating with a barrier) by also applying an administrative tactic (warning signs).

In a warehouse with high storage shelves, forklift trucks are used to place and remove pallets of materials on the shelves. There is a recognized risk of material falling from higher shelves and hitting a forklift operator. A standard practice is to use forklift vehicles with falling object protective structures—an overhead barrier that separates the operator from falling materials.

People and hazards may also be separated by time. This is the method used for some mechanical power presses (the positive-clutch type). The work cycle is arranged so the operator can reach to the point of operation only while the press ram is up. After loading the point of operation, the operator must retract both hands and use them to press a pair of actuator buttons to make the ram complete a stroke. When set up correctly, the operator's hands are separated from the point of operation during the part of the cycle when the ram is closing with tremendous KE and momentum.

Improving the resistance of that needing protection (Strategy 6) is commonly used for equipment. The designers write specifications for the equipment housing that include foreseeable impacts; as a result, machine tools for use in manufacturing environments have heavy steel housings. Most laptop computers have housing with limited capability for impact protection. For some military needs, special portable computers called "ruggedized computers" are available. The specifications for these require capability to withstand impacts with a hard floor when dropped from a specified height. Not only is the housing impact resistant, their internal components are made to tolerate the rather severe decelerations that occur from impacts.

The seventh strategy is to help people perform safely. An example might be a crane operator workspace with the best human engineering features to help the operator perform without errors, with precision, and within load capacity.

For the large, heavy equipment used in the mining and construction industries, the operators need to climb to reach their cab. During these climbs, falls can be prevented by helping the climbers control their personal gravitational energy using well-designed stairs and short ladders with handrails or handholds located for convenient and effective grasping.

Several types of PPE provide millions of workers a last line of defense from kinetic energy hazards. Perhaps the most prominent of these Strategy 8 devices are hard hats, which serve to mitigate the harm of a falling object hitting a worker's head. Although manufacturers design hard hats to meet downward impact standards, the hats also provide some protection from low-energy side impacts. Most common industrial hard hats have a brim to provide some protection of the eyes and face from small falling objects. Underground miners wear a particular type of head protection known as bump caps. These protect the miner's head when it bumps a protrusion from the mine roof.

Eye protection PPE offers considerable protection from projectiles traveling laterally. Older styles offered protection only from projectiles traveling directly toward the face. Experience proved that many eye injuries occur from objects reaching the eye from the side. This led to the availability of side-shield attachments to the old frames. Today, many safety glasses feature wrap around side protection, and they come in a wide range of sizes or adjustable features to accommodate different temple lengths and nose widths. The improved sizing, styling, and lighter weight of today's safety glasses help overcome the weaknesses of the older product and help employees tolerate wearing safety glasses for long periods (Strategy 7).

Footwear with impact protection is used in many jobs where employees handle heavy objects that might drop on their feet. The common impact footwear covers the toes with a shield of steel or other material. Some offer metatarsal protection as well. Safety footwear, like eyewear, now comes in a variety of styles. The soles also come in diverse materials and patterns to meet the needs of workers in varied worksites.

For many tasks performed at elevation, there are several methods for protection. One involves wearing a body harness connected by a lanyard to a secure object or lifeline located above the work area. If the worker falls, the lanyard will fully extend and stop the descending worker from a long fall. Like other kinds of PPE, effectiveness depends on workers using it at the proper time and in the proper manner.

The ninth strategy involves rescue, first aid, other immediate responses, recovery, and rehabilitation. The discussion in the previous chapter explained the value of developing plans for a post-incident response. One consideration unique to KE is planning for rescuing a worker who fell and was saved by personal fall protection gear. If the rescue takes too long, the suspended person will suffer the ill effects of suspension trauma.

LEARNING EXERCISES

1. Suppose we closed all the colleges and universities in the world for two years.
   (a) Do you think a student somewhere in the world would spend the two years thinking like Newton did?
   (b) Do you think having access to the Internet would help or hinder such a thinking process?
   (c) What Bloom level (see chapter 1) did Newton use to develop his three laws?

2. Imagine you coach a track team. You have three athletes who compete in the shot put event. All three have good technique and achieve similar trajectory angles. One can put the shot farther than the others. Using the concepts of energy, work, and power, explain why he can outperform the others.

3. Consider two objects of equal mass moving in a straight line. Object A has a velocity of 4 m/s. Object B has half the energy of object A. What is the velocity of object B?

4. Two boxes are sitting on shelves. Each has a mass of 8 kg. The shelves supporting box A and box B are 3 and 2 m, respectively. Object A has more gravitational energy than object B. What is the ratio of their gravitational energies (A to B)?

5. A spring with $k = 800$ N/m is compressed by 0.05 m. What is the energy of the spring? Include in your answer the SI units of energy.

6. Suppose you are a close friend of a physicist. You happen to share with her some occupational injury data regarding falls. She comments that it seems silly to have two categories of falls (falls on the same level and falls from elevation). She points out that both occur due to losing control of gravitational

energy, a descent, and an impact with the ground. She suggest it would be more logical to have one category called loss of gravitational support. How could you explain why having two categories is useful?

7. In this chapter, a dead-man actuator control is offered as an example of eliminating the hazard of KE by bringing velocity to zero. However, the KE is not eliminated instantly. There is a delay between releasing the actuator and when the motion stops. During this brief time (the after-reach time), what other strategies are used to protect the worker?

8. On the topic of tethering tools used on scaffolds and elevated platforms, attaching a tether to a railing of a guardrail is not recommended by the companies that make guardrails. Why do you think they object to this use of their railing?

TECHNICAL TERMS

Active state   The state of some type of energy while doing work, in contrast to the potential state of being dormant or inactive.

Energy   Ability to do work.

Gravitational energy   Potential kinetic energy of an object when located at a higher elevation than the base elevation.

Kinetic energy (KE)   Energy possessed by a moving object.

Potential energy   The state of some type of energy not actively doing work.

Potential energy of a spring (PE$_{Sprg}$)   Capability of a compressed or stretched spring to move something.

Power   Rate of doing work.

Work   Moving an object by applying a force.

REFERENCES

1. U.S. National Institute of Standards and, Technology. Fundamental concepts of force. Gaithersburg, MD: NIST. Available at http://www.nist.gov/mel/mmd/mf/whatis-force.cfm.
2. Hazen RM, Trefil, J. Science matters: Achieving scientific literacy, 2nd ed. New York: Anchor; 2009. p. 26–43.
3. Serway RA, Faughn JS, Vuille C, Bennett CA. College physics, 7th ed. Belmont CA: Thompson Brooks/Cole; 2006. p. 81–92.
4. Jensen RC. How to use workers' compensation data to identify high risk groups. In: Slote L, editor. Handbook for occupational safety and health. New York: Wiley; 1987. p. 364–403.
5. Salentine J. Tethering tools when working at heights. Professional Safety. 2011;56(9):66–68.

Chapter 15

Electric Energy Hazards

Electrical energy is essential to modern business and very useful for daily living. From an OSH perspective, the electrical energy used in workplaces may be regarded as a hazard source made safe by using appropriate hazard controls. Failure of the controls can release electrical energy that can then cause harm. This chapter has three main sections. The first provides a general background on electrical energy, the second explains the means by which electricity can harm, and the third provides examples of strategies and tactics for reducing risks associated with electrical energy hazards. A related topic, the electromagnetic spectrum, is discussed in another chapter.

15.1 ELECTRICAL ENERGY AS A SOURCE OF HAZARD

The major forms of electrical energy discussed in this section are static electricity, lightning, and electric current in circuits. The references of this chapter are well suited for OSH professionals seeking more insight into electrical hazards and safe practices.[1-3]

15.1.1 Static Electricity

Static electricity involves a rapid transfer of electrons between dissimilar materials. This transfer creates heat, possibly enough to ignite fuels. Of particular concern for OSH professionals is the possible ignition of air containing fuel vapors or combustible dusts. Static electricity is a widely recognized hazard in an environment where flammable and combustible liquids are present.[1] When these materials are flowing in pipes, they create areas where positively charged particles concentrate and other areas where negatively charged particles concentrate. Similarly, when a liquid fuel is poured from one container to another container, areas develop with different concentrations of charged particles. There comes a point at which the concentrations differ enough to generate a sudden transfer of the charged particles in the form of sparks. Another source is the rubber tires on vehicles. The vehicle body tends to develop an electrical potential different from the ground. A person standing on the

Risk-Reduction Methods for Occupational Safety and Health, First Edition. Roger C. Jensen.
© 2012 John Wiley & Sons, Inc. Published 2012 by John Wiley & Sons, Inc.

ground will have a different charge than the vehicle body. If the person touches the vehicle body, at first contact a tiny spark will occur, resulting in a slight sensation for the person, and more important, heat that may be sufficient to ignite flammable vapors in the air.

15.1.2 Lightning

Lightning is a sudden transfer of *ions* from one location to another. The common transfers are between a cloud and earth, two clouds, and one part of a cloud to another part of the same cloud. The cloud to earth strikes last less than 1 s, but create tremendous heat. Points on landmasses commonly hit by lightning are tall trees, mountain ridges, and buildings. Ship poles also get many hits. Damage at the sites hit results from the intense heat created as the ions transfer and from the shock wave created by the hot expanding air.

15.1.3 Electric Current in Circuits

Electricity in electrical circuits is capable of harming people and igniting fires. The harm to people occurs via electric shock, burning of skin and internal tissue, and *electrocution*. The ignition aspects are generally associated with faulty wiring, which can cause arcs or hot components with potential to ignite flammable or combustible materials. To understand electrical safety, a person needs to understand the basics of circuits. The following explanation is provided to help readers who are not very familiar with the terms and properties of electrical circuits.

The commonly used terms *current*, *voltage*, and *resistance* refer to the basic properties of circuits. The current (I) flowing in a circuit relates to voltage (V) potential and resistance (R) of the load according to Ohm's Law:

$$I = \frac{V}{R}. \tag{15.1}$$

Current is measured in amperes (A), voltage potential in volts (V), and resistance in ohms (Ω).

The Ohm's Law equation may also be expressed in two alternative forms: $V = IR$ or $R = V/I$. To illustrate these relationships, consider a simple circuit with a 6 V battery connected with copper wires to an appliance with a resistance of $3\,\Omega$. Using Equation 15.1, the current flowing through the appliance may be computed as follows: $I = V/R = 6\,\text{V}/3\,\Omega = 2\,\text{A}$.

A little thought tells us the current will increase if we lower the resistance. That fact explains why electric circuits use wires made of metals that have low resistance to electric current. Low resistance corresponds to high conductivity. The common metals for wires used in electric power distribution systems are copper and aluminum. More expensive materials, with almost no resistance, are known as superconductors.

Also needing explanation are the concepts of electrical energy, work, and power. In the *Système International (SI)*, joule (J) is the unit for electrical energy and various other types of energy. It expresses the ability of an electrical source to do work.

Recall that the definition of work is using force to move something. In the case of electrical energy, the force comes from the voltage potential in a circuit, and it is used to move electrical charges from one place in a circuit to another. The SI unit of both electrical energy and work is the joule. This may not seem intuitive at first, but consider that since energy is the ability to do some amount of work, using that energy to do that amount of work logically uses the same unit of measurement.

Electrical power is the time rate of doing work. Since work is in joules and the standard unit of time is second (s), the logical unit for power is J/s. However, the official SI unit of electrical power is watt (W), where 1 W equals 1 J/s. The watt is also used in the customary system of units used in the United States. Table 15.1 summarizes the units for energy, work, and power.

Another important property of electric current is the type of current flow—direct or alternating current. For a circuit supplied by a battery, the current flows in one direction and is called direct current (DC). Current is considered flowing from the positive terminal around the circuit and back to the negative terminal.

For a circuit supplied by the power company, the current flows in alternating directions. Conceptually, think of the free electrons in the conducting wires following a cycle of shifting slightly to the right and then slightly to the left. The pattern repeats rapidly and the movement in each direction contributes equally to the work. In the United States, typical houses and businesses are supplied with alternating current (AC) cycling 60 times per second or 60 Hz.

The application of Ohm's Law is straightforward for DC circuits. However, for AC circuits, the voltage and current follow constantly changing levels. If plotted against a time line, voltage changes according to a sine wave. The same is true of the current level. Thus, what value of voltage and amperage should be used for Ohm's

Table 15.1   Important Terms of Electrical Power

Term	SI name	SI letter	Explanation
Energy	Joule	J	Energies in various forms can be expressed in joules
Work	Joule	J	Think of a joule of energy as the amount of work that the amount of energy can do in the future
			Think of a joule of work as the amount of work that the energy has already done
Power	Watt	W	Work is power used for some time (Δt).
			Power (P) is time rate of doing work: $P = \text{work}/\Delta t$
	Kilowatt	kW	$1\,\text{kW} = 1000\,\text{W}$ and $1\,\text{W} = 1\,\text{J/s}$.
			Calculation formulas: $P = VI$; $P = I^2R$; $P = V^2/R$
	Kilowatt hour	kWh	The unit for selling and buying electric power.

Law calculations? To answer this, recognize that in each cycle both voltage and amperage have a peak value on the positive part of the sine wave and another peak on the negative side. Both positive and negative contribute to the work, so instead of treating one as adding and the other as subtracting, we treat both values as contributing equally. By doing so, the effective values may be determined mathematically in terms of the peak value. The effective voltage is 0.707 of the peak voltage and the effective current is 0.707 of the peak current. The electrical engineer or architect designing a circuit will use the effective voltage for Ohm's Law computations.

Arc welding is a special application of electrical circuits. An arc welding machine creates a large voltage potential between the hot wire and the neutral wire. The neutral is connected to the metal to be heated (known as the "work") and the hot wire connects the arc welding machine to an electrode. When the welder brings the electrode close enough to the work, an arc jumps across the gap. At that point, the potential electrical energy becomes active electrical energy as current flows across the gap. Much of the electrical energy is transformed into heat energy due to the high resistance of the air in the gap.

15.1.4 Electrical Power Supply and Usage

The final part in the basics of electrical circuits is understanding the connection between AC electricity in the power lines and the electric circuits in buildings. The power company provides a line from its distribution network to each customer. Within the customer's premises, the supply line connects to a junction box that provides connections to each of the numerous circuits in the facility. Each circuit is designed and installed according to particular specifications for voltage and current. Receptacles are wired into the circuits to allow use of the electricity. Figure 15.1 illustrates the wiring for a power tool plugged into a three-prong receptacle.

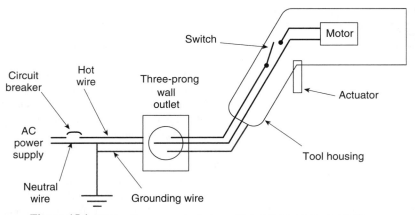

Figure 15.1 Wiring for a power tool plugged into a three-prong receptacle.

When selling and buying electrical power, the kilowatt hour (kWh) is used. To illustrate how fees are determined, consider a small business operating out of a rented office. Say the office used 20 A for 10 h, and the voltage supplied is 110 V. The power is calculated using the first equation for power ($P = VI$). Using the data, $P = (110\,V)(20\,A) = 2400\,W$.

Since the business used electric power for 10 h, it will owe the power company for using 2400 W (10 h) = 24 000 Wh. For convenience, the power companies prefer to sell power in 1000 W units, so the business will be charged for 24 kWh.

Although electricity within an engineered electrical circuit can injure people and ignite fires, circuits designed and installed according to applicable codes present very little risk. However, not all electrical installations comply with codes. Common reasons include degraded or damaged original equipment and electrical usage that exceeds the original design specifications. In industrial facilities, while process changes are in the planning stage, a person with electrical expertise should check the proposed changes to make sure the circuits are adequate for the process needs. This is the standard practice for organizations with a management of change program.

15.2 MECHANISMS OF HARMING

Through numerous mechanisms, electrical energy can damage electronic equipment, harm people, and provide the heat to ignite a fire.

15.2.1 Damage to Electronic Equipment

Many electronic devices are designed to use direct current. Their internal components would be ruined if connected to AC circuit. Because electricity supplied by the power company and distributed throughout our structures is AC, the current must be changed to DC before it enters electronic devices. This is accomplished by an AC adapter. It may be inside the housing of the electronic device or located external to the device. The externally located adapter is typically in a black box found between the power cord from the wall socket and the second power cord to the device. Adapters not only convert AC to DC, but they also protect the internal components by outputting the direct current at a voltage level matched to the device. The adapters may be thought of as a safety device for the electronic equipment.

Another cause of damage to electronic equipment is the occasional power surge, which may result from a lightning strike somewhere along the distribution lines serving the facility. The commonly used surge protectors are safety devices for moderating the excessive power before it reaches the electronic device.

15.2.2 Harm to People

The primary threat to humans is current in electrical circuits. Our entire nervous system functions by using electrical current. If current from an external source enters the body, it overwhelms the healthy functioning of the nervous system. In some circumstances, the external electricity will cause muscles to grip relentlessly—many people have reported being unable to let go of current carrying items such as power tools and aluminum ladders. In other circumstances, the electric current passes through the heart. The heart muscles, which normally contract rhythmically, can be so disrupted as to cease coordinated contractions and begin convulsing in an ineffective manner known as fibrillation. When in fibrillation, the heart no longer pumps blood through the normal distribution system. The effects are devastating. The fibrillating heart muscles need a lot of fresh oxygen, but fail to get it, resulting in permanent damage. Other body muscles and the brain also fail to get the oxygen they need. In some instances, current entering the chest causes the diaphragm to tense up, stopping respiration and leading to asphyxiation. Current flowing through one or more body parts can also cause the tissue temperature to rise rapidly, resulting in burning of the tissue.

The relationship between the level of current and the resulting health effects is complicated by multiple factors, including duration of contact with the current, body parts through which the current flows, skin wetness, gender, and frequency. Nevertheless, some general guidelines are available regarding expected health effects at various levels of AC current at a frequency of 60 Hz. The guidelines listed below indicate what to expect for contact with current at various levels, expressed in milliamperes (mA).

≤ 1 mA, no sensation

> 3 mA, painful shock

> 5 to 20 mA, muscle spasms

< 20 mA, not fatal

20 to 75 mA, sometimes fatal, especially as exposure time lengthens

> 75 to 300 mA, usually fatal

These exposure–effect data have two implications. First, all effects noted above occur with less than 1 A of current. The ordinary electric circuits in houses and industrial buildings have a circuit breaker or fuse at the junction box for automatically opening the circuit if current exceeds the circuit capacity. A circuit designed for 15 A will have a 15 A fuse or breaker and copper wiring sized to carry 15 A of current without appreciably heating up. If appliances and other loads supplied by the circuit draw more than 15 A, the fuse will blow or the breaker will open. This will prevent heat building up in the wires, which in turn will protect the circuit materials and prevent ignition of a fire. As useful as these devices are for protecting the circuit, they are not designed to protect a person who contacts the current. A person contacting the current in certain ways could get as much as 15 A passing through their body—at levels that far exceed the range from >75 to 300 mA that is associated with fatalities.

The second implication of the exposure–effect data concerns protection of people. As indicated, current levels of 5 mA or less are below the levels known to cause death or muscle spasms. To set a threshold for acceptable current level is complicated, but we can say that keeping exposures less than 5 mA should neither kill anyone nor cause muscle spasms that prevent letting go. The 5 mA level has been adopted for the design of a type of safety device known as a *ground-fault circuit interrupter (GFCI)*. When a GFCI is incorporated into an AC circuit, it monitors the system for equal current flowing in the hot wire and the neutral wire. The two currents should be the same. If the difference between the current in the hot wire and that in the neutral wire starts to exceed 5 mA, the device causes the circuit to open and cease conducting current. The response takes only 1 ms.

Why might the currents differ? Most likely, there is a leak, or short, somewhere in the circuit. It means the current has found an unwanted pathway out of the proper circuitry. A person contacting an item in that path can provide a path to ground. Often, that item is a power tool in which a path between the circuit wires and the tool housing allows some current to energize the tool housing. When the person holding the tool housing squeezes the actuator switch, current can flow from the hot wire to the housing, through the hand, through the arm, through the chest, down through a leg, and into the ground. With a GFCI in the circuit, the loss of current will be detected. If the loss exceeds 5 mA, the circuit will rapidly open. The person will feel a shock but not suffer any lasting harm. Without the GFCI in the circuit, the person could be exposed to a harmful level of current.

In addition to the above discussion on harm to people from the current in electrical circuits, lightning can harm people and property. Lightning bolts from cloud to cloud and within a cloud can damage airplanes, but the major concern for OSH is damage to people and property. Property not protected by lightning rods or similar devices is vulnerable to being struck. If struck, the heat can start the structure on fire, destroy electronic equipment, and electrocute personnel. For example, people have been electrocuted while bathing or showering since the lightning current tends to follow a path through water pipes leading to the ground. Likewise, people are advised not to use landline telephones during a lightning storm because the phone line provides a path to the ground.

Personnel working outdoors are at risk of being struck. Lightning can deliver immediate death to a person hit directly or nearly hit. Even if the person is not killed, his or her hearing can be permanently damaged from the extremely loud burst of acoustic energy. Personnel working outdoors can be electrocuted if their body becomes a path for current to flow to the ground. The person may be touching a piece of heavy equipment when it gets hit by lightning or when it comes into contact with a power line—an event that occurs much too often with equipment that can be extended vertically, such as cranes.

15.2.3 Ignition of Unwanted Fire

The last means by which electrical energy can cause harm involves ignition. An active electrical circuit can create heat that, in some circumstance, can ignite fuel. One way is if the load on a circuit is too much for the wiring. This is avoidable if the circuit has a properly sized circuit breaker or fuse. However, sometimes people get tired of a circuit shutting down frequently. Instead of having it fixed or replaced by an electrician, they choose to replace the fuse with one having higher capacity. Thus, a 15 A fuse might be replaced by a 20 A fuse, allowing more current than the wiring was designed to handle. Now the wires may be loaded with up to 20 A. The wires are too small to shed the heat generated by the current, so heat will build up in the wires. This scenario is the cause of many home fires because in older homes the circuits were designed for fewer appliances than we use today, and the wires run through wall spaces where accumulations of flammable and combustible materials may be present.

A second ignition concern is with electric *arcs*. These occur when an active circuit is switched *off*. Although these small arcs are usually out of sight, one can perform a simple experiment to see one. Arcs are visible when you unplug an active appliance. For example, if a clothes iron or lamp is plugged in and turned *on*, you can see an arc by dimming the room lights and pulling the plug. The arc will appear as a blue light between the wall socket and the prongs of the cord. The light is caused by current jumping across the small air gap separating the two conductors. It is essentially a tiny version of the arc created deliberately when performing electric arc welding. The little experiment mentioned should be tried only in an environment where the air is free of flammable vapors; as a rule, the safe practice is to prevent these arcs by using the appliance power switch to turn it *off* before pulling the plug.

The s*parks* developed by static electricity are another ignition source. The major concern for static electricity is when the environment might contain flammable vapors within the flammable range. A two-pronged prevention strategy involves (1) limiting the production of sparks and arcs and (2) keeping the flammable vapor concentration well outside the flammable range of the material. Limiting the production of sparks may be accomplished by *bonding* surfaces that might otherwise develop dissimilar charges. Bonding involves connecting a conducting cable between the surfaces. The practice is very common for transferring flammable liquids from one container to another. Without bonding, sparks would develop, and some might have enough energy to ignite vapors.

In summary, electrical energy can damage equipment, harm people, and provide the heat to ignite an unwanted fire. In the next section, we examine strategies and tactics for reducing risks of these harmful events.

15.3 STRATEGIES AND TACTICS FOR ELECTRICAL ENERGY

The strategies for reducing risks associated with electrical energy are listed in the left column of Table 15.2. On the right are examples of applicable tactics.

Eliminating the hazards is always the first strategy to consider, but when it comes to electricity, eliminating the hazard can be difficult or impractical. In modern commerce, we need electricity for computers, communications, lighting, powered equipment work, and many other applications. Another option noted in Table 15.2 is replacing electricity with a different form of energy. Maintenance personnel use the

Table 15.2   Strategies and Tactics for Hazards of Electrical Energy

Risk-reduction strategy	Risk-reduction tactics
1. Eliminate the hazard	1a. Use no energy
	1b. Use another form of energy
	1c. Before working on equipment, disconnect electrical circuits and apply locks and tags
2. Moderate the hazard	2a. Use minimum voltage to meet needs
	2b. Use minimum current to meet needs
	2c. For transferring flammable liquids in pipes, use pipe made of materials that do not promote buildup of static electricity
3. Avoid releasing the hazard	3a. Design and build high reliability into all connections in a circuit
	3b. Enclose sites where arcs may occur
	3c. Keep circuitry dry
	3d. Use lockout and tagout procedures
4. Modify release of the hazard	4a. Provide grounding for circuits
	4b. Provide a fuse or breaker for each circuit
	4c. Include in circuit a GFCI
5. Separate the hazard from that needing protection	5a. Ensure integrity of component insulation and proper connections
	5b. Use double-insulated housing for power tools
	5c. Isolate electrical equipment from people spaces
6. Improve the resistance of that needing protection	Use fire-resistant materials for items at risk of ignition from electricity
7. Help people perform safely	7a. Use human factors guidelines when designing equipment and user manuals
	7b. Provide appropriate warning decals on equipment
	7c. Perform usability testing of equipment, manuals, and warning decals
8. Use personal protective equipment	8a. Provide appropriate PPE for personnel exposed to potentially hazardous electrical energy
	8b. Provide quality training regarding the use of PPE
	8c. Maintain integrity of PPE
9. Expedite recovery	Include emergency preparedness plans for first response to victim of electrical contact

elimination strategy when they need to work on equipment. Before starting the work, they disconnect all circuits serving the equipment and lock and tag the switches.

The second strategy is to moderate the hazard. Basic tactics are to use the minimum necessary voltage and current to accomplish the task. For areas with flammable vapors, the use of printed circuit boards, semiconductors, and other devices requiring very low current can keep energy levels below that required for ignition. In operations involving the transfer of flammable liquids, the designers of equipment for transferring flammable liquids can specify the pipes and containers should be made of low-spark generating material such a stainless steel.

Strategy 3—to avoid releasing the hazard—finds many applications in electrical circuitry. The current in electrical wiring is normally contained within one or more layers of insulation, which effectively prevents the current from being released. Each end of an electrical wire is connected to another part of the circuit, so ensuring proper and secure connections will avoid current leaking at the juncture points. By assuring good and reliable connections, we can prevent the hazards associated with leaking current.

Keeping circuitry dry is a useful tactic for avoiding release of the hazard. Moisture can penetrate small spaces at junctions, creating a conducting path out of the designed circuit. People have died in bathtubs when using a powered device that became wet. Portable power tools soaked by rain might conduct some current to the housing.

The fourth strategy is to modify release of the hazard. This is the most useful of the strategies for reducing risks associated with electrical energy. Providing *grounding* for electrical circuits has long been recognized as an effective tactic for channeling the flow of errant current. If current occurs outside the designed path for a circuit, the grounding wire should carry the current directly to the ground where it immediately dissipates into the enormous mass of the earth. The rapid flow of the current from the hot wire to the ground will quickly exceed the threshold of the fuse or circuit breaker, resulting in opening the circuit path and shutting down the flow of current. Two types of grounding are used. System grounding is established by the power company at the transformer connecting the customer to the distribution lines. Grounding is also installed in circuits that supply current to appliances within the facility. The combination of grounding outside and inside provides two lines of defense against voltage surges from lightning and other events. In addition to system grounding, device-specific grounding is used for many power tools and appliances (see Figure 15.1). The power cord provided with commercial equipment should have the type of wall plug appropriate for the device. Beware of advice offered by anyone who thinks modifying the plug is a good idea.

Another common tactic for modifying the release of electrical energy is to incorporate a GFCI into a circuit. As explained earlier, it monitors the difference in current between the hot wire and the neutral wire. If the difference exceeds 5 mA, the GFCI will almost instantly cause the fuse or circuit breaker to open the circuit. GFCIs are used where leaks in a circuit are more likely, such as in areas with water.

The fifth strategy is to separate the hazard from that needing protection. This tactic is applied to high-voltage electrical installations by locating electrical equipment in a fenced yard, room, or vault with access restricted to certain qualified electricians. Similarly, power lines hung on high towers use this strategy to separate the hazard from people.

Using double-insulated power tools is another example of tactic for keeping people separated from the current. The idea behind these tools is that if a short develops inside the tool, the current will need to pass through two barriers of insulation in order to energize the tool housing and contact the person.

Another tactic for separating people from electric current is isolation. Using ladders made of nonconducting materials (e.g., fiberglass) while working on or near electrical equipment is one means. The public generally believes wooden ladders are nonconducting, but this belief is not entirely correct. A very dry wooden ladder may be nonconductive, but a wooden ladder wet from rain will conduct electricity. Even the wood in a new wooden ladder contains moisture that facilitates some degree of conductivity through the fibers. So the safe approach is to issue personnel ladders that are certified by a recognized testing laboratory as nonconductive.

The sixth strategy is to improve the resistance of that needing protection. The word "resistance" used in this strategy must not be confused with the concept of electrical resistance. This strategy is to make the item of concern more resistant to being harmed by electrical current, for example, by using fire-resistant materials to make items that are at risk of ignition from electricity.

Strategy 7—to help people perform safely—applies to equipment controls and displays. Providing labels for switches and other electrical control devices can prevent errors. Providing signal lights to indicate equipment status (e.g., whether in the *on* or *off* position) can help personnel avoid errors. For readers seeking additional information on displays and controls, several ergonomics books have chapters with guidance on selecting appropriate displays and controls. A book by Konz and Johnson and a book by The Eastman Kodak Company are recommended.[4,5]

The eighth strategy is to use personal protective equipment. Utility employees often use insulated gloves when working with energized power lines. They also use head protection made to minimize conduction from a current source to the head. Footwear is available with low-conductivity soles. Full-body protection is used by electricians working with high-voltage installations.

Planning and preparing to expedite recovery begins with emergency planning for a worksite (Strategy 9). If electrical incidents are foreseeable, plans should clarify capabilities needed for an effective first response. Some potential injuries are severe burns to skin, burns to internal tissues in an appendage, fibrillation, and inability to breathe. Other types of incidents are fires and explosions ignited by a spark, arc, or heat from an electrical circuit.

LEARNING EXERCISES

1. A residence in the United States is supplied with alternating current at 120 V. Electric current consumed during the day (12 h) averaged 20 A/h. What is the hourly average power consumption? What is the total power consumed that day?

2. Explain the difference between DC and AC currents.

3. How does an effective management of change program contribute to electrical safety?

4. Explain the reason for setting the trigger point of GFCI devices at 5 mA.

5. Why does arc welding generate intense heat?

6. Explain the reason for having a grounding wire on portable power tools.

7. Give an example of Strategy 7 applied to electrical energy.

8. Explain why wooden ladders should not be trusted to provide electrical insulation.

9. Suppose you are conducting a training session for supervisors. One of them says, "When I was 9 years old, I was electrocuted when I stuck a screwdriver into a wall socket." How would you explain why you are sure his facts are wrong?

TECHNICAL TERMS

Arcs	Visible current jumping the gap in an electrical circuit. Arcs can be continuous (as in arc welding) or single flashes (as when unplugging an appliance while it is turned on).
Bonding	Electrically connecting two or more items with a common conductor in order to equalize the electrical potentials of the items. The common conductor, generally called a bonding wire, is not part of an electrical circuit.
Current	The flow of electrical charge within a circuit.
Electrocution	Death resulting from electrical energy.
Ground-fault circuit interrupter (GFCI)	A shock protection device for an electrical circuit. A GFCI detects the difference in current in the hot wire and the neutral wire. If the difference exceeds a trigger level of about 5 mA, the device causes the circuit breaker to open and stop the flow of electricity.
Grounding	The practice of electrically connecting an item with the earth (ground). The connection uses good conductors such as copper or aluminum with one end attached to the item and the other end attached to a grounding rod or other conductor rooted in the earth.
Ions	An atom or group of atoms that has developed a positive or negative charge as a result of gaining or losing an electron.
Resistance	A force opposing current flow.
Spark	A tiny, hot, glowing particle. Some common spark producers are static electricity, welding, and friction.
Système International (SI)	The International System of Units, abbreviated SI from the French Système International d'Unités.[6]

Voltage   The effective difference in potential between any two points in a circuit; that difference establishes an electromotive force that moves electrical charges between those points.

REFERENCES

1. Hammer W, Price D. Occupational safety management and engineering, 5th ed. Upper Saddle River, NJ: Prentice Hall; 2001. p. 341–373.
2. Brauer RL. Safety and health for engineers, 2nd ed. Hoboken, NJ: Wiley; 2006. p. 172–173.
3. Wiggins JH. Managing electrical safety. Rockville MD: Government Institutes; 2001.
4. Konz S, Johnson S. Work design: Occupational ergonomics, 7th ed. Scottsdale, AZ: Holcomb Hathaway; 2008. p. 293–329.
5. The Eastman Kodak Company. Kodak's ergonomic design for people at work, 2nd ed. Hoboken, NJ; 2004. p.269–363.
6. Rowlett R. The international system of units (SI). Available at http://www.unc.edu/~rowlett/units/sipm.html.

Chapter 16

Acoustic Energy and Vibration Hazards

The discussion of acoustic energy and vibration is combined into one chapter because the hazards have similar origins and cyclical patterns. In OSH, acoustic energy is more commonly referred to as *noise* and is widely recognized for its adverse effects on hearing. Vibration is known in OSH as the main cause of occupational hand-arm vibration syndrome and as a contributor to the spinal wear and tear damage experienced by people who spend thousands of hours sitting in a vibrating seat.

16.1 BACKGROUND ON NOISE AND VIBRATION

In workplaces, most noise and vibration originates from the movement of solid objects. Another source is moving water such as in streams, ocean waves, and through the penstock and turbines of a hydroelectric dam. Numerous industrial processes have fluids flowing rapidly through pipes that occasionally cause the pipes to shake and emit noise, as well as through valves that can produce a whistling noise.

If the solid object is in contact with air, the kinetic energy of the object moving transfers some acoustic energy to the air. If the solid object is in contact with a fluid, the kinetic energy of movement imparts pressure to the fluid. In both situations, the back and forth motions of the solid object create waves of higher and lower pressure in the affected air or fluid. If the solid object is in contact with another solid object, it will transfer the kinetic energy of the motion directly to the second object. The OSH profession is concerned with a human body part being the second object and being subjected directly to vibration.

16.1.1 Characteristics of Noise and Vibration

Figure 16.1 is a graph for understanding cycles that follow a sinusoidal pattern. The vertical axis is the value of a sine wave with values ranging from -1.0 to plus $+1.0$.

Risk-Reduction Methods for Occupational Safety and Health, First Edition. Roger C. Jensen.
© 2012 John Wiley & Sons, Inc. Published 2012 by John Wiley & Sons, Inc.

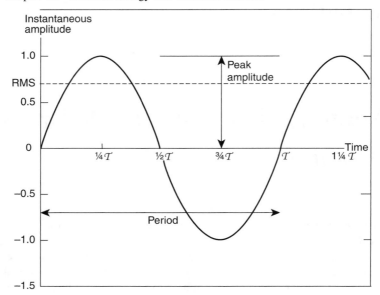

Figure 16.1   Instantaneous amplitude of a pure tone or vibration plotted against time.

When working with noise and vibration, we replace the sine wave variable with a more meaningful measure of the physical parameter involved.

For noise, the most common variables are intensity and air pressure. For vibration, commonly plotted variables are displacement, velocity, and acceleration. The horizontal axis may be time or distance. These various plots help visualize the key wave attributes: *period* (denoted T), *frequency* (f), *wavelength* (λ), *instantaneous amplitude* (the vertical axis), *amplitude* (A), and root-mean-square (RMS) value. The RMS serves as the effective amplitude of the plotted variable. Table 16.1 provides three fundamental equations relating some of these variables.

The variable in the vertical axis is something that can be measured with an appropriate instrument. In the case of sound, data for such a graph would be from an instrument capable of continuously measuring air pressure and located at a fixed site away from the source. For a single pure tone generated by a tuning fork, a graph of this *sound pressure* would show a pattern similar to that shown in Figure 16.1.

In the case of vibration, the vertical axis of the graph is often used for the instantaneous value of acceleration, which is the measurable parameter associated with adverse physiological effects on humans. Two other commonly plotted variables are instantaneous displacement from the neutral, or midposition, and instantaneous velocity of the motion.

Expressions for the intensity of sound and vibration are based on a ratio of the measured amplitude to a reference value. This general ratio approach is applied to several measurable quantities. For vibration, the common quantities are displacement, velocity, and acceleration. For noise, the most common quantity is the sound

Table 16.1   Some of the Common Equations for Noise and Vibration

Relationship	Equation
Sound pressure level in air is a function of the ratio of instantaneous measured air pressure (p) to the reference value p_o ($20 \, \mu$Pa), with results in decibel units	$20 \log(p/p_o)$
The decibel level of vibration acceleration is a function of the ratio of instantaneous acceleration (a) to the reference value a_o ($1 \, \mu$m/s^2)	$20 \log(a/a_o)$
The product of cycle time and frequency equals one cycle	$T \, f = 1 \, \text{cycle}$
Wavelength equals the speed of sound (c) divided by frequency	$\lambda = c/f$
The value of the root-mean-square amplitude is 70.7% of the peak amplitude	$\text{RMS} = 0.707 \, A$

pressure. Taking the logarithm of the ratio creates a linear numeric called the *decibel* (*dB*). Standard formulas are established for each of the quantities.[1] Two of these formulas are presented in Table 16.1, with the left column stating the relationship in sentence form and the right column showing the formula. The table also provides some common formulas applicable to the oscillatory parameters labeled in Figure 16.1.

16.1.2 Human Aspects of Noise and Vibration

Human perception of sound does not correspond directly to the attributes of sound already mentioned. What we call "pitch" is a sensation based largely on the frequency of sound, but is also influenced by the waveform. What we sense as "loudness" is largely based on the effective sound level, but is also dependent on frequency. The interplay between perceived loudness and frequency is presented in graphs showing lines of equal loudness perception across a range of frequencies. In order to have a practical scale for comparing levels of loudness, we use a *decibel scale*.

Occupational noise exposure standards and measuring instruments are readily available. When an employee's exposure regularly exceeds a standard, the expectation is that over a working career, the employee will experience more hearing loss than would occur naturally due to aging. The mechanisms of how noise exposure connects in a causative manner to hearing loss are discussed in section 16.2.

Occupational vibration involves two exposure categories, whole-body vibration and hand-arm vibration. Whole-body vibration occurs in jobs involving standing on a vibrating floor or sitting on a vibrating seat. The standing exposures are less of a concern because the legs partially dampen the vibration before being transmitted to the pelvis and spine. Hand-arm vibration occurs when manually handling vibrating power tools.

16.2 MECHANISMS OF HARMING

The mechanisms of harm for occupational exposure to noise, whole-body vibration, and hand-arm vibration may be conceptualized in terms of the following three-element model.

Source → mechanism for transfer → human receptor

The following three sections explain how this model applies to noise, whole-body vibration, and hand-arm vibration, respectively.

16.2.1 Effects of Noise Exposure

Occupational exposure to noise can cause permanent loss of hearing capability. Human hearing is optimal at the age when most people enter the workforce. People can expect some reduction in hearing capability between ages 20 and 50, with more pronounced declines during their fifties and beyond. Occupational exposure to noise can accelerate the hearing loss. The mechanism of harm is the acoustic energy encountered in the workplace—the cycles of varying air pressure enter the ear canal and deliver acoustic power to the eardrum. The resulting vibration of the eardrum is mechanically transmitted to the inner ear through the middle ear by three connected bones. Within the inner ear, special hair-like cells respond to the various frequencies. These cells interface with nerves belonging to the auditory system to produce nerve impulses recognizable by the brain. The hairs in the inner ear have considerable tolerance for noise levels, but the tolerance can be exceeded by some combinations of high noise levels and duration of exposure.

An extremely loud noise (e.g., an explosion) can permanently damage the eardrum and the inner ear. A couple hours of exposure to loud noise can cause a reduction in capability of detecting sound at particular frequencies. This temporary threshold shift is similar to muscular fatigue in that rest leads to full recovery. Sometimes a worker does not get enough quiet time for the hair cells to fully recover from the fatigue that caused the temporary threshold shift. With continued exposure to the noise, the hair cell damage can become permanent.

Because the damage affects a cluster of hair cells responsible for sensing certain sound frequencies, people with occupational hearing loss differ in the sounds they can hear. Results of audiometric testing are presented in a chart displaying the threshold (in decibels) of hearing throughout the range of hearing frequencies. Sometimes the frequencies showing the greatest loss can be related back to the same frequencies encountered on the job or in military service. The individual's audiometric tests may show a detriment in hearing threshold at the same frequency as their prior exposure. Such clear-cut links between exposure and harm are often obscured in today's work world because few people keep the same job with the same noise exposure for an entire career. More commonly, by the time a worker has suffered noticeable hearing loss, identifying a specific cause can be challenging because of the multiple factors associated with hearing loss—aging, military exposures, off-work activities, and various lifetime workplace exposures.

16.2.2 Effects of Whole-Body Vibration

Occupational exposure to whole-body vibration is part of many jobs. Some jobs involve standing on vibrating floors and others involve sitting on vibrating seats. Biomechanical modeling and several laboratory studies help us understand how a vibrating source transmits harmful energy to the human spine. Less is known about the way in which whole-body vibration can damage soft tissue in the torso and neck.

Numerous jobs include exposures to whole-body vibration. In transportation, railroad personnel riding trains stand or sit for a substantial portion of their shift. Personnel working on ferryboats and barges are similarly exposed. Numerous sitting jobs involve substantial vibration. Truck drivers are exposed to vibrations from uneven road surfaces as well as the vehicle motor. Roadway vibrations are transmitted through the tires, wheels, and suspension system to the truck body and from the truck body to the attachment points of the frame to the seat. Other jobs involving considerable vibration while sitting include driving farming equipment and city buses, operating heavy construction equipment, flying helicopters, and driving haul trucks in open-pit mines. Coal miners in underground mines are exposed to whole-body vibration driving continuous mining machines and ore hauling vehicles. Increased recognition of the long-term harmful effects of the vibration exposure and advances in technology have led to improved seating for many of the workers in such jobs.

Some whole-body vibration studies of jobs involving extended sitting on a vibrating seat (such as driving a truck, bus, or tractor) provide evidence of a relationship between measures of exposure and measures of back pain and other health outcomes.[1,2] One study found that the main contributors to low back pain among occupational drivers were the following factor interactions: posture plus vibration, posture plus manual materials handling, or all three.[2]

Sitting exposures to vibration are more consequential than standing exposures. Consider a person supported by a vibrating surface. If standing, some of the vibration is dampened by the shoe soles and more by the feet, ankles, lower legs, knees, upper legs, and hip joints. As a result, standing exposure to whole-body vibration has not been strongly associated with spinal damage. The story is different for prolonged sitting on a vibrating seat.

The vibrations in the seat exert alternating movements directly connected with the body of the person. The movements may involve up–down cycles, side to side cycles, and front to rear cycles. In three-dimensional coordinates, these cycles are referred to as the z-axis, y-axis, and x-axis, respectively. In each direction, the seat movement transmits corresponding movement to the person. These movements involve displacement in position and the accelerations inherent to changing direction with every cycle. In addition, rotational movements may be taking place. These are characterized using the nautical terms pitch, yaw, and roll. Occupational exposure standards use one graph for up–down vibration and another graph for side to side and front to rear movement.[3]

The forces of the up–down vibrations are transmitted to the sitting person's pelvis and from there to the lumbar spine. During the upward phase of each cycle, the upward force from the pelvis to the L5/S1 disc is opposed by an equal downward force from the weight of the upper body. These opposing forces squeeze the lumbar discs, making them flatter. When the motion shifts from upward to downward direction, the lumbar discs are subjected to pulling forces causing increased disc height. These repetitive cycles occur thousands of times per hour, and after many hours of exposure, the discs suffer fatigue damage as any other material does when subjected to repeated mechanical stress. The cartilage end plates of the discs are believed to incur small tears that never fully heal. Over time, the end-plate damage leads to reduced fluid in the discs, and the discs become permanently flatter, drier, and more rigid. Once in that state, the discs are more susceptible to becoming prolapsed or herniated and impinging on a nerve to cause pain (for a more comprehensive discussion of disc-related and other harmful outcomes, see Ref. 3).

The lumbar region of the spine is the most vulnerable part of the spine. Laboratory studies have provided information about the transmissibility of vibration from the vehicle seat to the driver's pelvis and up the spine. Most of the transmissibility affects the lower lumbar discs, subjecting them to the repetitive stress of alternating compression and tension. Thus, concerns about spinal discs damage should focus on the lower lumbar discs.

The vibration frequency is a major factor in disc damage. A *resonant frequency* occurs in the range of 4.5–5.5 Hz for the up–down motion of a sitting person.[3] That means the respective frequencies of the vibration loading and the spinal reaction become synchronized to create a much worse effect than occurring outside the resonance range. A fatal effect has been shown with monkeys subjected to vibration at resonant frequency for a few hours. A spinal effect on humans has been shown in laboratory studies. At the resonant frequency, laboratory measurements have shown that the upper spine actually moves vertically more than the pelvis. In the process, the spinal discs are subjected to an exaggerated, stressful loading.

16.2.3 Effects of Hand-Arm Vibration

Hand-arm vibration has been linked to work involving extensive use of vibratory hand tools. Effects may be felt by anyone who has spent a few hours using a power tool. Some power tools known for their vibratory effects are chisels, chain saws, sanders, grinders, riveters, drills, jackhammers, compactors, sharpeners, and shapers.[4] The forearm muscles get fatigued from extended gripping, and the vibration combined with gripping causes a feeling of numbness or insensitivity in the forearm. This tends to induce a tighter grip. The tighter grip facilitates transmission of vibratory motion from the tool to the hand, wrist, and forearm. Early work on hand-arm vibration stemmed from concerns among Scandinavian forest harvesters using chain saws in cold weather. That work extended to other occupations and led to development of standards for measurement, exposure assessment, and control.[4–7]

The original sources of vibrations—the tool motor and moving parts—cause the tool housing to vibrate. The vibrating surface of the tool housing transmits the vibration to the skin of the hand. Between the skin surface and hand bones, the vibration is damped a bit, but a large portion of the vibration transmits into the bones. The hand bones transmit the vibration through the carpal bones to the forearm bones. Although the etiology of what is now called *hand-arm vibration syndrome(HAVS)* is not precisely understood, the resulting symptoms have been strongly associated with prior exposure to vibrating hand tools.[3] Older literature refers to the syndrome as Raynaud's phenomena of occupational origin and vibration white finger. Symptoms of HAVS include white fingers following cold exposure; reduced function of sensory and motor nerves; and disturbances in muscles, bone, and joints.[3] If the symptoms are detected early, the person may fully recover by avoiding additional hand-arm vibration exposure. However, if the symptoms develop too far, the syndrome becomes permanent and the person may have substantial restrictions on tasks involving manual handling or dexterity. For the person with HAVS, rest may moderate the symptoms temporarily, but the symptoms can recur rapidly when the person uses a vibrating hand tool. Working in a cold environment exacerbates the response.

Most adverse health effects of noise, whole-body vibration, and hand-arm vibration are the cumulative result of extended exposures. All three sources have established methods for measuring occupational exposures and evaluating against a standard.

16.3 STRATEGIES AND TACTICS FOR NOISE AND VIBRATION

The strategies and tactics for mitigating the risks of noise, whole-body vibration, and hand-arm vibration have much in common but differ in details. Engineering approaches for both noise and vibration can become quite technical. This section presents risk-reduction tactics familiar to the OSH community. Readers seeking more technical information may find it in books on the subject.[8,9]

16.3.1 Risk-Reduction Tactics for Noise Exposure

The strategies for reducing the risk of hearing damage from noise exposure are listed in the left column of Table 16.2. On the right are examples of applicable tactics. Other sources provide more detailed information on the various tactics.[9–13]

The first strategy is to eliminate the hazard. Common noise sources in occupational environments include production machines (e.g., mechanical power presses), power tools, and the high-velocity flow of gases and liquids in pipes. In the design phase for an industrial process, avoid putting these or similar noise sources in areas where employees will be working. If the facility is already operating with a particular

Table 16.2   Strategies and Tactics for Reducing the Risks of Noise Exposure

Risk-reduction strategy	Strategies and tactics
1. Eliminate the hazard	1a. Avoid locating noise sources in employee areas
	1b. Remove existing noise sources from employee areas
2. Moderate the hazard	2a. Purchase quieter equipment
	2b. Substitute quieter for noisier equipment
	2c. Replace surfaces that reflect noise with sound absorbing materials
	2d. Use technology to moderate noise-generating sources
3. Avoid releasing the hazard	Enclose noise sources
4. Modify release of the hazard	4a. Locate noise sources in corners with sound absorbing materials for ceiling and adjacent walls
	4b. Direct generated noise away from employee areas
5. Separate the hazard from that needing protection	Separate noisy areas from employee areas by barriers or distance
6. Improve the resistance of that needing protection	Not feasible
7. Help people perform safely	7a. Avoid noise that interferes with interpersonal verbal communication
	7b. Provide auditory warning that can be distinguished from other sounds in the environment
	7c. Provide work environments that do not require employees to use hearing protection or respirators
8. Use personal protective equipment	8a. Provide employees with hearing protection suited for their noise exposure
	8b. Provide employees with the means and encouragement to properly use their hearing protective devices
9. Expedite recovery	Transfer personnel who show signs of hearing loss from noisy work to a low-noise job

noise source, eliminating the source would require removing it from the area where personnel work.

The second strategy is to moderate the hazard. In the design phase of an industrial process, equipment purchasing specifications can set a noise maximum and make noise reducing devices part of the deliverables. In existing facilities, substituting a quieter alternative for a noisy machine or hand tool might be a feasible option.

Damping is a widely used engineering approach for reducing noise and vibration. The vibrating mass that creates noise and vibration has some elasticity, which allows it to slightly displace from its neutral position. When it is displaced, it will exert a force tending to return to the neutral. Once it starts returning to the neutral, it will go too far and displace in the opposite direction. If there were no energy losses, the vibration would continue indefinitely, but there are losses to friction and perhaps to heat, resulting to a gradual reduction in the swing. This reduction in the energy of oscillation is known as damping.

Engineers often use damping to lessen the energy of a noise or vibration source. Figure 16.2 displays two cyclical patterns of amplitude. Figure 16.2a shows the pattern for an oscillatory motion with zero damping. Figure 16.2b shows how damping affects the oscillation by both reducing the peak amplitude with each cycle and shortening the time it takes to bring the motions to a trivial level. *Noise damping* reduces the sound level and *vibration damping* reduces the displacement. This decay in amplitude can be aided by having the vibratory motion go through a damping material. The damping serves to reduce the amplitude of each successive cycle.

If noisy flows through pipes and valves are anticipated, engineering technology can be used to moderate noise level. For example, could a slower flow rate be used? Can metal parts and pipes be secured with vibration damping devices? Noise damping and isolating technology may also be used for noise generated by the vibrations of motors, fans, and other moving apparatus. Materials falling into metal chutes can create considerable noise. Can the metal chute be replaced or fortified with wood or plastic to reduce the noise?

Strategy 3 is to avoid releasing the hazard. The main tactic is to completely enclose the source. Brauer provides some basic guidelines on making an effective enclosure.[10] At a minimum, the enclosure should be complete; even small openings will allow considerable noise energy to escape. In addition, the inner surfaces of the enclosure should be composed of or covered with sound absorbing material to avoid the buildup of reverberant noise.

The fourth strategy is to modify release of the hazard. In some situations, the noise from a source can be influenced so as to minimize adverse effects. One tactic is to locate the machine, process, or other noise source in a corner of the facility; position it so that the greater part of the noise is directed into the corner walls and install sound absorbing materials on these walls and the ceiling. Brauer describes several engineering approaches for moderating surfaces to limit reflection or increase absorption of sound energy.[10] Similarly, it may be feasible to arrange some noise sources so the noise

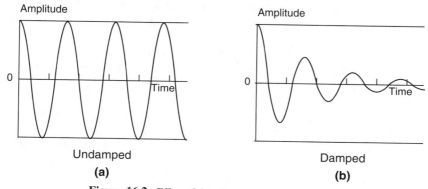

Figure 16.2 Effect of damping on vibratory motion.

energy is directed outside the facility or into a large area away from employees. Mufflers on engine exhaust systems are designed to allow exhaust gases to pass through while moderating the loud engine noises to acceptable levels.

The fifth strategy is to separate the hazard from that needing protection. Noise sources may be separated from employees by barriers or by distance. Barriers can reduce sound somewhat; however, sound moves easily around barriers, especially higher frequency sound. A shield-like barrier may seem as a great idea, but actual noise reduction can be less than might be intuitively expected. To be effective, a barrier should be close to the noise source, of material that limits noise transmission through it, and as tall as feasible. Using distance to separate a noise source from personnel will reduce the noise exposure level. If the noise source is moved sufficiently far from personnel to eliminate noise exposure, it would be part of the first strategy, eliminating the hazard.

Strategy 6, to improve the resistance of that needing protection, is not applicable to noise. It is not possible to make the hearing capability of employees more resistant to damage from excessive noise exposure. The same may be said of vibration.

The seventh strategy is to help people perform safely. In many situations, verbal communication plays an important role in safety. Background noise can interfere with this communication, so limiting background noise can help employees perform safely. Also, auditory warnings must be distinguishable from background noises in the environment. An example is the backup beeper on trucks and other large vehicles. If used in a noisy environment, the beeping may not be heard or may go unnoticed due to the worker being so accustomed to the generally loud work environment that no single sound stands out enough to draw attention. Another concern arises when personal hearing protection is used due to the loud background noise. The combination of loud background noise and use of hearing protection makes hearing safety information more difficult.

The use of respirators also interferes with transmitting spoken message to fellow employees. During the 1979 crisis at the Three Mile Island Unit 2 nuclear power station, control room personnel had significant communication problems because the respirators they had to wear partially muffled their voices and their hearing was impeded by the multiple conversations going on among people not normally in the control room. The operators had to pull their respirators aside to speak to others in the room. Thus, the ideal way is to design work so personnel do not need to use hearing protection or respirators.

The eighth strategy is to use personal protective equipment. While this strategy has a lower priority than the engineering control strategies, it serves the needs of millions of workers. Many workplaces have areas with noise levels considered too high based on company policy, a voluntary standard, or a mandatory standard. Whatever the reason, the use of hearing protective devices provides a last line of defense for protecting hearing. If an employer determines its employees need personal hearing protection, it should establish a hearing conservation program that includes acquisition of the PPE, training, maintenance, signage, workplace noise monitoring, periodic audiometric testing, and medical oversight.

Hearing protective devices are commercially available in several forms. Basic types are earplugs (ear inserts) and earmuffs, with several variations available for each type. In extra noisy worksites, using both earplugs and earmuffs is necessary to cut the noise reaching the eardrum down to a tolerable level. The use of personal hearing protection can make verbal communication more difficult. In extra noisy areas where workers wear earplugs and earmuffs for hearing protection, verbal communicate is ineffective, so they must use other forms of communication. Some employers have a standard set of hand gestures for these situations.

The ninth strategy is to expedite recovery. Periodic audiometric testing provides valuable information for early detection of hearing loss. If audiometric tests reveal an employee is losing hearing, a determination is needed as to the cause. The medical adviser may need to estimate the portion of hearing loss due to aging, disease, and workplace noise exposure. If noise exposure is suspected, the employee can be transferred to a job involving minimal noise exposure. This will not restore hearing capability, but it should end further loss due to employment. A transfer policy may prove beneficial for a couple of reasons. One is that, without such a policy, other employees may feel entitled to the low-noise job because of their seniority. In addition, a policy can establish how a transferred employee's wage rate will be affected. Having such a policy in advance can help avoid the appearance of being arbitrary or retaliatory.

16.3.2   Risk-Reduction Tactics for Vibration Exposure

Table 16.3 summarizes strategies and tactics for reducing risks associated with whole-body vibration. Because the tactics are self-explanatory and similar to those discussed in the preceding section, narrative accompanying the table is considered unnecessary. However, one point worth mentioning concerns seats available for vehicle drivers and heavy equipment operators. The so-called air ride seats effectively prevent driver exposures to resonant frequency. This enables the drivers and operators to avoid the most harmful aspects of whole-body vibration. Other technological advances in seat design also contribute to effective vibration damping.

Table 16.4 summarizes strategies and tactics for the hazards of hand-arm vibration. An extended narrative accompanying the table is considered unnecessary because the tactics in the table are sufficiently self-explanatory and similar to those used for noise. One observation worth mentioning is that the manufacturers of handheld power tools have made substantial progress reducing vibration or avoiding significant vibration in the resonant range. This is particularly useful because it makes unnecessary the use of antivibration gloves and antivibration coverings for the handles of power tools.

LEARNING EXERCISES

1. Perform the computations below using information in Table 16.1.
   (a) If a pure sound has a frequency of 3000 Hz, what is the value of the period (T)?

Table 16.3   Strategies and Tactics for Reducing the Risks of Whole-Body Vibration Exposure

Risk-reduction strategy	Tactics for whole-body vibration
1. Eliminate the hazard	1a. Avoid bringing vibrating equipment to the workplace
	1b. Fix equipment that vibrates inappropriately
	1c. Eliminate need for employee to sit on vibrating surface
2. Moderate the hazard	2a. When purchasing heavy equipment include specs for limited vibration at operator seat
	2b. Replace older operator seats with newer vibration damping seats so that exposure is not in resonant frequency range
3. Avoid releasing the hazard	Isolate vibration sources from media that can transmit vibration to humans
4. Modify release of the hazard	4a. Install vibration damping materials between source and transmission media
	4b. Install springs between source and transmission media to reduce vibratory accelerations
5. Separate the hazard from that needing protection	Use facility design to separate personnel and vibration sources
6. Improve the resistance of that needing protection	Not feasible
7. Help people perform safely	7a. Provide workstations that allow operator to alternate between sitting and standing
	7b. For sitting jobs requiring precise manual manipulations, provide operators with a nonvibrating seat
8. Use personal protective equipment	For vibrating floor, provide employees with footwear having soles suited for damping vibration
9. Expedite recovery	Actively manage medical care of personnel with a low back medical condition

(b) If a vibrating mass has a peak displacement of 1.3 mm, what is the value of the RMS displacement?

(c) What is the sound pressure level value (L_p) for a sound with an RMS value for $p = 14 \, \text{N/m}^2$?

2. Say an employee comes back from an audiometric test with results showing poor hearing at 4000 Hz. This was the only frequency showing a loss. The employee has been with the company for six years, always in the same job. What would be the basic steps for the OSH staff to take?

3. Say a haul truck driver in an open-pit mine comes with a report from his doctor that says the driver's lumbar spine shows signs of significant degeneration. He has been driving a haul truck for 22 years at your mine. If you were the

Table 16.4 Strategies and Tactics for Reducing the Risks of Hand-Arm Vibration Exposure

Risk-reduction strategy	Risk-reduction tactics
1. Eliminate the hazard	1a. Avoid bringing vibrating power tools to the workplace
	1b. Fix power tools that vibrate inappropriately
	1c. Eliminate need for employees to use handheld power tools
2. Moderate the hazard	2a. When purchasing power tools include specs for limited handle vibration
	2b. Choose tools powered with motors with an even number of cylinders[a]
3. Avoid releasing the hazard	3. Isolate vibration sources from media that can transmit vibration to humans
4. Modify release of the hazard	Purchase power tools with vibration damping materials built into the handle[b]
5. Separate the hazard from that needing protection	Have tool held by remote manipulator separated from human controller
6. Improve the resistance of that needing protection	Not feasible
7. Help people perform safely	7a. Provide workers with hand tools having handles suited for their hands and the tasks they perform
	7b. For assembly line workstations, use tool suspension devices to hang heavier tools
	7c. For power hand tools with substantial vibration, provide warning on product and in user manual
8. Use personal protective equipment	Provide power tool users with vibration damping gloves[c]
9. Expedite recovery	9a. Maintain medical surveillance in order to identify symptomatic employees before they develop permanent HAVS. Reassign them to jobs without hand-arm vibration
	9b. For employees with permanent HAVS, assign work with no hand-arm vibration exposure

[a] Motors with an odd number of cylinders tend to generate more vibration.
[b] Wrapping hard surface tool handles with vibration damping tape is not recommended because it increases the diameter of the handle, requiring the user to grip with greater force (see Ref. 12).
[c] Full finger gloves are recommended (see Ref. 12).

company medical adviser, and you agreed with the medical report, make two recommendations, one engineering and one administrative?

4. Explain the effects of damping on oscillatory waves.

5. Explain and compare the role of rest (i.e., nonexposure) as a prescription for HAVS, muscle fatigue, and temporary threshold shift in hearing.

6. One tactic for reducing noise exposure is to set up a barrier between the source and the personnel. What characteristics of an effective barrier are mentioned in the chapter?

7. Suppose a manufacturing facility is preparing to install a heavy new machine that will produce substantial vibrations. Engineers have identified two options for minimizing the effects of vibrations on the floor. One is to mount the machine on a thick pad of vibration damping material. The other option is to cut a hole in the floor and build an independent platform from the ground below up to the level of the floor. That would prevent the machine and floor it sits on from touching the room floor. Which strategy in Table 16.4 applies to each option?

8. Say a chain saw manufacturer makes chain saws for professional timber harvesters (lumberjacks). The present chain saw motor has three cylinders. One engineer suggests changing to a four-cylinder motor. Why?

TECHNICAL TERMS

Amplitude (A)	The maximum (or peak) deviation of a physical quantity from its equilibrium position. It is one of the two special values of the instantaneous deviation of the quantity. The other special value is the RMS.
Damping	Reducing the energy level of an oscillating phenomenon over a time by use of some sort of energy absorbing mechanism or material.
Decibel (dB)	A dimensionless variable, computed from the ratio of measured air pressure (p) to a reference air pressure (p_o). The formula is $20 \log (p/p_o)$.
Decibel scale	A unit for expressing level of noise or vibration relative to a low reference value.
Frequency (f)	The cycle rate, with units of cycles per second. Frequency is a fundamental attribute of oscillating phenomena and very useful for describing a pure tone, a vibrating mass, alternating electric current, and electromagnetic energy.
Noise	Undesired sounds because of either interfering with communication or appearing unpleasant to the listener. Also used for any factor that interferes with transmission of electronic signals.
Noise damping	Reducing noise energy by passing through a sound absorbing material.
Instantaneous amplitude	The deviation of a physical quantity from its equilibrium position at any point in time. Two special values of instantaneous amplitude are the peak amplitude (A) and the RMS amplitude.

Isolation	Separating the source of noise or vibration from a transporting medium.
Period (τ)	The time between corresponding points on successive waves of an oscillating phenomenon.
Resonant frequency	A small vibration frequency range within which the external oscillations create an exaggerated responding vibration in the subject, making the effect much worse.
Sound pressure (p)	The pressure of sound at any moment measured at a point some distance from the source. Usually measured in $\mu N/m^2$ or μPa.
Sound pressure level	The decibel expression for the pressure of sound at any moment measured at a point some distance from the source (see Table 16.2).
Vibration damping	A reducing effect on vibration energy by using a viscoelastic material to remove some of the vibration energy.
Wavelength (λ)	The distance between corresponding points on successive sound waves.

REFERENCES

1. Bovenzi M. Metrics of whole-body vibration and exposure–response relationship for low back pain in professional drivers: A prospective cohort study. International Archives Occupational Environmental Health. 2009;82:893–917.
2. Okunribido OO, Magnusson M, Pope MH. The role of whole body vibration, posture and manual materials handling as risk factors for low back pain in occupational drivers. Ergonomics. 2008;51: 308–329.
3. Wilder DG, Wasserman DE, Wasserman J. Occupational vibration exposure. In: Wald PH, Stave GM, editors. Physical and biological hazards of the workplace, 2nd ed. New York: Wiley; 2002. p. 79–104.
4. National Institute for Occupational Safety and Health. Criteria for a recommended standard: Occupational exposure to hand-arm vibration. DHHS(NIOSH) Publication No. 89-106. Cincinnati, OH: The Institute; 1989.
5. International Standards Organization. Guide for the measurement and the assessment of human exposure to hand transmitted vibration. Geneva: The Organization, ISO 5349; 1986.
6. American Conference of Governmental Industrial Hygienists. Hand-arm (segmental), vibration, in the 2010 TLVs and BEIs based on the documentation of the threshold limit values for chemical substances and physical agents & biological exposure indices. Cincinnati, OH: ACGIH; 2010. p. 188–191.
7. American National Standards Institute. Guide for the measurement and evaluation of human exposure to vibration transmitted to the hand. New York: The Institute, ANSI S3.34; 1986.
8. Ver IL, Beranek LL. Noise and vibration control engineering, 2nd ed. Hoboken, NJ: Wiley; 2006.
9. Crocker MJ, editor. Handbook of noise and vibration control. Hoboken, NJ: Wiley; 2007.
10. Brauer RL. Safety and health for engineers, 2nd ed. Hoboken, NJ: Wiley; 2006. p. 411–433.
11. Konz S, Johnson S. Work design: Occupational ergonomics, 7th ed. Scottsdale, AZ: Holcomb Hathaway; 2008. p. 426–433.
12. Bruce RD, Bommer AS, Moritz CT. Noise, vibration, and ultrasound. In: Denardi R, editor. The occupational environment: It's evaluation and control. Fairfax, VA: American Industrial Hygiene Association; 1997. p. 425–488.
13. Berger EH, Royster LH, Royster, JD, Driscoll DP, Layne M. The noise manual, 5th ed. Fairfax, VA: American Industrial Hygiene Association; 2003.

Chapter 17

Thermal Hazards: Heat and Cold

17.1 BACKGROUND ON THERMAL HAZARDS

Thermal energy concerns for the OSH community involve having too much heat or not enough heat. The concerns about too much heat are burns and heat disorders. This chapter summarizes the mechanisms of heat exchange between people and the environment, reviews indices used for evaluating exposures, explains the various disorders, and demonstrates how the risk-reduction strategies apply to thermal hazards.

17.1.1 Fundamentals of Thermal Energy for OSH

Heat as a "source" of burn injury is typically the result of another type of energy being transformed into heat. For example, the burners on a stove are a hazardous form of heat energy. The heat in the common burner is the result of electrical energy flowing through coils of relatively high resistance. If a person were to touch a heating element on a stove and get burned, the most proximal cause of the injury would be heat, not electric current.

A basic principle is that thermal energy flows from the warmer to the cooler region.[1] The extensive equations for quantifying this heat-transfer process are taught in university courses called heat transfer and thermodynamics. Older books on these subjects used several units for heat energy, work, and power such as calories, British thermal units, and now joules and watts. The current SI units are summarized in Table 17.1.

A common mode of heat transfer is by direct contact between the skin and a hot surface. Known as conduction, it occurs when skin directly contacts an object that is hotter or cooler than the skin. The radiant heat mode occurs when the skin is exposed to sunlight, flame, or a nearby hot surface. Radiant heat transfer also occurs in cold

Risk-Reduction Methods for Occupational Safety and Health, First Edition. Roger C. Jensen.
© 2012 John Wiley & Sons, Inc. Published 2012 by John Wiley & Sons, Inc.

Table 17.1   Important Terms for Thermal Energy

Term	SI name	SI letter	Explanation
Energy	Joule	J	Thermal energy that could perform work
Work	Joule	J	Thermal energy that was already used to perform work. $1\,J = 2.39 \times 10^{-4}$ kcal. The kilocalorie (kcal) unit is also used in the heat stress literature
Power	Watt	W	$1\,W = 1\,J/s$. Also, metabolic rate is often expressed as kcal/h. $1\,kcal/h = 4184\,J/h = 1.162\,W$

environments where one's skin and clothing will transfer heat to cooler objects in the area. A third kind of heat transfer, convection, is implicated in burns from contact with scalding hot water or other fluid. Convection also applies to the exchange of heat between the skin and surrounding air.

A less obvious type of heat transfer occurs when the sweat on skin and clothing evaporates. The evaporation process carries heat away from the skin. This mode of heat transfer plays an important role in heat stress and a moderate role in cold stress.

These modes of heat transfer provide the basis for a general mathematical model to explain the various factors that enable people to maintain a relatively constant body temperature even when the surrounding air feels unpleasantly hot or cold. Their model begins with the premise that human heat balance (*thermoregulation*) depends on balancing the heat added to the body and the heat removed. By quantifying each factor in the same units, we should be able to sum the heat additions and the heat losses to arrive at a value near zero. The summative equation for change in body heat content (ΔS) during some period of time is

$$\Delta S = (M - \text{Wk}) \pm C_V \pm R \pm C_D - E, \qquad (17.1)$$

where

M is metabolic energy,

Wk is external work performed,

C_V is heat gained or lost by convection,

R is heat gained or lost by radiant heat transfer,

C_D is heat gained or lost by conduction, and

E is heat lost by evaporative cooling.

Each heat transfer term in Equation 17.1 has a mathematical model (equation) for quantifying the effect on overall heat exchange. OSH professionals may find these equations helpful when working on control measures for particular heat sources. Readers looking for these equations are referred to book chapters referenced at the end of this chapter.[2–4]

For heat stress, the *conductive heat transfer (C_D)* term may be dropped from Equation 17.1 because conductive heat transfer contributes very little to heat balance. For cold stress, C_D is useful when the worker stands on frozen ground or on a cold floor (e.g., in a refrigerated room). The contact between the cold surface and a worker's boot can draw a significant amount of heat from the feet. Similarly, a job that requires manually handling cold objects, even with gloves, can remove considerable heat from the hands.

The *metabolism (M)* term in Equation 17.1 always adds to the total body heat, but not all metabolism becomes heat. Some portion of metabolic energy may be used to perform external work. For example, walking up a stairway uses metabolism for raising the body upward (the work), as well as generating heat. The part doing external work does not add heat. For that reason, the work term (Wk) in Equation 17.1 is subtracted from total metabolism. The result is the true metabolic heat load. However, for practical occupational exposure assessment, the work term is generally ignored because it is a small portion (<10%) of metabolic heat.

Convective heat transfer (C_V) may either add or take away heat from the total body heat content. Heat is transferred to the body when the surrounding environment is warmer than the skin, and heat moves out of the body when the skin is warmer than the surrounding environment. The rate of convective heat transfer is slower when the air-to-skin contact is static, as opposed to when the air moves over the skin. More generally, the temperature difference and the air velocity relative to the body surface affect the rate of convective heat transfer.

In OSH, we usually think of the surrounding environment as air, but in some occupational situations, the surrounding environment is water or other liquid. This can occur in many work situations such as falling into cold water while working on a bridge or boat dock. Some divers work underwater on the hulls of docked ships. Sometimes rescue personnel find themselves in cold water, intentionally or unintentionally, when attempting to rescue someone from a body of water. A worker surrounded by cold water will rapidly lose heat, leading to loss of ability to swim.

Radiant heat transfer (R) may either add or subtract heat from the total body heat content. Heat is transferred to the body when an external surface is hotter than the surface of the body facing the hot object. This effect is noticeable when working near a furnace, oven, heater, or flame, and becomes more pronounced when closer to the heat source. Another key factor in radiant heat transfer is the magnitude of the temperature difference between the external object and a person's skin and clothing. A greater difference means more heat transfer in the same time period. Radiant heat transfer also occurs in a cold environment, but the heat transfers from the person to nearby cooler objects. The primary factors that affect radiant heat transfer in the cold stress situation are the same as those in the heat stress situation: separation distance and temperature difference.

Evaporative heat loss (E) provides the primary means for workers in hot jobs to maintain their normal body temperature. The physiological mechanism is to increase sweat production. More sweat on the skin and clothing supports an increased rate of evaporative heat transfer from the body. The rate of this heat transfer depends on the difference in the water vapor pressure in the air and on the skin surface, as well as the air velocity. These two factors explain a basic control tactic for work in hot, humid worksites—using standing floor fans to help speed up E. Without the fans, the difference in vapor pressure between the skin and air will be rather small because the static air next to the worker's skin becomes fully saturated. The result is almost no difference in vapor pressure and almost no evaporation, and the sweat will drip from the worker's skin to the floor. In contrast, with a fan, the air next to the worker's skin is constantly moving. The air from the fan will have a lower vapor pressure than the air immediately next to the sweaty skin and clothing, and this difference in vapor pressure accelerates the evaporation of sweat. In most situations, the sweat will evaporate rather than drip onto the floor.

The human thermoregulation model in Equation 17.1 contains terms for the various factors involved in maintaining heat balance. Experts in cold stress and heat stress have attempted to find sound and practical indices to enable prediction of how various combinations affect humans. This task has led to some practical indices for OSH applications.

17.1.2 Indices for Cold Stress

Basic indices for cold stress account for air temperature and wind speed. Using a table format, various combinations of air temperature and wind conditions are presented in terms of equivalent temperature under calm conditions. Equivalent temperature scales are found on the websites of weather services in countries with colder winters.

Another index is the ACGIH TLV for cold stress.[5] Typically, wind chill charts used by weather services reflect the equivalent cooling effect on bare skin of the various air temperature and wind speed combinations. These charts should be regarded as very general guides because of multiple factors influencing acceptable exposures. For example, workers in cold jobs normally wear warm clothing, and their clothing varies greatly (e.g., type of coat, gloves/mittens, head cover, ear protection, footwear insulation, and nose cover). In addition, considerable variation is introduced by different amounts of metabolic heat generated by the work. Thus, tables based on exposed bare flesh provide only a rough basis for comparing cold conditions. The numerous variables explain why the world does not yet have a widely agreed upon exposure limit for workplace cold exposure.

17.1.3 Indices for Heat Stress

Standards and guidelines for occupational heat exposure are somewhat more useful than those for cold. Several indices for heat stress have been proposed. Excellent explanations and commentary on these indices are provided by Ramsey and Bishop as well as Bernard.[4,6] The index receiving the most support is the value obtained with a wet bulb globe thermometer (WBGT). It is a composite measure that accounts for the effect of air temperature, humidity, wind speed, and radiant heat transfer. It is based on three temperature measurements:

1. Black globe temperature (T_G) is measured by a thermometer in the middle of a black, metal globe. The matt black globe surface absorbs radiant heat from hot objects in the area and reflects almost none.

2. Natural wet bulb temperature (T_{NWB}) is measured by a thermometer with a wick around the sensor. The wick is wetted prior to taking a measurement. During the measurement, the water on the wick evaporates and cools. It is called "natural wet bulb" to distinguish it from the aspirated wet bulb temperature used by engineers for thermodynamic calculations. T_{NWB} reflects the effect of environmental humidity on the E factor in Equation 17.1. In a humid environment, the evaporative cooling effect lowers the thermometer very little, while in a dry environment, the wick lowers the temperature more. Air movement also contributes to cooling the thermometer by accelerating evaporation from the wick.

3. Dry bulb temperature (T_{DB}) is measured by a thermometer exposed to the air, and it accounts for the effect of air temperature on convective heat transfer.

Taken together, these temperature measurements provide values suitable for computing the WBGT. Two forms of the WBGT are used. For workers directly exposed to the sun, Equation 17.2a is used, and for those not exposed to the sun, Equation 17.2b is used.

$$WBGT = 0.7T_{NWB} + 0.2T_G + 0.1T_{DB} \qquad (17.2a)$$

$$WBGT = 0.7T_{NWB} + 0.3T_G \qquad (17.2b)$$

Heat stress exposure standards account for both environmental conditions using WBGT and physical activity level using estimates of *metabolic rate (MR)*. A comparison of five different heat stress standards showed agreement in recommended WBGT within 1 or 2 °C.[4] These standards use a table or graph to indicate their respective recommendations. Figure 17.1 illustrates a two-curve approach for evaluating measured heat exposures. The lower curve applies to workers who are not heat acclimatized; the upper one applies to workers fully acclimated to the hot conditions. The two curves are approximately the same as those found in the ACGIH TLV for heat stress, but in the TLV the two lines are labeled "action limit" and "TLV," respectively.[7] If exposures are above the action limit, various precautions need implementation. Exposures should not exceed the upper line. Points on the horizontal axis of Figure 17.1 are based on a time-weighted average metabolic rate (TWA MR) for 1 h. Using time expressed in minutes, the formula is

$$TWA\ MR = (1/60) \sum (MR_i)(t_i). \qquad (17.3)$$

Astute readers may notice that the metabolic heat rate for this exposure model ignores the external work term (Wk) in Equation 17.1. This is to simplify the process of

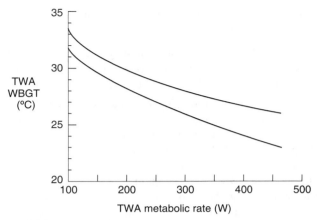

Figure 17.1   Example of heat exposure action levels and exposure limits.

obtaining a reasonable value for MR, and is based on the assumption that nearly all the energy from metabolism is in the form of heat (i.e., external work can be ignored). The corresponding formula for time-weighted average exposure, using WBGT, is

$$\text{TWA WBGT} = (1/60) \sum (\text{WBGT}_i)(t_i). \tag{17.4}$$

Obtaining values for these two equations requires monitoring the employee for 1 h. During that time, the employee may perform work at different metabolic rates. Therefore, the time (t_i) spent at each rate is required for Equation 17.3. Similarly, if the employee spends time in more than one work area, both the WBGT and the time in each work area are required for Equation 17.4. In practice, the metabolic rate values are less precise than the WBGT values. The common way to get metabolic rate values is to watch the worker, break down the hour into sessions of similar work level, and use metabolic rate values found in data tables for estimating MR for each session. Data tables are available with metabolic rates shown in various units, most commonly in kcal/min, kcal/h, or watts. Figure 17.1 uses watts.

The ACGIH TLV also has a table with guidance for sorting exposure levels based on the proportion of time a worker spends working in the heat versus resting in the same heat. On initial impression, these guidelines appear to simplify the application. However, that can be misleading because the guidelines assume the rest area and the work area have the same environmental conditions. Only in rare cases does a worker need to rest in the same hot area as he or she works. Even when working outdoors in the sun, workers can usually find a shady place to recover. In addition, the work–rest categories assume the work is at the upper end. For example, the 50–75% category assumes the worker worked 75% and rested 25%. If the worker actually worked 55%, the guidelines are misleading because the indicated exposure line is for 75% work. Thus, this author discourages using the work–rest guidelines approach. It is better, and no more difficult, to compute the time-weighted values using Equations 17.3

and 17.4. With those values and knowledge of the worker's acclimatization, employee's exposure can be compared to the curved lines in the ACGIH TLV.

17.2 MECHANISMS OF HARMING

This section explains the relationships between occupational exposure to thermal hazards and health effects—burns, heat disorders, and cold disorders. Also summarized are the effects of the workplace thermal conditions on the safety-related behavior of the workers.

The most common burns result from skin contacting with a hot object or fluid. Burns can also result from excessive contact with cold objects. The severity of heat burns is determined by two main factors: (1) the difference in temperature of skin and the contacted object or fluid, and (2) the duration of the contact. We all know from life experiences that if our finger contacts a hot object and we quickly withdraw, we can limit the severity of a burn or avoid any burn at all. We also know that for the same duration of contact, touching a hotter object will cause a more severe burn than touching an object that is not as hot. In addition to the burns incurred through conduction, we can be burned from exposure to radiant heat. Two common sources of radiant heat associated with occupational burns are sunlight and welding. Both can cause severe burns.

Burns from cold also involve conduction. Both important factors applicable to heat burns also apply to cold conduction burns—the colder the object, and the longer the contact, the greater the severity. Another factor is moisture. Moisture between the cold object and the skin will freeze and adhere to both surfaces. A curious child who touches his tongue to a cold metal pipe will soon find that his tongue is frozen to the pipe.

Humans are warm blooded and must maintain a core temperature within a narrow margin (the prescriptive zone). Too much heat can cause the core body temperature to rise above the normal, healthy level (approximately 38 °C), and too little heat can cause the core body temperature to sink below the normal, healthy range. For repeated occupational exposures to cold working conditions, it is undesirable to allow core temperatures below 36 °C. Lower core temperatures are associated with reduced vigilance and reduced manual dexterity. In addition, when the core body temperature gets low, the circulatory system reduces blood flow to the extremities, making fingers, toes, ear lobes, and the nose vulnerable to frostbite.

Disorders associated with heat are described in material by the U.S. National Institute for Occupational Safety and Health, and available at www.cdc.gov/niosh/topics/heatstress. It describes the heat disorders—heat stroke, heat exhaustion, heat cramps, fainting, heat rash, and transient heat fatigue—listing the symptoms and appropriate first-aid procedures. Similar material on cold stress is available at www.cdc.gov/niosh/topics/coldstress. It provides information about the cold disorders—hypothermia, frostbite, trench foot, and chilblains.

Information about the occupational groups most affected by heat and cold disorders comes from a retrospective review of workers' compensation records in the United States.[9] The occupational categories with the largest proportion of claims for heat disorders (heat exhaustion and heat stroke) were farm laborers (7.5%), firefighters (6.8%), miscellaneous laborers (5.6%), construction laborers (5.2%), miscellaneous operatives (4.7%), truck drivers (4.3%), laborers not specified (3.8%), and gardeners/groundskeepers (3.4%). The occupational categories with the largest proportion of claims for cold disorders (mainly frostbite) were miscellaneous laborers (10.4%), truck drivers (7.6%), construction laborers (4.3%), firefighters (3.6%), garbage collectors (3.4%), police and detectives (3.3%), and farm laborers (3.1%). An indication that most cases arose in outdoor conditions comes from the finding that 75% of the cold cases occurred in the coldest months, January and February. The body parts affected by cold were finger (24.5%); foot, not toes (14.4%); hand (14.0%); toe (12.9%); multiple body parts (5.1%); ear (4.2%); upper extremity (3.4%); respiratory system (3.1%); lower extremities (2.5%); and nose (1.4%). For the year studied, the total number of claims was 762 for heat disorders and 645 for cold disorders.

Occupational heat and cold exposure not only causes heat and cold disorders, but also affects safety-related behavior. A 14-month observational study was conducted in a metal products manufacturing plant and a foundry.[9] Thermal environment data and observations of safety-related behavior were obtained. Results showed the unsafe behavior rates were lowest in the mid-range of conditions (approximately 17–23 °C). The proportion of unsafe behaviors increased on both the cooler and the hotter ends of the range.

17.3 STRATEGIES AND TACTICS FOR THERMAL HAZARDS

The strategies for reducing risk associated with workplace heat are listed in the left column of Table 17.2. On the right are major groups of applicable tactics. Table 17.3 lists similar strategies and tactics for cold work environments. The text below clarifies or supplements the tactics in the table.

The first strategy is to eliminate the hazard either during the facility design phase or during the operation phase. The design phase of a facility may afford opportunities to eliminate from occupied rooms equipment having surfaces either hot enough or cold enough to harm personnel. Once operational, similar opportunities might be found for removing the same type of equipment from an unoccupied work area. It may also be feasible in some situations to make arrangement so that employees to perform work in a very hot or very cold area can do the work in a more moderate area.

The second strategy is to moderate the hazard. On building construction projects during winter, workers doing interior work often use a portable heater to make the room temperature more tolerable. Similarly, when work is required in a hot area, the use of spot cooling devices can reduce the heat level. Using fans to blow air directly on workers in hot, humid areas helps them maintain thermal balance by accelerating evaporative cooling from sweat. However, fans have limited value in dry climates

Table 17.2 Strategies and Tactics for Reducing the Risks of Workplace Heat

Risk-reduction strategy	Risk-reduction tactics
1. Eliminate the hazard	1a. Avoid hot surfaces in work areas
	1b. Eliminate existing hot surfaces
	1c. Eliminate need for personnel to work in hot area by having them perform the task in cooler area
2. Moderate the hazard	2a. For hot surfaces, use minimum heat needed
	2b. Provide portable fans or spot cooling to accelerate evaporative cooling rate
	2c. For work in the sun, provide some shade
	2d. Manage individual heat load to avoid having an extra high heat load hour during a shift by scheduling heavy tasks for cooler part of day, mechanizing heavy tasks, or assigning more personnel to share the heavy work
3. Avoid releasing the hazard	3a. For pipes and containers with hot materials under pressure, assure integrity of connection points to avoid releasing hot materials
	3b. For hot material stored or transported under high pressure, avoid an explosive release by installing a suitable pressure-relief valve to avoid an internal pressure buildup
4. Modify release of the hazard	4a. Fully enclose heat source except for in–out ventilation ports to take heat away
	4b. For very hot materials stored under pressure, install overpressure devices so that if internal pressure exceeds a set point, the released material will be sent in the least harmful direction
	4c. For pipes and containers with steam or other hot material under pressure, provide a shield around pipe nodes so that if a leak develops, the hot material spray will be interrupted and redirected
5. Separate the hazard from that needing protection	5a. For radiant heat sources, provide a barrier between personnel and source of radiant heat. Insulation around hot pipes and containers also helps by exposing a cooler outer surface to the workers
	5b. For heat generating equipment, locate heat source away from employee areas
	5c. For hot objects in workplaces, install insulation to separate the hot surfaces from personnel
6. Improve the resistance of that needing protection	Personnel can improve their resistance to heat stress, but this is classified as part of Strategy 7 because the change is not permanent.
7. Help people perform safely	7a. Make and use an acclimatization program for new workers and those returning from several days away from work.

(continued)

TABLE 17.2 (*Continued*)

Risk-reduction strategy	Risk-reduction tactics
	7b. Support a program to encourage employees who work under heat stress conditions to maintain or improve their physical fitness
	7c. Train employees exposed to heat stress on lifestyle, clothing, precautionary behavior, self-monitoring, and monitoring fellow workers
	7d. Make sanitary drinking water readily available for those exposed to heat stress
	7e. Provide rest areas suitable for cooling off
	7f. Encourage frequent breaks for those exposed to heat stress
8. Use personal protective equipment	8a. Provide personal cooling vests to help with briefer exposures to very hot work areas
	8b. For workers using full-body chemical protective gear, provide extra airflow through the suit to remove heated, humid air
9. Expedite recovery	9a. Maintain a cool space for any employee who shows signs of heat cramps or heat exhaustion
	9b. Maintain first-aid capability appropriate for treating burns and heat disorders

because the sweat already evaporates rapidly. Also, when the air temperature is well above skin temperature (e.g., above 40 °C), blowing hot air on the workers adds more heat through convection than it removes through evaporative cooling. To decide if fans will help, watch the workers to see if sweat accumulates on their clothing. If their clothing gets drenched, fans will probably help. If their clothes appear dry, adding fans will be ineffective since the sweat is already evaporating quickly without a fan. If work is performed in the sun, any form of full or partial shade will help reduce the radiant heat load.

Another tactic for moderating heat load is to use one of the following means for managing the heat load from heavy manual work. One approach is to schedule tasks so that the highest metabolic rate tasks are performed during the coolest part of the shift. A second is to assign extra personnel to share the heavy work performed in a hot location so that no individual is exposed to an excessive heat level. A third is to mechanize the heavy manual tasks.

Strategy 3 is to avoid releasing the hazard. The release of hot material under pressure (e.g., steam) from pipes and containers is prevented by designing, installing, and maintaining the interfaces/nodes so that no leaks develop. For hot material stored or transported under high pressure, an internal pressure buildup leading to an explosive release of hot material can be prevented by installing and maintaining suitable pressure-relief devices.

For some normally gaseous materials kept in a liquid state by maintaining them at an extremely cold temperature, release can be prevented by designing, building, installing, and maintaining quality storage vessel and transport tanks. Especially important are cryogenic materials—those stored at temperatures below $-180\,°C$ (93 K). Some materials stored in cold temperatures above the cryogenic level are also highly hazardous if the refrigeration fails. Managing such materials requires best-practice aspirations; anything less is simply unacceptable for liquefied materials that can expand rapidly and disperse into personnel areas. Best practices include having a first rate process safety management program.

The fourth strategy is to modify release of the hazard. For a specific source of heat, such as a machine or heating process, it may be feasible to enclose the source except for openings to allow ventilation ducts to bring in cooler air and exhaust the heated air. For pipes and containers with steam or other pressurized hot materials, leaks can develop at connection points. An engineering approach for this potential hazard is to install shields around pipe joints/nodes so that if a leak develops, the hot material spray will be interrupted and redirected for minimal harm. For storing or transporting liquefied gases, incorporating overpressure devices is standard practice. These devices work by releasing small to moderate amounts of material when internal pressure exceeds a set point.

The fifth strategy is to separate the hazard from that needing protection. Some basic practices are noted in the two tables. Locate heat sources away from employee areas. For pipes and containers with steam or other hot materials, install insulation around the pipes and containers to avoid having surfaces in the work area capable of causing burn injuries. In some work areas, radiant heat transfer contributes substantially to the total heat load by radiating to the worker's clothing and skin, as well as heating up various objects in the work area. A radiant heat shield between the radiating surface and work areas can effectively protect workers from the radiant heat.

For pipes and containers with steam or other hot materials, the potential hazard of a leak shooting hot material at nearby workers can be prevented by designing facilities so personnel will not be exposed. Containers with cold materials also benefit from insulation and placement away from employee areas. An example of Strategy 5 so routine that it may go unnoticed is the practice of putting thermal insulation on the handles on snow shovel and numerous other metal-handled tools used in cold environments.

The sixth strategy, to improve the resistance of that needing protection, is not applicable. Heat acclimatization is not a permanent change so it belongs in Strategy 7–help people perform safely. For work in hot environments, some employers have an *acclimatization* program for new workers and those returning from several days away from their hot jobs. Adaptations associated with acclimatization to heat include increased sweating, less salt in the sweat, and improved sense of how much water is needed to maintain hydration. Tactics listed in Table 17.3 for cold stress do not include acclimatization because, unlike working in the heat, working in the cold does not lead to significant physiological acclimatization. Both tables include the use of medical prescreening for personnel prior to assignment to a job with significant heat stress or

Table 17.3   Strategies and Tactics for Reducing the Risks of Cold Objects and Work Environments

Risk-reduction strategy	Risk-reduction tactics
1. Eliminate the hazard	1a. Avoid putting freezing surfaces in working areas
	1b. Eliminate existing freezing surfaces from working areas
	1c. Eliminate need for personnel to work in cold area by having them perform the work in warmer area
2. Moderate the hazard	Provide a portable heater to add warmth to a cold room
3. Avoid releasing the hazard	For very cold materials (e.g., cryogenics), design, build, install, and maintain storage vessel and transport tanks using best practices
4. Modify release of the hazard	For storing or transporting liquefied gases, place overpressure devices so that if internal pressure exceeds a set point, the expanding gas will be sent in the least harmful direction
5. Separate the hazard from that needing protection	5a. When outdoor weather is cold, provide personnel with comfortably warm indoor work areas
	5b. Locate freezing objects away from employee areas
	5c. When personnel must work outdoors in cold weather, provide wind-breaking barriers
6. Improve the resistance of that needing protection	This strategy does not apply because changes in people are not permanent.
7. Help people perform safely	7a. Support a program to encourage employees who work under cold stress conditions to maintain or improve their physical fitness
	7b. Train employees exposed to cold stress on lifestyle, clothing, precautionary behavior, self-monitoring, and monitoring fellow workers
	7c. Make sanitary drinking water and warm beverages readily available for workers exposed to cold stress
	7d. Provide rest areas suitable for warming up
	7e. Establish a work/rest schedule suited to the cold work exposure
8. Use personal protective equipment	8a. Provide personal heating devices to workers exposed to very cold areas
	8b. Provide cold-appropriate personal protective gear for hands, feet, and head
	8c. Provide some financial assistance for workers to buy cold-appropriate pants and coat
9. Expedite recovery	9a. Maintain a warm space for any employee who shows signs of cold stress such as disorientation and shivering
	9b. Maintain capability to treat an employee who develops a cold-related disorder

cold stress, and both tables list supporting a program to encourage employees with thermally stressful jobs to maintain or improve their physical fitness.

Several other tactics for Strategy 7 are noted in the two tables. Several tactics apply to work in both hot and cold environments. Among these is training workers on lifestyle, clothing, precautionary practices, self-monitoring, and monitoring fellow workers. Another is making water and other hydrating fluids readily available. Cooled drinks are desirable for work in hot environments, and warm to hot drinks are best for work in cold environments. A suggested policy for encouraging adequate hydration is to respect the preferences of individuals by making available diverse nonalcoholic beverages.

Attention to rest breaks is a third shared tactic. Obviously, a cool break area is desirable for employees who work in hot environments, while a warm break is needed for those who work in cold environments. In addition, more frequent, shorter breaks are preferred to less frequent, longer breaks. Some employers have a self-determination break policy—allowing each worker to decide when to take a break. Such a policy can succeed if the personnel share an adequate work ethic.

Strategy 8 is to use personal protective equipment. The tactics listed in Tables 17.2 and 17.3 are widely recognized and need no clarification. In addition to these widely used practices, some employers provide employees an allowance to purchase work clothing appropriate to the thermal conditions of the workplace.

The ninth strategy is to expedite recovery. Two similar tactics are listed for both hot and cold workplaces. For hot working conditions, a basic tactic is to maintain a cool space for any employee who shows signs of a heat disorder such as heat cramps or heat exhaustion. A second tactic is to maintain the capability to treat an employee who develops heat stroke. A worker suffering from heat stroke needs immediate and aggressive action to lower his or her core body temperature. It could be a fatal mistake to leave a heat stroke victim lying in a cool room or in the shade while waiting for an ambulance. They need aggressive cooling such as immersion in a bath of cold water.

For cold work conditions, workers may develop frostbite on the tip of their nose or on an exposed earlobe without being aware of it. A part of training for work in cold environments should include the role of each team member to watch fellow employees for frostbite or signs of early hypothermia such as disorientation, confusion, shivering, loss of coordination, and fatigue. Catching these developing conditions early, and providing appropriate care, can head off more severe outcomes.

LEARNING EXERCISES

1. Employers have diverse policies on paying for personal protective equipment. Imagine you are asked to consult for a large farming operation with laborers employed during cold winter months and hot summer months. For each of the items listed, do you think the PPE should be purchased by the employer, the employee, or should they share the costs?
   (a) Hearing protection inserts.
   (b) Gloves.

(c) Work boots.
(d) Work pants.
(e) Work coat.
(f) Head protection.
(g) Safety glasses (nonprescription).
(h) Safety glasses (prescription).

2. For the same farming operation, what policy differences would you recommend for the seasonal employees?

3. Two OSH professionals spent a hot summer day collecting data on employee exposures to heat stress. During the afternoon, they monitored a particular worker from 2 pm until 3 pm. One person set up the thermometer stand near the worker and recorded temperatures frequently. The other person observed the worker's activities and estimated metabolic heat using tables. The worker's activities were broken down into four segments of similar metabolic rate. Results are summarized below.

Work segment	1	2	3	4
Duration (min)	16	11	13	18
T_{NWB} (°C)	28	28	22	32
T_G (°C)	32	32	22	35
Metabolic heat (W)	340	280	100	300

(a) What value is the TWA WBGT for the hour?
(b) What value is the TWA MR for the hour?
(c) Compare the TWA values computed with the curves in Figure 17.1. What is the conclusion about heat exposure?

4. A topic of some controversy for heat stress work concerns whether to recommend that workers in these jobs increase their dietary salt intake. Associated with this issue is whether an employer should provide drinks containing sodium and potassium salts ("sports drinks"). Answer the following questions.
(a) What do you think is the rationale for encouraging increased salt intake.
(b) What do you think is the rationale for not encouraging increased salt intake.

5. There is widespread agreement that employers should provide sanitary drinking water for employees. Drinking fountains are a common means of delivery. In construction sites and agriculture workplaces, drinking fountains are generally not feasible and water jugs are common. Do some research on the Internet or in sanitation books to learn about sanitation requirements for drinking water. What minimum sanitation practices should be followed when using water jugs?

TECHNICAL TERMS

Acclimatization	The physiological changes that occur in response to a succession of days of exposure to environmental heat stress that reduces the strain caused by the overall heat stress.
Conductive heat transfer (C_D)	The net heat exchange by conduction between an individual and objects in contact with the individual during a specified period.
Convective heat transfer (C_V)	The net heat exchange by convection between an individual and the environment during a specified period of time.[10]
Evaporative heat loss (E)	Body heat loss by evaporation of water (sweat) from the skin during a specified period of time, expressed as kcal, Btu, or W.[10]
Metabolic rate (MR)	Time rate of energy produced by metabolism. Example units are kcal/min and W/h.
Metabolism (M)	Energy produced by oxidation in muscles. When used in Equation 17.1, *M* is total metabolic energy of the person during a specified period of time.
Radiant heat transfer (R)	Heat exchange by radiation between two radiant surfaces of different temperatures.[10]
Thermoregulation	The maintenance of core body temperature within a narrow range.

REFERENCES

1. Hazen RM, Trefil J. Science matters: Achieving scientific literacy, 2nd ed. New York: Anchor; 2009. p. 33–41.
2. National Institute for Occupational Safety and, Health. Criteria for a recommended standard: Occupational exposure to hot environments, revised criteria 1986 DHHS(NIOSH) Publication No. 86-113. Cincinnati, OH: The Institute; 1986.p. 18–23 and Appendix B.
3. Brauer RL. Safety and health for engineers, 2nd ed. Hoboken, NJ: Wiley; 2006.p. 337–357.
4. Ramsey JD, Bishop PA. Hot and cold environments. In: DiNardi SR, editor. The occupational environment: Its evaluation and control, 2nd ed. Fairfax, VA: American Industrial Hygiene Association; 2003.p. 613–642.
5. American Conference of Governmental Industrial, Hygienists. Cold stress. In: 2010 TLVs and BEIs based on the documentation of the threshold limit values for chemical substances and physical agents & biological exposure indices. Cincinnati, OH: The ACGIH Association; 2010.p. 200–208.
6. Bernard, TE Occupational heat stress. In: Bhattacharya A, McGlothlin J, editors. Occupational ergonomics. New York: Marcel Dekker; 1996.p. 195–218.
7. American Conference of Governmental Industrial Hygienists. Heat stress and heat strain. In: 2010 TLVs and BEIs based on the documentation of the threshold limit values for chemical substances and physical agents & biological exposure indices. Cincinnati, OH: The ACGIH Association; 2010. p. 209–218.
8. Jensen RC. Workers' compensation claims relating to heat and cold exposure. Professional Safety. 1983;28(9):19–24.

9. Ramsey JD, Buford CL, Beshir MY, Jensen RC. Effects of workplace thermal conditions on safe work behavior. J Safety Research. 1983;14(3):105–114.

10. National Institute for Occupational Safety and, Health. Criteria for a recommended standard: Occupational exposure to hot environments, revised criteria 1986 DHHS(NIOSH) Publication No. 86-113. Cincinnati, OH: The Institute; 1986.Appendix A, Glossary.

Chapter 18

Fire Hazards

18.1 FUNDAMENTALS OF FIRE

Fires cause considerable harm, including death, injuries, air pollution, property damage from the fire, and property damage occurring during firefighting. Financial losses continue after a fire from business interruption, time to recover lost documents, medical costs for those harmed, temporary relocation costs, salaries paid to personnel who are doing recovery work rather than work contributing to business goals, and damaged reputation in the community. Not all of these diverse consequences fit neatly into the mainstream duties assigned to OSH professionals.

OSH professionals tend to focus on employee safety and health, but responsibilities often extend to other related business needs. Common collateral duties include security, product safety, workers' compensation, environmental pollution, waste management, visitor safety, liability insurance, and property/fire insurance. Thus, facility *fire protection* is one of many knowledge areas the OSH professional may need during a career. This chapter discusses how fires start, are sustained, and spread; describes how fires cause harm; and provides numerous examples of how the nine risk-reduction strategies are used.

18.1.1 Elements for Starting and Sustaining Fires

Fires are transition processes for states of energy. The transition involves changing potential chemical energy into other forms of energy through combustion. In combustion, a fuel containing stored chemical energy is oxidized. In most cases, an ignition source triggers the initial chemical reaction to release the stored energy. Fire processes may be modeled simply as

$$\text{Initiating elements} \rightarrow \text{Fire} \rightarrow \text{Outputs.}$$

Risk-Reduction Methods for Occupational Safety and Health, First Edition. Roger C. Jensen.
© 2012 John Wiley & Sons, Inc. Published 2012 by John Wiley & Sons, Inc.

The initiating elements consist of a fuel, an oxidizing agent, and heat for ignition. A *fire* may be defined as rapid oxidation of material during which heat and light are emitted.[1] Other common outputs are flames, smoke, and gases.

Many fires start and soon die out. What it takes for continued burning is known as a *chemical chain reaction*, which combines with the three ignition elements to sustain a fire:

1. Fuel,
2. Heat,
3. Oxidizing agent, and
4. Chemical chain reaction.

These four elements are sometimes depicted graphically as a four-sided geometric shape (tetrahedron or rectangle). Figure 18.1 illustrates the difference in the elements for igniting a fire and for sustaining a fire. Basic background on each of the four elements is provided below.

18.1.1.1 Fuel

Fuels come in so many forms that a standard classification system is used for everyday practice. The fire insurance industry uses these classifications when assessing the fuels in an insured property. Fire protection engineers use the classifications when designing fire sprinkler systems for buildings. Building owners use the classifications to select portable extinguishers to match the types of fuels in the area. The classifications used in the United States are listed in Table 18.1. All are based on the fuel burned except Class C, which consists of energized electrical equipment. It differs from the other classes in that electrical fires are the result of extreme heat from current flowing across gaps in the circuit, as discussed in chapter 15. The extreme heat can burn insulation and create a constant source of arcs that sustains nearby burning.

The physical state of the fuel affects its ability to ignite and to sustain a fire. The solid materials (mostly Classes A and D) can be found in large pieces as well as fine

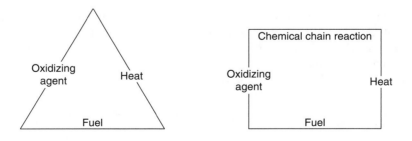

Fire ignition elements   Sustained fire elements

Figure 18.1   Elements for igniting a fire and for sustaining a fire.

Table 18.1   Fire Classes A, B, C, D, and K

Class	Contents	Features
A	Wood, paper, cloth	Relatively slow burning in initial stages. These fires leave ashes
B	Flammable and combustible liquids and gases	Fire develops rapidly. Examples are propane and gasoline
C	Energized electrical equipment	Equipment that is plugged in is considered energized even if the power switch is set to the off position. Examples are motors, appliances, and machinery
D	Combustible metals	Main concern is when in fine particle form. Special chemicals required to extinguish. Examples are magnesium, titanium, aluminum, and zirconium
K	Cooking oils	Newest of the fire classes. Formerly cooking oils were in Class B

dusts. A solid wooden beam is relatively difficult to ignite, whereas wood dust is relatively easy to ignite. The difference is explained by the ratio of surface area to mass—a higher ratio being easier to ignite. A second major factor is whether the fuel is in a gaseous, liquid, or solid state—gaseous being most easily ignited and solid being least. A third factor that plays an important role in the ignitability of fuels is temperature—a higher temperature makes a fuel easier to ignite, and the fire will burn with greater intensity.

The Class B fuels get extensive attention in the fire codes and in the practice of OSH. The obvious reason is that most of these materials are relatively easy to ignite—many having been produced to serve society as fuels. Most flammable and combustible (F&C) materials encountered in general industry are found in a liquid or gaseous state. Some are found in a solid state, such as those used as rocket fuel propellants. Industries have learned ways to transport, store, and dispense F&C materials, so the risk of unintentional ignition is quite low but not zero. When an unintentional fuel ignition occurs, it is usually the result of multiple failings in the established controls.

The actual burning of solids and liquids takes place within vapors on the fringes of the material. For example, when wood is heated, it gives off vapors. If the heat is sufficient, the vapors ignite, the combustion creates flame, the flame heats other parts of the wood, more vapors are produced, and the fire spreads. With F&C in liquid state, there is a layer of vapors just above the top surface. The concentration of this vapor is higher at the surface and declines with distance. Somewhere in this range, the concentration of vapor is within the *flammable range*. For each material, this range is defined by a lower flammable limit (LFL) and an upper flammable limit (UFL). When the vapors are ignited, the heat produced warms the liquid in the top layer of the liquid F&C material, causing an increased rate of liquid changing to vapor. If the temperature of the liquid is above its ignition temperature, the liquid continues to burn, constantly feeding the fire until all the liquid available is consumed.

18.1.1.2 Heat

Ignition sources are forms of heat energy. They may be grouped into four categories: mechanical, electrical, chemical, and nuclear. Figure 18.2 is a fault tree showing one way to visualize the role of the four fuel types in a sustained fire. Being under an OR gate means any one or various combinations of the fuels can sustain a fire. Each type of heat energy is discussed below.

Mechanical heat can develop from friction or compression. Fans in ventilation ducts provide an example of friction heat. The friction of rotating metal-to-metal parts can heat the metal enough to ignite flammable vapors in the ventilation duct. Diesel engines provide an example of compression heat. In the cylinders, each cycle involves compressing a mixture of air and diesel fuel until it explodes.

Electrical heating can develop in four ways: (1) resistance—too much current through the circuit causes heat buildup; (2) arcing—current flowing across a gap in the circuit; (3) static—electron flow between two surfaces having different charges; and (4) lightning—intense flow of electrons between a cloud and the earth.

Chemical heating can develop in four ways: (1) combustion—heat generated by a fire igniting nearby fuels; (2) decomposition—pile of materials subjected to an external heat source decomposes and generates heat; (3) spontaneous heating—decomposition rate of materials becomes rapid enough to initiate fire with no external source of heat; and (4) solution—mixing certain chemicals generates enough heat to start a fire.

Nuclear energy comes from changes in the nuclei of atoms. Both fission and fusion processes can occur in controlled conditions in which the heat generated by the reaction is continuously removed to maintain a heat level in the reactor core in the safe tolerance range. Without an effective cooling system, the heat would rapidly rise to a level high enough to ignite many objects in the area.

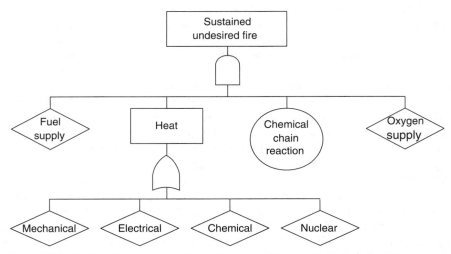

Figure 18.2 Fault tree depicting how a fire can be sustained by any of four alternative heat sources.

Regardless of which type of heat is involved, a fire will not start unless the amount of heat energy is sufficient. Under controlled laboratory conditions, a material is tested to determine the minimum temperature required to ignite it. The *ignition temperature* is useful for comparing different materials being considered for buildings, furniture, and consumer products. The amount of heat energy required to raise the temperature of a material to its ignition temperature is not so easily determined outside a laboratory. For example, consider that while you are at a station pumping gas into your car, you see another customer filling his car while smoking a cigarette. Should you worry? Is the amount of heat in his cigarette sufficient to ignite gasoline vapors?

18.1.1.3   Oxidizing Agent

Oxidation is an exothermic chemical reaction that releases the potential energy stored in a fuel. Oxygen is the most common oxidizing agent; other chemicals that function as oxidizers include chlorine, fluorine, hydrogen peroxide, nitric acid, sulfuric acid, and hydrofluoric acid.[2] These chemicals react by taking on electrons from a fuel and releasing oxygen during the combustion process. Common oxidation processes release energy at different rates, as follows:[3]

- Rusting is a slow form of oxidation.
- Metabolism is a faster form of oxidation than rusting.
- Fire releases energy at a much faster rate than metabolism.
- Combustion explosions release large amounts of energy in a very brief time.

18.1.1.4   Chemical Chain Reaction

The fourth element of a sustained fire is a chemical chain reaction. Think about being on a camping trip. You find a nice campsite, set up your tent, find some twigs and logs, and select a suitable fire pit. It helps to have some dry paper you can wad up for the base of the fire. On top of the paper, you lay your twigs. You continue by arranging on top of the twigs some thin-sized branches and small logs. You use a match to ignite the paper, creating a flame that heats the twigs. The twigs catch fire quickly, and you hope the branches will also ignite. With the base materials burning, the logs get hotter and begin to emit flammable vapors. If the materials were dry and arranged well, you will soon have a sustained campfire.

18.1.2 Fire Spread

Fires sustained by the four elements discussed above tend to spread from the burning fuel to unburned fuel. Spreading occurs in several ways, one being through direct flame contact (as in the above example, where the campfire spreads from the paper to the twigs and upward). Fire investigators looking for signs of arson seek visual signs of flames spreading upward from a single spot. Once they identify the spot of origin, they can hypothesize about how and why a fire could have started at that point. Other forms of heat transfer were discussed in chapter 17 on thermal energy but are revisited here from a different perspective.

When a building is on fire, convective heat transfer occurs when the heated air rises and fills the space near the ceiling. If the heated air in a burning room travels laterally to an adjacent room, it will carry the hot air to that room and possibly raise the room temperature enough to ignite materials, thereby spreading the fire. If the lateral flow is blocked, the rising hot air will push the cooler air down, setting up a circular flow within a space.

Radiant heat transfer occurs when two objects are separated by open space. The direction of the heat transfer is from the hotter surface to the cooler surface, like a campfire warming the campers. In hotter building fires, the hot wall of a burning building radiates substantial heat to adjacent buildings. The actual heat transfer is the result of electromagnetic waves initiated by the warmer object flowing to the surface of the cooler object. When the electromagnetic waves reach the surface of the cooler object, the energy converts to heat.

Forest fires spread through radiant and convective heat transfer. They radiate heat from burning trees to trees not yet burning, heating up and drying out the tree leaves, pine needles, pinecones, and branches. Air heated in the burning section spreads laterally and upward. The lateral spread heats up adjacent trees, and essentially prepares them for ignition. The vertical convective currents carry small embers from burning trees that, after reaching a plateau, tend to descend ahead of the fire spread onto dry tree parts that promptly ignite.

Conductive heat transfer occurs through a solid object or between two objects touching one another. This form of heat transfer plays a role in spreading fires within a building. Metal pipes are good heat conductors, so pipes between rooms can transfer heat from a room with fire to an adjacent room. The heat level can become sufficient to ignite materials with relatively low ignition temperatures. Of particular concern are flammable vapors. Thus, in rooms where flammable vapors might exist within their flammable range, metal pipes are recognized as a potential ignition source in need of appropriate controls.

18.1.3 Flashovers and Backdrafts: Threats to Firefighters

When fighting structural fires, firefighters face numerous threats to their safety and health, two of which are *flashovers* and *backdrafts*. Their respective progressions are briefly explained below.

Flashovers start with a fire in an enclosed room. Heat builds up until the ignition temperature of major materials in the room is reached. Suddenly, many things in the room ignite in a flash. The ignition creates a burst of expanding, hot air that can be fatal to an exposed firefighter.

Another special form of fire-originated explosion is known as a backdraft. Briefly, this starts with a fire in an enclosed room. All the room oxygen is consumed, leaving the space filled with smoke and hot gases with a high concentration of carbon monoxide—a flammable gas. An opening to the space occurs, perhaps when a firefighter opens a door. Oxygen enters the room. Very rapid burning of the fresh oxygen occurs, raising the room temperature enough to ignite the carbon monoxide and causing room gases to expand with explosive force. Such an event was dramatically depicted in the movie *Backdraft*.

18.2 MECHANISMS OF HARMING

Harm from fires starts with destruction of the fuel burned. Fire causes further harm by destroying valued property, causing smoke damage, and injuring people. In addition, when firefighters respond to a structural fire, they often need to break glass, chop openings in roofs, and spray powerful water streams into rooms, all of which cause property damage.

The fire by-products of heat, light, flame, smoke, and fire gases can be hazardous to both building occupants and firefighters. The respective mechanisms of harm are outlined below.

Heat. The heat from a fire transfers heat to surrounding areas, facilitating the fire spread. For human safety, heat is a problem if the fire develops so rapidly that some people are unable to avoid exposure to the intense heat. This scenario sometimes happens when people are trapped on one of the upper floors of a building. Heat from fires is also a key element in flashovers and backdrafts.

Light. The light produced by fire is not considered a serious threat to people.

Flame. Flame spreads fire and can cause extremely serious burns. An unfortunate but too-common scenario involves a person getting sprayed or soaked in a flammable liquid that ignites.

Smoke. Smoke damages the interior of a burning building and pollutes the air. More importantly for building occupants and firefighters, smoke reduces visibility. In many industrial fires, the fire burns materials like hydraulic fluid, with a by-product being very thick, dark smoke. The smoke initially rises to the ceiling. If there is no place for it to go laterally, the dark smoke fills the room from the top downward, thereby obscuring visibility to such an extent that people in the room cannot find an exit.

Fire Gases. Fire gases account for the largest number of fire fatalities. Two particularly common fire gases are carbon monoxide (CO) and carbon dioxide (CO_2). Exposure to CO disrupts the effective supply of oxygen to the brain and other body parts, which can lead to poor judgment and mental errors that make escape less likely. Exposure to CO_2 increases the rate and depth of a person's breathing, contributing to a faster rate of CO uptake.[3] These two gases, plus reduced oxygen concentration in the air, can cause unconsciousness and death from asphyxiation before the person manages to exit to a fresh air environment. Numerous other toxic gases can develop in fires as well, depending on the materials burned.

18.3 STRATEGIES AND TACTICS FOR FIRES

Risk reduction for fires involves the full spectrum of Haddon's three phases. During the pre-event phase, there are extensive methods for preventing a fire. This phase is the priority, but the subsequent two phases also afford useful opportunities for reducing risks. In the during-event phase, damage can be heavily influenced by an installed response system, evacuation procedures, and firefighter responses. During the post-event phase, there are many opportunities to minimize the financial losses to property and business. Also, getting high-quality medical care and rehabilitation services for anyone harmed in a fire can help limit the consequences for the victim and for any entity with financial liability. Including all three of the Haddon phases in a fire protection plan provides the most effective approach to minimize total risk.

Hot-work permit programs encompass several of the nine risk-reduction strategies. These are used in industries in which specific work operations generate heat capable of igniting a fire (e.g., welding and torch cutting). Some standards and codes list specific processes to include, but each company may include a broader list in its hot-work program. The idea behind hot-work permits is the requirement that jobs be planned before they are started. Planning includes inspecting the site, specifying the heat generating processes, removing or covering combustible materials, avoiding work if flammable vapors are present, using a person to stand by with a fire extinguisher and watch for any start of a fire, and following specific steps after completing the work. For the most part, the tactics incorporated into hot-work programs are the same as those used for fire protection generally, but some of the tactics listed in Table 18.2 are unique to hot work.

Risk-reduction tactics for fires and explosions are far too numerous to cover in a single chapter of a book. A more feasible approach is to refer readers to more extensive sources of information, particularly building codes, fire codes, and electrical codes used throughout the world. Some industrial sectors with high potential for major fires (e.g., chemical processing and petroleum refining) have created guidelines specific to the industry. In the United States, fire codes and electrical codes are developed and distributed by a voluntary organization—the National Fire Protection Association (NFPA)—while other countries may have their own codes and code-writing organizations. Whatever the source, these organizations are outstanding sources for information about what should be done to prevent fires, to protect structures, and to facilitate the safe exit of occupants. Some fundamental features in these codes warrant mention in this book on risk-reduction strategies for OSH.

A valuable approach in the fire codes is the classification of buildings according to their use and occupancy, which helps building owners to easily locate the provisions applicable to their buildings. For example, provisions concerning exits and related life-safety requirements differ considerably for industrial buildings, elementary schools, nursing homes, prisons, and high-rise office buildings. The classification approach is also used to classify areas within facilities. For rooms where flammable or explosive atmospheres might develop (i.e., hazardous locations), the engineering and architectural requirements spelled out in the codes are particularly rigorous. By

Table 18.2   Strategies and Tactics for Reducing the Risks of Fires

Risk-reduction strategy	Risk-reduction tactics
1. Eliminate the hazard	1a. Eliminate fuels
	1b. Eliminate heat sources
	1c. Eliminate oxidizers. Monitor to ensure goals are met
2. Moderate the hazard	2a. Limit fuel quantities
	2b. Substitute less flammable materials for more flammable materials
3. Avoid releasing the hazard	3a. Design and build systems to minimize risk of fuels, heat, and oxidizer coexisting
	3b. Install lightning protection on structures
	3c. Maintain very low oxygen content in a space
	3d. Store F&C materials well below or well above their flammable range. Monitor to ensure goals are met
	3e. Avoid static spark ignition by grounding and bonding containers involved in F&C transfers
4. Modify release of the hazard	4a. Provide automatic detection and response systems
	4b. Design buildings with venting system for fire gases and smoke
	4c. For hot work, provide a fire watch to spot and extinguish any fire before it grows
	4d. Have a pre-fire plan for how fire fighters will attack fire in different parts of the site
5. Separate the hazard from that needing protection	5a. Use space and/or barriers to separate operations of higher fire risk so that any fire damage will be confined
	5b. Maintain separation between F&C compressed gases and high-oxygen-content gases using distance and/or barriers (e.g., acetylene and oxygen cylinders)
	5c. Store small quantities of F&C materials in approved storage cabinets
	5d. Build firewalls to separate areas of fire risk from areas for people and valued property
	5e. Cover ignitable items in areas where hot work will be performed
6. Improve the resistance of that needing protection	6a. Construct facilities with fire-resistant materials
	6b. Cover structural beams with heat insulation to limit damage from heat of a fire
7. Help people perform safely	7a. Train personnel on fire prevention tactics
	7b. Conduct fire evacuation drills
	7c. Ensure fire exits are well marked and maintained
	7d. Provide remote controls for valves and other critical devices that might be needed during fire response

(*continued*)

TABLE 18.2 (*Continued*)

Risk-reduction strategy	Risk-reduction tactics
	7e. Provide appropriate training and equipment for personnel who may help with fire response
8. Use personal protective equipment	8a. Provide fire-resistant clothing for personnel who work with F&C materials
	8b. Provide appropriate PPE for personnel who may participate in response to a fire
9. Expedite recovery	9a. Establish and maintain a business continuity plan
	9b. Be prepared to render effective first aid to personnel affected by a fire

initially determining which classification applies to each room, the owner can have rooms designed and constructed according to code. Readers seeking more on the NFPA codes most applicable to OSH are referred to a book by Ferguson and Janicak.[2]

The strategies and tactics for reducing risks associated with fires are listed in Table 18.2. The numerous tactics for preventing unwanted fires in the pre-event phase may be summed up with the following principle: Avoid the coexistence of fuel, heat, and an oxidizing agent. Thousands of tactics for implementing this principle are found in the fire codes, building codes, and electrical codes. Table 18.2 lists some commonly used tactics for risk reduction. Supplemental comments on the tactics are provided in the following discussion.

The first strategy is to eliminate the hazard. For a particular building, vehicle, home, or other location, eliminating a fuel source is the most straightforward tactic. Eliminating the fuel during the initial design is obviously effective, and eliminating from an existing location achieves the same end. Eliminating heat sources can avoid risk of ignition. Eliminating the presence of oxygen is a tactic used in some confined spaces containing fuel. It involves replacing the oxidizing agent with an inert gas. This tactic is listed in Strategy 1 because it eliminates the risk of ignition. However, maintaining a zero-oxidizer environment is less reliable than employing the other two tactics in Strategy 1 because doing so requires several affirmative steps. The failure of any of the essential controls may result in a failure to maintain the desired atmosphere. Thus, this third tactic should be recognized as an elimination tactic contingent upon the maintenance of the unusual atmosphere within the space.

The second strategy is to moderate the hazard. The overarching tactic involves limiting fuel quantities through engineering and administrative processes. An example of the engineering approach is designing fuel storage facilities to limit the amount of fuel in occupied buildings. Preferably, large amounts of fuel are stored remotely from the occupied areas, with only enough for immediate needs stored in the occupied areas. Examples of the administrative approaches are to have and enforce housekeeping procedures for removing trash daily and providing self-closing containers for cloths contaminated with flammable organic liquids. Brauer provides a concise summary of the numerous specific engineering and administrative tactics found in the NFPA codes.[1] The second tactic for moderating the hazard involves substituting a less flammable material for a more flammable material.

Strategy 3—avoid releasing the hazard—equates to preventing a fire from starting. Five common tactics are listed in Table 18.2. The first is an overarching tactic calling for a facility design approach to minimizing the risk of fuels, heat, and an oxidizer coexisting. The second involves installing proper lightning protection, which will carry the current in a lightning strike into the ground, thereby avoiding the structure. Also, some lightning protection systems provide an invisible protective shield above the facility to make the facility less attractive to lightning. The third tactic involves maintaining the space so that the oxygen content stays below that needed for ignition. The fourth tactic is to store F&C materials so the vapor concentration is well above or well below the flammable range for the material. With this approach, a vapor monitoring device is usually installed to detect if the vapor content drifts outside the tolerance range. The fifth tactic is to avoid static spark ignition of F&C materials by grounding containers and following bonding practices when transferring F&C materials.

The fourth strategy is modifying release of the hazard. This broad strategy includes the various methods for responding to a fire. One of the most useful is installing automatic detection and response systems in areas with fire risks. Three commonly used approaches illustrate application of this tactic:

1. Automatic water sprinkler systems in facilities are useful for cooling a fire. These may actually extinguish a fire or limit heat buildup until qualified firefighters can respond.

2. Automatic detection and extinguishing systems are useful for protecting spaces where specific types of fires might start. These are designed to extinguish a foreseeable fire in the space by delivering a chemical agent that will kill the fire by removing one or more of the four elements of a sustained fire (see Figure 18.1).

3. Automatic detection–response systems in ducts containing flammable materials are intended to extinguish fires almost as soon as a fire starts. The system detects the start of a fire and automatically releases an extinguishing agent downstream to snuff out the fire.

Numerous other tactics are useful for modifying the release of the fire hazard. One is incorporating into facility designs some venting holes to allow smoke and fire gases to escape. Venting holes in roofs are effective for allowing hot gases and smoke to exhaust from the facility—making a safer environment for both evacuating the building and fighting the fire. Another tactic listed in Table 18.2 involves the use of a *fire watch* during hot work such as welding. The fire watch is a person responsible for monitoring the hot work to spot any unusual events. In particular, if a fire starts, the person is to promptly extinguish it with a portable extinguisher. Strategy 4 also includes the many tactics used by firefighters during a fire response. Instead of attempting to list all these tactics, a single tactic in Table 18.2 refers to having a pre-fire plan. The planning process requires working with the local fire department to mutually agree on how firefighters will attack a fire in various areas of the site. In the process, firefighting resource needs in the various areas should be identified and addressed.

The fifth strategy, to separate the hazard from that needing protection, is often implemented by providing space or a barrier. A commonly used tactic is to arrange industrial facilities so that processes posing higher risks of fire are kept separate from other facilities. Thus, if a fire develops, the damage will be confined to just one process area. Areas for filling the propane tanks used by forklifts are usually separated from the main facilities by space and a firewall. Workstations used for hot work are typically separated from common-use areas by space and barriers. Fuel product businesses design storage facilities (tank farms) with ample space between tanks so that one tank can burn without spreading the fire to neighboring tanks. Storing small quantities of F&C materials in approved F&C storage cabinets effectively separates the fuels from the heat that might develop if there is fire in the room. These cabinets provide an insulating barrier that keeps the temperature within the cabinet lower than the temperature of the external air. By this means, the air and fuels inside the cabinet will not reach ignition temperature as quickly as they would without the insulating barrier.

A second tactic is to store F&C compressed gases separately from oxygen cylinders, which greatly reduces the risk of a leaking fuel cylinder and leaking oxygen cylinder combining to create a highly flammable atmosphere. The most common example of this is the practice of separating stored acetylene cylinders from stored oxygen cylinders by distance, barrier, or both. This tactic also applies to thousands of industrial spaces in which flammable mixtures might be present. A major concern of these *hazardous locations*, as they are called in fire codes, is ignition by an electric arc or overheated wire. The codes classify the spaces based on the type of fuel in the room atmosphere. The more easily ignited the materials, the more rigorous the requirements for electrical equipment. If the room is in the easily ignited classification, electrical equipment needs to be enclosed appropriately (e.g., in a particular type of conduit) so that an arc or overheated wire will not directly contact the room air. If the room air is potentially explosive, the enclosures for electrical equipment need to account for a possible explosion within the enclosure. Appropriate equipment will allow the exploding gases to vent in a cooled condition to the outside atmosphere or other safe space.

Barriers are widely used tactics for Strategy 5. The firewalls in buildings function as barriers by resisting the spread of a fire in one room to an adjacent room. Wall materials are rated on the basis of how long they can continue their barrier function. The fire-resistant blankets used for hot work function as barriers by protecting combustible materials in the area from contact with sparks and other hot particles created by the hot work.

The sixth strategy is to improve the resistance of that needing protection. Architects use this tactic when specifying material for buildings and other structures. Many commercial materials are rated by nationally recognized laboratories for their fire-resistant properties. For larger buildings, constructed with steel beams, architects may specify that an insulating material be applied to the structural beams. This helps protect the beams from loss of strength due to the heat of a fire and can save on the cost of replacing beams after a fire. The insulation is not always adequate, as demonstrated in the 2001 World Trade Center disaster in New York City, when two commercial aircraft were intentionally flown into a pair of high-rise buildings. The building had insulated steel beams, but the intense heat of the fire from the burning jet fuel was too much. Before the fires ended, the beams softened and gave way, leading to the collapse of both buildings.[4]

The seventh strategy is to help people perform safely. The five tactics listed in Table 18.2 need no supplemental explanation.

The eighth strategy is to use personal protective equipment. In the oil and gas industries, many personnel are provided with outerwear made of fire-resistant material. This PPE provides protection from a brief fire for a substantial portion of skin. A second tactic involves providing PPE for personnel who may respond to a fire. Employee fire brigades are the usual group needing this level of PPE. Since these are company personnel, their safety and health is within the purview of the company OSH department. In the United States, an industrial fire brigade must meet several requirements to ensure that personnel are protected, which includes providing response team members with PPE suited for the type of responses they may need to make. If they are expected to enter burning structures to fight fire, they need gear equivalent to that provided to professional firefighters. If they will fight only small (incipient) fires, the PPE requirements are less demanding.

The ninth strategy is to expedite recovery. The most important tactic for expediting recovery involves developing and maintaining a business continuity plan. These plans are intended to help a company avoid bankruptcy, retain customers, and generally get back into business as soon as possible after a major disaster. A second tactic involves being prepared to render effective first aid for individuals injured in a fire, which should include plans for getting the more severely injured to appropriate medical care facilities.

LEARNING EXERCISES

1. List the four elements of a sustained fire.
2. Suppose you are at a gas station pumping gas into your car. You see another customer filling his car while smoking a cigarette. What information would you need to know in order to assess the risk of the cigarette starting a fire?
3. When a large, open tank of oil ignites, the flame spreads across the entire surface. An extinguishing method is to spread a foam agent all over the surface. Explain why this works.
4. Identify the three main sources of damage from a building fire.
5. Explain how smoke from a fire behaves.
6. Consider a room in an industrial facility containing several gas lines and gas cylinders. These normally operate without leaking any flammable material, so the room concentration is usually at zero. The room has an automatic

flammable vapor detector. The LFL of the material is 6%. The production manager recommends setting the automatic detector at 4.5% (three-fourths of the LFL). You learn from a colleague at another plant that they set theirs at 1.5% (one-fourth of the LFL). What are the pros and cons of each option?

7. What is the basic principle for preventing fire from occurring?

8. Which strategy applies to installed fire sprinkler systems?

9. Give two examples of using barriers for fire protection.

10. Explain the difference between fire prevention and fire protection.

TECHNICAL TERMS

Backdraft	A dangerous fire event resulting from a fire in an enclosed room consuming all the room oxygen, leaving the space filled with smoke and hot gases with a high concentration of carbon monoxide. An opening to the space occurs, allowing oxygen to enter the room. Very rapid burning of the fresh oxygen occurs, raising the room temperature enough to ignite the carbon monoxide and to expand the gas with explosive force.
Chemical chain reaction	The fourth element in a sustained fire. The chain reaction occurs within the material itself when the fuel is broken down by heat, producing chemically reactive free radicals, which then combine with oxidizers.[2]
Fire	Rapid oxidation of material during which heat and light are emitted.[1]
Fire protection	Various engineering approaches for helping protect people, property, and operations from fires and explosions.
Fire watch	A person assigned to continuously monitor hot work in progress, paying particular attention to any start of a fire or other deviations from normal.
Flammable range	Term used for F&C vapors to indicate the vapor concentration range in which burning can occur. Below this range, the vapor is too lean to burn. Above this range, the vapor concentration is too rich to burn. The flammable range is unique to each material.
Flashover	A dangerous fire event resulting from a fire in an enclosed room heating the room and its contents until the ignition temperature of the major materials in the room is reached. Suddenly, many things in the room ignite in a flash, creating a dangerous burst of expanding hot air.
Ignition temperature	The temperature above which a fuel will sustain combustion after being ignited by an external heat source.

REFERENCES

1. Brauer RL. Safety and health for engineers, 2nd ed. Hoboken, NJ: Wiley; 2006. p. 281–324.
2. Ferguson LH, Janicak CA. Fundamentals of fire protection for the safety professional. Toronto: Government Institutes; 2005.
3. Schroll RC. Industrial fire protection handbook, 2nd ed. Boca Raton, FL: CRC; 2002. p. 19–20.
4. Hamburger R, Baker W, Barnett J, Marrion C, James M, Nelson H. WTC1 and WTC2. In: McAllister T, editor. World Trade Center building performance study: Data collection, preliminary observations, and recommendations. Washington, DC: Federal Emergency Management Agency (FEMA 403); 2002. p. 2-1–2-12.

Chapter 19

Explosion Hazards

19.1 BACKGROUND ON EXPLOSIONS

A typical *explosion* involves a sudden expansion of material that produces outward pressure. The effects of the triggering event depend on the space available for expansion. If not fully contained, the material will rapidly expand, typically creating a loud noise and pushing hot gases outward. If fully contained, the explosion forces will exert great pressure on the inner walls of the containment vessel, which may hold firm, allow movement, or break and allow the gases to spew out in a violent manner.

Three common types of industrial explosions are (1) detonation of *explosives*, (2) unintended combustion explosions, and (3) mechanical explosions. This chapter concerns the first two. Chapter 20, on the hazards of pressure, addresses mechanical explosions such as pressure vessels rupturing and boiling-liquid expanding-vapor explosions (BLEVEs).

The detonation of explosives involves the use of explosive material and a means of detonating the explosion. It is used to break rock for mines, tunnels, and excavations and to demolish old structures. The whole process is highly regulated to ensure that the explosive materials do not fall into the hands of people with criminal intent. Regulations address manufacturing, transporting, selling, storing, using, and tracking.

Combustion explosions and ordinary fires have similar chemistry, but explosive events are much quicker and produce an outward pressure in the form of a high-pressure wave, expanding material, and heat. If the wave front has a velocity greater than the speed of sound, it is a true explosion; if it is less than the speed of sound, it is called a deflagration.

Dust explosions are a type of combustion explosion, the fuel being airborne particles of combustible material. Dust particles have a high ratio of surface area to mass, making ignition a quick process. These fires spread very rapidly from the origin space throughout the entire dust-containing space, raising the air temperature, causing

Risk-Reduction Methods for Occupational Safety and Health, First Edition. Roger C. Jensen.
© 2012 John Wiley & Sons, Inc. Published 2012 by John Wiley & Sons, Inc.

increased pressure, and creating turbulence. The turbulence often shakes loose dust that had settled on surfaces, adding airborne fuel that leads to a second explosion that is sometimes more harmful than the first.

Although combustion explosions have much in common with ordinary fires, some similarities and differences are worth noting. Both are exothermic chemical reactions. Both require the coexistence of three elements: fuel, heat for ignition, and an oxidizing agent. After initiation, a major difference is that deflagrations and dust explosions rapidly burn explosible material farther and farther from the origin. The resulting pressure wave and heat has destructive potential strongly influenced by the containment conditions, space available for dispersion, and fuel availability.

The focus of this chapter is on reducing the risk of explosions hurting employees and damaging business property. There is no discussion about the technology of creating explosives and explosive devices. Nor is there any discussion about reducing the risk of terrorists and other criminals using explosives to harm people or property. Also omitted from this chapter is a discussion of flashovers and backdrafts, because those were covered in the previous chapter on fires.

19.2 MECHANISMS OF HARMING

The result of an explosion depends a great deal on the containment. If in an open space or container, the expanding pressure and materials will travel outward, gradually decreasing in pressure and concentration. Typically, as the materials expand, they react with the air, causing a fireball. If in an enclosure, the expanding pressure and materials will exert great pressure on all internal surfaces of the enclosure. An enclosure designed and built to contain such explosions may do so without damage. In contrast, detonation of an explosive set within a drill hole for mining will break up the rock enclosure. Some enclosures have solid containment walls on all but one side. Such is the case with guns. The gunpowder detonation within the chamber creates an expanding gas that exerts intense pressure on the bullet, forcing it down the barrel. Some explosions in containment vessels break through a weak part of the vessel wall, allowing the expanding gases to spew out in a powerful and dangerous manner.

Explosions cause damage through the by-products of the exothermic chemical reaction. The main destructive by-products are a blast wave, a fireball, and hot gases. The blast wave consists of high-pressure, hot gases needing space for expansion. A fireball contains very hot gas expanding outward from the blast center. When the expanding gases are confined, the outward pressure of the blast wave can break through a weak part of the equipment. Once a small opening occurs, the rapid outflow of gases tends to expand the opening. The escaping gas can carry with it fragments of the broken equipment. The hot gas and fragments are propelled at high speed away from the reaction site. After that point, the potential for harm depends on who or what has the misfortune of being in the path.

Another hazard of explosions is damage to human hearing. A person close to an explosion can suffer damage to the eardrum and internal elements of his or her ear from the *acoustic trauma.*

19.3 STRATEGIES AND TACTICS FOR EXPLOSIONS

The NFPA identifies five strategies for controlling explosions,[1] which align with the strategies set forth in Table 19.1. The discussion in this section will supplement or clarify the tactics in Table 19.2. Because the focus herein is on strategies and tactics, readers seeking more detail may find it in other books for OSH professionals.[1–3]

The first strategy is to eliminate the hazard. For a particular location, eliminating a fuel that has the potential for explosion (e.g., explosives, other materials with known explosive potential) is the most straightforward tactic. It applies particularly to enclosed spaces where oxidation can produce a buildup of high-pressure gas that can be suddenly released when containment capacity is exceeded. Elimination during the initial design prevents the fuel from ever being introduced, and eliminating the fuel from an existing location achieves the same end. Eliminating or avoiding a heat source being near explosive material is another tactic listed in Table 19.2.

The second strategy is to moderate the hazard. The overarching tactic involves limiting fuel quantities through engineering and administrative processes. Two engineering approaches are to design fuel storage facilities to limit the amount of fuel in occupied buildings, and to design and build explosive material containers that, in the event of an explosion, will withstand the maximum pressure and contain the explosion. Preferably, large amounts of fuel are stored remotely from the occupied areas, and only enough for immediate needs are moved into the occupied areas. An example of the administrative approach is to enforce housekeeping procedures for removing trash daily and providing self-closing containers for cloths contaminated with flammable liquids. A second tactic involves regularly cleaning work areas, which limits the amount of dust available for combustion.

The third strategy is to avoid releasing the hazard. For explosives, a basic tactic is to strictly control the detonation. A related tactic is to ensure that explosive materials do not coexist with an ignition source at an unintended point in time. For dusts in confined spaces, putting inert gas into the air space above settled dust can effectively

Table 19.1 How the NFPA Control Strategies Match to the Risk-Reduction Strategies in This Book

NFPA control strategy	Strategy per this book
Containment	3. Avoid release of the hazard
Quenching	4. Modify release of the hazard
Dumping	4. Modify release of the hazard
Venting	4. Modify release of the hazard
Isolation	5. Separate the hazard from that needing protection

Table 19.2   Strategies and Tactics for Reducing the Risks of Explosions

Risk-reduction strategy	Risk-reduction tactics
1. Eliminate the hazard	1a. Avoid having explosive materials on-site
	1b. Avoid processes that temporarily create explosive mixtures
	1c. Eliminate a heat source near an explosive material
2. Moderate the hazard	2a. Limit fuel quantities by engineering and administrative processes
	2b. For combustible dusts, limit amount by frequent cleaning
3. Avoid releasing the hazard	3a. In vessels for containing explosive materials under specific pressure conditions, avoid an explosive release by having pressure-relieving devices on the vessel
	3b. Avoid static spark ignition by grounding and bonding containers involved in F&C transfers
	3c. Avoid ignition by hot objects by prohibiting specified objects in designated locations
	3d. Maintain oxygen concentrations of the air in contact with explosive materials at a low level by means such as displacement by an inert gas
	3e. For industrial vessels used to contain highly flammable gases at a pressure above boiling point, plan on the possibility of a leak releasing large amounts of material into the atmosphere. Avoid explosion by (1) a high level of monitoring vessel conditions and (2) isolating the vessel from heat sources
4. Modify release of the hazard	4a. Store explosive materials in vessels with vents designed to release upon internal pressure reaching a set point
	4b. Install explosion quenching systems capable of rapidly spraying water or dispersing a suppressant to stop further reaction
	4c. Design into equipment capability for dumping the reacting mixture into an area where it can be tolerated
5. Separate the hazard from that needing protection	5a. Store explosive materials in limited quantities separated by distance rather than storing a large amount in a single place
	5b. Separate multiple magazines for storing explosives according to standards
	5c. Locate far from populated areas any potentially explosive operations or storage facilities
	5d. For intentional explosions, use time and location to separate personnel and equipment from the products of the explosion

TABLE 19.2 (*Continued*)

Risk-reduction strategy	Risk-reduction tactics
	5e. Put barriers between explosive source and whatever needs protection
6. Improve the resistance of that needing protection	Design and construct items to withstand foreseeable explosive forces
7. Help people perform safely	7a. Provide remote controls for valves and other critical devices that personnel will need to use after an explosion
	7b. For crews setting explosives, provide SOPs, checklists, drug testing, and training
8. Use personal protective equipment	Provide personnel with body armor appropriate for foreseeable explosive projectiles
9. Expedite recovery	9a. Implement automated response systems before an explosion occurs
	9b. Implement procedures for personnel to follow after an explosion
	9c. Establish and maintain a business continuity plan

keep the partial pressure of oxygen below that needed for ignition. For dust that has settled on surfaces, avoiding turbulence helps keep the dust in a settled form. For airborne dusts, adding moisture to the air raises the ignition temperature.[1]

For flammable liquids, keeping vapor concentrations outside the flammable range will avoid a deflagration. The reliability of this tactic increases when automatic monitoring devices are used to detect and respond whenever the vapor concentration drifts outside the desired operating range, which is generally set well below or above the explosive range. Avoiding spark ignition in locations where flammable vapors may be present is a practice to emphasize. Some examples are worth mentioning. First, for many years it has been known that static sparks are common in flammable liquid transfer sites. A standard practice to avoid an explosion is to ground one or both containers and to bond the two containers. Second, a newer source of sparks has evolved with the use of electronic vehicle keys. In one case, a plumber completed a house call, returned to his van, pressed the remote door unlock button, and the van exploded. Apparently, while he was in the house working, a leaking acetylene tank in his van created an atmosphere of vapors in the flammable range. When he used the vehicle remote key to unlock the doors, a tiny spark in the vehicle was enough to ignite the acetylene.

Industrial vessels designed for containing flammable material capable of exploding commonly incorporate pressure-relieving devices. These normally hold the pressure within the system. When the pressure exceeds a set point, these devices open to allow some gases and vapors to escape. After letting out the excess material, most of these devices will reclose. The function of the open–reclose-type devices is to deal with deviations under normal operating conditions. A particular type of pressure-relieving device, known as *rupture disks*, is set to open at a higher pressure than the open–reclose type of pressure-relief devices. The rupture disks' opening pressure is close to that expected when an explosion is imminent. Thus, rupture disks may be considered a last line of defense against a full explosion.[4]

Some highly flammable materials that are gases under normal atmospheric conditions are stored and transported under pressure to keep the material in liquid form. A rare event needing attention is the possibility of a leak releasing large amounts of material into the atmosphere. Because these materials are kept in a liquid state by pressure, if a leak develops, the pressure is lost and the material escapes in a gaseous state. The cloud of material may float over a heat source such as a flare or hot stack. Two preventive tactics are

1. Conducting a high level of monitoring to prevent conditions that could lead to a leak (e.g., corrosion, abnormal stresses, mechanical shock), and

2. Isolating the vessel from heat sources by locating it far from flares and other heat sources.[5]

The fourth strategy is modifying release of the hazard. Tactics for this strategy involve preparing for the remote possibility that an explosion will occur. Appropriate planning and implementation will minimize the effects of an explosion. One way is to build into the container places for venting the material in acceptable directions. This venting tactic is useful for industrial buildings that house a process with potential for an explosion. Outward opening vent holes, skylights, windows, and doors can allow the hot expanding gases to escape and may prevent destruction of the structure. Saving the basic building structure will greatly reduce the time needed to repair the facility and return to operation.

Another common way to modify release of an explosion applies to air movement ducts that contain combustible dusts such as coal and wood dust. Experience has shown that fires occasionally break out in these ducts. The small initial fire can quickly develop into a full deflagration, blowing apart large sections of ductwork and associated processing equipment. Nearby people can be injured or killed by projected material. Fire detection and response systems are available commercially that provide monitors throughout the ducts. These explosion suppression systems have detection and response elements. Detection may be based on changes in emitted light characteristics increase in air pressure. If the detectors sense ignition in one part of a duct, a water spray or other extinguishing agent will be released downstream of the site within 0.1 s.[5] The release usually snuffs out the embryonic explosion before it fully develops.

The fifth strategy is to separate the hazard from that needing protection. Several widely recognized tactics fit this strategy. One is separating high-oxygen-content gases and fuels (e.g., flammable and combustible materials). The standard practice of storing cylinders of acetylene and oxygen either separately or with a solid barrier between the cylinders is an example. Strategy 5 includes many common practices for storing explosive materials, starting with a preference for storing these materials in smaller quantities at multiple sites, rather than in a large quantity at a single site. A second widely used tactic applies to storing munitions.

The storage sites, known as *magazines*, have specific standards to minimize the risks associated with explosions. In addition to provisions for minimizing explosions in individual magazines, the standards specify a minimum distance between the magazines in order to prevent an explosion in one from starting a chain reaction through other magazines on the site.

On a broader scale, decisions for locating industrial facilities regularly consider the potential for explosions and other major incidents. If a significant explosion is foreseeable (although considered unlikely), locating the facility far from a populated area is advisable. Barriers are often erected between industrial facilities that have a potential for explosion in order to avoid one explosion causing damage to adjacent buildings. The separate-by-location tactic, combined with timing, is used in mining. Between the time an explosive charge is set and detonated, time is provided for all miners to clear the area. Various procedures are followed to confirm the separation of all miners before detonation proceeds.

The sixth strategy is to improve the resistance of that needing protection. Any sort of man-made physical object that needs protection from an explosion may be designed with the goal of being able to withstand foreseeable explosive forces.

The seventh strategy is to help people perform safely. After an explosion, it is not uncommon to find that the resulting fire is being fed by flammable or combustible fluids coming out of busted lines. A tactic for helping people to perform safely applies to process designs that incorporate remote controls to allow personnel to control the flows without going into a dangerous location. Personnel who work with or around explosive materials—especially crews who work with intentional explosions such as those in mining, construction, and structural demolition—need awareness training and specific training for performing their work properly. These crews depend on each other, so numerous procedures and regulations govern their jobs. It makes sense that these regulated jobs include qualification standards including a criminal background check and substance testing. The employers help these crews perform safely by providing standard operating procedures, task checklists, and extensive training.

The eighth strategy is to use personal protective equipment. If personnel may be at risk because of their proximity to an explosive device, body armor can provide some protection. This tactic is used in war zones where soldiers might encounter exploding grenades, roadside bombs, or other explosive devices that spray shrapnel. For applications outside military operations, projectiles from explosions often hit personnel with an energy that exceeds the design limits of ordinary hard hats and safety glasses. Nevertheless, these safety devices may somewhat reduce the severity of injury from small explosions nearby and large explosions farther away. For example, the use of safety goggles in chemistry labs is standard practice because they provide protection in the event of a spill or small explosion. When developing a risk assessment, it may be tempting to note the use of head or eye PPE as a tactic for reducing the severity of harm from an explosion. However, the RA team should think through the actual value of such devices before assuming they will be effective for a particular explosion.

The ninth strategy is to expedite recovery. The tactics listed in Table 19.2 involve engineering approaches implemented before explosions occur. One is limiting fire damage. The fireball and hot air from an explosion often ignite other materials in the area. Having an installed sprinkler system in the location where an explosion is most foreseeable may prove useful for limiting the extent of damage from the core explosion. However, the sprinkler system may prove useless if significantly damaged by the explosion. Having installed sprinklers in adjacent areas might limit the fire damage by either cooling the area or extinguishing an incipient fire in those areas. Another tactic is an automated system for responding to the problems an explosion can create. An automated system could have an emergency power system, an automated emergency control system, and a system for qualified personnel to control affected operations.[5]

Other advance plans and systems for limiting damage after an explosion and ensuing fire are noted in the previous chapter on fires. Some notable ones identified by Baasel include installing remote control valves to allow personnel to isolate equipment and areas of the plant, blowdown tanks in remote areas of the plant for transferring materials removed from fire areas, drainage systems to remove liquid spills, and a system of interlocks to prevent materials from flowing in the wrong direction at any time.[5] Emergency plans for the response of personnel should include the response to an explosion. Another fundamental tactic is to have a business continuity plan ready to implement right after the explosion. All these tactics provide ways to prepare in advance for mitigating the damage from an explosion and facilitating recovery.

LEARNING EXERCISES

1. What are the two common types of industrial explosions discussed in this chapter?

2. What are three common by-products of combustion explosions?

3. Ductwork explosion suppression systems for avoiding dust explosions have great value, but they are not free of problems. For many systems, after each discharge, the system needs to be shut down for cleaning and discarding the affected material, which interrupts production. Many production supervisors order that the devices be turned off because of the numerous interruptions. If you were the OSH manager at such a facility, what would you recommend as a better solution?

4. A problem surfaced with the U.S. military operating in Afghanistan. Roadside bombs were exploding under vehicles carrying troops (Humvees). Troops were sustaining fatalities, loss of lower limbs, and other serious injuries from projectiles coming upward through the floor. Investigators recommended that the vehicles be equipped with a metal plate under the passengers to protect them. Into which strategy does this tactic belong?

5. Of the nine strategies in Table 19.2, which take effect in the Haddon phase called the "during-event" phase?

6. Look up the words *deflagration* and *detonation*. What characteristics make them different?

TECHNICAL TERMS

Acoustic trauma	Injury to the sensorineural elements of the inner ear from a single noise burst or by direct trauma to the head or ear.
Explosion	A sudden change in material that produces outward pressure from the material expanding, a pressure wave, and heat. The effects of the expanding pressure depend on the space available for expansion.
Explosives	Substances containing a large amount of stored energy and used to create explosions.
Magazines	Places for storing ammunition, bombs, and explosives.
Rupture disks	A metal membrane designed and manufactured to burst at a certain pressure and temperature to prevent overpressurization of the attached vessel.[4]

REFERENCES

1. Ferguson LH, Janicak CA. Fundamentals of fire protection for the safety professional. Toronto: Government Institutes; 2005. p. 35.
2. Brauer RL. Safety and health for engineers, 2nd ed. Hoboken, NJ: Wiley; 2006. p. 325–335.
3. Hammer W, Price D. Occupational safety management and engineering, 5th ed. Upper Saddle River, NJ: Prentice Hall; 2001. p. 425–439.
4. Wilson AT. Troubleshooting field failures of rupture disks. Chemical Engineering. 2006;113(13):34–36.
5. Baasel WD. Preliminary chemical engineering plant design, 2nd ed. London: Van Nostrand Reinhold International; 1990. p. 95–96.

Chapter 20

Pressure Hazards

20.1 OVERVIEW OF PRESSURE HAZARDS

In the course of studying this chapter, astute readers will notice that pressure, fire, and explosions are interrelated. The three chapters on these topics (chapters 18–20) discuss numerous incidents where two or more of these hazards come into play. Flashovers and backdrafts originate in building fires, progress by increasing pressure, and culminate in an explosive blast. Some explosive events start with a fire under or beside a liquefied petroleum gas tank, causing the liquefied gas to expand, increasing internal pressure, and eventually blasting through a weak part of the tank walls. Thus, the author considered various options for organizing these three topics before deciding to have separate chapters on fires, explosions, and pressure hazards.

Pressure is a form of energy because it can perform work by raising a mass. Aspects of pressure energy relevant to OSH professionals are grouped into two categories for this chapter. The first group relates to the challenges of working in unusually high- or low-pressure atmospheres. The second group encompasses pressure systems used in industry, such as compressed gas cylinders, boilers, chemical processing vessels, hydraulic systems, and pneumatic systems.

20.1.1 Unusual Pressure Atmospheres

Conditions for people working under water use elevated air pressure to maintain a space capable of resisting the external pressure of the water. Without the high air pressure, external water tends to find pathways into the space. In addition, the potential for implosion cannot be ignored. People work under water for various reasons. For example, in the building of bridges over large rivers and inlets, abutments to support the spans must be set on a solid footing, which is often deep under the water level. That requires establishing a temporary underwater environment for the foundation work. Similarly, creating tunnels under waterways requires sealed work areas maintained at high pressure. Professional divers provide a third reason for

Risk-Reduction Methods for Occupational Safety and Health, First Edition. Roger C. Jensen.
© 2012 John Wiley & Sons, Inc. Published 2012 by John Wiley & Sons, Inc.

working under water. Those who dive deep need equipment capable of maintaining a pressurized microenvironment inside their diving suit.

When people change from a high-pressure to a low-pressure environment, the body needs to adjust slowly. Making the change too quickly can cause decompression sickness and similar disorders. Another hazard of high-pressure environments is the air mixture of oxygen and nitrogen. Only the proper mixture ensures that the workers have proper breathing gases.[1]

Work at very high elevations challenges the cardiovascular system to deliver enough oxygen to the brain and other vital organs. Examples include constructing roads and hiking paths in high mountains. Aircraft crews regularly work at high elevations, but the tactic of pressurizing the aircraft interior normally protects them from the effects that would otherwise exist.

20.1.2 Pressure Used in Industry

Many industrial processes include pressure vessels of various kinds. The material contained may be in a liquid or gaseous state, and maintaining that state within specified temperature and pressure ranges is essential. In the early days of boilers, there were many explosions due to internal pressure exceeding the capacity of the vessels. The field now known as safety engineering originated at that time primarily to fix the boiler explosion problem. Since those early days, many practical devices have been invented and incorporated into pressure-vessel standards. These devices and standards have greatly reduced the rate of pressure-vessel explosions; however, continued diligence is required to completely eliminate these explosive events.

Boilers are not the only kind of pressure vessels that can burst open from excessive internal pressure. Bursting vessel explosions can occur when the pressure inside any fully enclosed vessel exceeds the capacity of the weakest part of the vessel walls. When pressurized material breaks through the vessel walls and expands, damage can result from the pressure wave, heat, or projected fragments from the vessel.

If the pressure vessel contains highly flammable gases at pressure above the boiling point (usually maintained in liquid state), another hazard needs attention. A significantly large leak or discharge can send a cloud of flammable vapor into the air. Once airborne, the material may encounter an ignition source and explode. The larger the release, the greater the explosion.

The compressed gas cylinders used extensively in industry for a variety of gases come in several standard sizes and shapes. They have a portal to allow filling and dispensing. The most vulnerable part of compressed gas cylinders is the stem. If the stem breaks off a cylinder filled with compressed gas, the exiting gas rapidly contaminates the local air and propels the cylinder like a rocket. If a leak develops in the stem or related hardware, the escaping gas contaminates the local air and may go undetected.

Industry makes extensive use of pressure in the form of hydraulic and pneumatic systems. Both consist of a device to compress a fluid, an enclosed network of hoses or pipes, and various controls. The fluid in hydraulic systems is liquid, and that in pneumatic systems is gas. The hazards of the compressed fluid are normally controlled through proper design, connections, and maintenance. When these controls fail, the release of pressurized fluid can be quite harmful.

20.2 MECHANISMS OF HARMING

The mechanisms of harm resulting from pressure are described in three subsections: work in high- or low-pressure atmospheres, pressure used in industry, and direct pressure on the worker.

20.2.1 Working in Unusual Pressure Atmospheres

Working in environments with low atmospheric pressure presents concerns about oxygen delivery to the brain and other vital organs. The physiology is explained in many books on physiology, aviation medicine, and occupational health. Stated briefly, as the altitude of the workplace increases, the capacity of the blood to deliver oxygen throughout the body declines.[1] The effect on workers doing light to moderate work is rather minimal from sea level up to about 2000 m, but for heavy labor the effects are significant at lower elevations. Workers adjust to higher elevations by increasing their breathing rate and heart rate. Additional adjustment may come from slowing the pace of work and taking more breaks.

Working in environments with high atmospheric pressure presents concerns about decompression sickness and variations known as caisson disease, the bends, and dysbarism. This occurs when a person changes from a high- to a low-pressure atmosphere, as when someone who has been working under water rises toward the surface. If the change is too rapid, gas bubbles form in the blood and other tissues. The bubbles can cause joint pain and alter the normal functions of the central nervous system and circulatory system.[1,2]

Implosions occur when a large pressure difference between the pressure outside and inside a closed container exceeds the strength of the container. Usual situations for implosions are when the internal pressure is air or a near vacuum and the external pressure is water or other liquid. For underwater exploration and submarines, structural designs are for specified depths; going deeper increases the risk of an implosion.

20.2.2 Pressure Used in Industry

When a pressurized container bursts, damage may come from several sources. The expanding gas can directly affect the personnel who breathe it. The rapid airflow of the expanding gas can send objects flying, lift flammable dust from surfaces, and blow small particles into the eyes of workers. When the container breaks apart, bolts, rivets, and various fragments may become fast-moving projectiles capable of harming personnel and equipment. Finally, when the valve of a compressed gas cylinder breaks off, the cylinder becomes a heavy, rocket-like projectile.[2]

There are several distinct types of pressurized-container incidents. One type that every OSH professional should recognize is called a boiling-liquid expanding-vapor explosion (*BLEVE*).[3] These events can occur with a normally gaseous material that is held in liquid state within a pressurized container. The only reason the material stays liquid is the high pressure. The container could be a railroad tank car, a highway tanker truck, or a stationary storage vessel. The initiating event is a rise in temperature, often due to a fire outside the vessel. The heated material creates more vapor pressure in the container. At some point, the weakest part of the vessel wall fails. The vapor that escapes through the small initial opening will rip open more of the vessel wall and allow a gush of very high-velocity, high-pressure gas to shoot outward, devastating a large area in its path. The sequential model below may help readers remember the events:

$$\text{Boiling liquid} \rightarrow \text{Expanding vapor} \rightarrow \text{Explosion.}$$

The power for pneumatic and hydraulic equipment comes from gas and liquid pressure, respectively. Both are created by specialized equipment and distributed through solid lines or flexible hoses.[1,2] When flexible lines are used, the failure of a connection can lead to fluid flowing through the hose and causing the hose to whip about. The whipping action can injure a person or damage workplace hardware.

The air in *pneumatic systems* is pressurized by passing through an air compressor. It then flows through solid lines or flexible hoses to equipment. Compressors come in many capacities—some of which are placed in a fixed location and others are designed for portability to various worksites. Uses of compressed air are many, including spray painting, tire inflating, cleaning, rock drilling, and powering pneumatic tools. The lines can develop leaks, but the problem most often mentioned in the OSH literature is misapplication of the air through the nozzle. Workers have died from air being shot into a body orifice and expanding within the body.[1,2] Another recognized hazard involves cleaning surfaces by spraying them with pressurized air. This process makes small particles airborne, and the particles can get into eyes. Spray cleaning this way is allowed if appropriate precautions are taken to protect the eyes of the sprayer and coworkers.

The fluid in *hydraulic systems* is a liquid substance specifically made for that use. If a leak develops, the fluid squirts out with high pressure and velocity. Both velocity and pressure decline as the material gets further from the leak point. If a person is near the leak, the fluid can penetrate the skin and enter the body. From there, it can have serious effects on health and can even cause death. In addition, because the fluid is flammable, if it leaks onto or near an ignition source, it can start a fire. The leak will continue to feed the fire until pressure to the line is closed or all fuel is exhausted.

20.2.3 Direct Pressure on the Worker

Numerous occupational injuries and fatalities result from direct pressure on the worker's body. This occurs when a worker is close to an explosion or lightning strike.

The pressure waves can damage the worker's hearing. Another way this happens is when the human body gets squeezed between objects. Squeezing on the chest can prevent inhalation and cause death from asphyxiation. This occurs when, for example, a person's body is caught between a floor and a heavy object, in a chute such as at the bottom of a grain bin, or between a machine and a fixed object like a wall or ceiling.

A case investigated by the author involved a worker who was told to take a scissor lift from the third floor to the first floor. This lift was drivable from the platform. He rode it into the freight elevator, parked near the elevator controls, leaned over the rail, and pressed the elevator down button. In the process, his torso apparently pressed the up button of the lift. The man's chest became trapped between the rail and the elevator ceiling, and his torso blocked access to the up and down controls of the lift. The elevator reached the first floor and sat there briefly before a coworker on the first floor opened the elevator and found the victim still alive but unable to breathe, talk, or release the up button of the manlift. The coworker could not figure out how to release the pressure and lower the manlift and yelled for help. Other workers arrived, and none could figure out how to release the pressure. Someone tried to cut the side rails with a welding torch. It took too long, and the man suffocated.

20.3 STRATEGIES AND TACTICS FOR PRESSURE-RELATED HAZARDS

Strategies and tactics for reducing the risks of pressure-related hazards are presented in two parts. The first addresses unusual atmospheric pressures, and the second concerns pressure used in industry.

20.3.1 Work Under Unusual Atmospheric Pressure

Unusual atmospheric pressures include too much and too little pressure. Table 20.1 summarizes strategies and tactics used to minimize the risks of work under high or low atmospheric pressures.

Strategy 1 addresses efforts to eliminate human exposure to high or low atmospheric pressures. For deepwater exploration, technological advances have successfully eliminated the need for human presence in many diving vessels by using robotics and remote controls. Media coverage of the 2010 Deepwater Horizon oil spill disaster in the Gulf of Mexico provided the public with a glimpse of these capabilities. For construction workers doing underwater work in pressurized environments, the elimination strategy applies to work performed on the surface rather than in the high-pressure environment. The one low-pressure tactic noted in Table 20.1 is aircraft flying at low-enough altitudes to eliminate the need for cabin pressure and oxygen control.

The second strategy involves moderating the hazard. For underwater work, the pressure level inside a structure should not be greater than that necessary to meet engineering requirements for structural integrity and leak protection.

Table 20.1   Strategies and Tactics for Reducing the Risks of High- and Low-Pressure Atmospheres

Risk-reduction strategy	Risk-reduction tactics
1. Eliminate the hazard	1a. For deepwater exploration, eliminate use of humans by using robotics and remote controls
	1b. For caisson work, perform as much as feasible on the surface rather than in a pressurized environment
	1c. For air travel, fly at lower elevations where there is no need for cabin pressure and oxygen control
2. Moderate the hazard	For work in pressurized environments below water level, set pressure level no higher than that needed for engineering requirements
3. Avoid releasing the hazard	For high-altitude aviation, avoid cabin pressure release through engineering design, manufacturing quality, and maintenance
4. Modify release of the hazard	After work in high-pressure location, release gases in blood slowly by decompression procedures
5. Separate the hazard from that needing protection	Not applicable
6. Improve the resistance of that needing protection	Not applicable
7. Help people perform safely	7a. For work at high elevations, adjust work-pace expectations for new personnel to allow time for acclimation to the atmosphere
	7b. For work at high elevations, allow more rest time for heavy manual tasks
8. Use personal protective equipment	For high-altitude flying, provide emergency breathing air for crew and passengers in case of loss of cabin pressure
9. Expedite recovery	Monitor workers throughout decompression. If symptoms develop, put workers in a medical chamber

The third strategy is to avoid releasing the hazard. The tactic in Table 20.1 relates to the various engineering and manufacturing efforts that contribute to aircraft cabins being able to hold the air pressure at levels needed by the crew and passengers.

The fourth strategy is to modify release of the hazard. For workers in pressurized environments, decompression procedures have been established to moderate the hazard of gas bubbles forming rapidly in the blood when decompressing. Having workers proceed through decompression chambers slows their physiological adaptation and allows most workers to cope with the hazard.

The seventh strategy is to help people perform safely. Two tactics in Table 20.1 relate to work performed at high altitudes. The first tactic is to adjust the work-pace expectations for new personnel to allow time for acclimation to the atmosphere.

The acclimation process of increasing hemoglobin levels in the bloodstream takes time, with considerable progress made in the first week followed by modest progress during the subsequent three weeks. For manual labor-type work at high elevations, the employer may allow more frequent rest breaks than would be normal for work performed closer to sea level. This tactic applies to both new employees and those who have been working at the elevation long enough to become acclimated.

The eighth strategy is to use personal protective equipment. The one tactic listed in Table 20.1 applies to maintaining cabin pressure in high-flying aircraft. If that fails, the backup is to provide individual breathing masks fed by an appropriate mixture of oxygen and nitrogen.

The ninth strategy is to expedite recovery. The tactic noted in Table 20.1 involves monitoring individuals as they proceed from a pressurized environment through decompression. The monitoring should detect if someone is not decompressing properly and needs medically supervised decompression, such as recompression followed by extra-slow decompression.[2]

20.3.2 Pressure Used in Industry

Table 20.2 links strategies to tactics for reducing risks associated with the use of various pressures in industry. For the first strategy of eliminating the hazard, one tactic would be to consider eliminating from the worksite pressurized vessels such as compressed gas cylinders and pneumatic and hydraulic systems. The tactics listed for the second strategy, moderating the hazard, are self-explanatory.

For the third strategy, avoiding release of the hazard, seven tactics are listed in Table 20.2. The tactic of using solid lines rather than flexible lines where practical comes from Hammer and Price.[2] They note that solid lines are less prone to whipping when they come loose. However, because solid lines are not completely free of that hazard, they recommend securing solid lines at points where they are most vulnerable to breaking. Regular inspections and maintenance also reduce the risks of pressure releases from failure in hydraulic and pneumatic systems. The third tactic is to depressurize systems before working on them—a standard part of hazardous energy control (lockout/tagout) programs. Many serious injuries and fatalities have resulted among maintenance workers and others who have been surprised by a pressure release. Part of this tactic involves ensuring that all pressure has been released. Employers with high aspirations for safety include a requirement to formally test the system to confirm that all pressure has been removed. Another tactic, applicable to compressed gas cylinders, is to follow the gas industry procedures for working with gas cylinders. Those procedures emphasize the behavioral and administrative practices rather than the engineering approaches.[2] Most of the engineering work has already been done by the gas industry, so cylinder users need to focus on procedures.

The last tactic in Strategy 3 concerns the importance of monitoring industrial processes to detect any deviations from the design parameters. In some industrial

Table 20.2   Strategies and Tactics for Reducing the Risks of Pressure Used in Industry

Risk-reduction strategy	Risk-reduction tactics
1. Eliminate the hazard	Completely avoid having any pressurized gas cylinders, pressurized vessels, pneumatic power, or hydraulic power at the worksite
2. Moderate the hazard	2a. For air hoses with nozzles, install nozzle that limits pressure to safe level[a]
	2b. For the whipping hazard of flexible hoses, make the hoses only as long as necessary[2]
	2c. For gas cylinders containing toxic materials, bring on-site only the amount needed for short-term usage
3. Avoid releasing the hazard	3a. For the hose-whipping hazard, use solid rather than flexible lines where practical
	3b. For hydraulic and pneumatic lines, conduct regular inspections and maintenance
	3c. For pressurized systems of all kinds, avoid working on them unless sure that all pressure has been released[2]
	3d. For powerful equipment with extension capabilities (e.g., scissor lifts), install a flange around extension/retraction controls to avoid unintended activation
	3e. For gas cylinders, follow the gas industry procedures
	3f. For pressurized vessels in industrial processes, monitor to ensure operations are within design limits
4. Modify release of the hazard	4a. For hydraulic and pneumatic lines, install metal guards at locations most vulnerable to failure
	4b. For the whipping hazard of flexible hoses in pressurized systems, install restraints to limit whipping[2]
	4c. Allow air hoses for cleaning only when precautionary procedures are followed
	4d. For water hammer hazard to liquid lines, install an air chamber or accumulator in the line[2]
	4e. For pressurized vessels, incorporate pressure-relieving devices
5. Separate the hazard from that needing protection	5a. For cylinders containing fuels, store outdoors in area away from heat sources
	5b. For storing oxygen and acetylene cylinders, store with appropriate separation
6. Improve the resistance of that needing protection	Secure protective caps on gas cylinders when not in use or while being transported
7. Help people perform safely	Train personnel on safe procedures for working with pressurized systems
8. Use personal protective equipment	Provide PPE indicated in JHA or RA

TABLE 20.2 (*Continued*)

Risk-reduction strategy	Risk-reduction tactics
9. Expedite recovery	9a. Install remote controls for hydraulic and pneumatic lines to allow stopping the pressure without local exposure 9b. For machines that might squeeze a person against a solid object, install a device to sense abnormal resistance and interrupt the drive mechanism when needed

[a] Air pressure levels allowed for cleaning clothing and surfaces vary by jurisdiction. When used appropriately, the United States allows using air pressure if less than 30 psi (207 kPa). Some Canadian provinces disallow cleaning with air pressure, while others allow using air pressure if less than 10 psi (69 kPa).

processes, the concern is too much heat, too much pressure, or both. In others, the concern is too little heat, too little pressure, or both. Gas laws tell us that when a gaseous material is in a vessel, a rising temperature coincides with increasing pressure. For pressurized vessels, the main concern is too much pressure. Too much pressure can lead to a leak or an explosion. Having and following pressure limitations reduces the risk of an explosion. Although this is standard practice spelled out by the engineers who designed the vessel, things occasionally go wrong. The post-startup maintenance and modifications are often identified as primary causes of pressure-vessel problems. The technical tools are well known for ensuring that operations are within design limits, but, unfortunately, too often people develop overconfidence in the technology and become less diligent in the follow-up processes.

The fourth strategy is modifying release of the hazard. The first two tactics, concerning hydraulic and pneumatic lines, address risk reduction in case a line failure starts releasing high-pressure gas or fluid. The third concerns misusing air hoses to spray-clean surfaces (a usage that may be acceptable if appropriate precautions are followed). The water hammer hazard noted in Table 20.1 can occur when a flow valve in a fluid line is opened rapidly rather than slowly, which causes the line to shake severely, possibly damaging the connections. An engineering approach is to install an air chamber or *hydraulic accumulator* in the line to temporarily store suddenly released fluid.[2]

The last tactic in Strategy 4 concerns the numerous pressure-relieving devices, brief descriptions of which are provided in books by Brauer and by Hammer and Price.[1,2] Brauer's list includes safety valves, relief valves, safety-relief valves, frangible disks, fusible plugs, discharge, vacuum failures, freeze plugs, and temperature-limit devices. In a typical operation, the devices must hold the internal pressure up to some level but open when pressure exceeds the level. Temperature-limit devices operate similarly but respond when the temperature reaches a certain level. When the material released is an air pollutant or flammable material, the discharge needs an appropriate engineered control system.

The fifth strategy is to separate the hazard from that needing protection. The two tactics reflect, without going into specifics, standard practices for storing compressed gas cylinders.

The sixth strategy is to improve the resistance of that needing protection. The tactic applies to protecting the valve stem and associated hardware from impact by using a safety cap as a protective shield when the cylinder is in storage or transit. Obviously, the cap must be removed when the cylinder is in use.

The seventh strategy is to help people perform safely. The tactic listed is to provide training to personnel who will work with pressurized equipment. This is particularly important with regard to pressure because there are numerous hazards not readily apparent to workers. For example, they may not know the hazards of air pressure sprayed on their skin or into a body orifice, the hazards of working on equipment with retained pressure, the effects of mishandling gas cylinders, or the reasons for various practices for storing gas cylinders. Because many of the hazards involved with pressure are not obvious or visible, workers need awareness training that goes beyond work rules—they need to learn the reasons for the rules.

The eighth strategy is to use personal protective equipment. Because the choice of PPE is highly dependent on the potential hazards encountered, the tactic is stated generically and simply defers to the PPE requirements of the applicable job hazard analysis or risk assessment. If either of these analyses has already been completed, the documentation should indicate any PPE needed.

The ninth strategy is to expedite recovery. A very important tactic is to install, before an incident, remote controls for hydraulic lines. When a fire or explosion involves hydraulic lines and a flammable liquid, the response can have a tremendous effect on the ultimate damage. To avoid the scenario of an employee or professional firefighter acting heroically by entering a dangerous scene to shut off a valve in the hydraulic system, devices for shutting off flow should be installed in locations away from the site where lines might fail. Such forethought can pay off greatly in terms of personnel safety and reduced damage from the fire. A failure to install such devices can lead to the kind of tragic consequences that resulted from a disaster in a chicken processing plant in North Carolina.[4] A leak in a hydraulic line sprayed flammable fluid onto a nearby hot object, which ignited the fluid. The burning hydraulic fluid created a dense black smoke that spread rapidly throughout the facility. No employee seemed positioned to shut off the flow of fluid into the line, so it continued spraying flammable liquid into the fire. The thick black smoke spread rapidly and killed 25 employees.

LEARNING EXERCISES

1. Search the Internet to find an example report on a BLEVE. Avoid news media reports due to their shallow coverage. Answer these questions: What was the chemical? What was the boiling point temperature? What sort of vessel was involved? What event initiated the loss of pressure? What is known about the burst event? What effects were due to the shock wave? What effects were due to the chemical dispersion? Other than effects of the shock wave and chemical dispersion, what other harmful effects were in the report?

2. Regarding flashovers and backdrafts, these are explosive events that develop in a confined space and lead to an explosive release of hot gas. The author

chose to describe them in the chapter on fires. Write a concise argument in favor of including these incidents in the chapter on fires rather than in this chapter on pressure.

3. Fatalities have been reported when someone enters the top of a grain bin while grain is being dropped through the chute into a train car. The assumed reason for the person's action is that he was trying to break up a crust on the top that was slowing flow out the bottom. For whatever reason, the person steps on the top crust of the grain and finds it does not support his body weight. The person falls through and becomes wedged in the chute at the bottom of the bin. Other workers may see feet sticking out the bottom, but they have no apparent way to rescue the victim quickly enough to prevent suffocation. What OSH practice or policy should be followed to prevent such a tragedy?

4. The chapter mentions situations where a pressurized vessel containing flammable gas might rupture, sending a cloud of flammable gas floating into the atmosphere. What can you suggest as a measure to minimize risk of the cloud being set afire by an ignition source?

5. Firefighters train for holding a fire hose so that it will spray in the intended direction. What is similar about a fire hose and a loose hose of a hydraulic or pneumatic system?

6. The 2010 BP Deepwater Horizon disaster in the Gulf of Mexico may be examined in light of risk-reduction strategies. Underwater oil extraction makes use of the high pressure under the sea floor to push the oil upward. Under the planned and normal conditions, the released oil is completely contained as it travels to the surface. Thus, the release of the pressure is a planned part of the process. When containment failed, the pressure continued pushing crude oil upward into the Gulf waters. Think about the issues posed by the following statements. Do you think the following characterizations are fair?

   (a) Refer to the definition of *hazard* in chapter 2. The crude oil was the source with potential to harm the Gulf waters, aquatic life, the coastline, and the fishing and tourism industries.
   (b) We can confidently assume that the risk of harm was recognized prior to the incident.
   (c) The strategy used by BP for operating the well was to modify release of the hazard by completely enclosing the oil as it flowed from under the Gulf waters up to the surface.
   (d) The tactic involved using oil industry technology to construct and operate the containment system.
   (e) The tactic and strategy failed, allowing oil to gush upward without containment.
   (f) BP recognized the importance of having a backup containment system. It identified one in the application for approval to drill. The method had never been tested in deep water, so it was unproven at that time.

(g) When the backup containment system was deployed, it was ineffective.
(h) Strategy 9 (expedite recovery) was inadequately incorporated into the project.

TECHNICAL TERMS

BLEVE	Acronym for boiling-liquid expanding-vapor explosion.
Hydraulic accumulator	A pressure vessel that functions as a shock absorber by storing or absorbing excess fluid.
Hydraulic system	Collection of equipment and flow lines containing a fluid used to perform work by transferring pressurized liquid.
Implosion	Sudden collapse of a container due to external pressure exceeding internal pressure enough to crush the container walls. Also applies to building demolition in which explosives are used to break or weaken key structural elements, causing an inward collapse of the walls.
Pneumatic system	Collection of equipment and flow lines containing air or another gas used to perform work by transferring pressurized gas.

REFERENCES

1. Brauer RL. Safety and health for engineers, 2nd ed. Hoboken, NJ: Wiley; 2006. p. 359–369.
2. Hammer W, Price D. Occupational safety management and engineering, 5th ed. Upper Saddle River, NJ: Prentice Hall; 2001. p. 312–340.
3. Ferguson LH, Janicak CA. Fundamentals of fire protection for the safety professional. Toronto: Government Institutes; 2005. p. 104–106.
4. Schroll RC. Industrial fire protection handbook, 2nd ed. Boca Raton, FL: CRC; 2002. p. 5–7.

Chapter 21

Hazards of Electromagnetic Energies

21.1 FUNDAMENTALS OF ELECTROMAGNETIC ENERGY

We notice only a small part of the electromagnetic energy in the environment. Our senses allow us to detect colors and the feel of sunlight on our skin, but our senses are not developed to detect many other forms of electromagnetic energy. This chapter provides a brief overview of the forms of electromagnetic energy, with an emphasis on those considered health hazards in the workplace.

An understanding of electromagnetic energy requires developing an extensive vocabulary of the terms in this unique specialty. It would be unrealistic to include a complete review of electromagnetic energy in one chapter, and dedicating more than one chapter of this book to electromagnetic energy would be disproportionate to the importance of this topic. Therefore, this chapter addresses a few fundamentals about waveforms and the various types of electromagnetic energy (also called "radiation" by many authors).

Waveforms within a broad range of wavelengths and frequencies are traditionally included in the term electromagnetic energy. Figure 21.1 shows the attributes of typical electromagnetic waveforms. The figure is similar to one encountered in the chapter on noise and vibration. Both plot a sinusoidal wave pattern. The vertical axis applies to amplitude of the quantity measured. The difference between the vibro-acoustic wave graph and this one is the horizontal axis. For the vibroacoustic graph in Figure 16.1, the horizontal axis uses time. For the electromagnetic graph, the more useful variable is travel distance. By using distance in the graph, we can see the meaning of *wavelength*—the spacing between corresponding points on successive waves. It is usually denoted with the Greek letter lambda (λ).

The wavelengths of electromagnetic energies include a tremendous range from gigameters down to picometers. Very different causes and effects are found in regions

Risk-Reduction Methods for Occupational Safety and Health, First Edition. Roger C. Jensen.
© 2012 John Wiley & Sons, Inc. Published 2012 by John Wiley & Sons, Inc.

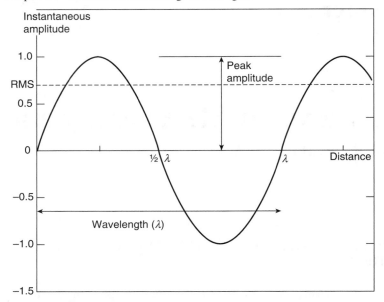

Figure 21.1 Instantaneous amplitude of an oscillating electromagnetic wave plotted against distance.

of this range. Figure 21.2 provides a graphic showing the categories of the spectrum typically discussed under the topic of electromagnetic energy or *electromagnetic radiation*. The category ranges are approximated by the bar graphics, using the wavelength scale on the left. Note that it uses SI prefixes to label every third point on the axis. Points between indicate the effect of moving the decimal one place. For example, when going from meters to millimeters, the points are 1.0 m, 0.1 m, 0.01 m, and 0.001 m (1 mm).

Wavelength is useful for describing and comparing all forms of electromagnetic energy. However, frequency is preferred for some categories. A frequency scale is included on the right-hand side of the figure for the applicable range. Frequencies (f) in any range may be calculated from the wavelength and the speed of light ($c \approx 3 \times 10^8$ m/s) using the equation $f = c/\lambda$. This calculation will yield a value of frequency with the units in cycles per second, or hertz.

The wavelength range included in Figure 21.2 includes the electromagnetic energies most applicable to the practice of OSH. Other electromagnetic energies exist outside the range shown. Those above the graphic are of interest to particle physicists, astrophysicists, and some research laboratories. Those below are of interest for technologists in the communication and power distribution fields. However, for the purpose of this chapter, these are considered outside the mainstream of OSH, and omitted from discussion.

Also not shown in Figure 21.2 is the range of *electromagnetic fields* (EMFs)— energy fields generated by electrical and magnetic equipment.[1] EMF exists around these devices in a static form (i.e., not varying with time). The wavelengths span the

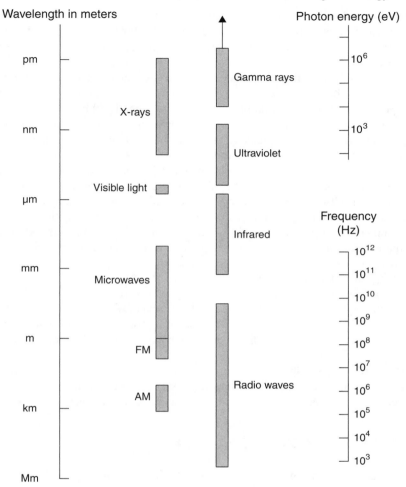

Figure 21.2 Forms of electromagnetic energy arranged by wavelength.

lower portion of Figure 21.2 (wavelengths >1 mm). We have the capability to measure the fields, but we lack clarity on the importance of EMF to the practice of OSH. Assertions of a causal link between EMFs and human health are controversial. Rather than waiting for results of further research, electrical equipment manufacturers have been proactive in efforts to limit the intensity of EMFs during design.[1] In addition, electronic product certification laboratories have criteria for EMF intensity. As for this book, further discussion of EMF energy is considered unnecessary.

The radio wave region of the electromagnetic spectrum includes a relatively broad span of frequencies. Within the broad span are some specific ranges used by our technical society. These include radio waves in the AM band, the FM band, and the broadcast band used for airwave transmission of television. Microwaves are in the short-wave portion of radio waves. These are useful for home cooking appliances, industrial heating processes, radar in aircraft navigation systems, and some specialized research laboratories.

Infrared (IR) radiation is produced by hot objects and molecules. These waves provide the link between the emitting source and the receiver. The receiver absorbs the heat. Some applications of infrared radiation warrant mention. Physical therapists use infrared emitting devices for heating tissues deeper than the skin. Some photography captures infrared light in the image while filtering out light in other wavelengths. This has numerous applications for research. The human eye does not sense the infrared waves directly, but devices have been developed to make it feasible. Night vision goggles and scopes are useful for military operations. Game hunters can locate animals during dark hours, but using these devices for actually targeting game is restricted by the hunting regulations of many governmental jurisdictions.

Human visual senses detect colors—the category just above the infrared category in Figure 21.2. Compared to the other types of electromagnetic energy, the range of visible light is very small, from 400 to 760 nm. But compared to human awareness of electromagnetic waves, visible light far exceeds the others.

Ultraviolet (UV) light includes electromagnetic waves with shorter wavelengths than visible light. UV is emitted by the sun, but fortunately most of it is filtered by the ozone in the stratosphere. The part of UV that gets through the stratosphere is in the region close to visible light. In this region, we encounter the following frequencies: UV-A, UV-B, and UV-C.

Proceeding up the spectrum depicted in Figure 21.2, X-rays and gamma rays are next. X-rays are created by machines and used for both medical and industrial applications. Gamma rays are emitted from the nuclei of radioactive atoms.

21.2 MECHANISMS OF HARMING

Mechanisms of harm depend on the type of radiation. *Nonionizing radiation* striking a human has minimal penetration of the skin, and some penetration of the eyes. Some forms of *ionizing radiation* travel through the skin, and others enter the body through ingestion and inhalation. Because of very different health effects, the two broad categories of electromagnetic energy are summarized in separate subsections.

21.2.1 Nonionizing Radiation

For several years, public health agencies have been informing the public about the adverse health effects of exposure to the sun and sunlamps.[2,3] Three adverse effects are summarized below.

1. Sunburn usually results from outdoor exposure to UV radiation from the sun. Excess exposure to a sunlamp, including models for home and tanning salons, also produces sunburn. Effects on skin can range from mild to intense pain.

For any particular exposure, individual factors also play a significant role in determining the effect.

2. Skin aging is accelerated by chronic exposure to UV.

3. Skin cancer risk is increased by exposure to sunlight. Individual factors also play major role, with fair-skinned people being most at risk. Ultraviolet light in all three regions (A, B, and C) is a risk factor for skin cancer.

Exposure to UV radiation also affects the eyes. Three major effects are listed below.

1. Arc eye is a condition resulting from eye exposure to the intense UV from arc welding. Also called "welder's flash," the condition can come from looking directly at the arc or from UV reflected off walls, the ceiling, or other surfaces around the welding environment.

2. Solar retinitis is a condition of considerable eye discomfort from UV-A penetrating the eye. The condition is also called "eclipse blindness" due to a cause being looking at a solar eclipse without an effective filtering lens.

3. Cataracts appear to have some association with UV radiation exposure. A clear causal relationship is not well established.

Laser light is another type of nonionizing radiation found in the wavelength range of IR–visible–UV.[4] Commercially available low-energy lasers are used extensively by lecturers and carpenters. Higher intensity lasers are used for machining a wide range of materials. Harmful effects include thermal burns, photochemical injuries, and eye injuries.[3] The primary health concern is for eyes.[3,4] Damage may occur in the cornea, lens, retina, or other part of the eye.[4] Eye damage depends on the beam intensity, wavelength, duration, source (direct or reflected), and direction entering the eye.[4]

21.2.2 Ionizing Radiation

By definition, ionizing radiation is electromagnetic or molecular radiation that, when it passes through matter, produces ions. Variations in ionization processes strongly influence the effects. This chapter provides a very brief summary of the topic. Numerous OSH reference books provide more details.[5–7] Readers with keen interest in this topic should look for books or courses in the field known as health physics.

Health effects depend on type of particle, the total dose, and the time over which the dose was acquired. Thus, this discussion begins with the classification of ionizing particles. The particles of concern are alpha particles, beta particles, neutrons, X-radiation, and gamma radiation.[5] The latter two are often referred to as X-rays and gamma rays.

Health effects of ionizing radiation stem from the ionization process that disrupts the electron balance of the affected molecule. Is that always unhealthy? There seems to be agreement among health physicists that a small number of ionized cells are tolerable. At some level, the number of ionized cells exceeds the tolerance of a particular organ, and adverse health effects develop. At the molecular level, an unhealthy molecule may no longer have the ability to reproduce or divide. It may damage a neighboring molecule by stripping electrons from that molecule, and it may cause mutations in other molecules. Different organs respond in different ways. Thus, the effects cannot be predicted precisely because of the numerous interactions among the type of ionizing particle, the type of healthy cell, the dose, the duration of exposure, and randomness. For individuals who work around ionizing radiation, tactics have been developed to keep their exposures within the tolerable level.

21.3 STRATEGIES AND TACTICS FOR ELECTROMAGNETIC HAZARDS

The strategies for reducing risk with the hazards of the electromagnetic spectrum are summarized in Table 21.1 for nonionizing radiation and Table 21.2 for ionizing radiation. On the right of each table are applicable tactics. Most of the tactics in the tables are self-explanatory; so the following supplemental explanations are brief.

The first strategy is to eliminate the hazard. Tactics for nonionizing radiation involve avoiding work under direct sunlight for most of a shift. Many highway construction projects schedule a majority of work at night. This reduces the disruptive effects on traffic flow during busy hours and eliminates worker exposure to sunlight. Consideration may be given to eliminating arc welding by choosing another process for the task or removing the arc welding from the site. For ionizing radiation, tactics involve eliminating sources from areas where people work. This tactic applies to not letting a source on the facility in the first place and removing sources already located on the site.

The second strategy is to moderate the hazard. For sunlight exposure, scheduling work for the evening hours reduces exposure because the intensity is decreased while the sun is low on the horizon. For laser application, using beams with only enough power required for the task moderates the potential exposure. Laser beams are categorized to reflect hazardousness.[3,4] Those in the more hazardous categories require more engineering and administrative controls. For ionizing radiation materials, diluting container contents may be an option. The dilution may change the classification of the contents and make disposal or storage less costly.

The third strategy is to avoid releasing the hazard. For sunlight and arc welding, the UV light emitted cannot be prevented from releasing. For ionizing radiation, there are several tactics. The first is to control everything leaving the facility to avoid spreading contamination to the community. That tactic includes inspecting and monitoring all outgoing personnel, shipped materials, waste water, and exhaust air. A second tactic applies to processes with potential to reach criticality (intense nuclear reaction). *Criticality controls* include monitoring closely to prevent deviations in work processes that could increase the risk of reaching criticality and releasing massive amounts of radiation. Steven Casey describes a sad case of failure to prevent deviations from the standard procedures for mixing materials used to make nuclear fuel pellets.[8] The third tactic applies to stored materials. A rigorous inspection

Table 21.1   Strategies and Tactics for Reducing the Risks of Nonionizing Radiation Hazards

Risk-reduction strategy	Risk-reduction tactics
1. Eliminate the hazard	1a. Instead of having workers construct something outdoors in the sun, have them perform some or all of the work in a building or otherwise protected from the sunlight
	1b. Schedule outdoor work for evenings and nights
	1c. Eliminate arc welding from a worksite
2. Moderate the hazard	2a. Schedule outdoor work for times when the sun is lower on the horizon rather than during midday hours
	2b. Use laser beams with only enough power required for the task
3. Avoid releasing the hazard	Use lockout and tagout procedures for high-energy lasers when personnel could be at risk
4. Modify release of the hazard	For arc welding and lasers, light reflected from walls, ceilings, and other surfaces can be modified by coating with low-reflectance paint
5. Separate the hazard from that needing protection	5a. Apply sunscreen to skin before exposure
	5b. Provide work areas with overhead shading devices (canopies)
	5c. For higher energy lasers, establish restricted areas with appropriate signage and limited access
6. Improve the resistance of that needing protection	Improve resistance to sunburn by gradually tanning skin. This does not change risk of skin cancer
7. Help people perform safely	7a. Encourage personnel to use plenty of sunscreen by providing instruction and supplying the lotion
	7b. Provide personnel with information about the hazards of sun exposure and precautionary practices
	7c. For higher energy lasers, post warnings and provide instructions for personnel who may interact with the equipment
8. Use personal protective equipment	8a. Ensure that workers exposed to arc welding or laser light protect their eyes by using light filtering lenses appropriate for light source
	8b. Workers exposed to sunlight can wear garments with built-in UV protection, a hat, gloves, and glasses with a UV filter
9. Expedite recovery	9a. Train personnel on early detection of skin cancer
	9b. For employees returning from skin cancer surgery, consider options for work not involving sun exposure

program may be used to detect any signs of container deterioration, and of course, deteriorated containers need replacement.

The fourth strategy is to modify release of the hazard. For arc welding, light reflected from walls and the ceiling can be modified by coating with low-reflectance

Table 21.2   Strategies and Tactics for Reducing the Risks of Ionizing Radiation Hazards

Risk-reduction strategy	Risk-reduction tactics
1. Eliminate the hazard	1a. For defined areas of facility, prohibit materials with ionizing radiation from entering, being used, or being stored
	1b. For facilities with known ionizing radiation sources, remove the sources from the site
2. Moderate the hazard	Dilute highly radioactive materials to lower radioactivity
3. Avoid releasing the hazard	3a. Inspect outgoing personnel and materials to detect and avoid releasing contaminated materials from facilities
	3b. Monitor exhaust air and water released from facility to detect contamination and avoid environmental damage
	3c. Establish and maintain criticality controls when processes have possibility of reaching criticality
	3d. Regularly inspect all containers of radioactive materials for signs of developing corrosion or structural weaknesses
4. Modify release of the hazard	4a. Designate as "*controlled access areas*" locations with high ionizing radiation. Limit employee exposures to these areas
	4b. Encase radioactive materials in sealed containers designed to limit leakage to zero or very small amounts
5. Separate the hazard from that needing protection	5a. Build radiation shields between radioactive sources and areas used by personnel
	5b. Provide remote controls for radiation-producing machinery
	5c. Provide remote controls for handling radioactive materials
	5d. Practice excellent housekeeping to prevent personnel from unintentionally acquiring radioactive materials that might be subsequently ingested or inhaled
6. Improve the resistance of that needing protection	No tactics. Cannot change human resistance to ionizing radiation
7. Help people perform safely	7a. Use SOPs to encourage employees to perform each task in the proper way
	7b. Train personnel to use the SOPs
	7c. Put radiation hazard label on all containers with radioactive materials
	7d. Have personnel carry a pocket dosimeter so that they can see if an exposure may have been excessive
8. Use personal protective equipment	8a. Have personnel use gloves and other protective clothing when possibly handling radioactive materials or containers
	8b. Have personnel wear respiratory protection based on a formal respiratory protection program
9. Expedite recovery	9a. Have in place a spill management system to limit harm
	9b. Monitor body fluids and excreta for early detection of overexposure

paint. This provides protection for personnel who are not welding but are exposed while passing by or working near the welding area. A tactic used for locations with ionizing radiation is to control the individual exposure time. To make this tactic effective, strict administrative procedures are used to make sure the exposure hours of individuals do not exceed a total allowed by health physics guidelines. A tactic used for storing and transporting radioactive materials is to encase the materials in sealed containers designed specifically for the material. Effective encasement should avoid releasing ionizing radiation or material from the container.

The fifth strategy is to separate the hazard from that protection. For UV light, sunscreen products applied to the skin before exposure partially block penetration of the harmful UV radiation. Although authorities agree that excessive exposure to the UV light should be avoided, there are some differences in opinion about how far to carry this advice. Some encourage avoiding sunlight as much as possible. Advice from others is to get enough sunlight to sustain vitamin D levels, but more than that is ill advised. All the experts agree that lotions with UV blocking agents should be applied to skin before exposures. They also agree that most people apply insufficient lotion, and they do not reapply often enough. Sunlight blocking agents are designated by the type of UV light for which they protect. Protection from UV-A and UV-B is most common.

The sixth strategy is to improve the resistance of that needing protection. Reducing risk of sunburn may be achieved by gradually acquiring a tan. This approach, however, does not reduce risk of skin cancer. Human susceptibility to health damage from ionizing radiation is not changeable.

The seventh strategy is to help people perform safely. Two tactics are listed in Table 21.1 for sun exposure. One is to encourage personnel to use plenty of sunscreen lotion. A fundamental way is for the employer to provide appropriate lotion to employees who work in the sun. The second tactic involves conducting periodic training sessions on hazards of sun exposure and precautionary practices for reducing risk of sunburn and skin cancer.

Tactics listed in Table 21.2 for ionizing radiation include four administrative practices. First, develop and use standard operating procedures (SOPs) to encourage employees to perform each task in the proper way. Second, training provides the connection between SOPs in a manual and the personnel who need to follow the SOPs. Third, put radiation hazard labels on all containers with radioactive materials and post signs at the entrances to protected areas. Fourth, have personnel carry a pocket dosimeter so that they can easily determine if a particular event exposed them to radiation. Employees may carry a pocket dosimeter as well as a more accurate monitoring device. The pocket dosimeter has the advantage of making quick feedback available to a worker who will then be enabled to take appropriate actions. These devices are less accurate than some other personal monitors.[6] Two other types are the film badge and the thermoluminescent detector. Carrying a pocket dosimeter plus one of the more accurate monitors permits both quick feedback and accurate exposure data determined somewhat later.

The eighth strategy is to use personal protective equipment. Welders use face shields to protect from splatter. Providing specific light filtering lenses to workers exposed to harmful lights is effective. Workers exposed to sunlight can wear garments with built-in UV protection, hard hats to protect their head, gloves to protect their hands, and sunglasses to protect their eyes. For ionizing radiation, personnel can use gloves and other protective clothing when possibly handling radioactive materials or containers. Finally, personnel can wear respiratory protection based on a formal respiratory protection program.

The ninth strategy is to expedite recovery. A common tactic is monitoring body fluids and excreta for early detection of overexposure. The discharge water from toilets needs monitoring to detect if radiation levels have increased to a troublesome level. If that happens, actions to prevent contaminated water from leaving the facility and getting into deep groundwater or surface water are essential, and investigating to find the source of the contaminant is a key step in the process of preventing future incidents. Another very important tactic for expediting recovery is to have in place a spill management system to limit harm. A spill management system can include multiple elements. Among these are automated systems to monitor and detect spills quickly, systems to seal off the area by means such as closing automatic doors, possibly applying a controlling agent, means for notifying personnel of the proper response, plans for evacuation known by personnel, and personnel trained and equipped to safely perform cleanup.

LEARNING EXERCISES

1. Out of the various forms of electromagnetic energy, which ones can humans detect without any scientific instrument?

2. Figure 21.1 looks similar to Figure 16.1. What is the difference in the horizontal axis? What is the difference in the spread between successive waves?

3. In most common radios, you can tune in AM as well as FM stations. Do the station numbers refer to the frequency or the wavelength?

4. Which type of electromagnetic radiation is most recognized for causing each of the following?
   (a) Sunburn, accelerated skin aging, and skin cancer.
   (b) Arc eye and solar retinitis.

5. What practices can be used at a nuclear power plant or nuclear research facility to protect the community outside the facility from ionizing radiation? Take into consideration the families of workers, dry cleaners, and community air and water.

6. In what ways can a nuclear power plant help people perform safely?

7. When machine shops are inspected, nuclear facilities take quite seriously accumulations of dust in corners. In contrast, inspections of machine shops in general industry (e.g., an automotive part supplier) regard dust in corners of minimal concern. Explain the reason for the difference in emphasis.

TECHNICAL TERMS

Controlled access area	A designated area in which the exposure of individuals to radiation or radioactive material is controlled.
Criticality controls	Engineering and administrative systems for making sure processes do not reach nuclear criticality.
Electromagnetic fields (EMFs)	Static electric and static magnetic fields associated with natural phenomena and some types of electromagnetic energy. The fields do not vary with time. The strength of a field at a point depends upon the distribution and behavior of the electrical charges involved.
Electromagnetic radiation	The propagation or transfer of energy through space and matter by time-varying electric and magnetic fields.[2]
Ionizing radiation	Electromagnetic or particulate radiation capable of producing ions when passing through matter.
Nonionizing radiation	The part of the electromagnetic spectrum where there is insufficient quantum energy to cause living matter to ionize. It includes radiant energy with wavelengths in the range of 100 nm to 300 000 km.[2]
Wavelength	The distance between corresponding points of two successive waves of a sinusoidal waveform such as electromagnetic energy. It is usually denoted with the Greek letter lambda (λ).

REFERENCES

1. Karpowicz J, Gryz K. Electromagnetic hazards in the workplace. In: Koradecka D, editor. Handbook of occupational safety and health. Boca Raton, FL: CRC; 2010. p. 199–214.
2. Hitchcock RT, Moss CE, Murray WE, Patterson RM, RockwellJr., Nonionizing radiation. In: DiNardi SR, editor. The occupational environment: Its evaluation and control, 2nd ed. Fairfax, VA: American Industrial Hygiene Association; 2003. p. 494–571.
3. Miller G, Yost M. Nonionizing radiation. In: Plog BA, Quinlan PJ, editors. Fundamentals of industrial hygiene, 5th ed. Chicago, IL: National Safety Council; 2002. p. 281–325.
4. Brauer RL. Safety and health for engineers, 2nd ed. Hoboken, NJ: Wiley; 2006. p. 390–393.
5. McCarthy ME, Thomas B. Ionizing radiation. In: DiNardi SR, editor. The occupational environment: Its evaluation and control, 2nd ed. Fairfax, VA: American Industrial Hygiene Association; 2003. p. 572–610.
6. Cheever CL. Ionizing radiation. In: Plog BA, Quinlan PJ, editors. Fundamentals of industrial hygiene, 5th ed. Chicago, IL: National Safety Council; 2002. p. 257–280.
7. Pachocki KA. Ionising radiation.In: Koradecka D,editor. Handbook of occupational safety and health. Boca Raton, FL: CRC; 2010. p. 297–325.
8. Casey S. The atomic chef and other true tales of design, technology, and human error. Santa Barbara, CA: Aegean; 2006. p. 13–26.

Chapter 22

Hazards of Severe Weather and Geological Events

22.1 BACKGROUND

This chapter discusses two hazard sources for organizational performance and employee safety. Significant weather events and major geological events can interrupt operations and harm personnel. These events are not preventable, and therefore, not viewed within the traditional framework of most OSH issues. For example, we have no standards to follow for compliance. The most prevalent view in the OSH field is to treat these events through emergency planning and preparation.

Major *weather events* include severe winter storms, hail, floods, torrential rains, mudslides, high winds, tornadoes, hurricanes, and nearby lightning strikes. These events arise from the natural transitions of air and water. The air movements are driven by temperature gradients constantly occurring as the heat from the sun is distributed. The pressure gradients are tied to the temperature gradients as the more dense cooler air pushes warmer air. A third factor adding considerable complications is the altitude. Air movements can vary in direction and velocity at different elevations. For these basic reasons, weather forecasts are based on probabilities rather than certainties.

Major *geological events* include earthquakes and volcanoes. Both arise from hot material deep under the surface exerting pressure on the more solid materials comprising the mantle. The heat comes from continued radioactive decay of materials deep in the Earth and from heat left over from the Earth's formation.[1] The mantle consists of large platonic plates that rest on a bed of molten material. Some plates contain continents such as the African Plate, the Eurasian Plate, and the South American Plate. Some plates, like the Pacific Plate, are the ocean floors with a few seamounts rising high enough to be islands.

Risk-Reduction Methods for Occupational Safety and Health, First Edition. Roger C. Jensen.
© 2012 John Wiley & Sons, Inc. Published 2012 by John Wiley & Sons, Inc.

Earthquakes originate from the junction of abutting platonic plates.[1] One type involves two plates sliding laterally relative to each other. This is the common case in California where the San Andreas Fault lies. The Pacific Plate moves northwest relative to the North American Plate. Movements occur in spurts, producing frequent minor tremors, and occasionally producing destructive shaking. A second type involves two plates separating. The Mid-Atlantic Ridge is formed by continental plates separating and allowing molten materials from below to push up into the Atlantic Ocean. During various geological periods, some of the resulting ridges grew peaks tall enough to form the present Atlantic islands. A third type is one plate sliding over another. The Andes Mountains are the result of the South American Plate being elevated by an ocean plate sliding under South America.

If the plates meet under the ocean, a very different result can occur. After years of two plates pushing against each other, a point is reached where the plates suddenly move, one sliding under the other. The rising plate creates an upward push on the water above it, while the descending plate pulls downward on the water above it. This pressure difference transmits all the way from the ocean floor to the surface, initiating a large wave in which the high water chases the low water across the vast ocean. These tsunamis (or tidal waves) travel great distances. When the wave reaches a shoreline, the low water arrives and causes an extreme low tide condition. The following high water wave then rushes ashore with the power to kill people, severely alter the natural environment, and destroy such human-built items as highways, bridges, buildings, and industrial facilities.

Volcanic eruptions arise in somewhat predictable locations. Some arise in a location where an ocean floor platonic plate meets a continental plate resulting in the ocean floor plate sliding under the continental plate. As the lower plate sinks further down it heats up, eventually becoming molten. The molten material breaks through a weak spot in the plate above it and spews out lava. This usually occurs in places where prior eruptions have occurred, such as old volcanic mountains. There are several of these volcanic mountains in British Columbia, Washington, and Oregon.

Another location for volcanoes is in a few "hot spots" around the Earth. These locations are above stationary openings located deeper than the plates. The platonic plates move over these hot spots very slowly but in a predictable direction. A major hot spot created the Hawaiian Islands. The oldest islands are in the northwest end of the chain. The newest are in the southeast end. The most recently created of the islands is the large island, Hawaii. These islands grew out of the Pacific Plate as it slowly moved northwest over the hot spot. Another hot spot is located in the North American Plate under Yellowstone National Park. From an OSH perspective, the important point is that the risk of a volcanic eruption is largely confined to geographic locations with a history of eruptions. Thus, location is predictable, but the timing of an eruption is speculative. Geologists may be able to predict that a volcanic eruption in a particular site will occur sometime in the next thousand years. Although advances in monitoring quakes continue to improve forecasts, considerable uncertainty remains. We simply need to accept the uncertainty and proceed with appropriate precautions to mitigate adverse effects and prepare for recovery.

22.2 MECHANISMS OF HARMING

Both major weather events and geological events can affect the people and functioning of an organization. Weather events can significantly affect employee commuting, especially for those using the roadways. Weather can also disrupt shipments into and out of the facilities. Both weather and geological events can interrupt access to utilities (e.g., electric power, fuel gas, and water) and communications (telephone, cell phone service, and Internet connections). Power surges from lightning can damage electronic systems. Major geological events like earthquakes can damage building structures and cause building materials to fall on personnel inside. Volcanic eruptions spew ash and gases into the air. The extended eruptions of a volcano in Iceland in 2010 caused cancellation of commercial air traffic throughout most of Europe for several days.

During the post-event phase, employees and external responders may become unknowingly exposed to hazards created by the weather or geological events. Examples include gas leaks, exposed electrical wires, unstable walls and ceilings, and damaged pneumatic or hydraulic lines. Hurricanes and ice storms often bring down power lines. Lines close to ground level are serious hazards for rescue workers and repair personnel.

Major weather events create situations that put people in need of rescue. Examples are people stranded by a severe snowstorm or flood, private boats being caught in a severe storm at sea, and people buried under rubble left by a tornado, hurricane, or earthquake. The people who volunteer to attempt rescue often find themselves subjected to dangerous situations. For example, a rescue at sea operation may expose rescue personnel to the hazards inherent in helicopter flight during bad weather and retrieving people from a sinking vessel. Job hazard analyses of these sorts of missions would reveal diverse specific hazards encountered during the planned steps, and for many of these the hazard control depends on timely and effective action by team members. A single decision error or mistake in execution can have serious consequences.

22.3 STRATEGIES AND TACTICS FOR WEATHER AND GEOLOGICAL EVENTS

The strategies for reducing risk with weather and geological events are listed in the left column of Table 22.1. On the right are some of the applicable tactics.

The first strategy is to eliminate the hazard. One tactic is to avoid building a new facility in a location known as high risk for significant weather or geological events. This may seem quite ivory tower, but companies looking for new facility sites include this risk among the numerous other site-selection criteria. A tactic for a rescue team is to postpone the mission if the situation would involve more risk to rescue personnel than is tolerable.

Strategies 2 through 4 have no tactics listed in Table 22.1. The author could not think of any way to affect weather or geological events by moderating the hazard,

Table 22.1   Strategies and Tactics for Reducing the Risks of Weather and Geological Events

Risk-reduction strategy	Risk-reduction tactics
1. Eliminate the hazard	1a. Avoid locating facilities in locations known as high risk for significant weather or geological events
	1b. Call off or postpone a rescue mission if conditions make it too dangerous
2. Moderate the hazard	Not feasible
3. Avoid releasing the hazard	Not avoidable
4. Modify release of the hazard	Not feasible
5. Separate the hazard from that which needs protection	Locate workplaces away from sites likely to experience a flood or mudslide
6. Improve the resistance of that which needs protection	6a. Construct buildings and other structures using technology for tolerating earthquakes
	6b. Establish redundant or backup systems for utility services and communications
7. Help people perform safely	7a. Provide shelters within facilities where employees may go during hurricanes and tornadoes
	7b. Conduct drills to help personnel learn emergency response procedures
	7c. Rigorously train and adequately equip personnel who will be performing rescue missions so that they can do so safely
	7d. Establish policies encouraging personnel to choose not to drive during severe weather
8. Use personal protective equipment	For outdoor workers, provide protective outerwear appropriate for foreseeable weather events
9. Expedite recovery	9a. Include in the business continuity plan those weather and geological events considered most likely
	9b. Include in emergency plans specifics about post-incident actions needed, responsible individuals, and required training and resources they will need
	9c. Have OSH trained people monitor the work of responders and repair personnel
	9d. Have repair teams develop a JHA or RA prior to starting tasks

avoiding release of the hazard, or modifying the release. A tactic for Strategy 5 is to locate workplaces away from areas likely to experience flooding or a mudslide.

The sixth strategy is to improve the resistance of that which needs protection. The governments of places with a history of destructive earthquakes have adopted requirements for new construction to mitigate effects of substantial tremors. Whether required or not, it makes sense to incorporate available earthquake technology into construction specifications for new structures located where the risk of earthquakes is substantial.

The seventh strategy is to help people perform safely. The first tactic listed is to provide shelters in building where employees may congregate during highly windy events like hurricanes and tornadoes. This should provide protection from glass and other objects thrown about by the destructive powers of the wind. In addition, the shelters may provide protection if the roof or walls collapse. Conducting emergency response drills is a well-known tactic for helping personnel respond appropriately to disruptive weather and geological events. A third tactic is to establish policies encouraging employees to choose not to drive during severe weather. Policies may consider both commuting employees and those who perform work-related driving.

The safety of people performing rescue operations makes use of Strategies 7 and 8. Performing rescue operations in the safest feasible manner requires extensive training and first class equipment. Through training and equipment, the team members will have opportunities to deal with obstacles and hazards encountered as they perform the rescue. Those performing this sort of work also benefit from personal protective equipment, for example, the self-contained breathing apparatus worn by firefighters when entering a burning building to rescue a child, and the bullet-resistant vests worn during some military and police operations.

For 24-hour operations, policies may clarify how decisions are made when severe weather is forecast. One such policy may address decisions about keeping some employees at work while having others stay home. For employees who drive as part of their job, policies can make clear the organization's support for individuals deciding to cease driving when they feel conditions are too risky. The costs to stop driving, and possibly pay for a motel room, are relatively small when weighed against the costs of roadway incident. Consider some of the potential costs. If an employee is involved in a roadway incident where others are seriously harmed, the potential for litigation is quite high. Lawyers commonly view large organizations as "deep-pocket" targets for legal action. The lawsuits, regardless of merit, can cost a great deal to defend, and settle. Damage to vehicles and roadside fixtures (e.g., signs, light poles, and guardrails) can be quite costly. Even small vehicular incidents like fender benders can lead to an employee using work time for visiting auto body shops to get repair estimates, dealing with insurance companies, and finding temporary transportation.

The eighth strategy, to use personal protective equipment, has only one tactic. It is to provide outdoor workers with protective outerwear appropriate for foreseeable weather events. The raincoats provided for police are an example. Rescue teams employ this tactic through rigorous training and providing quality equipment for personnel who will be performing the missions.

The ninth strategy is to expedite recovery. The first two tactics listed are discussed more extensively in Section 12.5. One is to include in the business continuity plan those weather and geological events considered most likely. The second is to include in emergency plans the responses and recovery activities needed after a weather or geological event. It is particularly important to identify individuals or positions charged with responsibilities to perform response tasks, and to provide those individuals with the training and specific resources needed to perform their assigned tasks. A third tactic is to have OSH trained people monitor the work of responders and repair personnel to provide an extra pair of eyes for spotting hazards. A fourth tactic is to have repair teams develop a job hazard analysis or risk assessment prior to starting a task. A tactic applicable to rescue teams is developing a plan and becoming prepared for responding in case the primary rescue or combat team needs help.

22.4 SUMMARY OF PART IV

Part IV has nine chapters about energy hazards—eight address forms of energy and this chapter addresses diverse forms of energy associated with weather and geological events. To put part IV in perspective, Figure 22.1 depicts two-dimensional matrix with a row for each of the seven hazard sources identified in chapter 2, and a column for each of the nine risk-reduction strategies. The first two rows apply to part IV. Cells marked with an X indicate for each source the strategies mentioned in part IV.

The first chapter begins with a section describing energy in general and explaining how energy frequently changes form. The main material in each chapter

Hazard Source

	1	2	3	4	5	6	7	8	9
Energy	X	X	X	X	X	X	X	X	X
Weather & Geologic Events	X				X	X	X	X	X
Hazardous Conditions									
Chemical Substances									
Biological Agents									
Musculoskeletal Stressors									
Violent Actions of People									

Risk-reduction Strategy

1. Eliminate the hazard.
2. Moderate the hazard.
3. Avoid releasing the hazard.
4. Modify release of the hazard.
5. Separate the hazard from that needing protection.
6. Improve the resistance of that needing protection.
7. Help people perform safely.
8. Use personal protective equipment.
9. Expedite recovery.

Figure 22.1   Matrix showing the risk-reduction strategies applicable to the two hazard sources addressed in part IV.

contains three sections starting with a background about the scientific aspects of the energy, followed by explanations of the mechanisms by which the hazardous energy can harm people or other valued items, and concluding with a section illustrating applications of risk-reduction strategies to the hazardous energy. A two-column table format is used to clearly map the various tactics to one of the nine strategies. Most tactics listed in the tables illustrate risk-reduction approaches found in OSH-related books, regulations, and voluntary standards. The author has attempted to include enough tactics to illustrate how the strategies are implemented.

The topics of the chapters in part IV are energy sources with potential to cause harm. These are kinetic and gravitational energy, electrical energy, acoustic energy, thermal energy, fires, explosions, pressure, electromagnetic energy, and weather and geological events. Due to the frequent change of energy from one form to another, some topics discussed in one chapter could have been discussed in another chapter. For example, some incidents start as fires, create high pressure, and explode. Explosions can create a shock wave that is a particular form of acoustic energy. Another example is earthquakes. Earthquakes result from the intense pressure between platonic plates being changed to kinetic energy when the plates move. If underwater, the kinetic energy transfers to the water and creates waves. If under a continent, the kinetic energy creates pressure waves that manifest as tremors or earthquakes. Thus, grouping all these energy topics into one of the major parts of the book emphasizes their interrelated character. The fifth major part of the book discusses occupational hazards not quite so clearly based on energy.

LEARNING EXERCISES

1. If this chapter on weather and geological events were left out of the book, into which energy chapter or chapters would you put (a) weather events and (b) geological events. Explain your rationale.

2. Do you think a tsunami should be considered a weather event or a geological event? Why?

3. Of the nine strategies in Table 22.1, which take effect in the Haddon phase called the "post-incident" phase?

TECHNICAL TERMS

Geological events   Earthquakes and volcanoes.
Weather events   Uncommon and disruptive weather conditions such as severe winter storms, hail, floods, torrential rains, mudslides, high winds, tornadoes, hurricanes, and local lightning strikes.

REFERENCE

1. Hazen RM, Trefil J. Science matters: Achieving scientific literacy. New York: Anchor Books; 2009. p. 215–250.

Part V

Risk Reduction for Other Than Energy Sources

Part V continues the discussion of the seven categories of hazard sources—forms of energy, weather or geological events, conditions, chemical substances, biological agents, musculoskeletal stressors, and the violent actions of people. Having addressed the first two in part IV, the remaining chapters address the other five hazard sources. Chapter 23 addresses conditions of workplace associated with increased risk. Due to the enormous variety of workplace conditions, the chapter focuses on a few that are both manageable and known to contribute to numerous occupational injuries, fatalities, and diseases—floors, stairs, ramps, confined spaces, and dusty working conditions. Chapters 24 through 27 provide background information and examples of risk-reduction strategies for the hazards of chemical substances, biological agents, musculoskeletal stressors, and hazards arising from the violent actions of people.

Throughout the chapters in this part, the emphasis is on the pre-event and during-event phases. The post-event phase is important, but it would be redundant to include it in the discussion of each hazard source. Readers will also notice distinctions based on different organizational aspirations for workplace safety and health. As explained in chapter 12, some organizations aspire to comply with applicable laws and standards, while others aspire to operate at a best-practice or world-class level. These organizations typically define best practice by referring to the latest voluntary standards applicable to their industry.

Chapter 23

Workplace Conditions

23.1 BACKGROUND

Workplaces have various physical features that can, if not managed appropriately, be hazards to employees. Perhaps all workplaces have walking and working surfaces, most have stairways and/or ramps, many have confined spaces, and some have dusty air. The condition of these features, and many other workplace features, can be made safe through attention to engineering controls and administrative practices. On the other hand, lack of attention can make these features hazardous conditions. This chapter addresses workplace conditions selected because they are found in so many workplaces, and because of their association with occupational injuries and diseases.

OSH professionals and the public generally think of walking surfaces with slippery spots as being hazardous to pedestrians. Thus, we can consider the slippery surface as a hazardous condition, waiting passively for a human to come along. If a person slips on the slippery spot and falls, we consider the slippery spot as the proximate cause of the injury. This is the view taken in this section. However, another perspective warrants acknowledgment. A physicist may take the view that the potential gravitational energy of the person's body is the hazard, and the slip caused failure of the base of support for the gravitational energy, resulting in release of potential energy, leading to impact with the floor. The physicist's perspective is entirely logical, but may seem a bit twisted for most of the OSH community and general public. The view that the hazard is the location rather than the energy is taken for this chapter because the opportunities to correct a hazardous location are abundant, whereas instructing employees to better control the gravitational energy of their body while walking is unlikely to change ambulatory habits. Similar differences in perspective are presented when the location is a confined space containing a potentially harmful chemical or energy.

Many spaces in industrial facilities contain air contaminants and various unseen energies with potential to harm anyone who might enter the space. Some industrial properties contain chemical-laden ponds that may look like swimming holes to

Risk-Reduction Methods for Occupational Safety and Health, First Edition. Roger C. Jensen.
© 2012 John Wiley & Sons, Inc. Published 2012 by John Wiley & Sons, Inc.

neighborhood children. The condition of such industrial spaces affects the risk of an injurious incident. Two other factors also contribute significantly to risk: frequency of exposure and the behavior of those exposed.

The importance of a hazardous condition increases as the extent of exposure increases. To emphasize this point, consider an industrial facility with two similar concrete stairways. Both have a chunk of concrete broken off the lower step. One leads to the basement and is used twice a month by a maintenance employee. The other is used by many employees every shift, averaging 800 descents per month. In an ideal world, the company would have both stairways repaired promptly. If, however, the company does not have an unlimited maintenance budget, and wants to commit funds to the most important matters, fixing both steps this year may not be feasible. A risk assessment would make clear that the stair used frequently has a much higher risk-reduction potential. Before saying this is obvious, think about the typical walk-through safety inspection. A small team walks through a work area and notes any conditions not conforming to standards. These sorts of checklist inspections consider both defects the same, with no concern for frequency of use.

To illustrate how significant this is, consider some numbers. Start with an estimate that the probability of a fall while descending either staircase once is 0.000 001. The probability of a fall on either staircase during a month may be calculated by multiplying the probability of a fall per descent by the monthly number of descents. Thus, for the frequently used staircase, the monthly probability is 0.000 001 falls/descent × 800 descents/month = 0.0008 falls/month. For the infre-quently used staircase, the probability is 0.000 001 falls/descent × 2 descents/month = 0.000 002 falls/month. The ratio of these risks (0.0008/0.000 002) indicates that the frequently used staircase has a risk 400 times greater than the infrequently used staircase. Basic calculations such as these illustrate how much more risk-reduction potential may be realized by allocating facility funds to repair the more frequently used staircase.

Throughout the following subsections about managing fixed-location hazards, risk-reduction opportunities include managing the condition itself, the frequency of use, and the behavior of those exposed.

23.2 FLOORS

The safety literature and codes agree that walking surfaces should be built and maintained for injury-free walking. This is an excellent goal for companies building new facilities and those with world-class aspirations for their OSH programs. However, there are many companies getting by in old facilities, and having less ambitious OSH goals. For these companies, obtaining funds for working surface safety projects may be very challenging. Thus, the following discussion attempts to address some considerations OSH managers may find helpful when prioritizing walking surface projects.

Public sidewalks, like the one shown in Figure 23.1, provide a good example of the need to consider more than just the hazardous condition. Other significant

Figure 23.1 A substantial crack in a sidewalk with low frequency of pedestrian traffic.

considerations include frequency of use, cost effectiveness, and human expectations. The frequency of use factor is obviously important and easily understood. The cost effectiveness factor comes into play as projects for OSH improvements compete for resources with each other and with proposals from other organizational units. The human expectation factor needs some explanation.

When people walk, they adjust their walking behavior to the walking surface they expect to encounter. Workers at construction sites do not expect a city-like walking surface; they know that peering at the surface ahead is required. People using public sidewalks provide another example. In a new neighborhood with new sidewalks, pedestrians develop expectations of level sidewalks. Along with that expectation, they walk without looking ahead at the sidewalk. In an older neighborhood with sidewalks having numerous cracks, broken concrete, and mismatched slabs, pedestrians do not expect level sidewalks and they understand the need to look ahead at the walking surface. If city officials do not have sufficient funds to maintain all sidewalks in tip-top shape, they will need to prioritize sidewalk projects, and in doing so, they are justified in considering human expectations along with frequency of use and cost effectiveness.

The following discussion begins with explanations of how humans interact with walking surfaces, and how falls sometimes occur. Following this discussion, risk-reduction tactics are noted.

23.2.1 How Pedestrians Fall

Hazardous locations associated with slipping involve interactions among several factors. Drawing the findings and conclusions of a few of the researchers in this area, the following explanation is provided.[1-4] People routinely walk on level surfaces without giving it any thought. Most walking is controlled by neural connections involving the lower extremities and the spinal cord. The brain is not actively involved in ordinary walking. When the person sees an obstacle or other variation in the walking surface ahead, the brain intervenes in the automated process and causes an adjustment in the walking pattern, or gait. This allows pedestrians to successfully traverse many types of walkways without falling. A common problem arises when the person encounters a surface variation they fail to see or notice. Why this is important requires a basic understanding of how people walk and occasionally fall.

As we walk, the leading foot makes initial contact with the heel. At that instant, the heel is exerting a forward shear force against the floor surface. For a fraction of a second, the heel slips forward (and a bit laterally). During a normal stride, the shear force (F_{motion}) is countered by force in the opposite direction (F_k) from the friction between the heel and floor. At first, $F_{\text{motion}} > F_k$. As more body weight is transferred to the floor, there comes a point where $F_{\text{motion}} < F_k$. After that, the heel motion decelerates, and usually comes to a stop soon. A slipping distance in ordinary walking less than 1 cm is not even noticed by the person. Slipping up to 3 cm is considered within the range of controlled walking for healthy people. As the length of the slip increases beyond 3 cm, concerns for slipping falls increase. As one can easily imagine, when the leading foot suddenly slides forward, we instinctively react. Sometimes the bodily reaction avoids a fall. The probability of a successful reaction declines as the length of the slide increases. For example, a relatively high proportion of recoveries from 4 cm slides are expected. But the proportion of recoveries from slides over 10 cm will be very low. In between this range, the proportion of successful recoveries declines.

Walking techniques that make sliding distance shorter are to walk slower and shorten stride length. Those of us living in northern climates learn this from walking outdoors during winter. When we encounter spots of ice, taking very short steps enables us to either avoid slipping altogether or limit the length of slipping. A similar approach applies when approaching a wet floor. If we know the floor is wet, we can either avoid it or walk on it slowly with very short strides. That technique reduces the reaction force required to control the initial heel slip and keeps the body's center of gravity approximately above the supporting feet. If one foot does slide, the other foot is not far from being positioned to support the upper body.

Falls initiated by slipping have some common fall patterns.[3] Two patterns follow slipping of the leading heel during straight-ahead walking. In pattern 1, the leading foot slides forward and becomes airborne. The trailing foot continues forward and also becomes airborne. The torso falls and the person hits the floor with their buttocks, possibly damaging their tailbone. The back of their head may also impact the floor. Pattern 2 starts similarly, but the trailing foot stays low and tucks under the leading leg. This fall pattern tends to result in floor impact involving one-sided buttock or hip impact, and often elbow impact. The side of impact is that of the trailing foot. Figure 23.2 is an event tree showing the paths to these two outcomes as well as paths to other patterns of falling from a slip. The probability values in the event tree are crude guesses provided for the third Learning Exercise at the end of the chapter.

When the person begins to change direction, a somewhat different fall pattern occurs. When the person starts a turn, the torso redirects in the direction of the turn,

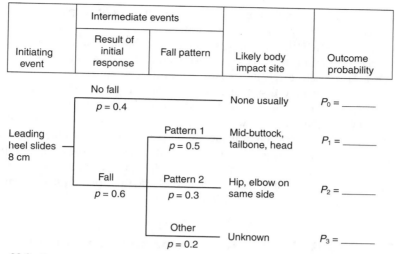

Figure 23.2   Event tree for a slip on the same level, showing paths of outcomes and crude estimates of branch probabilities.

and the leading foot demands more than usual floor reaction force to stop. In pattern 3, the leading foot starts sliding in the original direction of travel, but relative to the person's torso, the foot moves laterally. The trailing leg buckles, and the person lands on the side of their buttocks. Excellent graphics depicting these typical fall patterns are provided in Ref. 3.

A fourth slipping pattern is occasionally encountered.[3] The initiating event is a slip by the trailing foot as the toes push off. The trailing foot slides backward. Many people recover from this type of slip, but if recovery is unsuccessful, the typical fall pattern involves the knee of the trailing leg impacting the floor.[3] The secondary impact may be shared by the upper leg, hip, and arm on the side of the slipping foot.

The walking event known as *stumbling* involves interruption of the expected foot–floor contact. When we walk, the leading foot establishes contact with the floor (initially the heel, followed by a shifting to the toes). While the leading foot is in contact with the floor, the trailing foot leaves the floor and swings forward.[3] During this motion, the body's center of gravity continues moving forward. Normally the trailing foot ends the swing with a new heel contact in front of the center of gravity. In stumbling, the trailing foot is interrupted as it moves forward. Typically this involves the shoe sole brushing against the floor resulting in a slowing of the foot movement. Before assuming that only the most awkward among us would allow this to happen, recognize that the forward swing of most people provides a very small clearance with the floor. As people get on in years, they tend to have even less clearance. We have all seen older people shuffling along, almost dragging the following foot forward. Their clearance is so small that it takes only a slight elevation of the floor to catch the foot. The shoe sole can be easily slowed by a small bit of sticky surface such a gum deposited by an inconsiderate person days before. Fortunately, not all stumbling events lead to a fall. People in the workforce are generally fit enough to recover when the stumble only briefly interrupts their foot motion. Less fit people, especially the very old, are more likely to fall after stumbling.

A fall from stumbling resembles that from tripping. Loss of balance from tripping and stumbling produces a face-forward fall often accompanied by an extended arm impacting the floor. This response may soften the fall and avoid serious injury, or it may result in a broken bone. Fracture of the radius bone (a Colles' fracture) accounts for the largest portion of these fractures.[4]

A trip event occurs when a walking person's toe contacts an obstruction. The obstruction may be a protrusion from the walking surface or a loose object in the path. Examples are the curled edges of floor mats and wrinkles in rugs. These types of walking surface deviations are easily observed, and uncomplicated to correct. Employee awareness of the hazardous nature of such protrusions can provide the necessary monitoring for early detection.

23.2.2 Risk Reduction for Falls on the Same Level

Risk-reduction tactics for falls on the same level focus on preventing initiating events. Once the initiating event occurs, the outcome depends on the person's reaction. Aside from teaching people how to fall gently (e.g., skiers learn how to fall safely), there is nothing the OSH professional can do to affect the during-event phase of the incident. For the post-event phase, an effective first response can help reduce the ultimate severity of the injury.

The first priority is to eliminate the hazard. Eliminating slipping events involves providing consistently high friction throughout walkways. As explained earlier, a primary contributor to walkway slipping and stumbling is encountering an unexpected difference in the slipperiness of the surface.

The quantification of floor conditions from slippery to nonslippery uses the *coefficient of friction (COF)*. Values of COF, usually represented by the Greek letter μ, range from zero to one, with the low end for extremely slippery and the high end for extremely gripping. For an object sitting on a floor without moving, we use μ_s, the static value. For a moving object in contact with the floor, we use μ_k, the kinetic value. The flooring industry finds the static value useful as a guideline for quantifying slipperiness, with a value greater than 0.5 considered a rough dividing line between too little and adequate friction for flooring. The OSH community looks at the kinetic value as being more relevant because the most common slip-initiated falls occur during the early phase of heel contact with the floor—while the heel is moving forward. Figure 23.3 provides an explanation of the classic relationship among forces involved when a moving object is sliding on another solid object.

Researchers who study slipping while walking have concluded that the classic model for COF found in college physics textbooks contains assumptions unsuitable for directly characterizing shoe–floor friction due, in part, to the types of materials in the shoe soles, the flooring, and the contaminating material found on floors. These experts prefer to use a variable reflecting the friction required to counteract the shear force—the *required coefficient of friction (RCOF)*. The RCOF is the ratio of shear

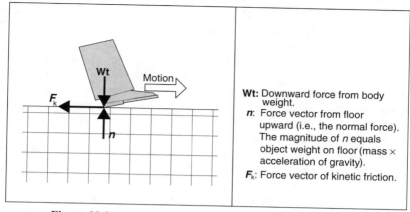

Wt: Downward force from body weight.

n: Force vector from floor upward (i.e., the normal force). The magnitude of *n* equals object weight on floor (mass × acceleration of gravity).

F_k: Force vector of kinetic friction.

Figure 23.3 Force vectors between heel and floor soon after contact.

force to the normal force. Normal force is the upward force from the floor pushing against the heel (see Figure 23.3). The way an RCOF variable may be used is illustrated by an example in Ref. 1. For a normally healthy person walking at a normal pace on a dry surface, an RCOF greater than 0.18 should be adequate to stop the initial heel slip quickly.[1] Thus, sometime in the future, there is hope that researchers will be able to use RCOF to define thresholds for floor friction.

One obstacle for standards development is the method for measuring μ_k. Various instruments have been developed for measuring the frictional properties of surfaces.[5,6] Unfortunately, no single instrument has achieved the status of "gold standard" for measuring the dynamic friction of floors in occupational facilities. Some progress has been made on standardizing surfaces that can be used to test the performance of slipperiness meters (tribometers) in terms of accuracy and reliability.[7] But more work lies ahead before instrument testing procedures are fully established and instruments meeting the standards become available for use by the OSH community. Even if that is accomplished, standards development committees will continue being challenged by the numerous factors required to validly differentiate acceptable from unacceptable floor slipperiness. A major factor is the ever-changing contamination of floors by dirt, oils, and water. Another is the incredible number of combinations of shoe sole materials and floor covering materials. Measuring the friction of a just cleaned and polished floor may seem like a good idea, but soon after people start walking on the floor, the measured levels will cease being representative of what the employees experience.

Given the present state of portable friction measurement instrumentation, employers may want to focus on making walking surfaces consistent, especially adjacent floors covered with different materials. Portable instruments are reasonably suited for identifying differences within the same environment. The difference approach would account for the underlying cause of slipping falls—an unexpected difference in COF. Also, measuring differences may help overcome some weaknesses in the different measuring instruments. Take, for example, two adjacent surfaces.

What if instrument A measured COF values of 0.7 and 0.4, while instrument B measured COF values of 0.6 and 0.3? These values do not match in absolute COF values, but they do provide the same difference (0.3). Perhaps, employers should start thinking about setting company guidelines for walkway surfaces based on avoiding abrupt changes in floor slipperiness as measured with existing portable devices. It would, of course, require making several measurements under varying conditions to obtain appropriate descriptive statistics (e.g., means and standard deviations). But that is no different from the professional practice of industrial hygiene that always emphasizes the importance of taking multiple measurements.

Another approach for eliminating falls on the same level involves housekeeping. Properly managed facilities designate walkways and fire egress pathways with painted lines. Rules are in place and enforced to maintain these floor spaces free of objects. Besides eliminating tripping hazards, maintenance of clear walkways sends a visual message to employees and visitors that the facility is committed to providing a safe environment.

When eliminating walkway hazards is not feasible, or impractical, there are ways to moderate the hazards. A common example involves the need to temporarily extend a power cord across a walkway. One approach is to eliminate the cord being a tripping hazard by hanging it above the walkway. Another way is to secure the cord to the floor with tape, under a carpet, or channeled within a commercially available device specifically intended for extension cords. With that approach the tripping hazard remains, but in a modified form with less risk than would otherwise be the case.

Cracks in walking surfaces present tripping hazards. The photo of a public sidewalk in Figure 23.1 shows a substantial crack on which a pedestrian could easily catch a foot and trip. If one side of the crack is elevated above the other, a person's toe can catch the raised part and initiate a trip. An unresolved issue concerns how great the elevation difference needs to be before we consider it a tripping hazard. One view is that a single elevation difference can be applied to all public places.[8,9] For example, a rule of practice may use a 10 mm or greater difference for categorizing a floor elevation change as being a tripping hazard. Another view is that the tolerable elevation difference should consider the flooring defect itself, the location, and the use pattern. This view finds support in the argument that worker expectations vary. For example, workers on construction sites and in underground mines, and railroad personnel doing track maintenance do not expect level walking surfaces. In contrast, visitors to shopping malls expect very level walking surfaces; after all, they are encouraged to look at merchandise displays rather than at the floor ahead. In between these extremes are the diverse workplaces in which a large percentage of people work. Thus, when it comes to walking surfaces having cracks with elevation differences, there are two views on how to distinguish between a crack large enough to warrant promptly taking countermeasures and one small enough to tolerate for the time being.

When substantial cracks develop in walkways, prompt repair is the preferred countermeasure. Less permanent alternatives involve modifying the hazard. One temporary modification is for the facility to mark the crack to draw the attention of approaching pedestrians. Marking may be as simple as painting the elevated side with orange paint. A warning sign may contribute to drawing attention to the hazard.

For some hazards on walking surfaces, redirecting pedestrians around the hazard provides a practical tactic. Workers needing to use part of a public sidewalk for their work routinely use this tactic. They may be repairing the sidewalk, loading/unloading materials, or working on a building exterior. Another common use of this strategy involves floor mopping. Janitors doing floor mopping routinely put up signs to redirect pedestrians away from the wet, slippery floor. In situations like these, the strategy is to separate the hazards of the work from the passing pedestrians.

While the conditions of level walking surfaces in workplaces and public places can present several hazards for pedestrians, the associated falls involve a relative short fall distance. The hazards of falling on a step or stairway present greater fall distances, and more severe injuries.

23.3 STAIRWAYS AND STEPS

Stairways may be thought of as inherently hazardous locations capable of being made acceptably safe by appropriate design, construction, and maintenance. Stairway characteristics affecting safety include the step dimensions, step uniformity, handrails, and guardrails.

Key step dimensions are riser height and tread depth. Not all standards agree on the details of how these parameters are defined and measured. Figure 23.4 illustrates a method based on effective dimensions. The terms *effective tread depth* and *effective riser height* indicate the measurements reflect dimensions critical to foot placement, and are made from the same point on a step to the corresponding point on the adjacent step. This approach accounts for any slope (wash) or irregularities in the step surface. In the illustration, the effective tread depth is 305 mm (12 inches), and the effective riser height is 176 mm (6.9 inches).

Taken together, the dimensions in Figure 23.4 make a 30° slope. Studies based on metabolic energy expenditures for ascending stairs indicate a preferred range of 25–30°. Studies of misstep frequency while descending indicate desirable tread

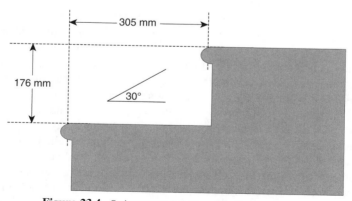

Figure 23.4   Stairway step height and depth measurement.

depths are over 305 mm (12 inches).[10] Taken together, most guidelines applicable to workplace stairways approve of dimensions somewhat like those in the illustration. However, guidelines do not specify exact dimensions because architects need some flexibility to design stairways to fit the floor space available and the height difference between the bottom and top.

The riser height in Figure 23.4 is appropriate for stairways in work settings. If a stairway is for general public use, including very young children or older/retired persons, the risers should be shorter (roughly in the range of 130–150 mm) to accommodate the needs of these diverse populations. The tread depth in Figure 23.4 provides a sufficient landing area for feet while descending. As tread depths get shorter than 280 mm, expect an increase in missteps during descent. Two patterns of missteps on descent are involved. One is the foot slipping on the nosing. The other, the more common one, involves overstepping. An overstep results in the heel of descending persons just catching the edge of the step, rather than landing firmly on the step. When only the heel catches, the foot bends downward rapidly, the knee buckles, and the body lunges forward.

Step uniformity is widely recognized as a component of safe stairways.[10–12] The reason requires an understanding of typical human behavior when using a stairway. Observations of people using stairways found patterns.[10] When people approach a stairway from above, they look at the upper steps. They tend to continue looking as they take the first step down. After one or two steps down, their eyes no longer look at the stairway. Because they are accustomed to stairways with uniform steps, they expect uniform steps and feel no need to continue looking downward. They adjust their stepping motions (gait) to match that of the first and second steps. If somewhere in the stairway a particular step has different dimensions, there is a mismatch between reality and human expectation. Thus, the reason stairway step variation is a risk factor for falls is that (1) people do not expect step variations, (2) they do not easily notice step variations, (3) if they are unaware of a step variation they will not adjust their gait as needed, and (4) they can easily misstep on the nonuniform step or an adjacent step. Once the misstep occurs, the person may fall or recover. If a fall occurs, the person often impacts the nosing of one or more steps as they descend. Very severe injuries and death can result from such falls.

There are ways to help people see deviations in step dimensions. One is to provide a visual contrast at the leading edge, or nosing, of each step. Many stairways have nosing strips for this purpose. The person standing at the top of a flight can look down and visually detect nonuniformity among these edge lines. Someone doing safety inspections may also use this simple technique. Opposite of making the edge lines easily visible is the poor choice of covering steps with heavily patterned flooring material or carpeting. To the person descending, the patterns act like camouflage, obscuring visual differences between adjacent steps. Another way to help people see deviations in steps is to provide uniform lighting throughout a flight. Uniform lighting avoids the shadows that can deceive the visual senses. A third approach is to avoid visual distractions within a flight. Investigations of people using stairs have found that visual changes, especially changes that attract the eyes to one side of the stairway, tend to draw the attention of the stairway users away from looking ahead. While looking elsewhere, the person has no chance of noticing anything unusual about the steps ahead, and they can easily misjudge when they have reached the last step.

Architects often specify nosing on the leading edge of steps (see Figure 23.4). Nosings help with ascending by making the step edge more visible and with descending by providing consistent friction at the edge of each step in the flight. Older stairs without nosings tend to have worn spots on the leading edge due to thousands of shoes rubbing the same spots. These spots become more rounded than other parts of the step, so the steps are no longer uniform.

Several recommendations for architects are provided in Ref. 12. These address such matters as planning for pedestrian traffic flow with the goal of minimizing stairway usage, avoiding traffic conflicts, and providing sufficient width to accommodate one person ascending while another person is descending. For example, in a workplace, if workers need to use a stairway to get between their work location and the restroom, they will be using the stairs frequently. A lower risk design will put the restroom on the same level as the work locations. In addition, architects may benefit from thinking of the top of a stairway as a hole in the floor. A hole in a floor is easily recognized as a danger for anyone who fails to notice it. For example, a janitor may back into a stairway while sweeping or mopping the floor; or a person may be walking toward a location, with their vision focused well above the opening, and never notice the stairway in their path. Thus, the location and design of the upper opening to a stairway needs consideration. One feature that helps make the stairwell visible is the handrail.

Stairway handrails contribute to both safety and usability. For ascent, people use a handrail to help pull their body upward. For descent, handrails are valuable to assist with postural orientation and to help recover from a misstep. The more serious fall injuries occur during descent, so placing a handrail on the side most used for descent makes sense. In the United States, the trend is to descend using the right side of the stairs. Thus, a handrail on the right side is a fundamental safety device. If the opposite trend is found in some other countries, the handrail recommendations need reversing. Having a handrail on the side used for ascent is less fundamental for safety, but useful for people ascending and for those descending who elect to use the handrail on that side. Handrails on both sides are useful for wide stairs and for stairs frequently used by older people.[12]

Handrails come in many shapes and sizes. To serve as effective safety devices, the handrail should be suitable for the descending person to easily slide their hand along the entire flight without interruption. It should be positioned away from the wall, and be shaped for easy gripping with a power grip. The preferred cross sections for power gripping are circular and oval shapes, with a horizontal diameter in the range of 38–50 mm (1.5–2 inches). The height of the handrail also needs consideration. Building codes typically specify or recommend a range for handrail heights. Since not all populations in the world have the same body size characteristics, the desirable range for handrail heights should consider the anthropometrics of the population.[13] For example, building codes in the United States provide a range of acceptable handrail heights (30–34 inches or 76–86 mm). This range would not be ideal for the populations found in many other countries. Within the range of acceptable heights, a higher placement is better for recovering from a misstep.

Many people do not distinguish between a stairway handrail and a *stairway guardrail*. A handrail is attached to a wall on one or both sides of a stairway. A stairway guardrail (also called a railing) is for an open side of a stairway with the primary function being protection from falling off the edge of the stairway. While the top rail can function as a handrail, the newer approach is to incorporate into the guardrail a separate handrail at the appropriate height. For workplaces, the standards for guardrail heights have been increasing in order to prevent persons from falling over the top rail. With these higher guardrails, worker may find the top rail feels awkward to use as a handrail during descent, especially the shorter workers. Many manufactured stairway guardrails are available with a top rail suited for fall protection and use by taller personnel, plus a lower rail suited for shorter people.

This discussion would be incomplete without pointing out the danger of short flights. Specifically, a single step down often goes unnoticed. Flights of two steps also have that problem. When the walker steps off the upper level, expecting their heel to land at the same level, they are completely unprepared when the foot only finds air. They tumble face first to the floor. They may instinctively extend an arm and break their radius bone. They may hit their face or forehead on the floor. Sometimes, a walker approaching a single step from the lower level will not notice the step, and trip on it. Recommendations by leading experts are consistent with the risk-reduction strategies listed in the previous chapter. First, architects may find a way to eliminate the need for a change in floor level by, for example, building the lower floor up to the same level as the upper floor. Second, substituting a ramp may be feasible. This is an excellent example of substituting a lesser hazard for a greater hazard. A third option is help the workers perform safely by making the change in elevation highly visible by installing features easily visible at heights well above the floor. A pair of handrails is a good start. A ceiling slope corresponding to the ramp or stair slope is another visual cue. Different lighting or different wall colors may also aid recognition.

23.4 RAMPS

Ramps provide another means for people to transition between floor levels. Issues associated with ramps include person's falling while walking on a ramp, usability for people in wheelchairs, tripping on ramp edges, falling off the side of ramps, and operating forklift vehicles on ramps. Building codes provide design guidelines and standards.

Persons falling while walking on a ramp occur most often during descent. One pattern is they unintentionally increase their walking speed to a point where they lose balance and tumble forward. Another pattern involves excessive slipping of the heel on contact with the floor. The slipping risk is greater than walking on a level surface because of the angle the heel makes with the floor. The RCOF is greater, and may not be met. For that reason, guidelines and standards specify giving the ramp surface a course texture to achieve a high COF. Also, the RCOF increases as the steepness of the ramp increases. Guidelines may specify a maximum slope for ramps intended for walking, but guidelines for wheelchair use specify less slope. Thus, the architects need information about the expected users. Providing a handrail, like with stairways, is a tactic for aiding with recovery from a slip.

People approaching a ramp from the side may not notice the rise near the bottom and trip on it. Therefore, making the lower part of the ramp highly visible when viewed from the side serves to help people avoid tripping. As the ramp rises, the potential fall distance gains importance. The usual safety device for this hazard is a guardrail along the open side of the ramp. A guardrail serves three safety purposes: it protects people from falling off the edge, it can serve as a handrail to aid people on both ascent and descent, and it provides visual cues of the ramp's existence to people walking near the ramp.

Forklift vehicles using ramps encounter a balance issue. Operating with the load on the downhill side can cause the vehicle to tip forward, with the front tires serving as a fulcrum. The rear wheels rise up and lose contact with the ramp surface. Since steering occurs through the rear wheels, the operator loses directional control. Therefore, when descending a ramp while carrying a load, the safe practice is to operate with forks in the uphill direction. Thus, having operators back down the ramp is the safe practice. When ascending a ramp while carrying a load, the safe practice is also to drive with forks in the uphill direction. If the load obstructs the operator's view ahead, a spotter assists the operator.

The various hazardous locations discussed above are working and walking surfaces that can be made safe for human use by using known risk-reduction strategies. The most common of these strategies is helping workers perform safely. We turn next to hazardous locations commonly known as "confined spaces."

23.5 CONFINED SPACES

The OSH community uses the term "confined space" to identify a class of locations for which special precautions are needed prior to entry. These locations are not intended for human occupancy, but may, infrequently, require a person to enter. Because many deaths have occurred in confined spaces, industrialized countries have regulations for confined space entry. Although these requirements vary somewhat in definitions of confined spaces and precautions required, their underlying goal is to prevent the sorts of fatal scenarios that killed workers in the past. As with other regulations, there will be some misinterpretations, misconceptions, and misapplications of the requirements (see paper by Taylor on the required precautions in the United States).[14]

A particularly troublesome scenario occurs when one member of a work crew enters a space, and collapses. A coworker sees what happened and enters the space with intent of a rescue.[15] The second worker also collapses. There have been cases of a three or more repeats of this deadly sequence.

The standards start by defining what is, and is not, a confined space. Definitions specify attributes of spaces included in the particular standard. Some attributes are the space (1) was not created for human occupancy; (2) has limited means of egress and ingress; and (3) lacks natural ventilation. Some definitions include an attribute about having a particular type of recognized hazard in the space. Some regulations define two categories of confined spaces based on level of threat to the entrant. This leads to a requirement to identify all spaces in a facility that fit the definition. Each space requires a sign posted at the entrance identifying it as a confined space, stating no entry unless authorized, and indicating the means for obtaining authorization. For the more hazardous spaces, someone designated by the organization must issue a permit after making sure the organization's procedures will be followed before, during, and after the entry. Among the procedures are requirements that the personnel involved have completed appropriate training. The paper permit is posted outside the space so that anyone walking by can check to see if the entry is an authorized entry.

Some common confined spaces are storage tanks, compartments of ships, process vessels, pits, silos, vats, wells, sewers, digesters, degreasers, reaction vessels, boilers, ventilation ducts, tunnels, underground utility vaults, and pipelines.[15] The most commonly encountered conditions that make confined spaces hazardous are

- Oxygen deficiency,
- Oxygen displacement (inert gases and simple asphyxiating substances),
- Flammable atmospheres,
- Toxic gases,
- Solvents,
- Engulfment, and
- Other physical hazards.

Some of the physical hazards in the last category warrant mention. The space may contain mechanical equipment or electrical connections that should be de-energized and locked out throughout the entry. The space may be connected to lines for material flow during normal operation. Before an entry, these connections need to be physically disconnected, blanked off, or emptied by procedures such as double blocking and bleeding out the material between the blocks. In spaces entered from above, there may be objects that could fall through the opening and hit the worker inside. There may be objects and surfaces hot enough to burn skin if touched. There may be moisture that could increase the hazards of electrical sources, and there may be a noise source in the space that gets amplified by reverberating off the walls in the space.

People in confined spaces have died from lack of oxygen resulting in several scenarios. One is the original atmosphere in the space has inadequate oxygen to sustain life. A person enters the space and after some time starts to experience symptoms of oxygen deprivation, losing ability to climb out of the space, and collapsing. A second scenario for asphyxiation occurs after purging a space with an inert gas to prevent a fire. If a person enters the space while it contains mostly inert gas, the shortage of oxygen will lead to oxygen deprivation, and eventually to asphyxiation. A third scenario occurs in containers for storing loose materials, like grain or coal dust. A common fatal scenario with loose materials is a person in the space above the materials falls into the materials. With their head fully engulfed, they desperately gasp for air, only to have their lungs filled with the material. The fourth asphyxiation scenario involves confined spaces with a chute at the bottom. A person falls into the space, becomes trapped in the chute, and suffocates.

Another fatal scenario in confined space is a fire or explosion. These occur in spaces normally used to store flammable material. If the task is to enter the space to perform cleaning or other maintenance, the first step is to empty the container of the flammable material. This inevitably leaves a residue of flammable vapors in the space. One procedure is to set up ventilation to pull the flammable vapors out while bringing fresh air in through another port. After terminating ventilation, there may be some vapors in the space with a vapor concentration within the flammable range. Fires have resulted when a person enters space, starts the task, and somehow causes a spark. The person has almost no chance of escaping. In these applications today, it is standard practice to test the atmosphere in the space using a flammable gas meter before allowing anyone to enter. Supplementing the precautions of venting and testing for flammable vapors, the employer can equip the maintenance person with non-sparking tools. A fire will ignite only if all three precautions fail.

Toxic atmospheres often occur in confined spaces. It is obviously preferred to avoid ever having a person enter a space contaminated with a toxic substance, but when it becomes necessary, providing the entrant with appropriate respiratory protection effectively separates their breathing zone from contaminated air inside the space. Appropriate breathing devices for entering confined spaces with unhealthy atmospheres are supplied air respirators and self-contained breathing apparatus.

The primary tactics for preventing deaths in confined spaces start with training the entrant and the person who will be monitoring the work so that they will be able to properly perform the work and deal with unexpected deviations in the process. On the day of a confined space entry, precautions include preparing the space by implementing the precautionary practices specified in the company's confined space entry program, and completing the task as planned. An entry that goes according to plans is part of Haddon's pre-event phase. If something happens that jeopardized the entrant's safety, the during-event phase would start and continue until the person exits the space. Haddon's third phase involves first aid and transport to a medical facility.

Further sources of information on confined spaces are readily available in OSH books, journals, and websites of governmental agencies in the applicable country. The next section addresses workplace conditions with dusty breathing air.

23.6 DUSTY AIR

Numerous kinds of airborne *dust* have properties with potential for harming exposed people. The main concern with these small particles is their effects stemming from breathing the particles. Some small particles encountered in workplaces have a long history of causing lung diseases named to reflect the cause. Some well-known diseases and agents are silicosis and silica dust, asbestosis and asbestos, black lung and coal dust, and brown lung and cotton dust. Some kinds of dust are associated with, or causes of, occupational diseases with long latency periods.

What happens after a particle enters the nose or mouth varies. Some small particles are trapped in the moist lining of the nose, mouth, throat, and windpipe (trachea). Some particles make it to the upper lungs only to get caught by the lining of the bronchus. Some of the smallest particles make it to the alveoli. Generally, the concentration of large particles reduces quickly in the upper passageways, mid-sized particles penetrate deeper before being caught, and the smallest particles reach the *alveoli*. This explains the recent fuss about nanoparticles because they are so small that they behave much like air contaminants in the gaseous state. Nanoparticles and gases distribute throughout the pulmonary system.

Once the inhaled particle reaches a stopping place in a respiratory area, numerous things may occur. Most of the larger *particulates* get stopped in the mucus lining of the upper airways, and expelled through a natural defensive mechanism. Particulates not expelled may lodge permanently in narrower passageways of the lungs. The smallest particulates and gaseous chemicals have the best chance of reaching the alveoli. Once settled deep in the lungs, the particle may just remain there, taking up space formerly used for normal lung functioning, or it may increase in size due to a physiological reaction to coat the foreign object with more material. With long-term exposure, thousands of small particles can accumulate, significantly decreasing the functional capacity of the lungs.

Strategies for protecting workers from airborne particulates emphasize ventilation and use of respirators. It is hard to imagine a situation where eliminating the hazard or substituting a less hazardous dust would be feasible. Avoiding the release of the hazard (Strategy 2) is a feasible strategy exemplified by watering dirt roads and other work areas where dust can be blown into the breathing zone of workers.

Buildings for industrial processes involving dust can be constructed so that personnel are separated from the dust (Strategy 5). Improving the resistance of that which needs protection (Strategy 6) is not a feasible option for employees, but it is feasible for coating surfaces of objects exposed to moving particulates. Many employers provide dust filtering respirators (Strategy 8) along with training for properly using and maintaining the respirators (Strategy 7).

An example of the ninth strategy, expedite recovery, is to provide exposed workers with pulmonary function tests in order to detect early signs of reduced lung capacity. A finding of reduced capacity could trigger medical consultation to make a diagnosis and establish an appropriate course of action.

LEARNING EXERCISES

1. Explain in your own words the author's point about the role exposure plays when assessing the importance of work floor surfaces being less than ideal.

2. Suppose in your workplace an employee slipped on a grease spot, fell, and injured her hip. A supervisor filled out an incident report in which the correction he recommended was that employees need to pay attention while walking. What could you say to educate the supervisor about the role of human expectations?

3. In the event tree shown in Figure 23.1, hypothetical probability values are provided for each intermediate event. Assume the initiating event occurs. What will be the computed probability values in the right column?

4. Summarize the obstacles in the way of developing a broadly acceptable standard for floor slipperiness.

5. Explain the author's suggestion to working around the obstacles.

6. What is the limitation in saying any crack in a floor surface with a difference in elevation of more than 10 mm is a hazard?

7. For stairways, explain what is meant by effective tread depth and effective tread height.

8. Explain why step uniformity is so important for stairways.

9. Why do many manufactured stairway guardrails have a built-in handrail instead of just letting the top of the guardrail serve the same purpose as the handrail?

10. If a facility has short flight of steps (one or two steps), what can be done to make it easily visible to employees approaching from above?

11. When a container normally used for a flammable liquid needs interior maintenance, the container is emptied and vented before allowing an employee to enter. Two precautions for fire protection are testing the vapor concentration prior to entry, and providing the employee with non-sparking tools. Use concepts from fault trees to explain why both precautions provide more safety than either alone.

TECHNICAL TERMS

Alveoli	Tiny sacs located at the ends of branches of the pulmonary system. Through the thin walls of the alveoli, chemicals are exchanged with nearby blood vessels—oxygen from inhaled air goes to the bloodstream, and carbon dioxide goes from the blood to the air in the alveoli where it is removed with exhaled air.
Coefficient of friction (COF)	A number between 0 and 1, reflecting the extent of friction of a surface and computed as the ratio of shear force to normal force. See Figure 23.3 for further explanation of force vectors involved.
Dust	Solid particles of organic or inorganic materials formed by breaking down larger collections of the same matter such as coal, wood, and grain. Airborne dusts tend to settle out of calm air under the force of gravity.

Effective riser height	The difference in height between adjacent steps in a stairway, determined from the same points on the leading edge of each. See Figure 23.4.
Effective tread depth	The horizontal distance between adjacent steps in a stairway, determined from the same points on the leading edge of each. See Figure 23.4.
Particulate	A small bit of solid or liquid matter. Aerosols and dusts are particulates.
Required coefficient of friction (RCOF)	The COF needed to counteract the shear force of an object sliding on the surface. If the RCOF is greater than the shear force, the object will decelerate and come to a stop.
Stairway guardrail	A fall protection barrier located on the side of a stairway where a fall might occur. Secondary uses include functioning as a handrail and providing a visual cue to make people aware of the stairway.
Stairway handrails	A bar along the wall side of a stairway that benefits those who use it by giving them something to help pull themselves up the stairs, improving their sense of postural orientation, and giving the person who has a misstep a contact point to help recover rather than fall. A handrail also provides a visual cue that aids postural orientation.
Stumbling	A type of misstep that occurs during locomotion. A stumble results from contact between the floor and the sole of the foot or shoe causing unexpected frictional resistance to the forward motion. The stumble event may lead to recovery or a forward fall.

REFERENCES

1. Chambers AJ, Margerum S, Redfern MS, Cham R. Kinematics of the foot during slips. Occupational Ergonomics. 2002/2003;3:225–234.
2. Redfern MS, Cham R, Gielo-Perczak K, Grönqvist, R, Hirvonen M, Lanshammar H, et al. Biomechanics of slips. Ergonomics. 2001;44:1138–1166.
3. Bakken GM, Cohen HH, Abele JR, Hyde AS, LaRue CA. Slips, trips, missteps and their consequences, 2nd ed. Tucson, AZ: Lawyers & Judges Publishing; 2007. p. 5–24.
4. Hyde AS. Fall injuries. In: Bakken GM, Cohen HH, Abele JR, Hyde AS, LaRue CA, editors. Slips, trips, missteps and their consequences, 2nd ed. Tucson, AZ: Lawyers & Judges Publishing; 2007. p. 127–132.
5. Brauer RL. Safety and health for engineers, 2nd ed. Hoboken, NJ: Wiley; 2006. p. 139–142.
6. Bakken GM, Cohen HH, Abele JR, Hyde AS, LaRue CA. Slips, trips, missteps and their consequences, 2nd ed. Tucson, AZ: Lawyers & Judges Publishing; 2007. p. 24–27.
7. ASTM F2508-11. Standard practice for validation and calibration of walkway tribometers using reference surfaces. Available at http://www.astm.org/Standards/F2508.htm.

8. Di Pella S. Slip, trip, and fall prevention: A practical handbook, 2nd ed. Boca Raton, FL: CRC; 2010. p. 1–2.

9. Bakken GM, Cohen HH, Abele JR, Hyde AS, LaRue CA. Slips, trips, missteps and their consequences, 2nd ed. Tucson, AZ: Lawyers & Judges Publishing; 2007. p. 34–36.

10. Templer JA, Mullet GM, Archea J, Margulis ST. An analysis of the behavior of stair users (NBSIR 78-1554). Gaithersburg, MD: National Bureau of Standards (now National Institute of Standards and Technology); 1978.

11. Cohen J, LaRue CA, Cohen HH. Stairway falls: An ergonomics analysis of 80 cases. Professional Safety. 2009;54(1):27–32.

12. Archea J, Collins B, Stahl F. Guidelines for stair safety (NBS Building Science Series 120). Gaithersburg, MD: National Bureau of Standards (now National Institute of Standards and Technology); 1979.

13. Kroemer KHE. Fitting the human: An introduction to ergonomics, 6th ed. Boca Raton, FL: CRC; 2008. p. 3–30.

14. Taylor B. Confined spaces: Common misconceptions & errors in complying with OSHA's standard. Professional Safety. 2011;56(7):42–46.

15. Pettit TA, Braddee R. Overview of confined-space hazards. In: Worker deaths in confined spaces. DHEW (NIOSH) No. 94-103. Cincinnati, OH: National Institute for Occupational Safety and Health; 1994. p. 5–10.

Chapter 24

Chemical Substances

Chemicals encountered in workplaces make many processes possible, and chemicals encountered outside of work make modern society possible. These, and many other beneficial uses of chemicals, are not without risks. Textbooks, reference books, websites, and scientific literature provide abundant information on chemicals. This chapter begins with a brief review of chemicals found in many workplaces, followed by a summary of the mechanisms of harm to workers, and ends with examples of strategies for limiting exposures.

24.1 MAJOR CATEGORIES OF CHEMICALS ENCOUNTERED AT WORK

The summaries in this section group chemicals into categories familiar to the OSH professionals. Some of these groups are based on similarity of use (e.g., solvents and pesticides), one by state of the compound (fumes from molten metal), and several by similarity of health effects.

Industrial solvents like turpentine and mineral spirits function effectively to dissolve unwanted materials such as adhesives, cleaning fluids, paints, and glues. The OSH community is particularly concerned with the health effects and flammability of industrial solvents.

Pesticides are compounds for killing pests, and to perform that function, pesticides must have poisonous properties. The toxicity levels for humans range from too toxic for use to low enough to be approved for selling to consumers without restrictions. The role of OSH professionals is to champion processes for minimizing risks for employees involved in the manufacturing, distribution, and application of pesticides.

Metals and metal compounds make up endless solid items like vessels, motors, mechanical energy transmission components, industrial machines, and heavy equipment. A major concern to OSH professionals is worker exposure to the fumes emitted from metals in the molten states such as during welding, ore processing, and metal production.

Risk-Reduction Methods for Occupational Safety and Health, First Edition. Roger C. Jensen.
© 2012 John Wiley & Sons, Inc. Published 2012 by John Wiley & Sons, Inc.

Asphyxiants harm by interfering with the essential function of supplying oxygen to various body parts. The basic process of breathing involves inhaling fresh air, extracting oxygen from it, and giving up carbon dioxide with the exhaled air. Chemicals that change the natural breathing process create problems. Normal fresh air is approximately 79% nitrogen, 21% oxygen, and very small amounts of other gases. Inhaled air with significant concentrations of carbon monoxide or carbon dioxide can alter the normal breathing process by reducing the supply of oxygen to organs and tissues. Some other well-known asphyxiants that interfere with normal breathing include cyanide gas, which prevents cells from using oxygen; nitrous oxide, which impairs the usual process of lungs to expel contaminants; and inert gases used to fill a space for fire prevention, which makes the space unacceptable for human respiration.

Corrosive substances react with and damage human skin as well as man-made materials. Two large groups of corrosive chemicals are acids and bases. Common acids are acetic acid, hydrochloric acid, nitric acid, chromic acid, hydrofluoric acid, perchloric acid, sulfuric acid, and fuming sulfuric acid.[1] Common bases are potassium hydroxide, sodium hydroxide, aluminum chloride, bromine, phosphorus trichloride, potassium bifluoride, sodium hypochlorite, and zinc chloride.[1]

Irritants are chemicals that irritate tissue by direct contact, causing an inflammatory response. The most common occupational exposures occur by inhalation of air containing irritant chemicals. Common irritants that affect the mouth, throat, and lungs are acetic acid, acrolein, formaldehyde, and formic acid.[1] Common eye irritants are ammonia, alkaline dusts and mists, hydrogen chloride, hydrogen fluoride, halogens, nitrogen dioxide, ozone, phosgene, and phosphorus chloride.[1]

Neurotoxic chemicals adversely affect the nervous system. Exposures can interfere with normal functions of the central nervous system, efferent nerves, peripheral nerves, and sensory organs. Some neurotoxins are methyl mercury, carbon disulfide, manganese, organic phosphorus insecticides, and tetraethyl lead.[1]

Oxidizing chemicals are capable of producing oxygen by spontaneously reacting with organic or combustible materials in environments with no more heat than ordinary room temperature. By producing oxygen, these reactions can contribute to fires and explosions. A few examples of oxidizing agents are perchloric acid, chromic acid, and inorganic peroxides.

Three chemical classifications defined by health effects—carcinogens, mutagens, and teratogens—warrant special attention. Carcinogens increase risk of cancer, mutagens increase risk of genetic mutations, and teratogens increase risk of birth defects in the offspring of those exposed. The scientific evidence for classifying a chemical into one or more of these categories comes from both epidemiological studies of humans and laboratory studies of mice and other laboratory animals. Standardization bodies examine the strength of this evidence and assign chemicals to categories such as "known" and "suspected" carcinogens, mutagens, or teratogens. As might be imagined, different committees of experts, looking at the same evidence, may disagree on which category to assign a particular chemical.

Fuels for fires and explosions consist of one or more chemicals. These materials are hazards because of their potential to be transformed from a state of potential energy to a fire capable of causing harmful consequences. In this book, separate chapters discuss fires and explosions. This chapter addresses only health effects of exposure to fuels.

24.2 MECHANISMS OF HARMING

The effects of chemical agents are complex because effects are chemical specific, organ specific, dose related, and affected by duration of influence on the organ. Specialists in the field of industrial toxicology are continuously contributing new knowledge to our understanding of the diverse combinations of chemicals, organs, and effects. The brief overview in this section provides a foundation for readers not already familiar with industrial hygiene.

Chemicals can harm the environment through emissions into the air, water, or soil. Chemicals in the air or water can slowly corrode metal structures and vessels through the process of rusting. Chemicals can harm people by entering the body through any of three routes—inhalation, ingestion, or absorption through the skin.

The route most commonly encountered in industrial workplaces is inhalation—the route through the mouth, throat, and lungs. Contaminants in gaseous state flow like the rest of the inhaled and exhaled air. Chemical contaminants in particulate form follow different patterns, largely determined by size but also influenced by other attributes. Some particles are trapped in the moist lining of the nose, mouth, throat, and windpipe (trachea). Some particles make it to the upper lungs only to be caught by the lining of the bronchus. Some of the smallest particles make it to the *alveoli*. Generally, the large particles penetrate least, mid-sized particles extend deeper into the lungs, and the smallest particles, including nanoparticles, reach the alveoli. In the alveoli, particulates may break down into their constituent chemicals. The chemicals in the deep lungs may diffuse across the alveoli walls, enter the bloodstream, and distribute throughout the body.

The ingestion route for chemicals starts with entry into the body. A common mechanism is purposefully eating food, drinking liquids, or taking oral drugs (lawfully prescribed or illegally obtained). Any of these ingested materials may have chemical contaminants. Inadvertent ingestion mechanism involves the person's hands transferring chemicals to food prior to consumption, and chemicals introduced to the mouth by contaminated hands touching the lips. Eating food containing potentially toxic properties is a common mechanism of ingestion. Consuming spoiled food can introduce biological agents that release harmful chemicals in the digestive tract. On rare occasions, a person may ingest food or drink another person (acting criminally) deliberately contaminated with a harmful chemical.

The third route of entry is by direct contact with the skin, eyes, or other exposed body parts. Skin contact by some chemicals leads to dermatitis (e.g., alkalis, acids, vapors, and irritant gases). A subset of the direct contact cases comes from employees becoming sensitized during their work around the chemical. *Occupational dermatitis* cases make up about half of the workers' compensation claims for compensable diseases. Eye contact with some chemicals causes symptoms such as irritation, red eye, and watering of the eyes.

Industrial solvents can harm by their inherent property of being able to dissolve other substances and materials. Inhalation and direct skin contact are the common routes of entry. Physiological effects vary considerably, depending on the specific solvents and extent of exposure.

Metals and metal compounds in a molten state produce fumes that workers may inhale. Depending on the exposure dose, health of the individual, and composition of the fumes, symptoms of certain diseases can develop. Many welders have developed a set of symptoms known as metal fume fever. These symptoms include fever, headache, cough, and a metallic taste.[1] For many welders working a traditional workweek, their symptoms are most intense on Monday mornings. The theory is they have some immunity to welding fumes that reduces over the weekend, so on Monday morning, they feel the effects, but soon reenergize their immunity and work the remainder of the week without being bothered. The immunity theory applies only to the acute symptoms mentioned. There is no reason to think this immunity provides any protection from the effects of long-term exposure to metal fumes. Some metal fumes have been causally linked to particular diseases, for example, copper fumes being linked to Wilson's disease and Indian childhood cirrhosis.[1] Exposures to some metal fumes have been linked to cancer.

Inhalation of metal fumes occurs regularly when welding without respiratory protection. The fumes are the result of molten metal giving off metallic vapor in gas form that regroup into tiny droplets of liquid metal and start to descend. Other chemicals are often part of the fumes. Welding fumes are primarily metals, but the fumes can contain other chemicals such as those from the coating material coming off the welded surface.

Pesticides come in numerous forms—some with much more harmful effects on humans than others. Several of the pesticides, with sufficient dosage, can interfere with a critical neurotransmitter chemical acetylcholine. And some, especially the organochlorine pesticides, have been banned from use because their persistent chemical makeup would, if usage continued, lead to ever-increasing concentrations in the groundwater, lakes, seas, and oceans. The water concentrations are picked up by the fish, and may end up on the human dinner plate. Ingestion of some of these chemicals may increase risk of cancer. Employees engaged in the manufacture and transportation of pesticides, as well as the agricultural workers, are potentially exposed.

24.3 STRATEGIES AND TACTICS FOR WORKPLACE CHEMICALS

The strategies for reducing risk with chemicals are listed in the left column of Table 24.1. On the right are major groups of applicable tactics.

Table 24.1 Strategies and Tactics for Reducing the Risks from Workplace Exposures to Chemicals

Risk-reduction strategy	Risk-reduction tactics
1. Eliminate the hazard	Eliminate the hazardous chemical from the workplace
2. Moderate the hazard	2a. Substitute a less hazardous chemical
	2b. Reduce the amount of the chemical
	2c. Reduce the concentration of the chemical
3. Avoid releasing the hazard	Enclose the chemical in a container that prevents any release
4. Modify release of the hazard	4a. Use local ventilation to control any released substance
	4b. Direct airflow to take process emissions away from breathing zone of workers
5. Separate the hazard from that needing protection	5a. House processes with hazardous chemicals in separate structures from those used by personnel
	5b. Use building walls to separate hazardous chemical locations from rooms used by personnel
6. Improve the resistance of that needing protection	6a. Apply protective coatings on materials used for equipment, tools, products, etc.
	6b. Apply protective ointments to skin prior to possible exposure to dermatitis-producing chemicals
7. Help people perform safely	7a. Instruct employees on safe procedures for working with or near hazardous chemicals
	7b. In rooms with hazardous chemicals, provide written material describing the hazardous properties of those chemicals
	7c. On chemical containers, provide warning labels with information for safe handling. To be effective, the text and symbols should fit the exposed population
	7d. Provide people with information about appropriate practices for avoiding ingestion of hazardous chemicals
8. Use personal protective equipment	8a. Provide and use whole-body protection by enclosing personnel in protective ensemble appropriate for the exposure conditions
	8b. Provide and use respiratory protection appropriate for the conditions.
	8c. Provide and use skin protection with a wearable chemical barrier such as chemical protective gloves
	8d. Administer a PPE program to ensure effective use of the PPE provided
9. Expedite recovery	9a. Provide safety showers and eyewash stations in appropriate locations

(continued)

TABLE 24.1 (*Continued*)

Risk-reduction strategy	Risk-reduction tactics
	9b. Include in first-aid courses instruction on procedures for responding to personnel who contact hazardous chemicals through ingestion, inhalation, or direct skin contact
	9c. Provide the phone number for a poison control center in rooms where hazardous chemicals are used

The first strategy is to eliminate the hazard. This strategy applies to the full range of system levels. At the workstations level, there are opportunities to remove hazardous chemicals by storing in other areas. At the facility level, this strategy applies to both stored chemicals and in-process chemicals. During plant design, or subsequent process modification, consider options for not having particular hazardous chemicals within the facility. This may necessitate having other firms perform the processes involving those chemicals. Although this solution carries ethical baggage, the outsourcing approach eliminates the hazardous chemical from one facility. At a broader system level, there may be opportunities for a large national or international organization to ban use of a particular chemical. This has occurred with certain pesticides.

Reducing the amount of a highly hazardous chemical on-site provides a feasible option. According to Baasel, the Union Carbide plant in Bhopal had three 15-metric ton storage tanks for methyl isocyanate.[2] Would it have been feasible to reduce that amount through process design? Apparently, Mitsubishi Chemical Industries had a similar plant in Japan that did not store any methyl isocyanate on-site. They also produced the chemical in an intermediate step, but afterward it went immediately into the next production process. Thus, there was no need to store any of the highly hazardous chemical.

The third strategy—avoid releasing the hazard—involves providing containers for chemicals that do not leak under normal conditions or during foreseeable abnormal conditions. Many opportunities for leaks and spills occur during transit. Mechanisms in modern rail tank cars and highway tankers provide multiple barriers, so a release will occur only if more than one protective mechanism fails.

The fourth strategy is to modify release of the hazard. If small amounts of a chemical enter a room, local and facility ventilation may be sufficient to control the concentration of chemicals in the workplace air. In pressure vessels, the use of a pressure-relief valve serves the function of modifying release of the stored chemical by emitting small amounts when the internal pressure exceeds the set point of the device.

The fifth strategy is to separate the hazard from that which needs protection. An approach widely used in the chemical industry is to protect populated communities by locating chemical plants in unpopulated sites. This requires a means of avoiding nearby property from becoming populated after the plant becomes operational.

For example, the Union Carbide plant in Bhopal started in a low-population area, but after starting operations, many employees moved their families into housing close to the plant.[2] This is one reason chemical plants tend to be located in industrial parks that have large areas set aside for industrial purposes, and the local government prohibits housing within the park. Similarly, within an industrial site, the processes and storage areas with hazardous chemicals may be located away from other structures.

Limited options are available for improving the resistance of that which needs protection (Strategy 6). Materials used for consumer products and industrial operations are usually coated to improve appearance and to protect the material from foreseeable damage (e.g., dings and corrosion). Another application of this strategy is having workers apply protective ointments on skin to protect themselves from dermatitis-producing chemicals and biological agents such as poison ivy.

The seventh strategy is to help people perform safely. The four tactics listed in Table 24.1 involve the usual practices for hazardous materials management such as providing material safety data sheets, labeling containers, and instructing workers on precautionary practices.

The eighth strategy is to use personal protective equipment. Common types of chemical PPE are respirators, full-body protective suites, and gloves. When these protective devices are used, the employer needs a program to assure proper use and maintenance of the equipment.

The ninth strategy is to expedite recovery. Three tactics listed in Table 24.1 are common in workplaces. Providing safety showers and eyewash stations contributes much to the capability for limiting harm from inadvertent chemical contact. Similarly, having personnel trained in proper first aid for victims of chemical contact and ingestion can be a life saver. It is also common to post the phone number of an organization with specialized information about poisons and first-aid procedures for those poisoned.

LEARNING EXERCISES

1. Chemicals may be classified according to several attributes. For classifications based on health effects, what are the chemical categories mentioned in Section 24.1?

2. Chemicals may enter the human body through three routes. What are the three routes?

3. Table 24.1 lists the nine risk-reduction strategies and various tactics applicable to each strategy. Can you think of a tactic not on the list?

TECHNICAL TERMS

Alveoli   Tiny sacs at the ends of branches of the pulmonary system. Through the thin walls of the alveoli, chemicals are exchanged with nearby blood vessels. The normal and

	healthy chemical exchange is oxygen from inhaled air goes to the bloodstream, and carbon dioxide goes from the blood to the air in the alveoli where it is removed with exhaled air.
Asphyxiants	Various chemicals that disturb the maintenance of adequate oxygen supply to different systems in the body.[1] Carbon dioxide exemplifies a simple asphyxiant; it reduces the oxygen getting to the lungs. Carbon monoxide is the most common chemical asphyxiant; it reduces the essential exchange and transportation of oxygen by chemically modifying the normal oxygen-carrying capacity of the hemoglobin.
Asphyxiate	To kill or make unconscious by inadequate oxygen, presence of noxious agents, or other obstruction to normal breathing.[3]
Occupational dermatitis	Inflammation of the skin caused or induced by a work-related exposure.
Suffocate	To stop respiration or deprive of oxygen. Also to die from being unable to breathe.[3]

REFERENCES

1. Dikshith TSS. Safe use of chemicals: A practical guide. Boca Raton, FL: CRC; 2009.
2. Baasel WD. Preliminary chemical engineering plant design, 2nd ed. London: Van Nostrand Reinhold International; 1990. p. 94.
3. Merriam Webster's Collegiate Dictionary, 10th ed. Springfield, MA: Merriam-Webster; 1995.

Chapter 25

Biological Agents

This chapter addresses several sources of harm from biological agents. These include plants, pets, livestock, wild animals, insects, mold, and blood-borne *pathogens*. As in other chapters, the discussion focuses on the occupational health and safety issues, and does not attempt to be exhaustive. The chapter organization is according to the source. For each source, the discussion includes a brief explanation of the hazard, those exposed, the mechanisms of harming, and some basic strategies and tactics for reducing risks.

25.1 PLANTS

The plants considered hazards to workers are those encountered by people working outdoors. These will vary in different regions of the world. In North America we encounter poison ivy, poison sumac, and poison oak. Examples of those exposed are tree trimmers, landscapers, gardeners, wildland firefighters, and people clearing vegetation for utility lines.

The plants mentioned above are inappropriately named "poisonous." A better term would be "allergenic" plants because their effect is an allergic reaction. Like many allergenic substances, people vary in their sensitivity. The extent of exposure also affects the allergic response. Touching the leaves with bare skin evokes the strongest response. Allowing the allergic substance to contact an open wound may increase the chance of a severe reaction spread over greater areas of skin. Getting the allergenic substance on clothing can extend the potential exposures from the outdoors into the home. Touching the affected clothing can transmit the substance to anyone who handles the clothing. Reactions vary from plant to plant, but tend to involve itchy skin followed by development of blisters.

In locations with known allergenic plants, and regular access by people, the possibility of using Strategy 1 is worth considering. Eliminating some of these plants is not a trivial task. The effort involves exposing personnel to the plants, using chemicals that can harm other plants, and there is no assurance of long-term success. In public parks and preserves, there will be objections to disrupting the natural

Risk-Reduction Methods for Occupational Safety and Health, First Edition. Roger C. Jensen.
© 2012 John Wiley & Sons, Inc. Published 2012 by John Wiley & Sons, Inc.

ecology of plant life. Discussions with experts on local plant life and various stakeholders are essential steps in the decision process. Thus, Strategy 1 may be an option to consider, but doing background research is advisable prior to making a final decision to attempt elimination.

The primary prevention approach is Strategy 5—keep skin and clothing separated from the plants. This requires that people have the ability to recognize the allergenic plants, pay attention while in vegetated areas, and avoid getting close. For employers of personnel potentially exposed, this starts with employee education to enable personnel to recognize the plants and know the precautionary practices they should take. Secondary prevention (Strategy 9) involves promptly and thoroughly washing any skin that may have contacted the plants. Washing with common hand soap and water is generally effective, but special washing products on the market may be more effective. Minimizing the handling of clothing that may have picked up some allergenic substances can avoid releasing the hazard from the affected material to other skin areas, as well as to the skin of other individuals (Strategy 3). It is best for the person who wore the clothing to take it directly to a washing machine, start the wash, and take a shower to remove any small traces of residue from skin that may have been missed by direct washing.

25.2 PETS

Domestic pets like dogs can be hazards to delivery personnel. Postal delivery personnel in the United States will deliver mail to mailboxes on porches of individual houses. Sometimes pet dogs can be very aggressive toward these letter carriers as well as other visitors. Pets other than dogs can also pose a threat to delivery personnel and visitors, but these are far less common than dogs. When bit by a dog, the treatment generally involves an unpleasant series of rabies shots.

For homes with a dog kept in a yard, the owner may put their mailbox outside the fenced area to allow the letter carrier to keep separated from the dog (Strategy 5). Owners may also use Strategy 4 (avoid releasing the hazard) by keeping the pet indoors during delivery hours, chaining the dog up to prevent reaching the letter carrier, and keeping the pet in a fenced area away from the path used by the letter carrier.

25.3 LIVESTOCK

Livestock on farms and ranches can be hazardous to personnel. People can suffer harm from mules kicking, herded animals trampling, goats biting, and camels spitting. The domesticated elephants in India can crush people against structures or underneath their great weight.

People can catch several diseases from livestock.[1] Anthrax is one of the diseases caused by the *zoonotic pathogens*. Workers at risk include those who work with animals such as veterinarians, ranchers, and farmers. Other workers are exposed during processing of animal products (hides, wool, and meat) including leather workers, butchers, wool workers, and even those working on wool carpets. Droppings from chickens and other birds are recognized as causes of histoplasmosis and ornithosis.

The basic approach for reducing risk to humans working around livestock is separating the animals from the humans as much as feasible (Strategy 5). Containing livestock in a pen, coop, or fenced area is a means of separating the livestock areas from people areas.

25.4 WILD ANIMALS

People performing work in rural areas occasionally encounter wild animals with capability to harm. Some of this work involves land and game management, surveying, and ranching. Hunting and fishing guides sometimes run into dangerous animals. Workers in public parks, zoos, and wild animal preserves routinely interact with the protected animals. Oil workers and other personnel working in the arctic regions encounter rabies-infected arctic foxes. Professional divers occasionally encounter sharks, electric eels, and various other aquatic life forms.

Through years of experience, and unfortunate incidents, people have developed guidance for reducing risks. It is not the place of this book to attempt to provide specific precautions for every species of animal in every corner of the world. Local wildlife agencies make precautionary information available for those who may need the information. The fundamental approach involves Strategy 5—separate the wild animals from the people. This requires individuals to use appropriate precautionary behaviors to respect the space around animal habitats. This may be complemented by efforts by wildlife and park personnel to discourage the most dangerous animals from locations commonly used by people. For example, bears can become habituated to raiding campgrounds for easy food left accessible by campers. Campsite managers work at getting campers to store food in a manner that does not attract bears. Getting all the campers to comply is more challenging than OSH staff trying to get all employees to comply with all work rules. In order to encourage compliance, campsite managers have found ways to make it easier for campers to dispose of food waste, limit aromas that attract bears, and store food appropriately (Strategy 7—help people perform safely).

Numerous species of snakes are poisonous. Of the many species, some well-known ones are the black mamba and saw-scaled vipers of Africa, the tiger snake of Australia, the rattlesnake of North America, and the Indian cobra of southern Asia. The strategy most suited for poisonous snakes is keeping people separated from the snakes (Strategy 5).

Outdoor workers often encounter bees, wasps, hornet, spiders, and other pests. Scorpions are found in many regions of the world. Spiders are found almost everywhere. Mosquitoes are both nuisances and vectors for disease. Useful information about the hazards presented and precautionary practices may be found on website of reputable governmental agencies in countries throughout the world. The agencies most likely to provide information on local insects are those involved in public health or agriculture. The NIOSH website (www.cdc.gov/niosh/topics), for example, has concise papers on the following topics: poisonous plants, venomous snakes, venomous spiders, West Nile virus, Lyme disease, tick-borne disease, fire ants, scorpions, bees, wasps, and hornets.

The principal strategy for reducing risk is Strategy 7—help people perform safely. This involves training personnel on three topics: (1) to recognize the animals likely encountered in the area, (2) to know appropriate precautionary practices, and (3) to know procedures for responding after being stung, bitten, or hurt in other ways.

25.5 MOLD

Mold has received considerable public attention in recent years. It is a concern in homes and workplaces due to people reporting symptoms they believe are caused by exposure to mold. Mold is a fungal growth existing both indoors and outdoors. The mold itself is not toxic, but some types produce toxins (mycotoxins) that affect some people.

Mold is a topic for which conflicting information is readily available. Readers are advised to question information based on the experiences of individuals who feel mold exposure caused them to suffer. This anecdotal information lacks scientific reliability. Better sources are found on the websites of governmental public health agencies. One is the NIOSH website (www.cdc.gov/niosh/topics) containing links to brief articles on several topics, including one on mold.

For occupational exposures, indoor mold is the primary concern. Mold tends to grow in moist, poorly ventilated places. Thus, logical approaches for reducing structural damage and mycotoxin production start with getting rid of the mold. Strategy 1 calls for eliminating the source of the hazard. Eliminating mold starts with eliminating the source of the moisture and getting rid of carpet and building materials already infiltrated by mold. Mold on hard surface materials can be removed by washing with household cleaners or a mixture of bleach and water. In areas inherently moist from climate or local water, the use of a dehumidifier and/or extra ventilation can moderate the growth of mold and concentration of mycotoxins (Strategy 2).

When employees report symptoms related to dampness and mold, Strategy 9 applies. The information found on the NIOSH website describes applicable symptoms for two conditions: (1) allergic responses, and (2) a type of lung inflammation known as hypersensitivity pneumonitis. NIOSH advises having symptomatic individuals examined by a physician for proper diagnosis and treatment. The treatment might include restrictions on working in locations known for the presence of mold.

25.6 PATHOGENS

Bacteria, viruses, and other pathogens present people with risks of numerous diseases. This section focuses on occupational concerns with pathogenic diseases associated with blood, with limited comments on other biological agent concerns.

Some occupational diseases result from microorganisms transmitted in blood. Well-known diseases from "blood-borne pathogens" are hepatitis B, hepatitis C, and human immunodeficiency virus (HIV). Hepatitis B produces varied symptoms that may eventually manifest as liver cancer, cirrhosis of the liver, and chronic hepatitis. Hepatitis C infection leads to chronic liver disease. HIV greatly reduces immunity to various diseases.

Workers most commonly exposed to human blood as part of their job include healthcare workers, laboratory workers, and first responders. Healthcare workers get exposed to blood in several ways. The common ones are through needlesticks or cuts from sharp instruments containing the blood of a patient. During surgery, the exchange of sharp surgical instruments between surgeons and surgical nurses occasionally results in cuts. Exposures also come from contact with any bodily fluid from an infected patient. Historically, the trash in medical facilities contained some used needles and other "sharps" with blood. Janitors were at risk of blood exposure from being stuck while collecting the trash. Fortunately, most exposures to contaminated blood do not result in infection. Various non-job exposures include needle sharing among drug users, having unprotected sex, and patients getting transfusions of infected blood.

Since these various means of transmission involve human behavior, people need information they can believe and translate into action (Strategy 7). An interesting experience during the early days of HIV provides some insight that OSH professionals can use for employee training. In the early 1980s, public health surveillance systems identified some new clusters of unusual cases that became known as acquired immunodeficiency syndrome (AIDS). The initial case clusters involved male homosexuals, leading to the general public impression that it was a disease confined to those who engaged in male-to-male sex. That public impression created difficulties for healthcare workers who acquired the disease. As public health surveillance reports began to show that the disease could be acquired by anyone, and research determined the transmission could be through any exchange of bodily fluid, not just sexual contact among homosexual men, the public needed some convincing. The U.S. Surgeon General, C. Everett Koop, saw the need to educate the public on what the public health community actually knew about AIDS.[2] In 1986, he wrote a 26-page report, in plain language, explaining the new epidemic, the means of transmission, the effects, and advice for reducing further spread of the disease. It was widely distributed. However, feedback indicated a great many people formed opinions about the contents without actually reading the full report. In particular, many moral idealists objected to having a government report talk about using condoms and providing sex education in elementary schools. This group included the White House staff under President Ronald Reagan. In spite of White House objections, Dr. Koop wrote a much shorter letter he hoped would be read by a larger portion of the public. Congress authorized a budget for mailing the letter to all households in the country. The authorization bill made clear the letter need only be approved by authorities in the Department of Health and Human Services (not the White House).[2] In early 1988, the letter was mailed to over 100 million households. The public response was generally positive, and negative feedback was remarkably limited. What can OSH professionals learn from this? When preparing employee training on topics with some potentially moral or religious implications, begin the course with the message that the course content is based on scientific evidence and best practices. Early in the course establish your credibility by presenting material that cannot be refuted. Inform trainees of the sources for your material, and explain that these sources can be trusted.

Risk reduction involves the first and third Haddon phases. Primary prevention for healthcare workers involves several practices known as *standard precautions in health care*.[3] These precautions emphasize hand hygiene for all, and use of PPE whenever a risk assessment indicates a possible exposure to body substances or contaminated surfaces. The PPE options include clean nonsterile gloves, clean nonsterile fluid-resistant gown, and mask and eye protection or a face shield. Respiratory hygiene and cough etiquette make a third component of the standard precautions in health care. Following these precautions contributes to the safety of healthcare workers and the patients. In addition to these valuable administrative practices, some engineering strategies also help with pathogen control.

A basic precaution is handling needles and other sharps using risk-minimizing techniques. For example, past practices revealed that many of the employee needle-sticks occurred when trying to place a cap on the needle before disposing of it. Modern healthcare facilities have eliminated capping (Strategy 1) by having personnel put the used, uncapped needles directly into a sharps disposal container located in rooms where needles are used. Some new designs for syringes make it easy to withdraw the needle into the syringe body without exposing the hands to the needle.[4] This Strategy 5 approach separates the hazard from the person. Protection of janitors also uses Strategy 5 in that the walls of the sharps containers provide a puncture-proof barrier between the used sharps and the hands of the person who collects the containers. The standard precaution requires caregivers to wear appropriate PPE (Strategy 8) that serves as a personalized shield between themselves and the blood of the patient.

Medical facilities provide or arrange the hepatitis B immunization of employees who might be exposed to patient blood. Immunization illustrates use of Strategy 6—improve the resistance of that which needs protection.

Once a healthcare provider becomes aware of being stuck by a needle, or otherwise exposed, there are procedures to follow. A reporting system is invaluable for knowing about the exposure. In case the person develops one of the diseases, the report will serve as documentation that the exposure occurred at work. This Strategy 9 tactic offers some value for those exposed to blood-borne pathogens. For hepatitis B, previously unvaccinated individuals can be treated effectively with the vaccination protocol. For HIV, there are treatment regimes available. The effectiveness of these regimes has increased substantially since the early days of the HIV epidemic in the 1980s and new developments are continuing.

Storing biological agents in biological safety cabinets avoids releasing the hazard (Strategy 3). Laboratory facilities for research on especially dangerous biological materials are operated throughout the world and classified into four levels based on their capabilities to contain these materials. A major strategy for these facilities is the use of containment zones. These zoned areas (some being separate buildings) have various features including independent ventilation systems. Access to zones is limited to those with proper training and a need for access. The Level 4 facilities are for biological hazards with potential for extremely severe effects (e.g., mass epidemics).

Table 25.1   Strategies and Tactics for Protecting Healthcare Personnel from Pathogens

Risk-reduction strategy	Risk-reduction tactics
1. Eliminate the hazard	Eliminate tasks of recapping used syringe needles
2. Moderate the hazard	Not applicable
3. Avoid releasing the hazard	3a. Design pathogen research labs with distinct containment zones
	3b. Use robots or remote manipulators to work with dangerous biological materials
4. Modify release of the hazard	4a. Use the standard precautions for health care to reduce spread of biological agents
	4b. Release pathogens by flowing through a filter containing a pathogen-killing substance
5. Separate the hazard from that needing protection	5a. Dispose of sharps in special containers
	5b. Collect sharps in sharps containers
6. Improve the resistance of that needing protection	Vaccinate healthcare employees for hepatitis B virus
7. Help people perform safely	7a. Place sharps disposal containers where most convenient for easy disposal after use
	7b. Mark all biohazard containers with biohazard symbols and text as appropriate
	7c. Provide training for personnel potentially exposed
8. Use personal protective equipment	Use PPE to shield workers from blood of patients
9. Expedite recovery	9a. For hepatitis B patients not already immunized, put them through the immunization process
	9b. For HIV patients, follow an evidence-based treatment protocol

For medical laboratories, research facilities, and some production facilities, personnel need to work with pathogens or other biological materials with potential to cause disease. If a biohazard cannot be eliminated, it may be feasible to contain it in specially built enclosures (Strategy 3). Another approach is to provide equipment such as robots and remote manipulators to help personnel get their work done while being separated from the hazardous biological agent (Strategy 5).

Table 25.1 summarizes risk-reduction strategies and tactics for personnel working around pathogens and other biological agents.

LEARNING EXERCISES

1. For contact with allergenic plants, how would you draw the line between the during-event and post-event phases?

2. Regarding pets, dogs are mentioned in the chapter. Identify another species of pet that may present personal risks for humans.

3. What types of venomous snakes are discussed on the NIOSH website?

4. In many worksites, when a worker has a small scratch or cut, they visit the OSH representative for first aid. If you are the OSH person, what words would you say to explain why you put on surgical gloves before cleaning and patching the wound?

5. Find the nearest biosafety Level 4 facility to your location. One place to look is on the following web page: www.answers.com/topic/biosafety-level.

TECHNICAL TERMS

Pathogen	An agent of disease. The term usually refers to infectious organisms such as viruses, bacteria, and fungi.
Standard precautions in health care	A set of precautionary practices meant to reduce the risk of transmission of blood-borne and other pathogens from both recognized and unrecognized sources.[3]
Zoonotic pathogens	Pathogens that transmit communicable disease from animals to humans.

REFERENCES

1. Brauer RL. Safety and health for engineers, 2nd ed. Hoboken, NJ: Wiley; 2006. p. 483–493.
2. Koop CE. Koop: The memoirs of America's family doctor. New York: Random House; 1991. p. 193–239.
3. World Health, Organization., Standard precautions in healthcare. Available at http://www.who.int/csr/resources/publications/standardprecautions/en.
4. Goetsch DL. Occupational safety and health for technologists, engineers, and managers. Upper Saddle River, NJ: Prentice Hall; 2011. p. 549–550.

Chapter 26

Musculoskeletal Stressors

26.1 BACKGROUND

This chapter addresses those musculoskeletal stressors of greatest concern to OSH professionals. To get started, a general concept needs explanation. Human physical activities might be characterized as lying on a continuum from completely inactive to intensely active. Health is related to level of physical activity as follows. The very sedentary life leads to reduced cardiorespiratory capacity, poorly toned muscles, and weight gain. In combination, these effects accelerate aging and shorten life. In contrast, being physically active supports cardiorespiratory capacity, tones muscles, and helps maintain a healthy weight. These positive effects occur when the activity matches the individual's capabilities and limitations. When the musculoskeletal stresses of the work exceed the musculoskeletal capability of the worker, the risk of developing a disorder becomes substantial. This chapter discusses these musculoskeletal stressors, mentions some common injuries and disorders, summarizes our understanding of causes and effects, and provides examples of risk-reduction strategies applied to musculoskeletal disorders.

Over 200 years ago, Italian physician Ramazzini wrote about health effects of work: "certain violent and irregular motions and unnatural postures of the body, by reason of which the natural structure of the vital machine is so impaired that serious diseases gradually develop therefrom." Today, much is known about the risk factors Dr. Ramazzini described as "irregular motions and unnatural postures." Occupational medicine specialists and ergonomists have developed a respectable understanding of the *risk factors* for the most common occupational *musculoskeletal disorders (MSDs)*. This section provides a brief review of some widely recognized occupational MSDs.

A number of musculoskeletal conditions develop gradually from repeated stress. When these are referred to collectively, various terms are used. Four worth mentioning are

1. Repetitive motion injury,
2. Repetitive strain injury,

Risk-Reduction Methods for Occupational Safety and Health, First Edition. Roger C. Jensen.
© 2012 John Wiley & Sons, Inc. Published 2012 by John Wiley & Sons, Inc.

3. Wear and tear disorders, and

4. Cumulative trauma disorders.

The label *cumulative trauma disorder(CTD)* fits many, but not all, of the musculo-skeletal disorders commonly associated with work. A closer look at the three words, cumulative trauma disorder, reveals much about their development. The word cumulative indicates a gradual buildup, trauma indicates a harmful event, and disorder indicates a medical condition. Each disorder has its unique etiology, but, in general, CTDs develop gradually from very small traumas that fail to self-repair and accumulate over time to produce a disorder.

Many musculoskeletal conditions result from a single incident like an automobile crash, fall, or excessive load on a body part. These *musculoskeletal injuries* as well as MSDs affect muscles, tendons, spinal discs, ligaments, nerves, cartilage, and joint surfaces. Both are included in this chapter.

26.2 MEANS BY WHICH MUSCULOSKELETAL STRESSORS CAN HARM

This section introduces some of the most common work-related MSDs and the musculoskeletal stressors known to cause, aggravate, or contribute to the condition. The discussion covers upper extremities first, followed by those of the back and spine.

26.2.1 Upper Extremity Conditions

The upper extremity discussion starts with the body system most directly affected—tendons, nerves, and neurovascular elements. *Tendinitis* is an inflammation of a tendon. Some causes are repeated exertion, vibration, and being in contact with a hard surface. People who need to work with hands over their heads sometimes get shoulder tendinitis. *Tenosynovitis* is irritation and inflammation of a tendon sheath. When the sheath swells, the inside diameter narrows, creating more frictional resistance to movement of the tendon inside. This is accompanied by reduced lubricant between the sheath and the tendon, further increasing the frictional resistance. Sometimes this leads to an inability to move the tendon. *De Quervain's disease* is a special case of tenosynovitis that occurs in the abductor and extensor tendons of the thumb where they share a common sheath. De Quervain's disease is associated with forceful gripping and hand rotation like wringing a wet cloth. *Tennis elbow* is the common term for *lateral epicondylitis*. The irritated tendons are those attached to the epicondyle (the lateral protrusion at the distal end of the humerus bone). It is most common among tennis players, and specifically associated with serving. In workplaces, tennis elbow can result from activities that involve forceful, rapid, and repeated forearm supination or pronation. An occupational example would be using a hand-powered screwdriver on an assembly line.[1]

The second group of MSDs includes those affecting nerves. *Carpal tunnel syndrome* is the result of compressing the median nerve as it goes through the carpal tunnel of the wrist. It is directly caused by swollen tendons in the carpal tunnel pinching the median nerve. The swollen tendons result from tasks involving repeatedly stressing the tendons in the carpal tunnel, especially by forceful exertions in nonneutral wrist postures, vibration, and high repetition of the same motion. Common symptoms are numbness and pain in the parts of the hand served by the median nerve.

The third group of MSDs includes neurovascular disorders. *Thoracic outlet syndrome* is a disorder resulting from compression of nerves and blood vessels between the clavicle and ribs at the brachial plexus, a site known as the thoracic outlet. Symptoms include numbness of the arm and may limit use of arm muscles. Occupational causes include working with arms elevated and any activity that puts pressure on the thoracic outlet. *Vibration white finger* involves insufficient blood supply in the hand due to narrowing of blood vessels. The narrowing results from vasospasm triggered by vibration. It is often associated with extensive use of vibrating hand tools such as chain saws and jackhammers. Using such tools in a cold environment significantly increases the risk of getting this disorder. See chapter 16 for more discussion of hand-arm vibration.

Each CTD has its own set of risk factors well described in a book by Freivalds.[2] However, most ergonomists agree that the major risk factors for CTDs in general are

- Highly repetitive movements,
- Awkward postures (or postures far from the neutral posture),
- High forces, and
- Lack of time for recovery from microtraumas.

Each of these risk factors increases risk of developing a CTD. The combination of two or more risk factors acts synergistically to greatly increase risk. Working as a cutter on assembly lines in meat and poultry processing plants exemplifies work with multiple risk factors, and their very high workers' compensation claim rates for non-impact wrist disorders confirm the expected effects.[3]

Data entry using a keyboard provides an example. Say a person uses her right hand to enter numbers on the numeric keypad for 8 h a day. This job has the first risk factor—highly repetitive movements. This factor alone may slightly increase her risk of developing carpal tunnel syndrome. If she also works with her wrist in a nonneutral alignment, the second risk factor, posture, would increase her risk considerably. If she tends to pound the keys unnecessarily hard, that would bring the third risk factor into the equation. If she works steadily, without breaks, the fourth risk factor would be present. The combination of three or four risk factors greatly increases risk of developing carpal tunnel syndrome.

Other commonly encountered risk factors for particular types of CTDs are

- Tools that irritate nerves in the palm,
- Carrying thin loads with a pinch grip,
- Low-frequency vibration, especially under cold conditions, and
- Working with the hands under cold conditions.

Various risk factors can be encountered at work and while off work. It is not always clear if the origin of a particular *cumulative MSD* was from activities at the present job, previous jobs, off-work activities, a personal characteristic of the individual, or some combination of these. Some individuals are more vulnerable to developing cumulative MSDs. Sometimes psychosocial factors appear to influence individual reaction after experiencing the symptoms; some quietly deal with the pain and continue working, others seek medical treatment and take time off work. Psychosocial factors also influence whether they report their condition as a work-related injury/illness. Due to the multifactor nature of causation, ergonomists say a *risk factor* is a factor that causes, aggravates, or contributes to a cumulative trauma disorder.

26.2.2 Spine and Back Conditions

The discussion now turns from MSDs of the extremities to MSDs and musculoskeletal injuries of the spine and back. A common symptom is back pain. Some people incorrectly use the term "back pain" as though it were a disease. It is actually a symptom associated with numerous medical conditions—some being soft tissue injuries and others being spinal disorders.

Common soft tissue injuries are torn muscles, often resulting from a motion or exertion. Torn muscles generally recover in 1–3 weeks. Often a slight muscle tear results in a tightening up response by adjacent muscles people refer to as a stiff back. Whether the person will miss work depends on the severity of the tear, the person's tolerance for pain, how much they like the job, and how physically demanding their job is.

Common spinal disorders occur in the lumbar and cervical regions. An unhealthy spinal disc can allow pinching a nerve, which results in pain radiating out to wherever the nerve goes. For example, if the sciatic nerve is pinched as it exits the lumbar spine, pain will radiate to the buttocks and down the leg. A similar pinching in the cervical spine can produce symptoms in the shoulders and arms.

The region of the spine associated with most occupational back disorders is the lumbar region, especially the L5/S1 and L4/L5 discs. These joints are subjected to greater compressive load than joints higher up the spine, and compressive loads are accepted by experts in occupational biomechanics as good indicators of overall biomechanical stress level on these spinal discs.[4,5] Experiments on cadaver spines show that these discs have limits for tolerating compressive loads, and when exceeded, cartilage end plates develop tears. When the person performs heavy manual work over a long period, more tears develop and the initial ones extend in length and scar over. Eventually, the accumulation of damage to the end plates no longer allows sufficient fluid diffusion to keep the discs filled with moisture, leading to the discs getting flatter and more rigid, and increasing risk of a pinched nerve.

The stressors that increase risk of back disorders may be found on the job and outside of work. Also important are individual differences in physical and psychological makeup. The major back stressors associated with increased risk of back pain and/or back disorders have been identified as follows:[6]

- Heavy lifting and heavy work.
- Frequent lifting.
- Lifting loads near one's strength capacity.
- Occasional very stressful load handling.
- Sudden unforeseen events (e.g., crashes, falls, impacts).
- Prolonged standing or sitting.
- Others back stressors include whole-body vibration, pushing, pulling, carrying, twisting, and bending.

Workers in jobs involving heavy, whole-body labor are at greatest risk of occupational back problems that involve medical care and/or time off work. Some of these jobs are nursing assistants working in nursing homes, garbage collectors, and laborers.[7]

26.3 STRATEGIES AND TACTICS FOR MUSCULOSKELETAL STRESSORS

The strategies for reducing risk from musculoskeletal stressors are listed in the left column of Table 26.2. On the right are major groups of applicable tactics.

The first strategy is to eliminate the hazard. This approach works for both tasks that stress the extremities and those that stress the back. Most often this involves replacing the human with a machine capable of performing the task. Occasionally, a task can be eliminated by changing the basic way of getting the work done. For example, a nursing home may have a procedure for weighing residents involving helping them transfer from a wheelchair onto a scale, and then back to the wheelchair. These two back-stressing transfers by a nursing assistant can be eliminated by having a larger scale onto which the wheelchair and resident can be rolled for weighing. After getting the weight, the wheelchair weight is subtracted from the result to obtain the person's weight. This approach eliminates two transfers involving exposure to shoulder and back stressors.

The second strategy is to moderate the hazard. For tasks identified as having risk factors for CTDs, modification of the risk factors should reduce risk of developing a CTD. For tasks involving manual lifting, the tasks can be assessed for known risk factors, and changes may be made to reduce risk factors associated with the task. NIOSH has guidelines for evaluating a manual lifting task to identify attributes furthest from optimal; so the most effective modifications will address those attributes.[5] The NIOSH method provides a numerical lifting index indicating how stressful the task is. The same index can be used to see how it would be if certain modifications were to be made.

Another approach often used in manufacturing is job rotation. The idea is to let workers divide their shift into two or more workstations. This can be useful if different body parts are stressed in each workstation. For example, one workstation might stress the person's lower back, while a second stresses their right wrist, and a third stresses their neck. The hope is that avoiding 8 h of stressing the same body part will avoid causing injury. Most ergonomists recognize the practical value of this approach, but regard it as a temporary solution—temporary until an engineering solution can be found to avoid overstressing the employee.

A method leading up to job modification is an employee survey known as a *body-part discomfort survey*. Employees are shown a sketch of a body and asked to express how each body part feels. Most forms use a rating scale for indicating the extent of discomfort. These surveys are used to identify mismatches between job demands and the employee's capabilities and limitations. In some instances of mismatch, a job modification can be made to provide a better fit.

The third, fifth, and sixth strategies do not appear suited for to the hazards of musculoskeletal stressors. The fourth strategy applies to some heavy exertion tasks. An example is changing the storage location of heavy object an employee needs to lift from being near the floor to being elevated to level between the knees and hips. This will effectively reduce the intensity of the musculoskeletal stressors on the lower back.

The seventh strategy is to help people perform safely. Three tactics for manual handling are noted in Table 26.1. The first is to provide training on work methods to minimize musculoskeletal stressors. This sort of training seeks to provide employees with the knowledge needed to recognize back-stressing tasks and choose a method that will minimize musculoskeletal stresses. Such training may also be viewed as fulfilling an employer's responsibility to inform employees of hazards and precautions appropriate for their job, like right-to-know training and warning signs. This training will usually be provided by a supervisor or experienced employee while showing the new worker how to perform the task. A different sort of training teaches employees to use proper body mechanics. A large cohort study of postal workers found this approach was ineffective for reducing incidence of back injuries.[8]

The second tactic is to make appropriate material-handling equipment available for use instead of brute force. This includes having the equipment where and when it is needed. Otherwise, workers may feel they can more easily do the task without the equipment.

Employers can encourage employees to engage in physical activity during nonwork hours to help maintain or improve their physical fitness. This could be most beneficial if acted on by employees inclined to being sedentary. It might not be helpful to employees who already engage in lots of high-intensity exercise in their off time. In a prospective cohort study of an occupational cohort, the incidence of low back pain was compared for employees sorted into three groups based on their physical activity level.[9] The highest incidence rates were for the lowest and highest exercise cohorts. The middle group, those who regularly engaged in moderate physical exercise, had the lowest back pain incidence. It was somewhat surprising that those who engaged in frequent, vigorous physical activity had higher rates than the moderate group. Perhaps some people are too zealous about their workouts.

Many employers have implemented programs to help employees prepare for their daily work. Examples include stretching and group calisthenics prior to commencing work. The scientific evidence on the efficacy of these programs has not clearly established benefits in terms of reduced rates of musculoskeletal disorders.[10]

Table 26.1   Strategies and Tactics for Reducing the Risks from Musculoskeletal Stressors

Risk-reduction strategy	Risk-reduction tactics
1. Eliminate the hazard	1a. Do not have a human do work involving unhealthy musculoskeletal stressors. Usually this involves having a machine do the work
	1b. Eliminate the need for the stressful work
2. Moderate the hazard	2a. Modify work by reducing risk factors
	2b. Modify the duration of exposure to stressful work using job rotation
	2c. Use body-part discomfort surveys to identify mismatches between employee and job demands; make appropriate changes
3. Avoid releasing the hazard	Not applicable
4. Modify release of the hazard	When performing a heavy manual task, position body optimally and use a smooth rather than jerking motion
5. Separate the hazard from that needing protection	Not applicable
6. Improve the resistance of that needing protection	Not applicable
7. Help people perform safely	7a. Provide job-specific training on work methods to minimize musculoskeletal stressors
	7b. Provide material-handling equipment for heavy tasks
	7c. Implement programs to increase the capabilities of exposed employees for resisting musculoskeletal stressors
8. Use personal protective equipment	Provide pads for body parts exposed to repetitive pressure
9. Expedite recovery	9a. Use symptom surveys to detect individuals with symptoms suggesting early stage of a CTD; change their work assignments appropriately to stop further development of CTD
	9b. Have employees with symptoms obtain medical care from an appropriate specialist
	9c. Support employees trying to recover from a CTD by providing a return-to-work program suited to their needs

However, this lack of evidence may be due to the difficulty of conducting a proper intervention study in a workplace.

The eighth strategy is to use personal protective equipment. The use of knee pads is a preventive practice for workers who stress their knees by impact, such as carpet installers who use their knee as a hammer to kick a carpet stretcher, and flooring installers who expose their patella to direct contact with hard surfaces. Workers who expose their palms to hard surfaces can benefit from a glove designed specifically to reduce compression forces that pinch nerves in the palm. Some employees wear a wrist brace during work, or only while sleeping. This is intended to keep symptoms of a wrist CTD from recurring or getting worse. Some employers provide manual handling employees a belt designed to provide back support when lifting. These are intended to stabilize the spine, reduce the stress on low back muscles, and hopefully reduce the risk of a back problem. Studies to determine if these industrial back belts actually reduce risk of back problem have been conducted with mixed findings. Consequently, the ergonomics community does not endorse the use of back belts for workers without a history of back pain.

The ninth strategy is to expedite recovery. The first and best tactic is to identify early symptoms of a CTD so that interventions can be implemented to head off a full-blown disorder. The means to accomplish this is to conduct a *symptom survey* of employees. Medical personnel evaluate the results and determine if intervention is needed. For ethical and perhaps legal reasons, symptom survey records should be treated as confidential medical records. Body-part discomfort surveys are easily confused with symptom surveys. Both call for using a rating scale linked to a specific body part shown on a diagram, and both rely on respondents to make a good faith effort to report how each body part feels. The tools are distinguished in purpose, rating scale, use, and confidentially as indicated in Table 26.2.

When a symptom survey indicates an employee experiences pain in a particular body part, further investigation is required to determine if the pain is work related. This might include a medical exam and an ergonomic assessment of the work. If the pain is work related, several responses may be considered. Interventions might involve changing the employee's tasks, changing to a different job, implementing job rotation, or prescribing medical interventions like anti-inflammatory drugs, using a wrist brace, or wearing a back brace or lifting belt. The second tactic comes into play when an employee has developed a diagnosed musculoskeletal disorder. It involves being proactive in finding a medical provider with qualifications appropriate for the

Table 26.2   Difference Between Two Similar Employee Surveys

Characteristic	Body-part discomfort survey	Symptom survey
Purpose	To identify task–employee mismatches	To identify early signs of a possible developing medical condition
Rating scale	Level of discomfort	Level of pain
Use	Used by ergonomists and industrial engineers with the goal of correcting the task–worker mismatch	Used by medical personnel with the goal of prescribing an intervention before a disorder matures
Confidentiality	Treat with same level of confidentiality as other nonmedical employee survey data	A medical record that should be treated with the same high-level confidentiality as other medical records

disorder. This has the potential of expediting recovery and return to work. The third tactic is to have an effective return-to-work program. Also called modified duty programs, these involve having the employee come into the workplace instead of sitting at home. This can avoid the malaise that tends to set in when a worker stays home while recovering, helps the individual sustain a feeling of being appreciated at the workplace, and often shortens the duration of temporary disability.

LEARNING EXERCISES

1. Sometimes in the popular press we find articles suggesting that ergonomists want to eliminate hard physical work. Considering the introduction to this chapter, what is a more correct characterization of the ergonomist view of hard physical work?

2. In the discussion of upper extremity cumulative trauma disorders, what are the three physiological systems used for organizing the discussion?

3. Give one example disorder from each of the groups identified in question 2.

4. List the four well-recognized risk factors for cumulative trauma disorders.

5. List the seven major risk factors for back pain.

6. Explain the different purposes of a body-part discomfort survey and a symptom survey.

7. Indicate what items of personal protective equipment may be useful for MSDs.

TECHNICAL TERMS

Body-part discomfort survey   A survey of employees asking them to indicate their level of discomfort in various body parts. The purpose of these surveys is to identify mismatches between the task demands and employee capabilities and limitations so that corrections may be made to reduce or eliminate job-related discomfort. Most rating scales are from no discomfort to intense discomfort.

Cumulative trauma disorder   A musculoskeletal disorder resulting from cumulative trauma.

Cumulative MSDs   Same as cumulative trauma disorders.

Musculoskeletal disorders   Medical conditions of the muscles, tendons, ligaments, joints, cartilage, nerves, or spinal discs that develop gradually. Some are work related.

Musculoskeletal injuries   Medical conditions of the muscles, tendons, ligaments, joints, cartilage, nerves, or spinal discs that result from a traumatic incident.

Symptom survey   A survey of employees asking them to indicate on a body-part-specific rating scale how it feels. The scales are usually labeled for level of pain ranging from no pain to intense pain. The purpose of these surveys is to identify early signs of a possibly developing musculoskeletal disorder so that medical intervention can be applied. Survey forms and results should be treated like other confidential medical records.

REFERENCES

1. Leclerc A, Landre M, Chastang J, Niedhanner I, Roquelaure Y. Upper-limb disorders in repetitive work. Scandinavian J. Work Environment Health. 2001;27: 268–278.
2. Freivalds A. Biomechanics of the upper limbs: Mechanics, modeling, and musculoskeletal injuries. Boca Raton, FL: CRC; 2004.
3. Jensen RC, Klein BP, Sanderson LM. Motion-related wrist disorders traced to industries, occupational groups. Monthly Labor Review. 1983;106(9):13–16.
4. Chaffin DB, Andersson GBJ, Martin BJ. Occupational biomechanics, 4th ed. Hoboken, NJ: Wiley; 2006. p. 130–146.
5. Waters TR, Putz-Anderson V, Garg A. Applications manual for the revised NIOSH lifting equation. Cincinnati, OH: National Institute for Occupational Safety and Health; 1994.
6. Jensen RC. Epidemiology of work-related back pain. Topics Acute Care Trauma Rehabilitation. 1988;2(3):1–15.
7. Klein BP, Jensen RC, Sanderson LM. Assessment of workers' compensation claims for back strains/sprains. J Occupational Medicine. 1984;26(6):443–448.
8. Daltroy LH, Iverson MD, Larson MG, et al. A controlled trial of an education program to prevent low back injuries. New England J Medicine. 1997;337(5):322–328.
9. Matthews T, Hegmann KT, Garg A, Porucznik C, Behrens T. The predictive relationship of physical activity on the incidence of low back pain in an occupational cohort. J Occupational Environmental Medicine. 2011;53(4):364–371.
10. Choi SD, Woletz T. Do stretching programs prevent work-related musculoskeletal disorders? J SH&E Research. 2010;2(3): Feature Article 2. Available at http://www.asse.org/publications.

Chapter 27

Violent Actions of People

The last hazard source category is one not generally recognized by authors of OSH books. It contains highly hazardous situations created when people become physically aggressive toward others and when terrorists attack a place of work. Once initiated, these situations can turn in many directions and end with outcomes ranging from no one being hurt to multiple deaths.

This hazard source category does not include careless conduct, negligent behavior, or failure to follow a safety rule. Take, for example, an employee who fails to use available and required eye protection. This behavior is clearly improper and undesirable because it takes away the last line of defense for the individual. It does not, however, create a new hazard for the misbehaving employee or for fellow employees. Likewise, an employee who operates a power tool with a power cord having damaged insulation is not creating a new hazard since the electrical hazard exists whether or not the cord has proper insulation; what the employee has done is not take advantage of a normally effective safety device. These examples of working with a *compromised hazard control* are offered to make clear that these behaviors are different from the violent actions of people. This chapter is limited to people-created, highly hazardous situations that imminently threaten the lives of people at work.

Strategies and tactics for the violent actions of people are summarized in Table 27.1. The discussion below begins with a section on workplace violence and ends with a section on attacks by terrorists.

27.1 WORKPLACE VIOLENCE

Everyone in the OSH field is aware of the potential for *workplace violence*. Four types of violent situations at workplaces reoccur often enough to form a pattern.[1,2]

27.1.1 Robbery and Other Criminal Acts

One repeated pattern occurs during a robbery or other criminal act.[1] These commonly take place in workplaces where there is cash, during late night or early morning hours,

Risk-Reduction Methods for Occupational Safety and Health, First Edition. Roger C. Jensen.
© 2012 John Wiley & Sons, Inc. Published 2012 by John Wiley & Sons, Inc.

Table 27.1   Strategies and Tactics for Workplace Violence and Terrorist Attacks

Risk-reduction strategy	Tactics for violent and outrageous conduct
1. Eliminate the hazard	No feasible means known to author
2. Moderate the hazard	Provide selected personnel training on de-escalation techniques
3. Avoid releasing the hazard	3a. High-risk businesses keep windows open enough for those outside to see inside
	3b. High-risk business have signs stating limited cash is kept in the premises
	3c. High-risk businesses have security camera system and signs indicating the place is protected
	3d. Provide good lighting in parking lots
4. Modify release of the hazard	Restrain or subdue the violent individual
5. Separate the hazard from that needing protection	5a. Use bullet-resistant glass to separate cashiers from would-be robbers
	5b. Limit workplace access to current employees and invited guests
	5c. Defend the air traveling public from car bombs by using most intense security measures in areas of greatest people density
	5d. In prisons, separate guards from prisoners by physical barriers
6. Improve the resistance of that needing protection	Not applicable
7. Help people perform safely	7a. Train employees of high-risk businesses on how to respond to robbers
	7b. Train service providers in high-risk jobs on how to anticipate and respond to the aggressive behavior of clients/patients
	7c. Treat employees fairly and let them know they are treated fairly
	7d. Train supervisors to recognize signs an employee might become violent
	7e. Provide assistance for employees feeling threatened by a nonemployee
8. Use personal protective equipment	Care providers may wear PPE for a specific patient (e.g., wear a face shield while caring for an individual prone to spitting)
9. Expedite recovery	For individuals who have been traumatized by the violent behavior of others, provide access to professional counseling

and where only one or two people are working. Common targets for these crimes are 24-hour convenience stores, fast food restaurants, and taxi drivers. Doing business in high-crime neighborhoods also increases likelihood of robbery. The initiators of these crimes typically do not know their victims and often use a gun to coerce compliance with demands. The most heartless of these criminals gets the valuables and then shoots the victim in order to eliminate the only person who could identify the perpetrator at trial. These situations account for the most workplace murders in the United States, except for year 2001 when terrorist, using hijacked commercial airliners, attacked the World Trade Center and the Pentagon killing over 2700 people working in the buildings and causing over 400 rescue workers to give their lives attempting to save others.

Strategies for addressing these hazardous situations start with discouraging the criminal minded person from targeting a particular person or establishment. Some tactics for implementing this Strategy 3 approach (avoid releasing the hazard) are used by many owners of convenience stores and fast food restaurants. They keep windows sufficiently free of advertising so that a person outside can see a robbery in progress, and they post signs near the entrance door informing would-be robbers that only small amounts of cash are kept in the premises and that surveillance cameras are in place.

An example of Strategy 5 used in high-cash businesses (e.g., banks, casinos, and racetracks) is separating the cashiers from the customers with a barrier of bullet-resistant glass. Similar barriers are used in some taxis. Training store clerks and taxi drivers how to respond during a robbery can improve their chance of surviving the incident (Strategy 7—help people perform safely). The approach is to politely treat the perpetrator like a customer, comply with all demands, and hope the perpetrator simply leaves the crime scene without killing the robbery victim.

27.1.2 Client Attacks on Service Provider

A second repeated pattern occurs when a service provider gets assaulted by the person being served.[1] These situations occur in healthcare facilities where a patient strikes a nurse or other care provider. Other situations involve a student assaulting a teacher and a welfare recipient assaulting a social worker. Fortunately, these assaults rarely result in death.[1]

Approaches for these assaults on service providers start with each provider attempting to recognize the potential of a particular person acting violently. If this sort of risk assessment identifies someone, the service provider can have a second person present during any period of contact with the person of concern. The presence of two people may influence the person to behave. The individuals who enter these sorts of service professions receive training on how to communicate effectively with those served. Part of that training teaches techniques for dealing with hostile and otherwise difficult people. Clearly, these Strategy 7 approaches can reduce the frequency but not prevent all cases.

27.1.3 Attacks by a Disgruntled Employee or Former Employee

The third recurring pattern involves an employee or former employee launching an attack at the workplace. Often the aggressor is someone who feels mistreated by the organization, a unit in the organization, a supervisor, or other employees. The outcomes of these assaults vary, with the worst result occurring when the aggressor brings a gun and shoots multiple employees.

It is highly desirable to remove motives for insider attacks by treating personnel fairly, having and following personnel procedures that assure all employees are treated the same, and convincing all employees that the organization and managers attempt to treat employees fairly. If this Strategy 3 does not succeed, and an employee remains convinced of having been mistreated, he/she could become the source of an attack.

Security measures that limit facility access to current employees will keep employees separated from a disgruntled former employee. A policy of no guns allowed in workplaces serves to deter gun attacks arising from a momentary fit of anger (Strategy 3). The idea is that the angry employee could only launch a gun attack by leaving the premises, getting a gun, and returning. This allows time for the person to cool off, and with a cooler head decide to call off the planned attack. It also provides time for the business to summon security personnel or police. Hiring practices that screen applicants using criminal background checks, contacting references, and conducting multiple personal interviews serve to avoid employing people with a history of violent behavior.

Training supervisors to recognize precursors of violent outbursts can avoid some of these incidents. Precursors include signs that an employee is having disputes with coworkers, being on the receiving end of bullying by fellow employees, or feeling distressed for reasons such as excessive workload, too much responsibility, or an overly controlling supervisor. Supervisors need to know who they should contact if they need assistance, perhaps someone with the human resources department or an employee assistance professional.[3]

27.1.4 Attacks Related to Domestic Squabbles

The fourth pattern occurs when a nonemployee enters the workplace with intent to hurt a specific employee. Usually the perpetuator's motive stems from a troubled domestic relationship with the employee (often the spouse or former spouse of the perpetuator). These situations account for 5% of the workplace murders in the United States.[2]

Various practices and procedures can reduce the risk of domestic violence occurring in the workplace. Having a building security system that limits access to the building can effectively separate the targeted employee from the would-be perpetrator. Having a security guard control the entrance to work areas serves a similar purpose. Having a receptionist at the entrance can discourage some would-be perpetrators, but not physically prevent them from entering the facility. Having policies that encourage employees to report when they feel threatened by a domestic partner, coupled with a plan to take appropriate precautions, can discourage violence at the workplace. Attacks in company parking lots can be discouraged by good lighting and making available to an employee who requests it an escort to and from their parked car.

For all these violent situations, training personnel on de-escalation techniques can moderate the perpetrator's behavior (Strategy 2). Angry people tend to think irrationally, so finding ways to slow things down allows the perpetrator time to cool off and start thinking rationally about the consequences of their actions. Time also provides an opportunity for security personnel or police to respond.

27.2 TERRORIST ATTACKS

Attacks by terrorists on workplaces are a source of highly hazardous situations threatening all affected employees. The motives of terrorists vary, but clearly differ from the four types of workplace violence discussed in the preceding section. The motives for some terrorist actions include a desire to draw attention to a cause, get revenge for a perceived injustice, and hurt people who do not share the religious or political views of the terrorists. The apparent motive for several previous airliner hijackings was to obtain ransom money or get associates freed from prison. Some terrorists attack a particular workplace because they see it as a symbol of something they distain (e.g., the Federal Building in Oklahoma City and the Twin Towers in New York City).

Governmental efforts to limit the weapons extremists can get are part of Strategy 2—moderate the hazard. A primary strategy employers can use is separating the workplace from outsiders with a barrier. Industrial facilities and military bases enclose the workplace with a security fence with gates for controlling access. Many western countries have fortified the perimeters and exterior walls of their embassy buildings to protect personnel and to improve the resistance of the buildings to car bombings. Airport security systems operate somewhat like gates to an industrial facility, separating the aircraft and legitimate passengers from persons who would like to highjack or destroy the airplane. All these tactics are applications of Strategy 5—separate the hazard from that which is to be protected.

Many airports have taken steps to address the car bomb threat to people. One is deliberately separating parking lots from high-density passenger areas; another is controlling traffic in arrival and departure areas to make it difficult for a terrorist to leave a car bomb parked in that busy area. These tactics will not prevent a determined terrorist from deploying a car bomb on airport grounds, but by making the likely location a parking lot—an area with few people at any point in time—the number of people killed will be less than it would be if exploded in a crowded area. These tactics modify release of the hazard (Strategy 4) by making the parking lot the most vulnerable location for deploying a car bomb.

Hiring practices that screen applicants using criminal background checks, contacting references, and conducting multiple personal interviews make it difficult for a terrorist organization to get their loyal followers a job in targeted industrial facilities where they would have opportunities to sabotage operations.

27.3 SUMMARY OF PART V

Part V consists of five chapters about risk reduction for hazards not classified as energy hazards. Figure 27.1 depicts two-dimensional matrix with the seven hazard sources in rows and the nine risk-reduction strategies in the columns. The last five rows apply to part V. Cells marked with an X indicate for each source the strategies mentioned in the applicable chapter.

Chapter 23 discusses some common hazardous conditions associated with workplace facilities—walking surfaces, stairways, ramps, confined spaces, and dusty air. The point is made that these locations exist harmlessly until a human comes along. Management decisions on resource allocations to improve the safety of these locations warrant risk assessment to account for both the frequency of human

Hazard Source	1	2	3	4	5	6	7	8	9
Energy									
Weather & Geologic Events									
Hazardous Conditions	X	X	X		X		X	X	X
Chemical Substances	X	X	X	X	X	X	X	X	X
Biological Agents	X	X	X	X	X	X	X	X	X
Musculoskeletal Stressors	X	X		X			X	X	X
Violent Actions of People		X	X	X	X	X	X		

Risk-reduction Strategy

1. Eliminate the hazard.
2. Moderate the hazard.
3. Avoid releasing the hazard.
4. Modify release of the hazard.
5. Separate the hazard from that which needs protection.
6. Improve the resistance of that which needs protection.
7. Help people perform safely.
8. Use personal protective equipment.
9. Expedite recovery.

Figure 27.1   Matrix showing the risk-reduction strategies applicable to the five hazard sources addressed in part V.

exposure to the location and the potential severity of harm. Regarding walking surfaces and stairways, the discussion emphasizes the importance of meeting the expectations of the people walking on working surfaces and using stairs.

The topic of chemicals encountered at work is discussed in chapter 24, beginning with a brief review of health effects, followed by comments on routes of entry, and ending with strategies and tactics for avoiding harmful exposures.

Chapter 25 discusses biological agents encountered at work. It has sections on plants, pets, livestock, wild animals, mold, and pathogens. Each section mentions some of the more common instances of occupational exposures to biological agents, and provides examples of risk-reduction tactics.

The discussion of the fourth topic in part V, musculoskeletal stressors, starts by recognizing that most physical activity is healthy and important for maintaining health. Problems arise when stressors on various joints and muscle groups exceed the capabilities of those body parts. Discussions along these lines are provided for the two most common areas affected—the low back and upper extremities. Two employee survey tools used by ergonomists and occupational medicine physicians—discomfort surveys and symptom surveys—are compared and contrasted. Strategies for addressing mismatches between a worker and the job demands emphasize adjusting the work demands to match the capabilities and limitations of the employee.

The final topic in part V concerns the violent actions of people. Topics in this chapter are workplace violence and terrorist attacks. The discussion of workplace violence describes four types and summarizes risk-reduction tactics for each. The section on terrorist attacks focuses on defensive tactics an employer can take prior to being attacked.

LEARNING EXERCISES

1. This chapter on the violent actions of people includes some actions and excludes others. It does not include the actions of a construction laborer who enters a deep trench knowing it lacks proper shoring. What rationale does the chapter author use to explain why this sort of behavior is not a hazard source?

2. For neighborhood convenience stores, list three factors that increase risk of being robbed.

3. Some nursing homes have a particular resident known for hitting care providers. What protective measures can be used?

4. What engineering approach mentioned in this chapter applies to both attacks by former employees and attacks stemming from domestic disputes?

TECHNICAL TERMS

Compromised hazard control   The failure of a worker to use normally effective and appropriate PPE or other available safety device or practice when performing work.

Workplace violence   Violent acts, including physical assaults and threats of assault, directed toward persons at work or on duty.[4]

REFERENCES

1. LeBlanc MM, Barling J. Workplace aggression. Current Directions Psychological Science. 2004;13 (1):9–12.
2. Peek-Asa C, Runyon CW, Zwerling C. The role of surveillance and evaluation research in the reduction of violence against workers. American J Preventive Medicine. 2001;20:141–148.
3. Anderson KR, Tyler MP, Jenkins EL. Preventing workplace violence. J Employee Assistance. 2004;34 (4):8–11.
4. Jenkins EL. Violence in the workplace: Risk factors and prevention strategies (Current Intelligence Bulletin 57 NIOSH No. 96–100) Washington, DC: U.S. Government Printing Office; 1996.

Index

Risk-Reduction Methods for Occupational Safety and Health, First Edition. Roger C. Jensen.
© 2012 John Wiley & Sons, Inc. Published 2012 by John Wiley & Sons, Inc.

THE GERMAN UNEMPLOYED

F4 MAR 1987

The German Unemployed

Experiences and Consequences of Mass Unemployment from the Weimar Republic to the Third Reich

Edited by
Richard J. Evans and Dick Geary

CROOM HELM
London & Sydney

© Richard J. Evans and Dick Geary 1987
Croom Helm Ltd, Provident House, Burrell Row
Beckenham, Kent BR3 1AT
Croom Helm Australia, 44-50 Waterloo Road,
North Ryde, 2113, New South Wales

British Library Cataloguing in Publication Data
The German unemployed.
  1. Unemployment—Germany—History—
20th century
I. Evans, Richard J.   II. Geary, Dick
331.13′7943 HD5779

ISBN 0–7099–0941–1

SHEFFIELD HALLAM UNIVERSITY
WL
331.137943
EV
ADSETTS CENTRE

Printed and bound in Great Britain
Mackays of Chatham Ltd, Kent

CONTENTS

LIST OF FIGURES

LIST OF TABLES

LIST OF ABBREVIATIONS

ADGB	Allgemeiner Deutscher Gewerkschaftsbund
AH	*Die Arbeitslosenhilfe*
APP	Archivum Panstwowe w Poznaniu
APS	Archivum Panstwowe w Szczecinie
ASte	Amt für Stadtentwicklung und Statistik, Augsburg
AVAVG	Gesetz über Arbeitsvermittlung und Arbeitslosenversicherung
AWO	*Arbeiterwohlfahrt*
BA	Bundesarchiv
BASF	Badische Anilin– und Sodafabrik
BWøB	*Berliner Wohlfahrtsblatt*
BzG	*Beiträge zur Geschichte der Arbeiterbewegung*
DHV	Deutschnationaler Handlungsgehilfenverband
DMV	Deutscher Metallarbeiter-Verband
DZW	*Deutsche Zeitschrift für Wohlfahrtspflege*
E	Evening edition
FAD	Freiwilliger Arbeitsdienst
FW	*Freie Wohlfahrtspflege*
GZ	*Gewerkschaftszeitung. Organ des Allgemeinen Deutschen Gewerkschaftsbundes*
H	Heft
HStA	Hauptstaatsarchiv
ILR	*International Labor Review*
IML/ZPA	Institut für Marxismus-Leninismus beim Zentralkomitee der SED
IWK	*Internationale Wissenschaftliche Korrespondenz zur Geschichte der deutschen Arbeiterbewegung*
KAH	Kirchenarchiv Hamburg
KJA	Kirchliches Jugendamt
KPD	Kommunistische Partei Deutschlands (Communists)
LA	Landesarchiv
LABln	Landesarchiv Berlin
M	Morning edition
ND	*Nachrichtendienst des Deutschen Vereins für öffentliche und private Fürsorge*
NF.	Neue Folge

NSDAP	Nationalsozialistische Deutsche Arbeiterpartei (Nazis)
RABl	*Reichsarbeitsblatt*
RAM	Reichsarbeitsministerium
RAVAV	Reichsanstalt für Arbeitsvermittlung und Arbeitslosenversicherung
RF	*Die Rote Fahne*
RFV	Reichsverordnung über Fürsorgepflicht (13 Feb. 1924)
RGBl	*Reichsgesetzblatt*
RGO	Rote Gewerkschaftsopposition
RGS	Reichsgrundsätze über Voraussetzung, Art und Mass der öffentlichen Fürsorge (4 Dec. 1924)
RM	Reichsmark
RMWD	Reichsministerium für wirtschaftliche Demobilmachung
SA	Sturmabteilung
SJDR	*Statistisches Jahrbuch für das Deutsche Reich*
SJH	*Statistisches Jahrbuch für die Freie und Hansestadt Hamburg*
SP	*Soziale Praxis*
SPD	Sozialdemokratische Partei Deutschlands (Social Democrats)
StA	Staatsarchiv
StAA	Stadtarchiv Augsburg
StAB	Staatsarchiv Bremen
StAH	Staatsarchiv Hamburg
StJB	*Statistisches Jahrbuch der Stadt Berlin*
USPD	Unabhängige Sozialdemokratische Partei Deutschlands
VfZG	*Vierteljahrshefte für Zeitgeschichte*
VO	Verordnung
VRT	*Stenographische Berichte über die Verhandlungen des Deutschen Reichstags*
VW	*Volkswohlfahrt*
ZB	*Zentralblatt der christlichen Gewerkschaften Deutschlands*
ZfG	*Zeitschrift für Geschichtswissenschaft*
ZfH	*Zeitschrift für das Heimatwesen. Amtliches Organ von Fürsorgeverbänden des Deutschen Reichs*
ZStA	Zentrales Staatsarchiv

PREFACE

Mass unemployment is the major social problem of the 1980s in the advanced industrial world. Politicians, economists, sociologists and academic analysts of all kinds are divided about its causes, its impact and its relevance to crime, rioting and political extremism. Historians too have a contribution to make to this debate. For our own society is not the first to have been affected by mass unemployment. Just half a century ago, in the Depression, an economic slump of even more dramatic proportions than that of the 1980s threw tens of millions of people out of their jobs right across Europe, causing misery and despair on a scale not even matched in the unemployment crisis of our own day. The most extreme depredations of unemployment were undoubtedly visited upon the ill-fated democracy of the Weimar Republic in Germany. Founded in the aftermath of defeat in the First World War, the Republic, thanks to the leading role played in it by the social reformers of the working-class Social Democratic Party and the Social Catholics of the Centre Party, in many ways had every right to be proud of its social legislation and its efforts to create a genuine welfare state. But these achievements in the end counted for little in the face of economic collapse and political extremism. The fate of the Weimar Republic, as it collapsed and gave way in 1933 to the Nazi dictatorship of Hitler's 'Third Reich', has stood ever since as a grim reminder of the fragility of democracy and the frightening ease with which it can be destroyed in a time of crisis.

Curiously, however, the very drama of the Republic's political collapse has diverted historians' attention away from the mass unemployment which did so much to bring it about. In Britain, where the political resonances of the Depression were far less severe, as was the impact of the economic crisis, the early 1930s live on in popular memory principally as a time of economic hardship and social misery, and it is these features of the crisis which historians have done most to illuminate, although recent research has also stressed that for those in employment this was a period of rising real wages. In Germany, however, it is the tramping of jackbooted stormtroopers, the roar of the Nazi crowd, and the bumbling intrigues of the incompetent political leaders of the late Weimar Republic that have attracted most attention. It is only the re-emergence of mass unemployment in our own day that has prompted historians to take a closer look at the same problem as it beset the Weimar Republic in the 1920s and 1930s. The availability of large quantities of published and unpublished documentation makes it possible for historians to study the problem at many levels in a depth and detail unattainable by the contemporary observer; while the distance in time from which we approach the problem enables us to set it in a longer-term perspective that is unavailable to those who study unemployment only in the present. History, of course, seldom repeats itself; Britain and Europe, and still more Germany, are very different today from what they were fifty or sixty years ago. Above all, the political context within which economic crises occur has changed. Nevertheless, the studies collected in this book offer some striking and sometimes unexpected parallels, as well as exposing some obvious, and perhaps also not so obvious, differences between the two periods.

Unemployment, especially in the dimensions which it eventually attained under the Weimar Republic, is a highly complex phenomenon, and its causes are still a matter for conjecture and dispute. The authors of this book are not so much concerned with the economic analysis of the origins of unemployment, a subject which belongs more properly to the often rather technical realm of economic and financial history, as with its social and political consequences. What groups were most severely affected by unemployment, and why? How did they react? How effective were welfare and job-creation schemes? Did unemployment fuel social instability and political extremism, and if so, in what ways? How far was unemployment a cause of the collapse of the Weimar Republic and the triumph of Hitler? Did the Nazis solve the unemployment problem by peaceful Keynesian measures or through other, less laudable means? In responding to these questions, the contributors to this volume do not seek easy answers. Above all, they do not neglect the unemployed themselves. It is all too easy to treat the jobless as mere statistics, or the anonymous objects of social welfare and political debate. But a wealth of sources exists through which historians can recapture something of the subjective experience of unemployed men and women. Only by trying to understand what it felt like to be jobless in Weimar Germany can we hope to comprehend why the unemployed reacted to their situation in the ways they did. A major focus of this book, therefore, is on reconstructing the experience of the German unemployed. The resulting revelations are often unexpected, and enable us to see the social and political crisis of the Weimar years in a new and often disturbing light.

Disturbing too are the parallels which emerge with our own time. In the pages of this book can be found a wide variety of devices through which successive governments in Germany, up to and above all including the Third Reich, fiddled the statistics so as to make it look as if unemployment rates were much lower than any reasonably fair definition would have made them out to be. The exposure, in several chapters of this book, of the statistical manipulations of the authorities in the early 1930s, serves as a reminder that similar manipulations being carried out by government today will not escape the attention of future historians. Then as now too, it was political expediency that dominated government thinking: and so-called job-creation schemes, as a number of the authors of this book show, mostly had little effect on long-term unemployment, providing only temporary relief at such low wage-rates that they had no stimulating effect on demand at all. Worse still, they were achieved only by various forms of coercion, under the Weimar regimes disguised in legal formulae and gift-wrapped in the language of welfare, in the Third Reich naked, brutal and overt. Apprentices were taken on only because they were cheap, and were replaced by other apprentices when they finished their training, not by full-time adult workers. 'Voluntary Labour Service' was less than voluntary, least of all after the Nazi seizure of power. It was rearmament and military expenditure that did most to bring a real increase in jobs, at a cost that was eventually to prove appalling in any terms, although other kinds of job-creation schemes had already begun to have an impact on unemployment levels under the last Weimar governments. Here too is a lesson we might ponder in our own day, as the welfare state is steadily dismantled while expenditure on armaments and 'law and order' continues to rise.

The graphic details which emerge from this book about the sufferings of the unemployed themselves are no less striking. In chapter after chapter, we shall see how they were exposed to continued and increasingly arbitrary official action, cuts in benefits, coercion and discrimination. The story of how they responded does not offer much comfort to socialists or indeed to democrats of any persuasion. Unemployment brought fatal divisions into the labour movement and the workforce. Those affected by it flocked in their millions to vote for the Communists, while fear of unemployment and the unemployed, and anger at the effect the crisis was having on business, led millions of other voters into the polling-booths to support the Nazis. Genuine political commitment among the jobless was relatively rare, and when they mobilised on the streets, no political party, not even the Communist, was able to control them. Apathy and indifference were the major psychological consequences of long-term unemployment, and those who sought a way out of the impasse, whether through crime, or through gang violence, or through political activism, remained a minority. The experience of unemployment was thus not only financially ruinous but also morally and psychologically debilitating.

The 1920s and 1930s were a time when full-time wage-labour was perhaps at the height of its historical career as the defining factor in people's expectations of life in countries such as Britain and Germany, although in Germany there was still a large artisanal and peasant sector that fell outside this category. It may be that with the growth of economic activity beyond employed wage labour, whether in the form of the 'black economy' of clandestine work, or in the shape of computer-based freelance domestic labour, its importance is now in a phase of secular decline. But however rapid the effects of technological and social change in our own day, we still have a long way to go before full-time waged employment ceases to be the basis on which millions of people in advanced industrial societies construct their lives. It is for the reader, perhaps, to draw from the following pages any lessons that might suggest themselves for social policy and political action to confront the problem of unemployment in our own time. What can be said, however, is that spurious solutions, whether based on statistical sleight of hand, on demagogic rhetoric, on forced labour, or on increased arms expenditure, bring about a 'reduction' of unemployment at a very high price. The dangers of avoiding a genuine confrontation of the problem should be as apparent in Britain and Germany in the 1980s as they were in the much more dramatic circumstances of half a century ago.

The following chapters attempt to trace the history of unemployment from the beginning of the Weimar Republic (Chapters 2 and 3), through its whole history from 1918 to 1933 (Chapter 5), and into the Third Reich (Chapter 11), so that the years of mass unemployment in 1929–33 can be seen in their proper perspective. But the concentration of the book (Chapters 4, and 6–10) is inevitably on the Depression which brought about the Republic's collapse. While some of the contributions (Chapters 3, 4, 6, 11) deal mainly with government and municipal policies towards the unemployed, some (Chapters 2, 5 and 7) with the social distribution of unemployment, and some (Chapters 8, 9, 10) with attempts to bring about the political mobilisation of the unemployed, none of them neglects the subjective experiences and aspirations of the unemployed themselves, and in some contributions (Chapters 5, 7, 8 and 9) these form indeed the central focus of study.

Throughout this book an attempt has been made to allow the German unemployed to speak to posterity, both directly, from interview material, and indirectly, through their actions and opinions as reported in contemporary sources. Material on which to base such an attempt is plentiful, above all on a local basis, and a number of the contributions, whether on Augsburg (Chapter 3), Frankfurt (Chapter 5), Hamburg (Chapters 6, 11), Berlin (Chapter 8) or Altona (Chapter 9) reveal the advantages of local studies in this respect. Inevitably, the development of unemployment legislation and welfare provisions forms an important part of the context, without which some of the points at issue are not easy to grasp: so the general introduction (Chapter 1) tries to provide an outline of basic information on this context, as well as to sketch in some of the main trends and structures of unemployment in the Weimar Republic, to show how they have been studied by historians, and to suggest some of the ways in which the contributions to this book come together to revise or extend the current state of knowledge on these problems.

The origins of this book go back to the seventh meeting of the Research Seminar Group on German Social History, held at the University of East Anglia in July 1983, when several of the chapters were presented as discussion papers. None of them has previously been published, though during the preparation of this book, some may have appeared in different versions in German. Thanks are due to the Nuffield Foundation for providing the financial support for the meeting, to the University of East Anglia for its assistance in organising it, and to the participants for providing a stimulating discussion which, as they will recognise, has left its traces in the Introduction (Chapter 1) at various points.

We would also like to express our gratitude to Cathleen S. Catt for translating Chapter 3, to Lynn Abrams for reading the proofs and helping with the translation of Chapter 4, to Marjan Bhavsar and Elvi Dobie for their work in assisting with the final preparation, and, as with previous volumes in this series,[1] to Richard Johnson for converting our often obscure graphs into readable artwork, and to Croom Helm and especially to Richard Stoneman for the editorial patience and tolerance which the very protracted period of gestation has been accorded.

Richard J. Evans
Norwich

Dick Geary
Lancaster

Note

1. Richard J. Evans and W. R. Lee (eds), *The German Family. Essays in the Social History of the Family in Nineteenth- and Twentieth-Century Germany* (Croom Helm, London, 1981); Richard J. Evans (ed.), *The German Working Class 1888–1933: The Politics of Everyday Life* (Croom Helm, London, 1982); Richard J. Evans and W. R. Lee (eds), *The German Peasantry: Conflict and Community in Rural Society from the Eighteenth to the Twentieth Centuries* (London, Croom Helm, 1986); and also Richard J. Evans (ed.) 'Religion and Society in Germany' (special issue of *European Studies Review*, vol. 12, no. 3, July 1982).

1 INTRODUCTION: The Experience of Unemployment in the Weimar Republic

Richard J. Evans

I

To be unemployed is to be without a job: yet the notion of a 'job', in the sense of regular, full-time employment is, historically speaking, quite new. In pre-industrial society, with its seasonally determined patterns of work and its fluid, often unarticulated and barely visible dividing-line between work and leisure, to be idle or inactive was almost unknown, for even leisure had its productive aspects. Only with the rise of industrial wage-labour did fixed hours of work, in a place away from home, on a regular, uninterrupted basis, five or six days a week, become the defining norm against which the concept of unemployment could establish itself. Yet even in the late nineteenth century, as Germany was still an industrial society in the making, millions of people were engaged in casual or domestic forms of labour; or changed from one job to another, indeed from one type of job to another, with a frequency that made the notion of a fixed, permanently employed workforce inapplicable; or migrated from country to town and back again at intervals during the year, as the demands of the harvest or the opportunities of seasonal employment dictated. The concept of unemployment was by no means absent from political discourse in the late nineteenth century: indeed, there were even demonstrations of the unemployed on one famous occasion in Berlin early in 1892. Some groups of workers were already coming to regard wage-labour in a chosen trade or place of abode as so essential to their existence that they could feel justifiably aggrieved if they were deprived of it. But the fluidity and instability of the workforce was still such that any attempt to calculate overall levels of 'unemployment' for this period would be premature.[1]

For the definition of unemployment is itself subjective. Official definitions are quite often relatively arbitrarily changed according to the political exigencies of the moment. As we shall see later in this book, the governments of the Weimar Republic were no exception to this rule (still less was that of the Third Reich). Government

1

statistics ultimately depend on the actions of bureaucratic agencies which register some people as unemployed, but refuse to register others. Correspondingly, however, individual workers themselves also vary in their willingness to 'sign on' as unemployed. Women, for example, have been in some situations notoriously reluctant to regard themselves as unemployed when dismissed from their jobs; although, as Helgard Kramer points out in her contribution to this volume, the reasons for this behaviour and its variations between different categories of women were more complex (and more interesting) than has often been supposed. Only when millions of workers — the majority of the workforce — had come to regard a full-time job as the normal, indispensable basis for their existence, and to reject any alternatives that might be proffered, was it possible for those selfsame millions, on losing their jobs, to become unemployed.

So unemployment was the product of the triumph of a particular kind of work ethic, and its defintion became — as it is today — the focus of political controversy. The state may operate one definition of unemployment: the unemployed themselves, even if refused recognition as such by the rest of the state, might operate a very different one. By the 1920s, certainly, the concept of unemployment had entered the centre of political discourse in a way that it had never managed to do before the First World War. Probably it was helped on its way there by the beginning of welfare legislation in the area. Whatever the social policy of Bismarck and his successors did in the way of creating a viable system of insurance against illness, accidents and old age, it never went so far as to insure people against the loss of their jobs. All that was available to them was the bounty of the Poor Law, which took away their civil rights, including the vote, and subjected them to a regime of demeaning inquisition and intrusive control.[2] As Richard Bessel and Merith Niehuss suggest in the first two contributions to this volume, it was the First World War that began the move away from this system and towards a real scheme of unemployment relief. Unemployment was not a serious social or political problem in Germany before the First World War, in contrast to Britain. But the dislocation of the economy caused by the switch from consumer goods to war production in 1914–15 brought about a substantial level of unemployment. The need to deal with this, to support dependents of soldiers at the Front, and then later, to provide those returning from the war in 1918–19 with the means of subsistence until they found a job: these were the political imperatives that lay behind the emergence of the unemployment benefit system that took on concrete form at the start of the Weimar era.

The 13 November 1918 saw the introduction of the first in a long series of ordinances establishing a system of support for 'those without an occupation' (*Erwerbslosenfürsorge*) which lasted, despite frequent amendment, until 1927.[3] Amendment was necessary not least because of the galloping inflation which was becoming the dominant economic fact of the early Weimar years, so that the rate payable had to be changed with ever-growing frequency. The financial burden of the payments was borne partly by the central state authorities in Berlin, which paid a half, the *Länder* or federated states (such as Bavaria, Baden or Hamburg), which paid a third, and the local city or district authorities, which paid the rest. In the inflationary conditions of the early 1920s these financial arrangements soon began to prove very costly. Already in November 1918, therefore, measures were taken to encourage those supported to find work. By 1919 an extensive system of state labour exchanges (*Arbeitsnachweise*) was being established, to be co-ordinated by a specially-created central Reich Office for Employment (*Reichsamt für Arbeitsvermittlung*) set up in May 1920, and the maximum period allowable for receipt of benefits • was being restricted to six months, or in some cases even less. What these and other regulations meant for those to whom they applied is shown in detail by Merith Niehuss, using the example of Augsburg, in Chapter 3. The situation in which these various laws and institutions were created to deal with, however, was only a temporary one. As Richard Bessel suggests in Chapter 2, fears of permanent mass unemployment in the wake of demobilisation proved to be exaggerated, and before long the returning troops had been largely re-absorbed into the labour market.[4]

Inflation meant full employment in the early Weimar years; until, that is, it became hyperinflation and got completely out of control. By 1923 the labour exchanges and benefit offices were in full swing again. The crisis of hyperinflation, as Merith Niehuss shows, had a much more serious and long-lasting effect on the labour market than the demobilisation in 1918–19 had done. It threw not only the unskilled out of their jobs, but also forced independent craftsmen to stop trading, and even professionals to seek welfare support. New kinds of claimants were thus coming into being, and as conditions reached their most chaotic in the late autumn of 1923, the existing system of supporting 'those without a livelihood' came close to collapse. The consequences for those receiving benefits — illustrated in graphic detail in Chapter 3 — were less than pleasant. As the crisis receded with the currency stabilisation of early 1924, and employment prospects started improving once more, government began to consider how to prevent such a near-collapse occurring again. Support for 'those without a livelihood' had originally been regarded as a temporary measure to deal with demobilisation. In the long run it seemed inadvisable for the central and regional authorities to bear so much of the cost of supporting the unemployed, when it was the local authorities who in reality were best placed to deal with them and find them jobs.

On numerous occasions from 1919 onwards, indeed, there had been attempts to introduce a new, permanent and comprehensive system of unemployment insurance. In the spring of 1924 the existing system of support for 'those without a livelihood' was restricted to employees who had contributed to the state health insurance scheme for a minimum of 13 weeks over the previous year. This not only introduced the principle of insurance but also excluded the long-term unemployed and lower-middle-class groups, such as white-collar workers and the self-employed, who did not participate in health insurance schemes.[5] The numbers of these excluded groups, as Merith Niehuss shows in Chapter 3, could be quite considerable. Meanwhile, the financial provisions of the system were also reformed, so that contributions came from employers, employees and local authorities instead of the Reich and the *Länder*. The maximum period of eligibility was set at 26 weeks, with the possibility of extension to 39. The rates paid, as before, were scarcely adequate even for the smaller number of people now covered. Even so, employers were inclined to object that they were so close to some wage-rates they paid that they weakened the will to work of those who actually were in employment.[6]

This scheme too soon ran into difficulties. True, unemployment declined rapidly during 1924 as the economy recovered in the first phases of currency stabilisation. But in the longer run, the consequences of stabilisation were less favourable for workers. Cost-cutting and rationalisation all round helped business, but by the winter of 1925–6 unemployment was rising sharply again,

largely as a consequence of the realignment of the economy that
followed the recovery of the mark in 1924. Soon the unemployed
were to be counted in millions again.[7] For the rest of its existence
the Weimar Republic had to live with mass unemployment. The
years 1924–9 are sometimes thought of as years of prosperity and
stability, a peaceful interlude between the inflation and political
crisis of 1918–23 and the Depression and political collapse of
1929–33. But from 1926 onwards they were years of hardship for
the millions of workers who were unable to get a job. In particular,
it quickly became apparent that something would have to be done
about the lower-middle-class groups and long-term unemployed
excluded from support for 'those without a livelihood' by the
reforms of 1924. The result was a fundamental series of laws
passed in 1926–7 which laid down the framework of unemployment support for the rest of the Republic's existence.

On 20 November 1926, after much debate, a new form of
support was established to help those excluded in 1924. This was
the system of so-called 'crisis benefits' (*Krisenunterstützung*).
When the period of support for 'those without a livelihood' — by
now extended to a year — ran out, claimants could receive 'crisis
benefits' for a period to be determined according to trade and
locality, and provided they passed a means test. Three-quarters of
the necessary sums were raised by central government, the rest by
local authorities. In 1927 a maximum length of 26 weeks was
imposed for the receipt of crisis benefits, extended to 39 weeks in
1928 (one year for white-collar workers over 40 years of age).
Although attempts were made to restrict crisis benefits to certain
trades, they were made general from June 1929 for all branches of
employment except seasonal trades. In the conditions of
permanent mass unemployment that began in 1929, however,
increasing numbers of the unemployed began to run through the
39 weeks allowable for receipt of crisis benefits. Their support
after this was a matter for poor relief, and soon the numbers of
these 'welfare unemployed' (*Wohlfahrtserwerbslose*) were growing
rapidly. Since crisis benefits were largely paid by the Reich, and
poor relief wholly by local authorities, this placed an ever-increasing burden on local authority finances as the Depression wore
on.[8]

Meanwhile, an even more significant legislative intervention
had taken place in the shape of a new law on labour exchanges and
unemployment insurance (*Gesetz über Arbeitsvermittlung und*

Arbeitslosenversicherung), which came into effect on 1 October 1927. This replaced the old system of support for 'those without a livelihood' (*Erwerbslosenfürsorge*), which finally ceased to exist in 1928, with a new system of unemployment insurance (*Arbeitslosenversicherung*) in which employers and employees were to pay contributions which were intended to finance unemployment benefits for a maximum of 26 weeks (in some cases 39). These benefits (*Arbeitslosenunterstützung*) were paid at levels that varied according to trade, locality and former earnings-levels, and were supplemented by family benefits where appropriate. All this was to be implemented by a new system of labour exchanges (*Arbeitsämter*) organised in 13 large districts which corresponded to economically determined geographical areas rather than already existing administrative structures. The whole package has rightly been regarded as one of the major achievements of the Weimar welfare state. For all that, however, it soon proved incapable of dealing with the unprecedented levels reached by unemployment during the Depression of 1929–33.[9]

For the numbers of unemployed increased to more than five million in the winter of 1930–1, and six million in 1931–2, without significantly declining thereafter. By January 1932 it was estimated that the unemployed, with their dependents, made up about a fifth of the entire population, some 12.86 million people. These numbers were substantially the same as late as March 1933. In this long period of mass unemployment, the insurance contributions paid by employers and employees soon proved unable to cover the vast sums being paid out in benefits. The problem of what to do with the unemployment insurance system now moved to the centre of the political arena. Central government became increasingly reluctant to cover the gap between contributions and benefits because of its commitment to solve the crisis by deflation and reducing government expenditure. Employers did not want to increase their contributions at a time when profits were falling. Trade unionists and workers were unwilling to see the level of benefits reduced at a time when it seemed more and more likely that they themselves would soon become dependent on them. Local authorities were anxious not to see a shift in the burden of support from the centre to the localities at a time when the numbers of 'welfare unemployed' supported by local authorities were rapidly increasing. The political institutions of the Republic were unable to reconcile these differences. From 1930 they were overriden by the 'government of experts' under Heinrich Brüning, which now began to issue a series of 'emergency decrees' whose cumulative effect was to bring about a gradual dismantling of the welfare state so painstakingly assembled in the years of relative prosperity.[10]

Figure 1.1: The Unemployment Rate in Germany 1921–38

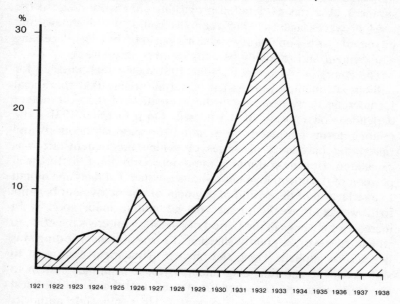

The graph shows registered unemployed persons as a percentage of dependent occupied persons.
Source: Dietmar Petzina *et al.* (eds), *Sozialgeschichtliches Arbeitsbuch* III (Munich, 1978), p. 119.

The Brüning government moved to deal with the situation by reducing benefits, increasing contributions and providing financial support from the Reich all at the same time, in the emergency decree of 26 July 1930. But no sooner were these adjustments made than unemployment reached fresh heights and brought the problem of financing benefits to a head once more. A gesture towards a solution was made on 6 October, when contributions were raised again, but little was achieved by this step and the situation grew steadily worse. On 5 June 1931, therefore, the

Brüning government went much further and reduced unemployment benefits by up to 14.3 per cent. The period claimable
was reduced to 20 weeks for seasonal labourers and the rate for
them fixed at the same as that for 'crisis benefits', which were also
reduced by up to 14.3 per cent. Claimants under 21 were disqualified unless they had a family to support, and married women were
subjected to a means test (thus breaching the principle of insurance). A series of detailed provisions effectively reduced the
level of wages against which benefits could be calculated, and
increased the obligatory waiting-period between ceasing
employment and receiving benefits to up to three weeks.

The emergency decree of 5 June 1931 meant real hardship for
millions of unemployed, and it is not surprising that the Social
Democratic Party made its further toleration of the government
dependent on revision of the decree. On 6 October 1931 some
minor reforms improved the situation for seasonal labourers and
raised the base-level of wages on which the benefit-rate was
calculated. But the continuing rapid deterioration of the financial
position of the unemployment insurance system led the same month
to a reduction of the period of payment of unemployment benefits
to 20 weeks (16 weeks for seasonal labourers), made good by an
increase of 6 weeks in the period of eligibility for crisis benefits, to
which, however, a greatly sharpened form of means testing was
henceforth applied. Thus, the burden of payment continued to
shift from the self-financing unemployment insurance system to
the Reich-financed crisis-benefit system, and thence to the local
authority-supported welfare system. Each responsible authority
was now imposing increasingly restrictive definitions of unemployment in order to reduce the number of claimants. Beyond
this, too, the Reich was increasingly using the indebtedness of local
authorities as a lever to force on them an acceptance of its own
deflationary policies, as Elizabeth Harvey shows, using the example of Hamburg, in Chapter 6.

By the winter of 1932–3, it has been estimated, there were over
a million people capable of work who were without either a job or
any form of support, so that the official figure of six million
unemployed was certainly a gross underestimate by almost any
definition except that of the responsible officials.[11] The variety of
ways in which the number of registered unemployed was held in
check by official action was virtually endless, as numerous examples in the following chapters will illustrate. By 1932 no more than

Figure 1.2: Recipients of Unemployment Insurance and Crisis Benefits 1927–37

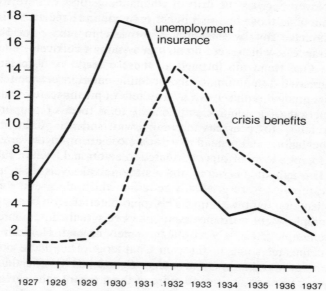

Recipients
(100,000 s)

The figure for 1927 covers October–December only.
Source: Dietmar Petzina *et al.* (eds), *Sozialgeschichtliches Arbeitsbuch* III
(Munich, 1978), p. 160.

a third of the unemployed were receiving unemployment benefits; most of the rest were on crisis benefits or welfare support, and a significant number were receiving nothing at all. The Brüning cabinet seemed incapable of solving the problem of financing these payments, and the burden falling on the local authorities continued to grow. In September 1932 there were only 618,000 people on unemployment pay, 1,230,000 on crisis benefits and 2,500,000 on welfare support. The right-wing Papen cabinet, which came to power in the spring of 1932, passed a decree on 14 June 1932 that was intended to rescue the finances of the unemployment insurance system by reducing the period claimable to a mere six weeks, and reducing the level of benefits by up to 23 per cent. After six weeks claimants were subjected to a means test. If they were allowed onto crisis benefits (which were reduced by 10 per cent) they could stay only for a maximum of one year. The reform meant yet more hardship for the unemployed, but it did relatively little for the near-bankrupt local authorities, even though the level of welfare payments was also reduced by 15 per cent. Only in November was the drift of claimants to local welfare stopped by allowing those on crisis benefits to continue receiving them indefinitely. But by then it was too late to rescue local authority finances, which were by now in a state of collapse.[12]

One trend ran through all these emergency decrees: the progressive dismantling of the unemployment insurance scheme and its gradual replacement by a system of means-tested benefits. The period of automatic entitlement to a level of support directly related to previous earnings was reduced until it became negligible, and instead more and more unemployed were forced to accept a level of support related to a notional level of subsistence. Throughout the whole process the actual levels of support in all categories were gradually reduced, and the criteria for granting benefits tightened up. This piecemeal destruction of the unemployment insurance system had very real consequences for the unemployed who were on the receiving end: Heidrun Homburg delineates some of these in Chapter 4. They were often at the mercy of definitions and categories imposed by officialdom, and as time went on, more and more of them fell into the category of the 'welfare unemployed'. They were increasingly exposed to arbitrary variations in practice between different local authorities. The growing financial crisis of the municipalities made their situation steadily worse as cuts and restrictions in welfare followed. The 'welfare unemployed', as a consequence, were discriminated against and marginalised as the financial burden of the crisis was now inexorably transferred from the insurance scheme to the Reich, from the Reich to the local authorities and finally, from the local authorities to the unemployed and their families themselves. Growing poverty and malnutrition and an increase in diseases like rickets and tuberculosis were the inevitable results.

II

Changing official definitions of unemployment were most marked perhaps in their application to women: but here even more than among other groups of the population, a crucial role was played by the willingness or otherwise of those thrown out of work to define themselves as unemployed. Arranging the dismissal of certain categories of women from their jobs in order to make way for men, and then refusing to regard these dismissed women as un-employed, was a method of 'reducing' unemployment which was relatively easy to implement, given the strength in German society of the belief that woman's place was in the home. Such measures were taken during demobilisation in 1918–19, as women drafted into jobs previously done by men were ousted, now that men were returning from the front.[13] In the stabilisation crisis of 1923–4, the authorities introduced special ordinances to reduce the employment of married women (so-called 'double earners' supported both by their husbands and by their own job). Again in 1931 the state officially decreed the removal of married women employees from the civil service, while in 1933–4 the Nazi regime actually offered loans to those women who gave up their jobs in order to get married. The effect of these measures is far from clear. As Helgard Kramer points out in Chapter 5, because male unemployment in the Depression rose faster than female, it is often assumed that women were simply not registering as un-employed when they lost their jobs (whether this was through official action or not). Yet there is a singular lack of evidence to back this assumption, and as Kramer shows, the reality was a good deal more complex.[14]

To begin with, by removing or drastically reducing the income of the male head of household, mass unemployment forced many previously non-working women to find jobs (even if only part-time) in order to make up the family wage. Thus, women coming onto the labour market went a long way towards balancing out those who were leaving it. Moreover, many unemployed women could ill-afford to do without benefits in this situation, and despite all the efforts of the authorities to deny women benefits (and indeed classification as unemployed) on the grounds that they were being supported by their families, those women who had had employment in a trained or skilled post such as white-collar work — and they were very numerous by the end of the 1920s — were very difficult to treat this way. Women often tended to be employed in sectors of the economy less severely hit by the Depression than the male-dominated heavy industrial sector. On the other hand, many were engaged in seasonal or part-time work (for example in the garment trade) which fell outside the statistics. Women workers were used to frequent job-changing and short-time or part-time work, and were more flexible and adaptable in their employment behaviour than men were. For this reason, and also because part-time women workers did not bring with them any obligation on the part of employers to pay insurance contributions on their behalf, many employers found it preferable to take them on during the Depression. As Helgard Kramer shows, the complexities of this situation were considerable, and varied from job to job and between different sectors of the economy. But all in all it seems more than possible that the relatively low rates of recorded female unemployment may have reflected a complex reality rather than a simple statistical illusion.

If women thus enjoyed a certain freedom of choice in how they responded to changes in the official definition of employment, the same was clearly less true in the case of young people. Here too the state undertook repeated attempts to reduce the financial burden of paying benefits by removing various categories of the unemployed from the register. First, apprentices were not obliged to pay contributions until near the end of their apprenticeship; then the under-21s were excluded from crisis benefits; and finally, from 1930, payment of unemployment insurance benefits to the under-17s (then from 5 June 1931 the under-21s) was means tested against the parents' income. As Elizabeth Harvey shows in Chapter 6, these tactics increasingly ushered young people without jobs into the ranks of the 'welfare unemployed'. Such policies reflected not least the fact that the young were particularly hard hit by unemployment in the Weimar years. As Detlev Peukert demonstrates in Chapter 7, this period saw the entry onto the labour market of the last substantial birth cohorts of the prewar years, before the continuous fall of the birth-rate that began at the turn of the century and reached dramatic proportions during and after the First World War took effect. An increase in the female labour supply further worsened the chances of young men. Many of them had got their jobs without training in the special circumstances of the First World War, and now faced trying to find employment without skills to offer. Apprenticeship schemes made little difference in the long run because employers simply replaced apprentices when they ended their apprenticeship, preferring a continuous supply of cheap temporary labour to the employment of full-time adult workers. And as Dick Geary shows in Chapter

10, the trade unions made things worse by their policy of agreeing to the dismissal of young workers in order to keep older workers and heads of families in their jobs.[15]

While Weimar governments complacently assumed that women dismissed from their employment as 'double earners' would return to a safe and non-threatening existence as mothers and housewives, they were far more concerned about the reactions of unemployed youth. Throughout the Weimar Republic, therefore, they tried strenuously to prevent them becoming a source of crime and disorder. Central to these efforts were the numerous attempts of the authorities to direct young people into specially created work schemes where they could be supervised and controlled. For governments, of course, did not rest at juggling with statistics and definitions of unemployment in their efforts to 'reduce' the unemployment rate. It is often thought that job-creation schemes in Germany were the invention of the Nazis. But it is a mistake to see the advent of the Third Reich as being followed by a Keynesian bolt from the blue, in which the jobless were quickly put to work on building autobahns and the like. To begin with, the old poor-relief system of the Imperial period and the new system of benefits for 'those without livelihood' of the immediate postwar years, both gave priority to finding work for claimants over paying them benefits. Indeed, claimants were virtually obliged to take on any job assigned to them by the authorities. Only when all possibilities had been explored did they finally receive their payments. The authorities did their best to see that this point was never reached. As soon as the war ended the government launched a state-funded emergency works scheme in which the unemployed were offered jobs on road, canal and dam construction and other specially created civil-engineering projects. In 1924 the new law on unemployment benefits strengthened the obligation to carry out 'compulsory labour' (*Pflichtarbeit*). Even the unemployment insurance scheme of 1927 allowed the authorities to assign jobs to claimants who were under 21. As Elizabeth Harvey shows in Chapter 6, the authorities enrolled large numbers of young people in compulsory vocational training courses, rural work programmes, emergency labour schemes and so-called 'voluntary labour service'. By the time the Nazis came to power in January 1933, as Birgit Wulff demonstrates in Chapter 11, a whole range of job-creation schemes was in existence, including massive road-building, agricultural improvement and housing-construction

Figure 1.3: Persons on State-supported Short-time and Emergency Work 1928–38

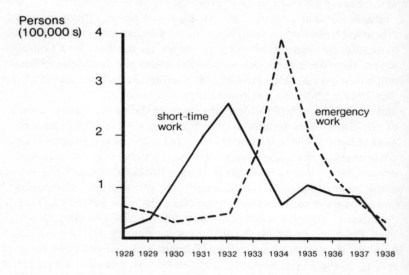

Source: Dietmar Petzina *et al.* (eds), *Sozialgeschichtliches Arbeitsbuch* III (Munich, 1978), p. 122.

projects planned by Brüning, effected under Papen and considerably expanded by the short-lived government of Kurt von Schleicher.

These schemes were not popular with the unemployed. The postwar civil engineering projects were poorly paid, and workers were not given adequate clothing and resented having to stay overnight on construction sites far from home in accommodation that was often utterly miserable. Even less popular, as Merith Niehuss shows in Chapter 3, were compulsory labour schemes in which the unemployed were put to clearing snow, chopping wood, cleaning schools and so on, on pain of withdrawal of benefits. The element of control in these schemes was obvious, as the authorities sought to keep the unemployed busy as a diversion from the dangers of idleness. It was there too in the job-creation programmes of the final years of the Republic. As Elizabeth Harvey shows in Chapter 6, both compulsory municipal labour schemes and so-called 'voluntary labour service' (*Freiwilliger Arbeitsdienst*), introduced by the Brüning government in June

1931, had strong disciplinary elements and were intended not simply to reduce unemployment but also to instil moral backbone into those who were enrolled in them.[16] It was not surprising that when harsh and restrictive conditions and close supervision were added to low pay, workers responded by passive resistance, evasion, low productivity and even sometimes strikes. It was only the coercive measures introduced by the Nazis after Hitler became Chancellor in January 1933 that enabled such schemes to be operated effectively and even expanded.

There were worse things than unemployment. Recognition of the hardship and suffering engendered by unemployment should not lead to a romanticisation or glorification of work. Under some circumstances it might even be preferable to be without a job. As Helgard Kramer points out in Chapter 5, many women were only too glad to leave the physically demanding and often dangerous industries into which they had been drafted during the war. In addition, the conventions of the sexual division of labour meant that men may have been reluctant to take on 'women's jobs', such as in clerical and secretarial work, except in sectors such as machine construction and building, which were male-dominated and where unions and employers agreed to dismiss women. Moreover, as Elizabeth Harvey shows in Chapter 6, the disciplinary element involved in job-creation schemes, when coupled with low pay and poor conditions, made them an unattractive alternative to unemployment. However, the freedom of choice of the unemployed to escape from their plight and yet remain within the bounds of the law was steadily diminishing, and the possibilities of manipulating the support system in their own interests grew steadily less with the growing power of the authorities to categorise them, assign them to job creation and other schemes, and take decisions about their lives with no real possibility of appeal, as Heidrun Homburg illustrates for the case of the 'welfare unemployed' in Chapter 4. The results for the stability and survival chances of the Weimar Republic were little short of catastrophic.

III

Unemployment, especially as it became long-term, engendered increasing apathy and indifference in those who had to suffer it. In Chapter 7 Detlev Peukert quotes from a number of contemporary sociological and psychological investigations which stressed not only the resignation of the long-term unemployed but also their emotional volatility.[17] The stigma attached to unemployment could not be thrown off by engagement in officially sponsored labour schemes, youth centres and the like. The only viable way out was to choose an alternative that generated its own kind of self-respect. Crime was one such escape-route for some, though not as many as might be supposed. Others formed street gangs such as the so-called 'cliques' of Berlin.[18] But the most attractive way out of the boredom, indifference and loss of self-respect that unemployment caused was through active involvement in radical politics. Here we come to what was in many ways the most crucial consequence of mass unemployment in the last years of the Weimar Republic. For there can be no doubt that the Nazi vote rose as unemployment rose. The two curves follow each other with a parallel motion too obvious to ignore.

Yet as Ross McKibbin argued as long ago as 1969,[19] this does not mean that the unemployed simply voted for the Nazis. Once we begin to look at voting patterns on a regionally differentiated basis, a very different picture soon emerges. Recent quantitative studies have shown 'that the NSDAP could only register relatively small electoral successes in areas with high unemployment rates'. Indeed, Jürgen Falter and his collaborators, who have conducted the most sophisticated electoral analysis so far available for this period, have concluded that

> It can hardly be doubted any longer that there is a *negative* statistical correlation between the level of unemployment and the electoral successes of the NSDAP in the Reichstag elections of 1932 and 1933. In terms both of its absolute strength and of the growth of its electoral support, the NSDAP tended to be more successful in areas where unemployment was below the Reich average, and less successful where it was above it.[20]

It seems reasonable to conclude, as Falter and his group do, therefore, that the Nazi vote grew because of general despair and disillusion at the failure of the Republic to solve the crisis, perhaps also in response to a growing fear among those who were not unemployed of the political consequences of the growing radicalisation of the unemployed themselves.[21]

For there seems little doubt that it was above all the Communists who gained most from the political mobilisation of the jobless. As the unemployment rate rose through 1930, 1931 and 1932, so too did the Communist vote. It even continued to rise in November 1932, as unemployment continued to rise, but the Nazi vote was beginning to fall. And as Eve Rosenhaft shows in Chapter 8, the connections between unemployment and Communism were even more obvious on the streets than in the polling booths. This was indeed one reason for the growing attempts of the state to discipline the young unemployed. As the factories emptied, radical politics moved from the workplace to the neighbourhood. Unemployed men and youths, reluctant to stay at home all day, spent much of their time on the streets or in pubs, bars, railway stations and other public places, as both Chapter 7 and Chapter 8 illustrate. And as a consequence, the neighbourhood became a central element in the political struggle, as mass unemployment opened up sharp divisions between social classes (for example, working-class families with declining incomes on the one hand, shopkeepers and landlords with falling takings or unpaid rents on the other). As the social fabric of community solidarity began to crumble in what Eve Rosenhaft calls 'the scramble for cash in the neighbourhoods', people began to turn to extremist political parties — the NSDAP and the KPD — for help; help both particular and concrete — for example, in struggles over rent payments, or over cuts in the education budget and problems of school management — and generally, in bringing an end to the political impasse in which they found themselves.

The unemployed, therefore, were not simply the passive instruments of political demagogy. On the contrary, their political mobilisation had its own dynamic. As Anthony McElligott points out in Chapter 9, the KPD was energetic in its attempts to mobilise the unemployed virtually from the very beginning of its existence. But for most of the period of the Weimar Republic it merely attempted to use them as auxiliaries in the struggles of those who were still in work. Not surprisingly, therefore, they tended to gravitate towards anarcho-syndicalist organisations instead. It was when the phenomenon of mass unemployment at the end of the Republic led to a downgrading of the political importance of the workplace that KPD attempts to mobilise the jobless, now built around issues of direct concern to them (such as benefit-levels, welfare payments and the like) began to meet with some success. Even here, however, the Communists' intention of channelling the energies of the unemployed into disciplined political activity frequently failed, as unemployed demonstrators allowed their anger and frustration to spill over into spontaneous violence against the police and the agencies of the state.

Nevertheless, the fact remains that the KPD by the end of the Weimar Republic was overwhelmingly a party of the unemployed, as both Anthony McElligott (Chapter 9) and Dick Geary (Chapter 10) clearly show. The party's gains tended to be made very much at the expense of the older-established, more moderate Social Democratic Party. Tension between the two parties, as Dick Geary shows, existed not merely at a political level, but at a social level too. Social Democratic unionists protected their jobs at the expense of the Communists; older workers, mostly in the SPD, connived at the dismissal of younger ones, mostly in the KPD; and 'respectable' working-class housing estates which were strongholds of the SPD increasingly lost touch with the 'rough' inner-city slums which accordingly transferred their loyalty to the KPD. Tensions between those who had jobs and those who had none added to these antagonisms. So another major political consequence of mass unemployment was a fatal weakening of working-class solidarity, a division of the labour movement which made it all the more difficult for the working class to mount any effective resistance to the Nazi seizure of power and the destruction of working-class organisations in the first six months of 1933.

The disillusion and hopelessness of the unemployed did not, on the whole, lead them into the Nazi Party, any more than it led them to vote for Hitler. Certainly, as Dick Geary points out, rural labourers, domestic workers, workers from industrial villages, and some others with little or no contact to labour movement traditions gravitated towards the Nazis. But there was little real movement of supporters between the socialist camps of the SPD and KPD on the one hand and the National Socialist camp on the other.[22] The effect of unemployment on labour-movement traditions and allegiances was much more long-term. For whatever their level of support for the KPD at the ballot box, the unemployed remained in their majority relatively indifferent in terms of real political activism. The street violence described in Anthony McElligott's contribution (Chapter 9) was the work of a relatively small minority, even perhaps as time went on an increasingly professionalised minority. As Detlev Peukert notes (Chapter 7), long-term unemployment left virtually a whole generation of working-class youth without any experience of full-time employment, cut off from labour-movement and socialist traditions as they were transmitted through the workplace. The consequences of this only really became apparent after the Second World War.

In retrospect, however, it is not the decline of working-class solidarity and labour-movement traditions that has remained in the popular memory as the outstanding feature of the un-employment crisis of the early 1930s, but rather the achievement of the Nazis in finding an eventual solution to the crisis. As Birgit Wulff argues in Chapter 11, however, the solution was largely a spurious one. To begin with, most of the Nazi job-creation schemes did little more than build on plans already laid and even to some extent implemented under the last cabinets of the Weimar Republic. As in so many other areas, Hitler had few concrete ideas of his own, and relied on improvisation covered in rhetoric. In the second place, the Third Reich proved even more ruthless and adept at cooking the books than its predecessors had done.[23] It soon removed people on relief work, rural job schemes and the voluntary labour service from the unemployment statistics, even though these people continued to be supported by local authorities as members of the welfare unemployed. The regime cut working hours (thus also effectively cutting wages) and drastically stepped up the coercion of young people to joint the Voluntary Labour Service and similar schemes. Unemployment rates were further reduced by the removal of women from the workforce by offering them marriage loans on condition they left their jobs. None of this had much to do with real job creation, and in fact the lack of any genuine rise in employment was reflected in the absence of any corresponding rise in living standards or consumer purchasing.

Hitler's 'solution', as Birgit Wulff shows, was in fact achieved by force. Existing job creation schemes had little real effect in re-ducing long-term unemployment and were paid for by cuts which had a negative effect on employment elsewhere. Real jobs were created only by rearmament. Road-building and similar schemes were undertaken not least for strategic reasons. Money was poured into heavy industry, machinery, engineering, automobile, ship and aircraft production, all the areas of the economy most central to the armaments industry, and the idea of a job-creation programme spread evenly across the major sectors of un-employment was abandoned. Instead, remaining areas of unemployment were tackled by massively coercive measures, only possible after the removal of trade unions and the creation of the Nazi police state, in which the unemployed were drafted into forced-labour schemes or into the army. The conquest of un-employment achieved in the Third Reich was thus a spurious one, linked from the beginning to preparations for a war which would inevitably bring death and destruction not only on Germany itself, but on the whole of Europe as well.

This is not, of course, to make unemployment in Germany the major cause of the Second World War or even the coming of the Third Reich. In the tangled skein of national and international politics it was merely a single thread. Yet, the essays collected in this volume do suggest that it was a really major problem which deserves more attention in the history of German politics and society in the interwar years than perhaps it has received so far. Germany was not, and is not, Britain, of course: the effects of mass unemployment were stronger than in Britain partly because the unemployment rate was higher (33 per cent as against 25 per cent at its peak), long-term unemployment was much more pervasive and the young were proportionately more seriously affected (partly because of differing demographic structures). The regional structure of employment was more varied in Britain, so the unemployed had somewhere to move to if they were willing to pay the social and psychological price; in Germany the million or so young people moving around the country in the early 1930s had nowhere to go except into paramilitary movements or, ultimately, into forced-labour schemes.[24]

The labour movement was more able to remain united in Britain because it was not responsible, as the SPD was, for welfare administration and policing; although the Labour Government of 1929–31 certainly gave little evidence of being able to solve the unemployment problem, it did manage to avoid much of the responsibility for it, and Labour-led local authorities were able to take acceptable measures to alleviate it. In comparison to Germany, welfare in Britain was relatively effective and living standards did not suffer so much. Of course, Britain lacked a mass Communist Party such as had existed in Germany since the early 1920s, and levels of political mobilisation had been higher in Germany since before the turn of the century. The experience of revolution, so decisive in 1918–19 for the future of the Weimar Republic, was absent in Britain, and the provisional nature of the

Republic in the eyes of many of its citizens raised the stakes in the political game far beyond those that were being played for in Britain. Unemployment was not only more serious in Germany, it took place in a political context that was already far more volatile and violent than in the United Kingdom, and so its effects, as it were, became magnified as a consequence. In many ways they are still with us, both as a reminder of the difficulties that long-term mass unemployment presents, and as a warning of what can happen if they are not tackled with the honesty and determination that they, and those who suffer most directly from them, the unemployed themselves, deserve.

Notes

I am grateful to Dick Geary for his helpful comments and suggestions on an earlier draft of this Introduction.

1. For attempts to gauge the 'real' level of unemployment in the late 19th century, see e.g. Hermann Schäfer's contribution to the collective volume of studies edited by Werner Conze and Ulrich Engelhardt, *Arbeiterexistenz im 19. Jahrhundert* (Stuttgart, 1981). See also Bernd Balkenhol, *Armut und Arbeitslosigkeit in der Industrialisierung, dargestellt am Beispiel Düsseldorf 1850–1900* (Düsseldorf, 1976); more generally, cf. Frank Niess, *Geschichte der Arbeitslosigkeit. Ökonomische Ursachen und politische Kämpfe: ein Kapitel deutscher Sozialgeschichte* (Cologne, 1979), and John A. Garraty, *Unemployment in History: Economic Thought and Public Policy* (New York, 1978).

2. See Hansjoachim Henning, 'Arbeitslosenversicherung vor 1914: Das Genfer System und seine Übernahme in Deutschland' in Hermann Kellenbenz (ed.), *Wirtschaftspolitik und Arbeitsmarkt* (Munich, 1974), pp. 271–87, and Anselm Faust, 'State and Unemployment in Germany 1890–1918 (Labour Exchanges, Job Creation and Unemployment Insurance)' in Wolfgang J. Mommsen (ed.), *The Emergence of the Welfare State in Britain and Germany* (London, 1981), pp. 150–63.

3. Ludwig Preller, *Sozialpolitik in der Weimarer Republik* (Düsseldorf, 1949, repr. 1978), p. 236.

4. For unemployment in the early years of the Weimar Republic, see also Merith Niehuss, *Arbeiterschaft in Krieg und Revolution. Soziale Schichtung und Lage der Arbeiter in Augsburg und Linz 1910 bis 1925* (Berlin and New York, 1985); the same author's 'Arbeitslosigkeit in Augsburg und Linz 1914–1924', *Archiv für Sozialgeschichte*, XXII (1982), pp. 133–58; Lothar Wentzel, *Inflation und Arbeitslosigkeit. Gewerkschaftliche Kämpfe und ihre Grenzen am Beispiel des Deutschen Metallarbeiter-Verbandes 1919–1924* (Hanover, 1981).

5. Preller, *Sozialpolitik*, pp. 276–82.

6. Ibid., pp. 364–5; see also Albin Gladen, 'Probleme staatlicher Sozialpolitik in der Weimarer Republik' in Hans Mommsen *et al.* (eds), *Industrielles System und politische Entwicklung in der Weimarer Republik*, (Dusseldorf, 1974), pp. 248–58; and F. Syrup, *Hundert Jahre staatliche Sozialpolitik* (Stuttgart, 1957).

7. See generally Dietmar Petzina *et al.*, *Sozialgeschichtliches Arbeitsbuch III. Materialien zur Statistik des Deutschen Reiches 1914–1945* (Munich, 1978).

8. Preller, *Sozialpolitik*, pp. 420–1.

9. Ibid., p. 374.

10. Ibid., p. 440, and for the following.

11. Ibid., p. 165.

12. Ibid., p. 448.

13. See also Richard Bessel, '"Eine nicht allzu grosse Beunruhigung des Arbeitsmarktes". Frauenarbeit und Demobilmachung in Deutschland nach dem Ersten Weltkrieg', *Geschichte und Gesellschaft*, IX (1983), 2, 211–29.

14. See also Karin Jurcyk, *Frauenarbeit und Frauenrolle. Zum Zusammenhang von Familienpolitik und Frauenerwerbstätigkeit in Deutschland 1918–1933* (Frankfurt, 1976) and Stefan Bajohr, *Die Hälfte der Fabrik. Geschichte der Frauenarbeit in Deutschland 1914–1945* (Marburg, 1979).

15. See also Dick Geary, 'Jugend, Arbeitslosigkeit und politischer Radikalismus am Ende der Weimarer Republik', *Gewerkschaftliche Monatshefte*, 34 (1983), 304–9.

16. Wolfgang Benz, 'Vom Freiwilligen Arbeitsdienst zur Arbeitsdienstpflicht', *Vierteljahreshefte für Zeitgeschichte* 16 (1968), pp. 317–46; Hellmut Lessing and Manfred Liebel, 'Jungen vor dem Faschismus. Proletarische Jugendcliquen und Arbeitsdienst am Ende der Weimarer Republik', in Johannes Beck *et al.* (eds), *Terror und Hoffnung in Deutschland 1933–1945* (Reinbek, 1980), pp. 391–420; Joachim Bartz and Dagmar Mor, 'Der Weg in die Jugendzwangsarbeit. Massnahmen gegen Jugendarbeitslosigkeit zwischen 1925 und 1935' in G. Lenhardt (ed.), *Der hilflose Sozialstaat. Jugendarbeitslosigkeit und Politik* (Frankfurt, 1979), pp. 28–94.

17. The most celebrated contemporary study was probably Marie Jahoda *et al.*, *Die Arbeitslosen von Marienthal* (1933, 4th ed., Frankfurt, 1982).

18. See Eve Rosenhaft, 'Organising the "Lumpenproletariat": Cliques and Communists in Berlin during the Weimar Republic' in Richard J. Evans (ed.), *The German Working Class* (London, 1982), pp. 174–219.

19. Ross McKibbin, 'The Myth of the Unemployed. Who voted for Hitler?' *Australian Journal of Political History* 15 (1969), 25–40.

20. Jürgen Falter *et al.*, 'Arbeitslosigkeit und Nationalsozialismus. Eine empirische Analyse des Beitrags der Massenerwerbslosigkeit zu den Wahlerfolgen der NSDAP 1932 und 1933', *Kölner Zeitschrift für Soziologie und Sozialpsychologie* 35 (1983), 525–54. See also Thomas Childers, *The Nazi Voter* (Chapel Hill, 1984) and Richard F. Hamilton, *Who Voted for Hitler?* (Princeton, 1982).

21. For a general survey, see Richard Bessel, 'Germany's Unemployment and the Rise of the Nazis', *The Times Higher Education Supplement*, 4 Feb. 1983.

22. See Dick Geary, 'Nazis and Workers: A Response to Conan Fischer's "Class Enemies or Class Brothers"', *European History Quarterly* 15 (1985), 453–64; for Fischer's part in the exchange, see ibid., 15/3 (1985), 259–79 and 15/4, 465–71.

23. For work-creation schemes during the crisis, see generally Michael Wolffsohn, *Industrie und Handwerk im Konflikt mit staatlicher Wirtschaftspolitik? Studien zur Politik der Arbeitsbeschaffung in Deutschland 1930–1934* (Berlin, 1977); and Michael Schneider, *Das Arbeitsbeschaffungsprogramm des ADGB* (Bonn, 1975).

24. See Stephen Constantine, *Unemployment in Britain between the Wars* (London, 1980).

2 UNEMPLOYMENT AND DEMOBILISATION IN GERMANY AFTER THE FIRST WORLD WAR[1]

Richard Bessel

I

The purpose of this essay is exploratory: to delineate the extent and character of unemployment in Germany in the aftermath of the First World War; to consider how it shaped and was shaped by the processes of demobilisation; and finally, to suggest how it affected the social and political upheavals which marked the early years of the Weimar Republic. As such, this forms part of an attempt to examine the processes by which a modern industrial society made the transition from war to peace.

In attempting to construct a general outline of the unemployment problem in Germany after the First World War, a number of preliminary points need to be borne in mind. First, before 1918 a range of economic statistics now taken almost for granted — including unemployment statistics — were not yet recorded systematically by the state, and the economic statistics generated during the demobilisation period often were imprecise.[2] Secondly, faced with the problems of the economic demobilisation the German state assumed a much more active role with regard to the labour market than had been imagined in 1913, with the first efforts of the Reich government to gain a comprehensive picture of unemployment and the first coherent attempts by the state to regulate the peacetime labour market.[3] The administration of unemployment relief was linked with the labour exchanges; guidelines were issued by the state authorities for the hiring and firing of different categories of employees; during the winter of 1919–20 an administrative section was set up in the Reich Labour Ministry to co-ordinate unemployment relief and the filling of vacant posts; in May 1920 a Reich Office overseeing the labour exchanges, the *Reichsamt für Arbeitsvermittlung*, was set up; and employers were required to report job vacancies, planned dismissals and planned factory closures to the state authorities. Thirdly, aggregate national figures for unemployment hide at least as much as they reveal, particularly with regard to part-time and home working and to regional and sectoral variations. The changes in home working among women, the unemployment generated by the Allied occupation of parts of western Germany after the armistice, the effects of the loss of Alsace-Lorraine and territories in eastern Prussia (Posen, West Prussia, Upper Silesia), the impact of war industries upon particular local labour markets, and the differences in the demand for and availability of labour in different sectors (for example, as between textiles and agriculture) are only some of the things buried beneath the aggregate figures. And fourthly, it is by no means clear precisely what was meant by unemployment during the demobilisation period and who in fact was being counted. At a time when millions of men were returning from the armed forces, when many ex-soldiers displayed a disinclination to return to their old communities and to jobs on the land, when hundreds of thousands of women were being pressured out of paid employment,[4] when shortages of materials and short-time working affected many factories, and when labour turnover reached very high levels,[5] it was more than usually difficult to apply a precise definition of 'unemployment'.

That said, a glance at unemployment levels recorded among trade union members (Table 2.1) gives a rough indication of the changing nature of the labour market between 1913 and 1924 — that is, from the last full year of peace to the currency stabilisation (although, it should be noted, the trade union figures for the war years 1915–18 are open to question).[6]

Table 2.1: Unemployment Among Members of Trade Unions in Germany, 1913–24 (percentages)[7]

	1913	1914	1915	1916	1917	1918	1919	1920	1921	1922	1923	1924
January	3.2	4.7	6.5	2.6	1.7	0.9	6.6	3.4	4.5	3.3	4.2	26.5
February	2.9	3.7	5.1	2.8	1.6	0.8	6.0	2.9	4.7	2.7	5.2	25.1
March	2.3	2.8	3.3	2.2	1.3	0.9	3.9	1.9	3.7	1.1	5.6	16.6
April	2.3	2.8	2.9	2.3	1.0	0.8	5.2	1.9	3.9	0.9	7.0	10.4
May	2.5	2.8	2.9	2.5	1.0	0.8	3.8	2.7	3.7	0.7	6.2	8.6
June	2.7	2.5	2.5	2.5	0.9	0.8	2.5	4.0	3.0	0.6	4.1	10.5
July	2.9	2.9	2.7	2.4	0.8	0.7	3.1	6.0	2.6	0.6	3.5	12.5
August	2.8	22.4	2.6	2.2	0.8	0.7	3.1	5.9	2.2	0.7	6.3	12.4
September	2.7	15.7	2.6	2.1	0.8	0.8	2.2	4.5	1.4	0.8	9.9	10.9
October	2.8	10.9	2.5	2.0	0.7	0.7	2.6	4.2	1.2	1.4	19.1	8.4
November	3.1	8.2	2.5	1.7	0.7	1.8	2.9	3.9	1.4	2.0	23.4	7.3
December	4.8	7.2	2.6	1.6	0.9	5.1	2.9	4.1	1.6	2.8	28.2	8.1

From these figures the following general picture of un-employment after the First World War emerges: during late 1918 and early 1919, when the demobilisation of the soldiers (and the removal of women from employment) had the greatest impact, unemployment peaked at between roughly 6 and 7 per cent of the labour force. This is roughly corroborated by the fact that about 1.1 million unemployed people were receiving relief in Germany in February 1919.[8] In the second half of 1919 this figure was approximately halved;[9] unemployment rose again during the summer of 1920, but by late 1921 there were only about 150,000 people receiving unemployment benefit in the entire Reich; and in the summer of 1922 the number of those receiving unemployment benefit was a mere 12,000.[10] In other words, at the peak of the inflationary boom of the early 1920s there was virtually no un-employment in Germany whatsoever. In 1923 things changed, however, and the number of people receiving unemployment ben-efit jumped from about 150,000 in the summer to roughly 1.5 million at the year's end.[11]

These figures suggest that the direct effects of the military demobilisation upon unemployment levels were relatively small and short-lived. Altogether, roughly six million men were dis-charged from (or discharged themselves from) the German military during the last two months of 1918 and the first four months of 1919,[12] a time when unemployment threatened large numbers of people within the Reich since (as Joseph Koeth, who headed the Reich Office — then Ministry — for Economic De-mobilisation from November 1918 until its dissolution in April 1919, later put it) 'the armaments industry — and that was more or less the whole of German industry — was paralysed by the sudden end of the war'.[13] More precise outlines of the German labour market during the crucial demobilisation period — and of its absorption of the ex-soldiers — are provided by figures published by the labour exchanges (Table 2.2) of people registered as looking for work.

These figures show clearly that the major impact of the demobili-sation upon the labour market took place between December 1918 and March 1919. But what is particularly striking is not so much the rise in unemployment in early 1919 as the speed with which the unemployed appear to have found work. Thus, in January 1919 there are 775,588 new registrations of men looking for work, and one month later the *total* number of men so registered was less

Table 2.2: People Registered at Labour Exchanges as Looking for Work in Germany, December 1918–November 1919[14]

| | Total | | New registrations | |
	Men	Women	Men	Women
December 1918	650,446	196,621	596,443	169,501
January 1919	1,068,923	381,735	775,588	291,226
February 1919	978,774	387,714	557,673	224,083
March 1919	840,395	353,657	458,378	173,910
April 1919	695,325	321,761	368,714	166,823
May 1919	756,869	346,558	380,929	178,098
June 1919	740,739	329,499	391,202	168,583
July 1919	708,948	305,142	417,578	168,873
August 1919	738,144	311,990	432,277	172,316
September 1919	731,023	299,800	455,076	178,946
October 1919	734,616	284,311	477,142	179,225
November 1919	721,363	243,404	457,004	148,758

than a million, despite the fact that there were over half a million new male registrations in February! Very few of the men looking for work in early 1919 seem to have remained unemployed for long. And when one compares its effects upon the German labour market with those of the mobilisation in 1914 or of the currency stabilisation of late 1923 and early 1924, the military demobilisation does not appear to have caused a particularly severe unemployment problem. Most soldiers returned to find jobs fairly quickly, and the sudden shift of millions of men from field grey into mufti does not seem to have put the German labour market severely out of joint. No wonder then, that when reflecting on the efforts of his Demobilisation Office to put people to work against the approximately one million Germans receiving unemployment relief in early 1919, Joseph Koeth regarded this figure as 'not too disturbing'.[15]

It has been suggested that the trade union figures given in Table 2.1 greatly underestimate the unemployment problem caused by the demobilisation in early 1919.[16] For one thing, many of the returning soldiers were not yet organised in trade unions, and therefore would not appear in the trade union figures; for another, many employees — particularly women — may not have registered themselves as unemployed even though they were out of work. No doubt the figures understate unemployment to some extent, if only because of high unemployment in consumer-goods industries with relatively large proportions of female labour and relatively low levels of trade union membership. However, the claim that unemployment in early 1919 approached heights equal to those reached during the early 1930s can be traced back to a statistical error by Rolf Wagenfuhr in an important article published in 1933;[17] there is no evidence to suggest that when the returned soldiers joined trade unions by the millions during the course of 1919 this greatly increased the trade union unemployment figures; and it seems that it was not until the second half of 1919 that the authorities succeeded in removing from the relief rolls women unwilling to accept the undesirable jobs on offer, particularly positions in domestic service.[18]

Similar impressions of the unemployment problem during the demobilisation period may be found in Wladimir Woytinsky's examination of the German labour market between 1919 and 1929:

> During the period 1919 to 1922 the considerable economic disturbances had almost no effect upon the labour market. The destruction of hundreds of thousands of men on the battlefields made itself felt in the inadequate supply of human labour power. With cheap money and starvation wages the employers had no need to be especially thrifty with regard to labour. One hurried to repair the damage of the war period and to replace the factories which had been lost in the territories taken from Germany with new factories with even greater capacity . . . The demand for labour also was stimulated by the shortening of the working day. Of course this situation was not 'normal', but for the labour market (more precisely: for the development of unemployment) the abnormality of the situation consisted solely in the fact that a very high demand for labour was met by an inadequate supply and that as a result almost all those looking for work could find employment . . . The first really severe disruption of the labour market came at the end of 1923: the stabilisation crisis![19]

As Gerald Feldman has observed (while noting — in contrast to Woytinsky's comment about 'starvation wages' — that there was a considerable rise in real wages during 1919–20), 'unemployment remained at remarkably low levels given the difficulties of the period' immediately after the War.[20]

Three general points can be made here. First, Germany was not the only country where military demobilisation after the First World War appears to have had a relatively small effect upon unemployment. In Britain, for example, the return home of the heroes of the trenches seems to have had even less of an impact upon unemployment levels, which remained under 3 per cent for almost the whole of 1919; in the summer of 1920 the Ministry of Labour estimated that fewer than 200,000 British war veterans were unemployed.[21] Indeed, international comparisons suggest that military demobilisation *per se* had rather little to do with levels of postwar unemployment. In the Netherlands, which had kept out of the war, the unemployment level shot up from 9.2 per cent in September 1918 to 18.6 per cent in February 1919 — for reasons which obviously had nothing to do with ex-soldiers flooding the labour market.[22]

The second point also is a comparative one. Although the effects of the demobilisation upon the labour market were not fundamentally different in Germany than elsewhere in 1919, during the early 1920s the development of the German economy differed greatly from what occurred in other industrial countries. Whereas German unemployment remained quite low for most of the time until late 1923, during 1921 and 1922 other industrial economies — in particular the American and the British — were in the grip of a sharp depression characterised by falling prices and rising joblessness. In Britain over one-fifth of the labour force was out of work in June 1921.[23] In Germany, expansionist and inflationary economic policies helped to create a very different climate and aided the re-integration of the 'war generation' into the world of work.

The third point is that the unemployment which arose in Germany after the Great War was essentially short term. The low levels of joblessness which followed the demobilisation suggest that few of those thrown out of work in late 1918 and early 1919 need have remained unemployed for long. Thus, the unemployment which characterised the German economy at the outset of the Weimar period was very different, both in quality and quantity, from the unemployment which accompanied the Republic's destruction in the early 1930s. This is of crucial importance both in understanding the nature of the political upheavals and working-class protest of the early Weimar years and in appreciating what unemployment meant to the Germans who experienced it after the First World War. It also needs to be kept in mind when reading the reports of officials who understandably were extremely worried by the problems facing them during the demobilisation period.

II

Although with the benefit of hindsight we can say that unemployment was not really a major problem during the demobilisation period, things looked rather different at the time. Indeed, to many of those involved in wartime planning for the demobilisation the prospects seemed terrifying; and the spectre of mass unemployment once war production ceased and the soldiers returned played a major role in the erection of a nationwide system of unemployment relief, introduced on 13 November 1918.[24] And it was not only government officials who were frightened. For example, at a regional planning meeting in the Bamberg Town Hall on 18 October 1918 a trade union functionary cried out in near desperation:

> It is probably the greatest mistake of the Reich government that in drawing up the demobilisation plans one did not reckon that Germany might end the war as the defeated party. One always reckoned that the demobilisation would proceed slowly and constructed the new plans for the economy on that basis. As a result of the course of events the whole demobilisation planning has been thrown out the window, and we are confronted with the fact that as soon as the war is over the troops will have to be discharged. We probably will be forced to disarm immediately and to discharge the veterans back into Germany. This will mean that a huge number of workers will come back home and face destitution. I do not now believe that enough raw materials can be acquired in the foreseeable future in order to revive the economy completely everwhere and to employ the entire labour force returning from the battlefield.[25]

More sober observers, who realised that it was not the returning soldiers but the women who had taken up the soldiers' jobs during wartime who would be the first to face unemployment, were hardly more sanguine about the prospects. For example, in August

1918 the Prussian Factory Inspectorate in Düsseldorf expressed the fear that the women due for dismissal once peace broke out would face 'a quite extraordinarily unfavourable employment situation'.[26] It appeared self-evident that the end of the war would bring mass unemployment and that those out of work would face gloomy prospects for years to come.

It also appeared self-evident that this unemployment would pose serious political dangers. The prospect that millions of soldiers might remain jobless long after they returned from the front aroused great unease in government. In late 1918, when Germany's imminent. defeat became obvious, the Reich Economics Office justified the recommendation that public funds be used to provide employment by asserting that 'order will depend on whether we succeed, in the most unfavourable circumstances, to provide work and income for those discharged from army service and for those who no longer can be employed in the armaments industry'.[27] Unemployment, it was feared, threatened the fabric of German society and the future of the German state. To avoid a collapse of 'order' the large-scale use of public funds for make-work projects seemed a small price to pay. Employers jumped on the bandwagon, pointing to the threat posed by high unemployment to urge that state contracts come their way.[28]

The surge in unemployment immediately following the military collapse appeared at first sight to confirm wartime fears. The number of jobless rose from virtually nil in October 1918 to well over one million in early 1919, and the problem was particularly serious in large cities. There, as the demobilisation committee in Wiesbaden observed in February 1919, unemployment had reached 'horrendous proportions' with 'no improvement in sight'.[29] Indeed, the unemployment problem of early 1919 was very largely an urban problem (in that the unemployed mostly were to be found in cities and in that relief provision was more developed in cities), and one concentrated in particular areas. In January 1919, for example, of the 1,068,923 men registered by the labour exchanges as looking for work, 125,468 (11.7 per cent) were to be found in Greater Berlin, 157,909 (14.8 per cent) in the Rhineland and Hohenzollern and 125,681 (11.8 per cent) in Saxony; of the 381,735 women registered as seeking employment, 65,631 (17.2 per cent) were in Greater Berlin, 52,216 (13.7 per cent) in the Rhineland and Hohenzollern and 62,541 (16.4 per cent) in Saxony.[30] Other estimates for Berlin and Saxony were much higher. For example, in March 1919 the demobilisation authorities asserted that Greater Berlin accounted for roughly 275,000 of Germany's registered jobless.[31] (The fact that the Berlin labour exchange figures for January included only 3,153 people who had come into the city from elsewhere — at a time when by all accounts Berlin was attracting large numbers from outside its boundaries, most of whom could expect no help from the labour exchanges in finding work — also suggests higher unemployment figures than those officially recorded.) An almost equal number was claimed for Saxony, where unemployment in centres of the textile industry (such as Plauen) was especially severe: in April 1919 over 232,000 people were said to be jobless in Saxony.[32] In general, cities dependent upon heavy industry (for example, in the Ruhr and in Upper Silesia) had lower unemployment levels than those with a more mixed economy (such as Berlin or Hamburg) or those more dependent on the textile industry (such as Elberfeld, Barmen or Plauen).

It is worth noting here that many of the areas with high postwar unemployment had experienced especially sharp falls in their population during the war: between 1910 and 1917 Berlin's population fell by 15.8 per cent, Saxony's by 9.6 per cent and Plauen's by 24.3 per cent.[33] This suggests that high postwar unemployment in these places was due partly to the sudden return of a disproportionately large section of their prewar working populations. It is also worth noting that many of those areas in which the postwar political unrest was so pronounced (such as Berlin and Saxony) contained a disproportionately large number of Germany's unemployed.[34]

At the same time there were regions and sectors of the economy with severe labour shortages. The disparities in the labour market were painfully evident to Joseph Koeth, who observed in February 1919:

Unemployment is growing daily. The unemployed are piling up in the large cities. On the other hand there is such a shortage of labour in the countryside that the spring cultivation, and with it the nourishment of the German people, is being put into question. Mining likewise suffers from an acute labour shortage.[35]

It is clear from the reports of district demobilisation committees that even in February 1919, when the postwar unemployment peaked, there was no lack of work available in agricultural areas.[36] But, as the town council in Lissa, near Glogau, observed in May 1919, 'the oversupply of positions available in agriculture cannot be filled by the numerous unemployed from other occupations, since they cannot be moved to accept any kind of work on the land'.[37] Complaints abounded that 'the so-called unemployed' in the cities preferred drawing unemployment benefit to working in agriculture,[38] and attempts were made during 1919 to resettle some of the urban unemployed on the land.[39] However, such efforts met with a fair amount of resistance from the unemployed, in some cases resulting in 'stormy demonstrations'.[40] Understandably, few city-dwellers, even if they were without work and especially if they were married, had much desire to uproot themselves to do strenuous labour on the land at low wages.[41]

Nor, for that matter, did returning soldiers who previously had lived in rural areas necessarily display enthusiasm for taking up agricultural work again after the war was over. This had been a major worry during the war. Estimates that before 1914 roughly 45 per cent of recruits from farms did not return home after military service caused grave concern about the likely actions of the three to four million rural inhabitants (including about two million agricultural labourers) who had been drafted into the wartime army.[42] And once peace arrived it seemed that there had been good cause for worry. Thus, for example, the head of the labour exchange of the Silesian agricultural chamber in Breslau spoke in December 1918 of 'a certain disinclination to work' (*eine gewisse Arbeitsscheu*) and claimed that many soldiers who had come from the countryside 'do not even consider returning to the land'.[43] Clearly, the prospect of unemployment after the armistice did not fill people with sufficient dread to induce them to work on the farm.

It is possible to trace similar reactions to the emergency works projects set up throughout Germany to combat unemployment. For example, from Spandau — which bordered on Berlin, with its high unemployment, and which had been very dependent upon its armaments factories for employment during the war — it was reported in February 1919:

The emergency works projects planned by the city have been started in part. At present there is still a lack of workers. Most of the unemployed have little desire or inclination to accept emergency work, despite relatively high wages. There is a particular disinclination towards forestry work.[44]

The widespread unwillingness to take up this sort of work is an indication that the unemployment of the immediate postwar period was rather less serious than contemporaries believed. Furthermore, it suggests that it was not primarily the government-financed emergency works programmes — limited in any case by shortages of fuel, raw materials and money[45] — which solved the unemployment problem of early 1919. (Indeed, one purpose of the emergency works seems to have been weeding out the work-shy from those claiming unemployment benefit. This was made clear by the Demobilisation Commissar for the Thuringian States when, in late November 1918, he stressed that 'only the genuine unemployed' should receive benefit and that providing work was preferable to providing support: 'Therefore as much opportunity for work as possible must be created through emergency works in all places where the unemployed congregate.')[46]

That the postwar unemployment problem was perhaps not so serious also is suggested by the continued unsatisfied demand for domestic servants. Even at the time of the highest unemployment immediately after the war — when women were being removed in their hundreds of thousands from wartime jobs in offices, transport facilities and armaments factories — positions available for female domestic servants remained unfilled.[47] Thus, in January 1919 the Association of Pomeranian Labour Exchanges reported that 'the reluctance to accept positions in domestic service unfortunately has not diminished', and in February it was reported from Breslau that, despite high and rising female unemployment, women displayed 'great resistance' to being placed in jobs on the land or in domestic service.[48] Throughout 1919 labour exchanges reported great demand for domestic servants, even in areas where many people were out of work and at a time when unemployment rates among women were higher than among men.[49]

The unattractiveness of the work also exacerbated the labour shortages in the coalmining industry; many unemployed were reluctant to take up employment in the pits.[50] But the labour shortages in mining also were due to other causes: the introduction of a shorter working day, serious housing shortages in coalmining centres, sharp declines in labour productivity as well as a disin-

clination of some men to return to dirty and dangerous work.[51]
Housing shortages proved a particular problem in the Ruhr, where
they prevented additional workers from coming into the region
from outside in order to provide desperately needed labour.[52]
Even so, the numbers of people employed in the mining industry
rose in 1919 to levels considerably higher than those of 1913, which
explains why coal production did not decline by even greater
amounts than it actually did. In the Ruhr the numbers employed in
coalmining rose from 387,000 in 1913 to 471,000 in 1919 and in
Upper Silesia from 123,000 in 1913 to 192,000 in 1919; and in the
lignite fields employment increased by even greater amounts, from
about 59,000 in 1913 to 153,000 in 1919 (an increase of 260 per
cent, which explains why the production of lignite in 1919 actually
exceeded the 1913 figure — 92.2 million tonnes as against 87.2
million — despite sharp falls in productivity).[53] Yet while the
problems of the coalmining industry led to labour shortages in
mining centres, they also were among the more important reasons
for joblessness elsewhere. The contradictory effects of the coal
shortage upon the labour market were illustrated well in a report
of the Demobilisation Commissar for the Thuringian States, in
which in the same paragraph he observed that the coal shortage
caused the closure of 'numerous factories' and that 30,000 workers
were needed for the Zeitz-Weissenfels lignite field.[54] Postwar
shortages of coal had serious repercussions throughout the
German economy, for which coal was the main source of energy,
and during early 1919 coal shortages were cited repeatedly as a
main reason for dismissals and the rapid rise in unemployment.[55]

   Although agriculture and mining faced the greatest labour
shortages after the armistice, it would be mistaken to assume that
unemployment was a major problem everywhere else. In many
instances workers were taken on after the war by employers who
anticipated a postwar consumer-goods boom once controls were
dismantled, who were keen to replace the prisoner-of-war labour
which they had lost so suddenly, or who wanted to ensure that
when the expected economic upturn arrived they would not be
caught out for lack of labour. An illustration of this is provided by
a prognosis, written in mid-November 1918, of the local economy
of Graudenz, a garrison town and minor industrial centre in
eastern Germany which became Polish after the War.[56] Certainly,
it appeared that Graudenz, where roughly 20,000 soldiers were
sitting in the town's military installations when hostilities ended,

would face a considerable problem of demobilised soldiers looking for work. However, regional government officials considered there to be 'no shortage' of available work, since local industry needed to replace the prisoners of war; the local brickworks in particular were looking for labour in anticipation of rising demand from the building industry during the coming summer; and in any event the city was regarded as well prepared with emergency works projects to soak up any extra labour. The main problem, it was felt, was not the prospect of high unemployment but high wage demands, supported by the local workers' and soldiers' council, which together with the introduction of the eight-hour day put the 'economic viability' of firms 'into question'.

Table 2.3: Unemployment Among Male Members of Trade Unions in Germany, 1919–24 (percentages)[57]

	1919	1920	1921	1922	1923	1924
January	6.2	3.3	4.7	3.8	3.9	29.4
February	5.5	2.7	4.9	3.1	4.8	28.6
March	3.6	1.9	3.7	1.2	5.1	19.0
April	4.8	1.9	3.8	0.9	6.6	11.7
May	3.5	2.4	3.5	0.6	5.9	9.4
June	2.1	3.5	2.9	0.5	3.6	11.1
July	2.8	5.0	2.5	0.5	3.1	12.9
August	2.6	5.2	2.2	0.5	5.7	12.7
September	1.7	4.1	1.4	0.6	9.6	10.7
October	2.3	4.0	1.1	1.1	19.3	8.6
November	2.7	3.9	1.4	1.6	24.4	7.6
December	2.8	4.3	1.7	2.4	30.8	8.8

Table 2.4: Unemployment Among Female Members of Trade Unions in Germany 1919–24 (percentages)[58]

	1919	1920	1921	1922	1923	1924
January	7.9	3.6	3.7	1.7	5.3	17.1
February	8.0	3.6	4.1	1.7	6.4	13.3
March	4.8	2.1	3.7	0.9	7.2	7.9
April	6.8	2.2	4.4	0.8	8.5	6.0
May	5.0	3.8	4.4	0.8	7.3	5.6
June	3.8	5.9	3.4	1.0	5.3	8.1
July	4.2	10.0	2.8	0.8	4.7	11.3
August	4.8	8.7	2.3	1.1	7.9	11.5
September	4.1	5.9	1.4	1.4	10.8	9.8
October	3.9	4.9	1.4	2.4	18.4	7.5
November	3.8	3.8	1.2	3.4	20.1	5.9
December	3.6	3.4	1.3	4.3	19.6	5.5

A particularly revealing aspect of the postwar unemployment is how it affected men and women respectively. Since one of the most important postwar changes in the labour market was the removal of hundreds of thousands of women from their wartime jobs to make way for the returning men, during the demobilisation period unemployment levels among women were rather higher than among men — something probably understated in the available statistical data, because of the tendency of the authorities not to consider home workers (mostly women) as eligible for relief and to be more restrictive about granting women unemployment benefit.[59]

Not until 1923–4 was there a decisive shift in the relationship between male and female unemployment. Higher female unemployment had been a peculiar characteristic of the demobilisation period, a reflection of its extraordinary nature as a time of transition. It was during the stabilisation — that normalisation of economic relationships — that a greater percentage of men found themselves without work.

Of course the shape of the German labour market after the war was far more complex than even the above observations indicate. Not only were there significant differences in unemployment levels among men and women, in different regions and at different times; there also were wide differences in the job prospects for the skilled and unskilled, among manual labour and office staff, from occupation to occupation. Some indication of this complexity may be gathered from a look at the workings of the local job market during the demobilisation — in this case as reflected in data from Ludwigshafen. Although Ludwigshafen was peculiar in that the giant BASF chemical works played so dominant a role in the local economy, some general characteristics may be seen in figures compiled by the city's labour exchange (Table 2.5) of those whose applications it processed during 1919.[60]

According to the Ludwigshafen labour exchange, the beginning of 1919 saw a discrepancy between the number looking for work and situations vacant 'larger than ever before'; however, from March this began to ebb, and by the end of the year the number of unemployed was so low that the hope for an early ending of unemployment as a consequence of the war appeared justified — in large measure as a result of the recovery of the chemical industry. The great majority of the unemployed were unskilled (66 per cent) and single (63 per cent of the men and 80 per cent of the women). At the end of the year, when women largely had been removed from the relief rolls and the level of unemployment

Table 2.5: Applications for Jobs in Ludwigshafen, 1919, by Selected Occupations

Men	Job applicants	Situations vacant
Workers in chemical factories*	3,845	3,375
Workers in other factories	2,046	626
Bakers	173	42
Printers and compositors	108	17
Skilled metalworkers	858	221
Electrical engineers	175	49
Building workers	3,501	1,807
Drivers	330	114
Warehousemen	2,210	1,311
Commercial employees	1,076	538
Waiters	40	11
Agricultural workers	316	235
Apprentices	348	316
Machine-fitters, mechanics, boiler-makers	1,504	550
Bricklayers and plasterers	437	283
Butchers	91	2
Smiths	146	101
Tailors	63	99
Plumbers	276	109
Day labourers	2,251	1,122
(*essentially workers at BASF)		

Women	Job applicants	Situations vacant
Factory workers	4,370	411
In service (agriculture)	135	153
In service (domestic)	2,275	2,357
In service (public houses)	625	620
Commercial employees	824	133
Waitresses	898	642
Seamstresses	208	101
Cleaners and washerwomen	2,215	2,040
Chambermaids	114	82

had become 'relatively low', those who still could not be placed in jobs were described as 'mainly clerical staff, commercial employees, artisans, war wounded, disabled and other people for which in any case it is difficult to find positions'.

The job prospects of commercial employees, office workers and white-collar staff gave particular cause for concern during the demobilisation period. Whereas placing skilled workers in jobs in industry generally posed no great problem,[61] finding appropriate work for returned soldiers who previously had occupied white-collar positions was often difficult.[62] The demand for office work far outstripped supply in early 1919 and led to considerable strains on the job market.[63] Thus, it is not surprising that the sharpest conflicts over the continued employment of women during the demobilisation period arose over women working in offices, and that unemployment among office staff gave added stridency to the demands by the conservative *Deutschnationaler Handlungsgehilfen-Verband* that women be removed from such work.[64]

III

With regard to the labour market — as in so many other respects — the demobilisation period was a time of upheaval. Confronted simultaneously by a political revolution and massive and sudden changes in the German labour force, the state authorities, employers and most employees hoped to see order return as soon as possible. And considering the obstacles to such a return, they were remarkably successful. Contrary to most expectations, the aspect of the re-integration of the 'front generation' into civilian life which went most smoothly was the re-integration into the world of work. The worst fears about a postwar breakdown of order as a result of mass unemployment were proved unfounded. One might speculate that in a sense the German authorities got it wrong in 1918–19 — that far from defusing radical protest, the relative ease with which men found jobs after the war made involvement in the unrest less risky. Certainly, it is difficult to imagine the strength of the working-class protest in the aftermath of the war (or the success with which strike action was used to break the right-wing putsch attempt headed by Wolfgang Kapp and General von Lüttwitz in March 1920) had unemployment reached and remained at levels expected when the war ended. In another sense, however, they got it profoundly right. The fact that it proved relatively easy for the soldiers of the First World War to find jobs upon their return (as women were being pushed out of employment) made possible a widespread normalisation of social relationships — with the reconstitution of millions of households with male breadwinners at their heads — and an underpinning of the existing social order.

The unexpected success which Germany had in achieving relatively low levels of unemployment after the war came to a crashing end with the stabilisation crisis of 1923–4. Then Germany experienced an economic contraction which matched in severity those which affected Britain and the United States two years previously.

The delay — in comparison with Britain — in the coming of mass unemployment may have been an important reason why the Left was able to achieve as much as it did in Germany after 1918, even if its achievements were short lived. For mass unemployment does not usually pave the way for great political advances from the Left; and in Germany it provided one of the conditions for the destruction of democracy and the triumph of National Socialism.[65]

Notes

1. I would like to thank Dick Geary and Richard Evans for their helpful comments on an earlier draft, and to express my gratitude to the Alexander von Humboldt Foundation, the Wolfson Foundation and the Open University for providing the financial assistance which made the research for this essay possible.
2. See Peter-Christian Witt, 'Staatliche Wirtschaftspolitik in Deutschland 1918 bis 1923: Entwicklung und Zerstörung einer modernen wirtschaftspolitischen Strategie' in Gerald D. Feldman, Carl-Ludwig Holtfrerich and Peter-Christian Witt (eds), *Die deutsche Inflation. Eine Zwischenbilanz* (Berlin and New York, 1982), pp. 163–6; Peter-Christian Witt, 'Bemerkungen zur Wirtschaftspolitik in der "Übergangswirtschaft" in 1918/19' in Dirk Stegmann, Bernd-Jürgen Wendt and Peter-Christian Witt (eds), *Industrielle Gesellschaft und politisches System. Beiträge zur politischen Sozialgeschichte* (Bonn, 1978), p. 90.
3. See Ludwig Preller, *Sozialpolitik in der Weimarer Republik* (Stuttgart, 1949), pp. 236–7; Witt, 'Bemerkungen zur Wirtschaftspolitik', pp. 89–90. For an overview of the state's role in regulating unemployment in Germany before the First World War, see Anselm Faust, 'State and Unemployment in Germany 1890–1918 (Labour Exchanges, Job Creation and Unemployment Insurance)', in Wolfgang J. Mommsen (ed.), *The Emergence of the Welfare State in Britain and Germany* (London, 1981), pp. 150–63.
4. See Richard Bessel, '"Eine nicht allzu grosse Beunruhigung des Arbeitsmarktes". Frauenarbeit und Demobilmachung in Deutschland nach dem Ersten Weltkrieg', *Geschichte und Gesellschaft*, vol. ix, no. 2 (1983), 211–29.
5. One indication of this are the figures for turnover among the work force at BASF, given in Dieter Schiffmann, *Von der Revolution zum Neunstundentag. Arbeit und Konflikt bei BASF 1918–1924* (Frankfurt/Main and New York, 1983), p. 461, Table 5.
6. Wladimir Woytinsky, *Der deutsche Arbeitsmarkt. Ergebnisse der gewerkschaftlichen Arbeitslosenstatistik 1919 bis 1929* (Berlin, 1930), vol. i, p. 11.
7. *Statistisches Jahrbuch für das Deutsche Reich (SJDR) 1921/22* (Berlin, 1922) 'Anhang' ('Internationale Übersichten'), p. 78; ibid *1924/25* (Berlin, 1925), p. 296.
8. Preller, *Sozialpolitik*, p. 236. In Bavaria (which held slightly more than 10 per cent of the German population) registered unemployment peaked at about 113,000 in early 1919, the equivalent of between 8 and 11 per cent of the population employed in trade and industry (depending on whether figures for the labour force from 1910 or 1917 are used). See Kurt Königsberger, 'Die wirtschaftliche Demobilmachung in Bayern während der Zeit vom November 1918 bis Mai 1919', *Zeitschrift des Bayerischen Statistischen Landesamts*, vol. lii (1920), 210. Königsberger stressed how low this unemployment was considering the problems created by the demobilisation.
9. See Georg Gradnauer and Robert Schmidt, *Die deutsche Volkswirtschaft. Eine Einführung* (Berlin, 1921), p. 193.

10. Preller, *Sozialpolitik*, p. 164.

11. Ibid., p. 164.

12. The number of men in the German armed forces had shrunk from roughly seven million in May 1918 to six million in October. See Reichswehrministerium, *Sanitätsbericht über das Deutsche Heer im Weltkrieg 1914/1918*, vol. iii: *Die Krankenbewegung bei dem Deutschen Feld- und Besatzungsheer* (Berlin, 1934), section ii, p. 5; Wilhelm Deist, 'Bemerkungen zur militärischen Demobilmachung 1918' (paper delivered to the German Historical Institute and Open University Conference on 'Social Processes of Demobilisation after the First World War in Germany, France and Great Britain', London, May 1981), p. 4.

13. Joseph Koeth, 'Die wirtschaftliche Demobilmachung. Ihre Aufgabe und ihre Organe' in *Handbuch der Politik*, vol. iv, *Der wirtschaftliche Wiederaufbau* (Berlin and Leipzig, 1921), p. 165.

14. *RABL*, vol. xvii (1919), no. 1, 36–7; no. 2, 166–7; no. 3, 246–7; no. 4, 334–5; no. 5, 410–11; no. 6, 476–7; no. 7, 560–1; no. 8, 632–3; no. 9, 714–15; no. 10, 788–9; no. 11, 788–9; no. 12, 962–3. It should be noted that the month-by-month figures are not absolutely comparable, since in some months statistics from certain labour exchanges were unavailable. Thus, the December 1918 figures omit the unemployed in Württemberg and Hamburg; the January 1919 figures omit the unemployed in Württemberg; the February figures omit Bremen; the March, May and November figures omit Württemberg; and figures from the Prussian province of Posen are either completely or partially lacking from June 1919.

15. Koeth, 'Die wirtschaftliche Demobilmachung', p. 167.

16. See, for example, Merith Niehuss, *Arbeiterschaft in Krieg und Revolution. Soziale Schichtung und Lage der Arbeiter in Augsburg und Linz 1910 bis 1925* (Berlin and New York, 1985), p. 214.

17. Rolf Wagenfuhr, 'Die Industriewirtschaft. Entwicklungstendenzen der deutschen und internationalen Industrieproduktion 1860 bis 1932', *Vierteljahreshefte zur Konjunkturforschung*, Sonderheft 31 (Berlin, 1933), p. 24. Wagenfuhr took his figures mistakenly from a book published in 1921 by two SPD Reichstag deputies, Georg Gradnauer and Robert Schmidt. See Gradnauer and Schmidt, *Die deutsche Volkswirtschaft*, p. 193. Gradnauer and Schmidt took their figures from the same trade union statistics given above, and Wagenfuhr evidently confused 6.6 *per cent* of trade unionists with 6.6 *million* people unemployed.

18. See Statistisches Amt der Stadt Mannheim, *Verwaltungs-Bericht der badischen Hauptstadt Mannheim für 1919/20* (Mannheim, n.d.), pp. 194–6. In Mannheim both male and female unemployment rose considerably in late 1918 and early 1919. The figure for men declined rapidly thereafter, while that for women remained at nearly the January–February 1919 levels until September. Between September 1919 and March 1920, however, registered female unemployment fell to virtually nothing, while male unemployment remained fairly steady. Commenting on these figures, the Statistical Office in Mannheim reported: 'Special emphasis was given to referring gradually those female unemployed in some way suitable into domestic service . . . Thanks to unremitting efforts it has been possible to sift out the female unemployed in this regard so that in the end only an infinitesimal portion still was receiving support.' See also Chapter 5, below.

19. Woytinsky, *Der deutsche Arbeitsmarkt*, pp. 19–20.

20. Gerald D. Feldman, 'Socio-economic Structures in the Industrial Sector and Revolutionary Potentialities, 1917–22' in Charles L. Bertrand (ed.), *Revolutionary Situations in Europe, 1917–1922: Germany, Italy, Austria–Hungary* (Montreal, 1977), p. 163.

21. David Englander, 'Die Demobilmachung in Grossbritannien nach dem Ersten Weltkrieg', *Geschichte und Gesellschaft*, vol. ix, no. 2 (1983), 202–3; *SJDR 1921/22*, 'Anhang', p. 78.

22. Ibid., p. 78.

23. Ibid., p. 78.

24. The 'Decree on Unemployment Relief' of 13 November 1918 was the first piece of legislation to emerge from the Reich Office for Economic Demobilisation called into being the previous day. See *Reichsgesetzblatt*, 1918, pp. 1305–8. See also Preller, *Sozialpolitik*, p. 236. For evidence of government attitudes towards relief programmes on the eve of the armistice, see Zentrales Staatsarchiv (ZStA), Dienststelle Merseburg, Rep. 120, BB, Abt. VII, Fach 1, Nr. 3y, Band 1, ff. 15–21: 'Aufzeichnung zu der Besprechung (im Reichsarbeitsamt) über Erwerbslosen-fürsorge vom 19. Oktober 1918'.

25. Bayerisches Hauptstaatsarchiv, Abt. IV, MKr 14412: 'Aufzeichnung über die von der Kriegsamtstelle Würzburg einberufene Besprechung am 18. Oktober 1918 im Sitzungssaal des Rathauses zu Bamberg'.

26. HStADüsseldorf, Reg. Düsseldorf, Nr. 33557: Königliche Gewerbein-spektion Düsseldorf-Land to the Regierungspräsident, Düsseldorf, 22 Aug. 1918. It should be noted, however, that not all officials were so gloomy. In early November 1918 the Head of the Trade and Industry Section of the Bavarian Foreign Ministry expressed the belief that 'for Bavaria the picture for the employment possibilities of the workers returning home from the front is on the whole not unfavourable'. See Staatsarchiv (StA) Dresden, Ministerium für au-swärtige Angelegenheiten, Nr. 2494, Band I, ff. 77–8: Sächsische Gesandtschaft to the Minister für auswärtige Angelegenheiten, Munich, 7 Nov. 1918.

27. Ibid., ff. 44–5, and Hessisches Hauptstaatsarchiv (HStA) Wiesbaden, 405/6165, ff. 1–2; The Reichskanzler (Reichswirtschaftsamt) to the Minister des Innern, Berlin, 29 Oct. 1918.

28. See, for example, Niedersächsisches Hauptstaatsarchiv (HStA) Hannover, Hann. 80 Hann. II, Nr. 1981: Handwerkskammer zu Hannover to the Reg-ierungspräsident, Hannover, 19 Dec. 1918; ibid., Hann. 80, Hann. II, Nr. 1991: H. Wohlenberg, Drehbankfabrik und Eisengiesserei to the Demobilmachungskom-missar in Hannover, Hannover, 20 Dec. 1918.

29. HStA Wiesbaden, 405/6165, ff. 249–51: Demobilmachungsausschuss der Stadt Wiesbaden to the Demobilmachungskommissar, Wiesbaden, 15 Feb. 1919.

30. *RABL*, vol. xvii, no. 8 (1919), pp. 166–7. The high figures for the Rhineland were due partly to the effects of Allied occupation. In neighbouring Westphalia relatively few people were registered as looking for work: In January 1919 43,787 men (4.1 per cent of the total) and 10,211 women (2.7 per cent).

31. StA Potsdam, Rep. 2A Reg. Potsdam I SW, Nr. 798, f. 78: The Vorsitzender des Demobilmachungsausschusses des Kreises Teltow to the Regierungspräsident in Potsdam, Berlin, 14 Feb. 1920. See also ZStA Merseburg, Rep. 120, BB, Abt. VII, Fach 1, Nr. 30, Band 4, f. 413: Demobilmachungsausschuss Gross-Berlin to the Minister für Handel und Gewerbe, Berlin, 16 July 1919; Frauke Bey-Heard, *Hauptstadt und Staatsumwälzung. Berlin 1919. Problematik und Scheitern der Rätebewegung in der Berliner Kommunalverwaltung* (Stuttgart, 1969), p. 114.

32. StA Dresden, Gesandtschaft Berlin, Nr. 681, ff. 179–80: Gesandtschaft Berlin to the Reichspostministerium, Berlin, 19 May 1919. For the special prob-lems of Plauen, see StA Dresden, Kriegsarchiv/Kriegsministerium, Nr. 25012, f. 564: Beauftragter des Min. f. Mil. Wesen, Garn. Kdo. Plauen, Plauen, 13 Aug. 1919.

33. StA Dresden, Ministerium für auswärtige Angelegenheiten, Nr. 2494, Band I, ff. 118–20: Sächsisches Ministerium des Innern to the Ministerium der au-swärtigen Angelegenheiten, Dresden, 16 Nov. 1918.

34. See Dick Geary, 'Radicalism and the Worker: Metalworkers and Revolution 1914–23' in Richard J. Evans (ed.), *Society and Politics in Wilhelmine Germany* (London, 1978), p. 274.

35. StA Potsdam, Rep. 2A Reg. Potsdam I SW, Nr. 793, f. 114: Reichsamt für wirtschaftliche Demobilmachung, Berlin, 1 Feb. 1919.

36. See, for example, the reports of the demobilisation committees in rural districts in the *Regierungsbezirk* Potsdam during February 1919, in StA Potsdam, Rep. 2A Reg. Potsdam I SW, Nr. 796, ff. 161–72; and similar reports from the Regierungsbezirk Hannover, in HStA Hannover, Hann. 80 Hann II, Nr. 1981.

37. Archiwum Panstwowe w Poznaniu (APP), Landratsamt Lissa, Nr. 118, f. 49: The Magistrat to the Landrat, Lissa, 8 May 1919.

38. ZStA Potsdam, Reichsministerium für wirtschaftliche Demobilmachung (RMwD), Nr. 2, ff. 371–2: Arbeitsnachweis der Landwirtschaftskammer für die Provinz Schlesien to the Kriegswirtschaftsamt Breslau, Breslau, 4 Jan. 1919.

39. A good example of this is the efforts of the city of Spandau. See StA Potsdam, Rep. 2A Reg. Potsdam I SW, Nr. 796, ff. 131–2: The Magistrat to the Regierungspräsident in Potsdam, Spandau, 4 Feb. 1920.

40. See, for example, HStA Hannover, Hann. 80 Hann. II, Nr. 1981: The Vorsitzender des Kreisausschusses to the Regierungspräsident in Hannover, Neustadt a/Rbg., 5 Feb. 1919; ibid., Hann 122a/XXXIV, Nr. 368, ff. 440–1: Kriegsamtstelle to the Reichsamt für wirtschaftliche Demobilmachung, Hannover, 15 Mar. 1919. This latter report describes a demonstration in Hanover of young women protesting against being sent to work on the land. It also should be noted that the authorities were wary in early 1919 of sending out to the countryside urban unemployed who were 'unsuited' to agricultural labour or were contaminated with 'Bolshevism'. See, for example, StA Greifswald, Rep. 38b Loitz, Nr. 1010, ff. 32–4; Kriegsamtstelle Stettin and Zentralauskunftstelle Stettin to the Landräte, Vorsitzenden der landwirtschaftlichen Vereine, Magistrate and Arbeitsnachweise, Stettin, Feb. 1919.

41. See Archiwum Panstwowe w Szczecinie (APS), Oberpräsidium von Pommern, Nr. 3952: Zentralauskunftstelle Stettin, 'Wochenbericht über den Arbeitsmarkt der Provinz Pommern', Stettin, 25 Jan. 1919. According to a report about the labour market in Danzig in July 1920, when unemployment reached 13,000 in a city with about 200,000 inhabitants, it was chiefly among the *unmarried* unemployed that the authorities had success in inducing people to accept work outside the city. See ZStA Merseburg, Rep. 120, BB, Abt. VII, Fach 1, Nr. 30, Band 5, ff. 12–13: Magistrat der Stadt Danzig an das Auswärtige Amt in Berlin, Danzig, 19 July 1920.

42. For this estimate of the percentage of rural soldiers who failed to return home before the war, see ZStA Merseburg, Rep. 77, tit. 332d, Nr. 1, Band 2, ff. 166–7: Kriegsministerium to the Minister für Landwirtschaft, Domänen und Forsten, Berlin, 8 June 1918. For estimates of the numbers of wartime recruits from the countryside, see StA Dresden, Ministerium des Innern, Nr. 16074, ff. 211–12: 'Niederschrift über die Sitzung des Arbeitsausschusses der Kommission für Demobilmachung der Arbeiterschaft am 28. Oktober 1918, nachmittags 4 Uhr, unter Vorsitz des Unterstaatssekretärs Dr. Müller'; Jens Flemming, *Landwirtschaftliche Interessen und Demokratie. Ländliche Interessen, Agrarverbände und Staat 1890–1925* (Bonn, 1978), p. 82.

43. ZStA Merseburg, Rep. 120, BB, Abt. VII, Fach 1, Nr. 30, Band 2, ff. 123–4: Arbeitsnachweis der Landwirtschaftskammer für die Provinz Schlesien to Regierungsrat Jacques in the Breslau Oberpräsidium, Breslau, 18 Dec. 1918.

44. StA Potsdam, Rep. 2A Reg. Potsdam I SW, Nr. 796, ff. 104–8: Demobilmachungsausschuss to the Reichsamt für wirtschaftliche Demobilmachung, Spandau, 11 Feb. 1919.

45. Bey-Heard, *Hauptstadt und Staatsumwälzung*, pp. 117–18.

46. StA Weimar, Thüringisches Wirtschaftsministerium, Nr. 130, f. 13: The Staatskommissar für die Demobilmachung in den Thüringischen Staaten to the Demobilmachungsausschüsse, Weimar, 27 Nov. 1918.

47. Bessel, '"Eine nicht allzu grosse Beunruhigung des Arbeitsmarktes"', pp. 222–3.

48. APS, Oberpräsidium von Pommern, Nr. 3952: Pommerscher Arbeitsnachweisverband, 'Bericht über die Lage des Arbeitsmarktes der Provinz Pommern im Monat Dezember 1918', Stettin, 13 Jan. 1919; ZStA Potsdam, RMwD, Nr. 18/1, ff. 297–301: Kriegsamtstelle to the Reichsamt für wirtschaftliche Demobilmachung, Breslau, 8 Feb. 1919.

49. See tables 2.3 and 2.4.

50. StA Greifswald, Rep. 65c, Nr. 2891, ff. 3–4: Reichszentrale für Kriegs- und Zivilgefangene to the Kriegsgefangenenheimkehrstellen in Königsberg, Berlin, 30 Oct. 1919.

51. HStA Wiesbaden, 405/6182, ff. 73–4: Reichsamt für wirtschaftliche Demobilmachung, 'Richtlinien für die Tätigkeit der Gewerkschaftsfunktionäre', Berlin, 1 Feb. 1919; ZStA Merseburg, Rep. 120, BB, Abt. VII, Fach 1, Nr. 30, Band 5, ff. 57–9: The Minister für Handel und Gewerbe to the Reichskanzler, Berlin, 4 Sept. 1919.

52. Jürgen Tampke, *The Ruhr and Revolution. The Revolutionary Movement in the Rhenish-Westphalian Industrial Region 1912–1919* (London, 1979), p. 196.

53. See Gradnauer and Schmidt, *Die deutsche Volkswirtschaft*, pp. 196–7.

54. StA Weimar, Thüringisches Wirtschaftsministerium, Nr. 130, f. 57: Der Staatskommissar für wirtschaftliche Demobilmachung in den Thüringischen Staaten to all Demobilmachungsausschüsse, Weimar, 31 Dec. 1918.

55. See, for example, StA Dresden, Amtshauptmannschaft Pirna, Nr. 496, ff. 9–10: Amtshauptmannschaft Pirna to the (Saxon) Ministerium des Innern, 'Stimmungsbericht', Pirna, 16 May 1919; APS, Regierung Stettin, Nr. 13267. f. 66: Vereinigte Ganzstoff-Fabriken A.G., Fabrik 'Sydowsaue' to the Regierungspräsidenten, Sydowsaue b/Stettin, 5 Feb. 1919.

56. ZStA Merseburg, Rep. 120, BB, Abt. VII, Fach 1, Nr. 30, Band 2, ff. 35–6: The Regierungspräsident to the Minister für Handel und Gewerbe, Marienwerder, 18 Nov. 1918.

57. *SJDR 1924/25* (Berlin, 1925), p. 296.

58. Ibid., p. 296.

59. ZStA Merseburg, Rep. 120, BB, Abt. VII, Fach 1, Nr. 30, Band 5, ff. 197–203: 'Eingabe des Berufsverbandes der Katholischen Metallarbeiterinnen (Abteilung Grubenarbeiterinnen) und der Berufsorganisation der Textilarbeiterinnen (Heimarbeiterinnen) im Verbande Katholischer Vereine erwerbstätiger Frauen und Mädchen Deutschlands', Kattowitz, 22 July 1920. See also note 18.

60. Stadtarchiv Ludwigshafen, Nr. 6670: Städtisches Arbeitsamt, report for 1919.

61. For example, despite relatively high unemployment the labour exchange in Aachen reported in March 1919 that the demand for skilled labour in industry exceeded the supply. See HStA Düsseldorf, Reg. Aachen, Nr. 7760. f. 159: Bezirksstelle für Arbeitsnachweis für den Regierungsbezirk Aachen to the Regierungspräsident, Aachen, 25 March 1919.

62. See, for example, Generallandesarchiv Karlsruhe, 456/E.V.8, Bund 112: Minsterium für soziale Fürsorge, 'Arbeitsbeschaffung für kaufmännische und technische Angestellten', Karlsruhe, 8 Jan. 1919.

63. HStA Wiesbaden, 405/6181, f. 402: 'Zusammenstellung der Anzahl der Stellensuchenden, offene Stellen und besetzte Stellen in Frankfurt a. M. vom 11. January 1919'; ZStA Potsdam, RMwD, Nr. 18/1, ff. 91–3: Kriegsamtstelle Breslau to the Reichsamt für wirtschaftliche Demobilmachung, Breslau, 4 Jan. 1919; ibid., ff. 313–14: Kriegsamtstelle Magdeburg to the Reichsamt für die wirtschaftliche Demobilmachung, Magdeburg, 8 Feb. 1919.

64. For example, see the complaints of the DHV during the summer of 1919 in Bochum, in Stadtarchiv Bochum, B 265.

65. For a general discussion of unemployment and the Nazi triumph, see Richard Bessel, 'Germany's Unemployment and the Rise of the Nazis', *The Times Higher Education Supplement*, 4 Feb. 1983.

3 FROM WELFARE PROVISION TO SOCIAL INSURANCE: The Unemployed in Augsburg 1918–27

Merith Niehuss

I

Although welfare provision for 'those without a livelihood' (*Erwerbslose*) first came into existence in 1918, it was not actually the Revolution of 1918–19 which laid the first foundation stones for a system of social insurance for the unemployed. Rather, its origin must be sought in the policies of the old regime during the 1914–18 war. Characteristically, the plight of the unemployed created by the mobilisation crisis of 1914–15 did not galvanise the state into action; these unfortunates were left to their fate, and had to rely on help from the trade unions. And during the rest of the war the shortage of manpower ensured that the state paid only scant attention to the problem. In the end, it was the need, for political reasons, to provide for the dependents (i.e., the wives and children and later the widows and orphans) of soldiers and workers who had been involved in the war effort, which forced the state to reassess its welfare policies. After the 'civil truce' of the war years it would hardly have been politically viable to leave the dependents of war veterans to the mercy of the existing poor-relief system, which as in Britain in the nineteenth century was largely based on local charity. This move away from the previous poverty-relief system, where the poor had been reliant on the discriminatory practices of local welfare provision — exemplified by the existence of the workhouse — was an essential step on the road which led eventually to a legal right to support during periods of unemployment:

> The various welfare committees were meeting places for representatives from the state, local welfare institutions and industrial organisations, notably the employers' organisation and the trade unions. This had an emancipatory effect and was one facet of the 'civil truce'.[1]

44

At any rate, the road from a system of — fairly arbitrary — welfare institutions to a system of social insurance was still a long one. In the following pages I shall examine this road and the various stages along it, from general and special support for short-time work, through emergency and compulsory labour schemes to full unemployment benefits, using the town of Augsburg in South Germany as an example.

II

Support for those without a livelihood, instituted hastily after the end of the war in order to guarantee an initial income for the mass of returning soldiers, and thus to decrease the possibility of social unrest, was certainly not a system of social insurance for the unemployed. So there were no employment conditions attached to it.

The support for those without a livelihood was created as a local institution and was carried out locally on behalf of the Reich. Its aim was to support not only the unemployed but also those without the means to subsist. Ameliorating the poverty created by a total or partial loss of livelihood as a result of the war was its underlying rationale. Local councils were not allowed to regard this support as part of the poor relief system.[2]

The Reich bore half of the cost of this support, the regional council one third of the cost, and the local council the remaining one-sixth. Besides this particular benefit there were three other welfare institutions which the local council had to administer. These included support for the war wounded and dependents of the war dead; the Reich paid one-fifth of this; and the remainder was divided between the regional and local councils equally. Special support was also instituted for those on public pensions (*Sozialrentner*) from 1921 and for those on small pensions (*Kleinrentner*) from 1923. Four-fifths of these last two were paid by the Reich. Thus, the local councils were responsible for nearly all the administration of the welfare payments, and also had — to some extent — responsibility for setting the rates. However, in comparison with the regional councils and the Reich their financial burden was light. This meant that, amongst other things, in the initial stages guidelines for providing support for those without a livelihood could be relatively generous, as at first the number of people entitled to claim fell rapidly.

Figure 3.1 shows the monthly rate of those receiving this benefit in the south German town of Augsburg.[3] The recovery of the economy in Augsburg can be clearly seen in the sharp drop in numbers receiving the benefit up to the spring of 1920. Until the end of 1922 the number of those unable to support themselves remained low. In April and May 1922 it fell for the first time to below 100, and reached its lowest point in June 1922 with 19, when the Augsburg economy was experiencing an inflationary false recovery.[4] The benefit offices were closed in 1922 and the officials were engaged only occasionally in paying out the benefits. It was not until 27 November 1922 that the offices reopened.[5] The rise in unemployment figures at the beginning of 1923 was caused partly by the Ruhr crisis, and the peak reached in the winter of 1923–4 was caused by the collapse of the currency and the subsequent stabilisation crisis. During this period a new law governing support for the unemployed was passed (this will be dealt with in more detail in section IV below). The crisis during this winter was decisive for the employment market. It certainly overshadowed the crisis which had followed demobilisation, although relatively speaking it was of shorter duration. A permanent crisis for the employment market began towards the end of 1925. It reached a first peak in the winter of 1925–6. From this time onwards the unemployment figures fluctuated at a high level and welfare payments became a permanent fixture. Even the new legislation passed in 1924 proved to be insufficient — above all, the length of time for which a person was entitled to receive support had to be continually lengthened in order to alleviate the worst cases of need. The legislation of 1927 finally brought about a change in the situation, which will be dealt with in more depth later in this chapter.

At the beginning of the 1920s the conditions required for receipt of a benefit for those without a livelihood were, apart from need, fitness and willingness to work. House-ownership and savings disqualified the claimant. The council doctor decided who was fit to work. Whoever could not — or for reasons of age could no longer — take up employment, was turned over to the local welfare institutions, which were essentially the same as the poor-relief system which had existed before the war. The condition of

Figure 3.1: Numbers of People Without a Livelihood Receiving Support in Augsburg 1919–26

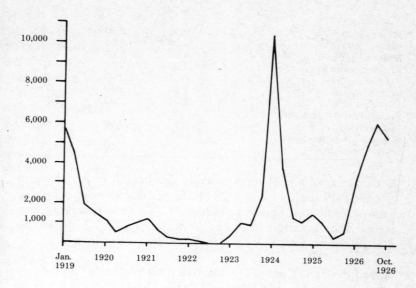

willingness to work meant that the claimant was under an obligation to accept work under certain conditions. In practice this meant that claimants had to accept virtually any work that they were physically capable of undertaking:

> A skilled worker is not allowed to decline unskilled work on the grounds that there are unskilled workers looking for work; a clerical worker cannot support a refusal to undertake physical labour with the fact that he is a clerical worker; a carpenter cannot refuse to undertake logging work during periodic unemployment on the grounds that the wage for this work is lower than his usual rate of payment.[6]

If these conditions were fulfilled the claimant — and youths from the age of 14 could claim this benefit — would receive a money payment based on the rates of the standard sickness insurance benefit. The payments varied greatly with respect to the age and sex of the claimant, as Table 3.1 shows.

Table 3.1: Daily Rates of Benefit in Three Large South German Towns on 27 January 1919 (in Marks)[7]

	Munich	Nuremberg	Augsburg
Men over 21	8.00	6.00	5.20
Women over 21	5.00	4.00	4.20
Men under 21	4.50	5.00	4.30
Women under 21	3.00	3.00	3.50
Men under 16	3.50	3.00	3.00
Women under 16	2.50	2.10	2.50
Supp. for wife	2.00	2.00	1.04
Support for dependents older than 17	2.00	1.50	0.52
Support for dependents younger than 17	1.00	1.00	0.52

A man without a livelihood claiming this benefit in Augsburg in January 1919 could receive 7.28 Marks per day at most, if he also claimed for his wife and two children. If his wife was working, this income was deducted from the payment. Table 3.2 shows the relationship of these rates of benefit to the cost of living.

Table 3.2: Daily Cost of Necessities for a Worker's Family Consisting of Man and Wife and Three Children, November 1919 (Augsburg Prices, given in Marks)[8]

Rationed food	4.66	
Unrationed food	4.17	(vegetables, fruit, beer)
Rent	1.17	(3 rooms, kitchen, no bath-WC)
Heating and light	1.13	(yearly average)
Clothing and sundries	5.00	(new, repairs, insurance, tax)
Total	16.13	

One can fairly safely assume that this daily budget represents very closely the minimum cost of living. Approximately 16 Marks per day were necessary for survival — and a man without a livelihood, with a wife and three children to support received 10.25 Marks. On 19 December these rates were reduced by a total of 1 Mark. It was only in July 1920 that the rates were increased.[9] The people receiving this benefit did not, at this time, receive any supplement in the form of food or fuel.

There is no doubt that it is very difficult to ascertain from these figures what conditions were actually like in the households of the claimants. The amounts and prices of the rationed foodstuffs vary too greatly, and in particular the possibilities open to people, and their capabilities to make ends meet, differed enormously. A further example from Augsburg makes this clear (Table 3.3).

Table 3.3: Weekly Budget of a Three-person Family in Augsburg, January 1920 (Prices in Marks)[10]

Rationed foodstuffs		Sundry expenses	
Bread	1.80	Soap, soap powder	3.75
Potatoes	0.72	Cleaning materials	1.00
Meat	2.65	Brushes etc.	2.00
Sausage	2.00	Coal, wood	15.00
Margarine	3.06	Gas, lighting	4.50
Butter	1.56	Rent	6.00
Eggs	0.69	Maintenance	2.50
Sugar	1.08	Taxes	10.00
Flour	1.08	Insurance, union dues	4.00
Ingredients for soup	2.25	Clothing, shoes	30.00
		Replacement utensils	2.00
Unrationed foodstuffs		Tobacco, newspapers etc.	9.25
Oil	4.00	Total	129.69
Coffee	5.00		
Coffee substitute	1.50		
Milk	2.10		
Pulses	5.20		
Spices	4.00		
Herrings	4.00		
Fruit	2.00		

At this date a man without a livelihood with a wife and one child received 7.45 Marks per day (i.e. 52.15 Marks per week).[11] Even if one reduces the very high figures given for sundry expenses, the fixed costs for rent, heating (in January!), lighting and for soap remain at a total cost of 29.25 Marks; these still take approximately half of the total income of a family subsisting on these payments. Thus, even during the relatively favourable economic conditions of the early years of the Weimar Republic these payments can only be regarded as a desperate short-term measure for supporting those in need.

With inflation increasing towards the end of 1921 a new system of welfare payments had to be introduced (as mentioned above). Those receiving public pensions from the health and clerical insurance schemes were in desperate economic straits as a result of the inflation.[12] Indeed, the inflation was already making itself felt by the middle of 1920. But this time welfare payments were keeping in step with price increases and rose with the cost-of-living index of the Reich, in 1920 and 1921 in fact rising even faster. Thus, if we take the cost-of-living index of February 1920 as the base — 1 — , in July it had reached 1.25, while the welfare payments for men over 21 had reached 1.35 in Augsburg; in November the ratio was 1.31 to 1.73, and with the next increase in welfare payments, in August 1921, the ratio was 1.57 to 1.82.[13] This high rate of increase was caused, amongst other things, by the relative ease with which local authorities could obtain money during the inflation, not least by simply printing it themselves under the emergency banknote printing rules.[14] However, this does not disguise the fact that the rates were now as before still very low. As a result, a series of new rules came into effect in July 1920. The two most important were those which limited the length of time and the amount which could be paid out in benefits.

The amount of time a claimant could receive benefit was generally limited to 26 weeks. Only in exceptional cases could support be extended to one year. There was moreover a stipulated period of one week which the claimant had to wait between signing on and receiving the first payment. The maximum amount a family could receive was now limited to two and a-half times the rate for a man over 21.[15] In fact, welfare payments now fell for a time below what the claimant would have received as sick pay in his former occupation, because rates of pay, and thus rates of sick pay, had risen with inflation. A report by the Augsburg town council expressed the fear that the unemployed would thus attempt to get themselves registered as sick in order to claim the higher rates.[16] At the same time as this development was taking place in the Augsburg labour market, the town magistrates attempted to get Augsburg classified on a higher rating. Towns were classified on a scale of A to C, according to the supposed rate of inflation. The rating A — for example Munich — had the highest rate of inflation, while the rating C — and Augsburg was still classified as such at this time — was supposed to reflect the lowest rate of inflation. However, as the magistrates ascertained, inflation already reached Munich levels in Augsburg during 1921. So they felt it incumbent on themselves to try and obtain for those claiming welfare benefits the higher rates of payment set by the Ministry of Social Welfare (*Ministerium für Soziale Fürsorge*) in Munich.[17] However, it was not until the beginning of 1923 that Augsburg was finally reclassified with an A rating.[18]

At the beginning of 1921, as a result of a decision by the town council, there was a tightening-up of controls. It was feared that the unemployed were engaging in illegal black market activities or that they were not 'willing to work'. They were therefore compelled to 'sign on' several times a day at the welfare offices, in order not to lose their payments. Until this point it had been customary to sign on once a day. This was not all. The limitation on the maximum amount that could now be paid out to a single family affected large families particularly badly. Until now, these families had received a relatively favourable level of support. Indeed, the welfare offices had initially had to deal with complaints that some families, as a result of the supplements paid to dependents, had been better off on welfare than if the head of the family had been in employment. The kind of organisational problems, apart from the purely financial, which this upper limit on payments caused the affected families, is exemplified in the following petition from the Augsburg magistrates to the Ministry of Social Welfare in Munich:

If, for example, both parents are employed, which is the general rule, the children must, while they are still little, be looked after by someone who is paid for this. The cost of this on average (in February 1921) is approximately 120–150 Marks. As the unemployed are generally anxious to find work as soon as possible, the children cannot be brought home, as if work is found, it is often not possible to find another satisfactory situation for the children immediately.

However, the rate of support for a child was only 97.50 Marks (and after the above-mentioned restrictions came into effect, it could be considerably less than this). In order to avoid hardship the town council petitioned that the excess above the support-rate should be counted against the income of the partner still in work.[19] The petition failed. The efforts of the Augsburg welfare offices naturally decreased as the numbers receiving support began to decline in 1922. For the few remaining claimants this was mainly noticeable in that the welfare payments no longer kept up with inflation at all. In March 1922 the cost-of-living index and the rate of welfare payments stood at approximately the same level (3.5; Feb. 1920 = 1). In August 1922 the cost-of-living index rose to 9.17, while the welfare payments were only five times what they had been in February 1920. Even when the numbers of unemployed began to rise again in the winter of 1922–3, the increase in the value of the welfare payments did not keep pace with the general level of inflation. (November: 52.7/27.0; December: 00.9/69.2; January: 132/115; February: 312/138).[20]

The second most signficant economic crisis after the First World War, the hyperinflation and stabilisation crisis of 1923, affected the labour market and the unemployed very much more drastically than had the demobilisation crisis of 1918–19. The changeover from wartime to peacetime production had in 1918–19 affected mainly the giant industrial plants associated with heavy industry; many small and middle-sized concerns were able to carry on virtually undisturbed. Even some branches of larger concerns, for example the Augsburg textile industry, were able, despite a dearth of raw materials and production problems, to continue to supply a wide section of their clientele. Finally, even during demobilisation many former employees had been able to live off their savings and thus had not needed — nor were they eligible — to claim welfare payments. The picture was very different in 1923 and the winter of 1923–4. The effects of inflation affected every branch of industry equally. While the giant concerns had to lay off large numbers of workers, or go over to short-time production, a large number of small and middle-sized businesses were forced to shut down completely.[21] In the final phase of the inflation two distinctly new classes of unemployed suddenly sprang into existence. Many independent craftsmen and professional people were forced, as a result of losing their working capital or their occupation, to apply to the welfare offices.[22] For those who, for one reason or another, were unable to offer themselves for other kinds of work, and were thus not eligible for support under the existing welfare arrangements, a third branch of welfare was erected, the so-called 'small pension supplement' (*Kleinrentnerfürsorge*). The Reich paid the lion's share of the cost for this.[23] As a result of the devaluation of the currency there were also a large number of people driven to seek assistance from the welfare system who had previously lived by trading or using the black market, or who had simply lived in the countryside, offering their labour in return for food. The worsening economic crisis towards the end of 1923 made it increasingly impossible to survive in this way.

The inflation not only brought about a rapid increase in the number of unemployed (see Figure 3.1), but also very severe organisational problems. At the beginning of 1923 the welfare office had decided to make payments twice weekly, instead of once a week. Even getting hold of enough money was difficult. Transferring money through the banking system, in view of the daily increase in inflation, took too long, so the money for the welfare office in Augsburg, which came from the Reich and from Munich, had to be personally collected from Munich by an official.

> For several months an official was sent twice a week to Munich where after collecting official passes from the regional council offices (*Landesamt*) he collected cheques and finally actual money from the Bavarian treasury and the State Bank. He brought the money back to Augsburg on the express train. When the emergency currency (*Notgeld*) was introduced at least this performance was ended.[24]

The rates of payment had to be continually increased to keep pace with inflation. In 1922 it was raised five times, but in 1923 it was increased 32 times. It went up twice in January; in February and March it was left temporarily at the same level; in April and May it was raised once each month, in June twice, in July three times, and in August six times. Until this point the Reich Office for Employment in Berlin had refused to link welfare payments automatically to increases in the cost-of-living index.[25] However, it now seemed unavoidable, and from September 1923 until the stabilisation of the currency the welfare payments increased weekly in line with the cost-of-living index.[26] The recipients of welfare payments, who mostly had to 'sign on' twice a day, and who usually collected their money twice a week, filled the welfare offices; endless queueing was a daily experience throughout the period of hyperinflation. As the money collected in the morning was worth less by even the same afternoon, long queues formed outside the offices very early in the morning,[27] as everyone wanted to collect their money as early as possible.

Many recipients of the benefit had been supported by welfare payments for a long period, and had been unable to afford to buy the necessary wood or coal for the winter. This inability to purchase necessities in advance, which lasted from this winter until the end of the world economic crisis, substantially increased the cost of living for the unemployed.[28] Even the fact that eventually welfare payments were automatically kept in line with inflation did not represent a financial improvement in conditions for those affected. It is true that recipients of other welfare payments —

those receiving payments from local funds, the war-wounded and the pensioners — were in a worse position than those receiving welfare payments for lack of livelihood, because their rates of payment could not keep in step with inflation to the same extent.

> It is an extraordinarily bad state of affairs and has already led to some unpleasantness, when the rates paid to those without a livelihood differ sharply from the rates received by the war wounded and pensioners. It is particularly difficult when they are far higher than what is paid to those injured in the war.[29]

Despite this, welfare payments for those without a livelihood were scarcely enough to live on. In April 1923, following an announcement by the Ministry of Social Welfare, the rates of support were further limited. The rate of payment (including supplements for dependents) was not to exceed three-quarters of the customary wage.[30] A family had, therefore, to exist on, at most 75 per cent of what the head of the family had previously normally earned. The length of time for which payments could be made was again curtailed. Previously the authorities had been able, in exceptional cases, to extend the right to payments to one year. Now the upper limit was set at 39 weeks. 'Exclusion' after 39 weeks meant that the claimant had to turn to the existing local welfare system, i.e. the poor-relief system. There is very little evidence indeed about the extent of assistance, in money or goods, that the claimant would receive from there. One can however assume, from the tenor of the official reports, that the assistance was very much lower than that received from the welfare offices.

The leader of the welfare office dealing with those without a livelihood in Augsburg, Lawyer (*Rechtsrat*) Kleindinst, made a serious attempt to increase the length of time a family could be supported. He stressed the financial aspects: 'It is very important to remember that after the 39 weeks have expired the burden of supporting those without a livelihood falls directly on the local council (*Gemeinde*) — they are not able to carry this financial burden as they do not have access to sufficient funds.'[31] He also complained about the rates of payment, which were felt to be too low. The rate paid for a man without a livelihood, it is true, rose with inflation, but the supplementary rates paid for dependants did not keep pace:

The rate of support for the week 29.10 to 3.11.23 enabled six children to be supported. However from the 5th November onwards, when inflation increased at a phenomenal rate (650.5% in one week) the family supplements fell so far short that only two children could be supported. It was only with the rates that came into effect on the 19th November that it became possible for four children to be supported.[32]

Indeed, from the beginning the rates paid to women without a livelihood had been between one quarter and one fifth below those paid to men. On the other hand, the supplements paid for dependents were subject to wide fluctuations. In January 1919 the supplement for a dependent spouse had been about 30 per cent and that for a dependent child roughly 10 per cent of the rate paid for the head of the family. These rates increased gradually and reached a peak in February 1923 with 46.7 per cent paid for a spouse and 40 per cent paid for a child. Finally, they fell rapidly to 35 and 30 per cent respectively, but remained more or less at this level, with occasional fluctuations until 19 November. In the following weeks, however, they fell again to 25 per cent for the spouse and 19 per cent for a child. As inflation had been halted at about this time, the rates remained at this level until the middle of December 1923, when the rates were again altered.[33]

In the week from 26 November to 2 December 1923 the following rates were paid[34] (the enormous sums reflect of course the staggering scale reached by hyperinflation at this time): for a single person over 21, without a livelihood, 4,680,000,000,000 Marks; for a married person with 1 child, 6,780,000,000,000 Marks; with 2 children, 7,680,000,000,000 Marks; with 3 children, 8,580,000,000,000 and with 4 children 9,360,000,000,000 Marks. Thus, in this particular week a family consisting of two adults and three children had to live on a budget of 8,580,000,000,000 Marks. The minimum weekly cost of subsistence for such a family, calculated on the basis of the cost of living index, comes to 36,809,000,000,000 Marks. This calculation does not include costs for consumer goods or taxes etc.[35] Thus, an Augsburg family supported by these welfare payments received approximately 25 per cent of the necessary sum. What this calculation also does not show is the effect of continuing hyperinflation. On the day on which half of the welfare payment was made (prices given are those for 26.11.23) the family could purchase the minimum amount of bread

(23.5 lbs = 6,157,000,000,000 Marks), 1 kg of fat (which does not represent even the minimum amount) for 2,423,000,000,000 Marks, and nothing else.[36] Thus the attempt to get the welfare payments increased also failed.

The sole success enjoyed by the Augsburg welfare offices was in getting the waiting period of one week before the first payment reduced to three days. However, this decision, which took effect from October 1923, was later rescinded soon after the currency was stabilised in March 1924.[37] The crisis of 1923 had shown all parties that this method of supporting those 'without a livelihood' was no longer financially viable. The local welfare system could not cope with the large numbers who were passed on to them after the 39 weeks had expired as well as having to support those ineligible for the benefit at all. The central and regional governments, now they were no longer able to print their own emergency money, were unable to pay out large sums as they had previously done.

> In 1921/22 large sections of the working population had demanded the continuing existence of the welfare system as it stood. This was based on a belief that society as a whole was responsible for unemployment and that the financial burden should be carried by the people as a whole. At that time, because of this attitude, people were prepared to accept support for unemployment which still bore the stamp of charity. However the stabilisation of the currency demanded the strictest possible restraints on public spending, with the result that statutory contributions by employers and employees were established instead. All the trade unions now joined together in demanding that the logical conclusion of this contract between employers and employees should be drawn, namely that a proper system of insurance against unemployment should be set up.[38]

There was legislative provision for establishing a new system to deal with support for the unemployed under the 'first enabling act, on which the clauses of 15 October 1923, relating to the provision of sufficient funds for supporting those without a livelihood are based'.[39] However, before we move on to a closer look at the new regulations governing support for the unemployed, it is necessary to round off the picture by casting a brief glance at the set of special measures which the authorities had introduced for 'those without a livelihood' during the period 1919–23.

III

Alongside the welfare payments which enabled large numbers of those without a livelihood to be supported a series of special measures was adopted, partly to eke out existing employment, partly to create new jobs and partly, in order to avoid a so-called 'moral decline' in recipients, to set up a kind of 'work therapy'. One of the most significant measures available to add to existing work was a regulation passed in 1920 governing the suspension of production. This law meant that businesses had, if possible, to introduce short-time working before they could lay people off. It was rescinded on 13 October 1923, and from this point on only industries which came under the jurisdiction of the demobilisation authorities could be ordered to introduce short-time work.[40] In Augsburg, as in many areas, short-time working was widespread, as Table 3.4 shows.

Table 3.4: Number of People on Short-time Work in Augsburg, October 1923–March 1924[41]

1923		1924	
1 Oct.	17,539	1 Jan.	5,096[43]
15 Oct.	18,592	15 Jan.	2,863
1 Nov.	19,000	1 Feb.	3,130
15 Nov.	19,930	15 Feb.	1,694
1 Dec.	19,851[42]	1 Mar.	810
15 Dec.	12,500	15 Mar.	240

The payment of wages for these workers on short time caused many problems for the authorities, employers and workers alike. Calculating them during a constant fall in currency values was so complicated that delays were experienced on all sides.

Large factories had to set up separate departments to calculate payments. Wage rates, the level of welfare support and the amount of time worked changed so quickly that these departments were scarcely in a position to get things sorted out by payday. Thus calculations often arrived late at the welfare offices, which had also had to set up special sections for dealing with the short-time workers. Smaller concerns were guaranteed advance payments from the welfare offices. Factories well-known to the offices were granted the payments after a number of test cases had been worked through. Others however had to have their books properly inspected. Despite the speed with which all departments worked, many businesses were damaged by the rapid inflation and necessity of introducing short-time work.[44]

Welfare supplements for short-time work were stopped in 1924, mainly because it was feared 'that they were developing into a form of subsidy for industries that were inefficient and were thus hindering the regeneration of the economy'.[45]

The Augsburg authorities gave other reasons for stopping these supplements; the welfare offices had already demanded the cessation of support for short-time work at the beginning of December 1923. This eventually happened on 1 April 1924. This desire to suspend support for short-time work occurred at precisely the point where there were the largest number of short-time workers in Augsburg (almost 2,000, as Table 3.4 indicates) and was based on the high cost. Up until November 1923 a working man on short time who was losing 16–18 hours a week, with a wife and two children might receive — after the complicated procedure outlined above — his full wage. It was not, of course, possible to receive any more.[46] He also did not have to 'sign on' at the welfare offices.[47] From November 1923 the calculations were somewhat simplified and the upper limit was set at five-sixths of the full-time wage.[48] The support for short-time work was high, and conversely the support for those totally unemployed so low, that short-time workers were better off in almost every case, even without the welfare supplements, than their unemployed counterparts, even if they worked only three days a week.[49]

One of the most comprehensive measures taken to create employment throughout the whole period of the Weimar Republic began directly after the war. This so-called *Produktive Erwerbslosenfürsorge* later became known simply as 'emergency employment'. It was a series of measures designed to create employment for the largest possible number of people, but offering the poorest possible wage. The work so created had to be 'of positive value for the economy'.[50] The projects had to prove that they could not be put into practice without a state grant. The work was mainly construction labour, above all civil engineering projects such as river regulation, the building of dams, road and canal construction. At first the financial backing for these projects was supposed to come from the welfare office funds, with the Reich paying half, regional governments one-third and local councils one-sixth. But as these sources were insufficient, the Reich instituted another way of financing these ventures. The state made funds available for each project which represented the saving on welfare payments, the so-called 'basic grant'. In addition, particularly 'valuable' projects[51] could be granted a further sum of money from the central treasury. This additional sum was, on average, two and a- half times the 'basic grant'.[52] This was administered differently in the different regions. In Bavaria the Ministry of Social Welfare usually only granted an additional matching grant at the same level as the 'basic grant'.

> One result of the emergency employment measures, clearly not intended by the original legislation, was that in Bavaria the funds of the regional government intended for the support of the unemployed were significantly increased as a result of the practice of only making matching grants. On the one hand the local councils had to put up with an almost intolerable drain on their resources as a result of the emergency employment measures, while, on the other, the regional government did very well out of the arrangement, [because] . . . if a person was receiving support from the welfare offices, sickness insurance contributions were paid by the regional government, but under the emergency employment schemes, the contributions were paid by the contractor.[53]

After a project had been approved and granted funds there was still a series of regulations that had to be complied with. The people taken on had to come from authorities where at least 3 per 1,000 of the population were receiving support from the welfare office. However, as in some towns and parishes no support was paid, there were cases where there were people able and willing to work, but who were refused employment in the schemes on these formal grounds of exclusion.[54] In addition, only people who had been receiving welfare for a period of at least four weeks before the commencement of a project were entitled to employment (in

1923 this was reduced to two weeks).[55] For those directly concerned, perhaps the hardest condition of all was that these
emergency employment schemes were in no sense voluntary.

> A state which has to provide for a large number of unemployed
> must use emergency employment as a test of 'willingness to
> work'. There were always people who refused to take work for
> the most varied reasons, or who supplemented their welfare
> payments with more or less regular work on an illegal basis. If
> an unemployed man refused work on an emergency
> employment scheme without a valid reason, his welfare
> payments were stopped. Thus the emergency employment
> schemes offered, in the face of massive unemployment, a way
> of testing the willingness of the unemployed to work, even if
> these schemes only affected a small proportion of the total
> number of unemployed (on average 10 per cent).[56]

Not surprisingly, the unemployed disliked these emergency
employment schemes. For a start the rates of pay for a working
week of 48 hours were very low. In January 1919 the Labour
Exchange in Augsburg noted

> According to the guidelines communicated orally to the
> employment office in Augsburg by the local planning de-
> partment (*Stadtbauamt*) we must offer the places on the
> emergency employment schemes in particular to local married
> people with large families . . . However, these guidelines are
> very difficult to put into practice. According to the wage rates
> the workers receive 7.20 Marks for an eight hour day. A worker
> with four children can end up worse off than if he was receiving
> welfare payments, which in this case would amount to 9 marks
> a day.[57]

Shortly afterwards the wages were raised to 10.80 Marks per day,
but the difference between wage-rates and welfare payments continued to seem far too small to those employed on the schemes.
Complaints about low wages were made continually through the
years. An additional reason why those employed on the
emergency employment schemes complained was the fact of insufficient clothing. Most of the work, as already mentioned above,
was concerned with civil engineering. Most of the larger

emergency employment schemes in Augsburg were concerned with river regulation. The workers had to spend most of the day on wet construction sites and did not receive any suitable protective clothing for this. As shoes had become almost a luxury article for the unemployed in the Weimar Republic, and outer garments could very seldom be replaced, the authorities did try to meet the workers' demands to some extent, in that they did not allow the construction work to continue in the late autumn and winter months. Finally, it was only in exceptional cases that the schemes were carried out near where the workers lived. In many cases the workers had to spend the week in barracks on the construction site, and were able to return to their families only at the weekends. Here as well, the welfare offices were concerned to try and mitigate the hardship in that it was generally single men who were sent to the most remote construction sites, and they generally tried to accommodate married men as near as possible to their homes.

Looked at overall, neither the number nor the nature of the complaints made by those forced into the emergency employment schemes altered substantially during the time studied. As the welfare offices could threaten to stop welfare payments for at least 4–6 weeks in the case of 'unreasonable refusal to comply',[58] there were very few cases of refusal. For example, in 1924, of approximately 250 cases per week, only in 6 per cent of cases did the worker concerned refuse to take the place on an emergency employment scheme offered to him.[59] Often the welfare office was quick to increase the wage-rates when the ratio to welfare payments was obviously out of step. In February 1923, for example, in response to the complaints of the workers the hourly rate was increased from 320 Marks to 520. The wage rates were as shown in Table 3.5.

In August 1924 a decision of the Reich Ministry of Labour set the maximum wage for a worker employed on the emergency schemes as 70 per cent of that of an unskilled clerical worker or a construction worker. A worker on one of the emergency schemes in Augsburg received at this time 6 Marks main support per week, 5.10 Marks family supplement with a wife and two children, and 7.20 Marks supplement for length of work (which was paid for work of more than 24 hours per week from January 1924). That came to 18.30 Marks per week. Seventy per cent of the net wage of an unskilled construction worker was 20.30 Marks, and 70 per cent of the net wage of an unskilled clerical worker with a wife and two

Table 3.5: Wages and Construction-site Prices for Emergency Employment Workers in Augsburg, February 1923[60]

	Hourly rate (Marks)	%
Immigrants to the area (married)	520	= 100
(single)	474	= 90
Local married men	474	= 90
Local single men	421	= 80
Construction-site prices	Marks	
Overnight accommodation and coffee	30–50	
Midday meal with or without meat	300–500	
Evening meal	100–250	
Bread, 2 kg	600	
Milk, 1 litre	240	

children living in a town with the same rating was 17.39 Marks. Thus, as a result of this decision the worker on the emergency employment scheme now received 17.39 Marks instead of 18.30.[61]

In order to avoid unrest the authorities generally followed up workers' complaints. The most common complaints, apart from those about wages and clothing, concerned accommodation arrangements. Complaints about rats and lice, which were often put forward by the workers as sufficient cause for leaving the emergency schemes, were not generally substantiated by the investigating officials. In most cases where workers had left the schemes without permission, their complaints were not accepted, and welfare payments were stopped. The following example is typical of the years 1922–7:

> Despite the fact that the unskilled workers had been demanding the setting up of an emergency employment scheme, it was very difficult to find 50 workers for Weilheim. It is true that the most efficient workers were sought, and to this end the doctor examined the entire workforce, pronouncing them fit for work. However there were no less than 145 cases of refusal to work. Workers under the age of 21 or over 55 were naturally not considered for the emergency employment scheme.

On the very next day 15 men arrived back from the construction site; the resultant investigation found their complaints unfounded.

'Made cautious by previous experiences the welfare office sent its manager to the construction site.' It was decided not to send such a large contingent of men to the construction site at the same time in future, but rather to send only small groups of 3–4 men at a time, in order to avoid similar disturbances occurring again.[62]

The welfare office not only had difficulties in sending workers to more remote emergency employment schemes, but also conversely it experienced problems in convincing construction firms to use these workers. According to a report by the town council, apparently the rural population harboured a distinct prejudice against the urban unemployed, which increased the problems of finding accommodation. The rural councils were by and large on the side of the local population in this, and tended not to notify urban authorities of vacancies. The contractors preferred to employ people from the local rural areas, first because they were often physically in better shape than the badly nourished urban unemployed, and secondly because they worked harder: often they were not working for the basic necessities of life as they could find food and fuel from other sources. The money from the emergency employment schemes represented an additional income.[63] As a result of the very limited resources available to finance the emergency employment schemes and the very high rate of unemployment, especially in the crisis year of 1923, the town council felt itself forced to 'invent' new emergency employment schemes continually, and to take advantage of every possible opportunity, including those offered on the open market, to find employment for the unemployed.

This example, from 1923, shows how difficult it could be to organise emergency employment schemes. Two construction firms were contracted to work on a particular stretch of canal. They ran into difficulties and claimed that they would not be able to complete the work without state aid under the emergency employment scheme. They applied for a grant of 250 million Marks.

This is where the difficulties started . . . The original workforce worked a 48-hour week in two shifts, and even put in overtime. To comply with the regulations for receiving state aid the firms must take on the local unemployed. They have therefore had to dismiss their original workforce. After eight days compulsory waiting, the workers then had to apply for welfare benefit for fourteen days, and then the firms are able to reemploy them under the emergency employment scheme. So in practical terms, construction had been halted for this period. Some of the engineers have left completely. All in all it has taken five weeks before construction has been able to begin again properly.

As the barracks for the construction workers had not been fitted out properly, more problems followed; and the firms discovered that men employed under the emergency employment schemes did not work as hard as the original workforce had; thus, in the end they were unable to complete their contract on time.[64]

Some commentators looked on these schemes quite favourably:

> Compulsory labour (*Pflichtarbeit*), created by the welfare offices without legislative justification, and initially regarded by central government with the greatest reserve, has become an important instrument for the authorities. It maintains discipline within the workforce, and makes it possible to avoid the deleterious effects of enforced leisure.[65]

But this idealistic view of the most unpleasant of the special measures taken to combat unemployment was not shared by many. There was no payment for compulsory labour; but if a person refused it, the welfare payments were stopped. As early as the winter of 1919 the welfare office in Augsburg had experimented with compulsory labour: the unemployed were set to clearing snow. In order to avoid the work many of the unemployed appeared at the offices after 10 a.m., when the work had already started.[66] After this short-lived attempt the welfare office obviously decided that there was no point in continuing this kind of special measure.[67] However, it was different in the winter of 1923–4 when, with the massive increase in unemployment constantly in view, the authorities turned to every possible measure to keep the unemployed occupied. Augsburg introduced compulsory labour officially on 15 October. Clearing snow, removal of household refuse, splitting wood, levelling ground, construction of paths, cleaning schools (for women) and unloading and distributing coal deliveries — these were only some of the many kinds of employment projected.[68]

The unemployed were divided into three-day shifts, with 6 hours work per day. An individual would be called on to work on average once every 4–6 weeks.[69] Refusal to work, which was higher here than on the emergency employment schemes (10 per cent as opposed to 6 per cent), was strictly dealt with: the offender's welfare payments were stopped until he or she worked. As a result there were public disturbances amongst the unemployed in the winter of 1923–4:

> Initially this kind of work, for which the individual received only the welfare payments, engendered a great deal of resistance on the part of the unemployed and of the general public (a higher percentage of which were unemployed themselves). There were even strikes and civil disturbances, so that the police had to be kept on alert . . . In evaluating the affair, it must be remembered that many of the unemployed had insufficient clothing . . . and there were many amongst them who had never handled a pick or shovel. These people find it particularly difficult, they have to rest more frequently, and often bloodblisters develop on their hands, which also hinders the work.[70]

Eventually, certain concessions were made to the unemployed. It is true that it was only in exceptional cases that they received clothing and shoes at reduced prices,[71] but for every shift (three days at 6 hours) they received a special extra allowance of food or coal. A married man would get 38 kg of coal, or 750 grams of fat or 3 kg of bread. A single person got 1½kg of bread.

The attitude of the unemployed to compulsory labour revealed important underlying assumptions about society. In principle all the unemployed were called on to perform compulsory labour. However, commercial and technical clerical workers as well as the few unemployed professionals were usually only called on to perform office work.[72] But even among the working class distinctions have to be made, because it was mainly the skilled workers who refused compulsory labour, which they felt discriminated against them. However, the welfare office continued to call on them to perform these tasks 'where they are seldom seen. If one agrees to these wishes they generally accept the work, if unenthusiastically. If not the work is often refused.'[73] All three measures designed to create work — short-time work, emergency employment schemes and compulsory labour — continued throughout the period investigated, and were almost completely unaffected by changes in the legislation relating to welfare provision. The constant changes in the level of welfare payments and length of time they were paid for, along with other changes, affected the position of the unemployed in a general way of course, but it was the new welfare regulations of April 1924 that first brought about fundamental changes.

IV

Unemployment support[74] formed the intermediate step between welfare support for those without a livelihood and unemployment benefits based on a modern system of social insurance. In its financial aspects the new insurance system of 1924 prefigured the system set up in 1927 in certain features, in that both employers and employees had to pay certain statutory contributions (the local council had also to contribute, but the Reich and the regional governments no longer contributed to the cost). On the other hand, the new regulations still had something of the same aspect as the welfare support, for the worker was subjected to a means test, and had no legal right to support. The most important of the new regulations limited still further the number of people eligible: the claimant had to have been properly employed and paying sickness insurance contributions for at least 12 weeks in the previous 12 months. This regulation turned what had been a general system of welfare support for those without a livelihood into a more limited system of support for the unemployed. It meant that all the self-employed, and unemployed professionals, were no longer covered. Furthermore, it also excluded the long-term unemployed. Though they might well have worked inter-mittently, they were not able to fulfil the condition of just three-months' continual employment. With a wildly fluctuating economy and continual redundancies and re-employment, it was of course quite possible for this to happen to individuals without their being in any way personally to blame. Finally, it was usually a case of 'last in, first out'.[75] The proportion of unemployed not eligible for support varied in 1924 between 8 and 22.8 per cent. Figure 3.2 shows the distribution of this during the crisis winter of 1923–4, week by week.[76]

The numbers affected were so large that it led to the assumption that in the future there would be 'two classes of the unemployed', in other words 'those who got an income from the unemployment

Figure 3.2: Numbers of Unemployed and Those Receiving Benefits in Augsburg 1923–4

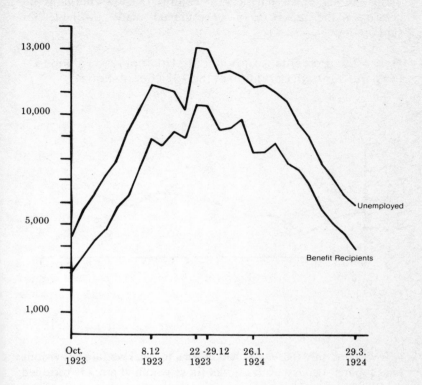

insurance provisions, and those who had to be supported by local welfare . . . The boundary between the two groups will be drawn by economic chance'.[77] In the following years the number of this second class of unemployed, with small fluctuations, increased continually. The Augsburg Labour Exchange, which in the meantime had taken over the job of administering the payments, was very concerned to alleviate at least the worst instances of hardship, and to try and avoid turning individuals over to local welfare. The long-term unemployed were the worst hit. They could be given short-term employment (often seasonal work on building sites), but they then afterwards lost their right to support, although they might not have been receiving that support for the full term. Marginal cases such as these received redress only in

1927, by means of a decree from the Reich Ministry for Employment whereby such short-term jobs were to be regarded as 'interruptions' in the payments.[78] Figure 3.3 shows the large increase in the numbers of unemployed made ineligible for support.

Figure 3.3: Percentage of Augsburg Unemployed Failing to Fulfil the 3-month Condition of the 1924 Regulations[79]

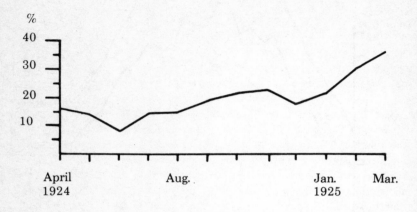

It was not only these who had to turn to local welfare provisions for support of course, but also all those who had already received the maximum support (39 weeks). As the previous employment crisis (1919) had been of relatively short duration, and the economic fluctuations between 1919 and 1923 not very violent, the first time the numbers of these excluded unemployed reached significant proportions was the spring of 1924. In the month of April 1924 alone, 365 unemployed had to be turned over to local welfare as they had been receiving support for the maximum length of time they were entitled to (Table 3.6).[80]

Even the authorities seemed to be surprised at the high figures, because it was by no means a case of 'ageing workers who have been out of the labour market since the stabilisation crisis, but also men in the prime of life . . .'[81] The new regulations were supposed to be a step in the direction of a proper system of social insurance. But they represented a definite step backwards for all of those

Table 3.6: Number of Unemployed Excluded from Welfare Payments Due to Length of Time Supported, Augsburg 1924

Month	Total unemployed	Excluded absolute figs.	%
April	3,683	385	10.4
May	1,362	99	7.2
June	1,240	144	11.6
July	1,264	143	11.3
August	1,489	212	14.2
September	1,395	231	16.5
October	1,099	209	19.0
November	883	118	13.4
December	1,108	61	5.5
January 1925	1,499	9	0.6

excluded under the new provisions. Even those who benefited under the new regulations could not always see them in terms of progress:

> The unemployed can't understand that after they have made statutory contributions to welfare support they might be ineligible for support, or at least be only partially eligible, if the family has another source of income. In the main, it is felt to be a great hardship in cases where the parents are fairly old and their dependents are of an age to work. Before the war these family members would have been able to move out of the parents home long since, but as a result of the housing shortage they are still forced to live with their parents.[82]

In such cases of hardship, which increased in number, the Employment Office offered to exclude the income of family members over thirty in their calculation of the rate of support.

Like the means test, the way in which the rate of support was calculated was based on the principles which had governed the welfare payments rather than on the principles underlying a system of social insurance. Thus, the rates were not differentiated on the basis of what a claimant had previously earned, but rather different rates were set for men, women and young workers. In 1925 the rates paid to women were increased so that they approached those paid to men, and from February 1925 the rates were supposed to reflect only differences in age; but even in this the rates paid to the youngest workers were made closer to that of the men.[83].

The underlying assumption was that welfare support should only cover the bare necessities, which for men, women and adolescents were not basically very different. From 1926 onwards, however, it became increasingly clear that even those in receipt of proper wages found it difficult to afford the basic essentials.[84] At any rate, this was one of the continual complaints of the clerical workers' unions. Rural labourers, people starting out on their careers and young workers all generally earned less than the rate of support.[85] The long-overdue 1927 legislation providing insurance for the unemployed, which gave them a firm legal right to support, and the removal of the means test signified the last decisive steps away from the discriminatory poor relief of prewar days. There was, however, a sequel, in which the financial problems of the Ministry of Labour during successive world economic crises after 1925 once more meant poverty and dependence on welfare for many unemployed.

Notes

This chapter was translated by Cathleen S. Catt

1. Stadtarchiv Augsburg (henceforth StAA). Nachlass Kleindinst, 7, MS., 'Wohlfahrtspflege und Arbeitslosenfürsorge', p. 2.

2. Ibid.

3. Figure 3.1 is based on figures from *Augsburger Statistisches Taschenbuch* (1st ed., 1927), p. 84. The discrepancies between Figures 3.1 and 3.2 are discussed in the text on pp. 66.

4. For a detailed description of the Augsburg economy and the effect of demobilisation on the labour market see: Merith Niehuss, 'Arbeitslosigkeit in Augsburg und Linz 1914–1924', in *Archiv für Sozialgeschichte*, vol. XXII (1982), 133–58.

5. StAA, Nachlass Kleindinst, 7, MS., on 'die Entwicklung der Erwerbslosenfürsorge'.

6. *Deutsche Sozialpolitik 1918–1928* (Erinnerungsschrift des Reichsarbeitsministeriums, Berlin 1929), p. 169.

7. StAA, Abg., 10, 1635, 27.1.1919.

8. StAA, ibid. This report was demanded by the Ministry for Social Welfare, Munich, from several towns.

9. StAA, Abg., 10, 1637, 30.1.24. Information on all the different rates paid by the welfare office, collected by the Erwerbslosenfürsorgestelle.

10. StAA Abg., 10, 1635, 20.1.1920.

11. StAA Abg., 10, 1637, see note 9.

12. StAA Nachlass Kleindinst, 7, MS., 'Wolhfahrtspflege und Arbeitslosenfürsorge', p. 2.

13. Collected and calculated from: StAA, Abg., 10, 1637.

14. Amt für Stadtentwicklung und Statistik, Augsburg (henceforth ASte). Verwaltungsbericht, MS, 1924/25, p. 442.

15. *Deutsche Sozialpolitik*, p. 105, and StAA, Abg., 10, 1636, 1.11.20.

16. StAA Abg., 10, 1636, report 16.7.20.

17. Ibid., report by the town council, 9.9.21.

18. Ibid. Suggestion by the town council concerning the new calculation of welfare-rates. 5.3.23.
19. Ibid. Request from the town council to the Ministerium für Soziale Fürsorge.
20. See note 13.
21. ASte, Verwaltungsbericht.
22. StAA, Nachlass Kleindinst, 7, MS., on *Erwerbslosenfürsorge*, p. 1.
23. Ibid., MS on *Wohlfahrtspflege*, p. 2.
24. Ibid., MS on *Erwerbslosenfürsorge*, p. 2.
25. For a more detailed exposition see: Stephan Leibfried: 'Existenzminimum und Fürsorge-Richtsätze in der Weimarer Republik', in Ch. Sachsse and F. Tennstedt (eds), *Jahrbuch der Sozialarbeit* 4, (Reinbek, 1981), pp. 469–523, here pp. 471–3.
26. ASte, Verwaltungsbericht, p. 440.
27. StAA, Nachlass Kleindinst, MS., on *Erwerbslosenfürsorge*, p. 3.
28. In more detail, M. Niehuss, 'Lebensweise und Familie in der Inflationszeit' in G. D. Feldman, C. L. Holtfrerich, G. A. Ritter, P. C. Witt (eds), *Folgen der Inflation* (Berlin, 1986).
29. StAA, Abg., 10, 1636, Rundschreiben des Reichsarbeitsministeriums, 16.3.23.
30. Ibid. Decision of the Stadtrat, 6.4.23. The 'customary wage' was an official rate per job calculated by the authorities as a basis for sick-pay rates.
31. Ibid., 1637, report from Kleindinst to the Min. für Soz. Fürsorge, 20.11.23.
32. Ibid., 6.12.23.
33. Calculated from StAA, Abg., 10, 1637.
34. Ibid.
35. Cf. M. Niehuss, *Arbeiterschaft in Krieg und Inflation. Soziale Schichtung und Lage der Arbeiter in Augsburg und Linz a.D. 1910–1925* (Berlin, 1985).
36. StAA, Abg., 10, 1637, Kleindinst's report, 6.12.23.
37. Ibid. Decision by the Min. für Soziale Fürsorge, 18.10.23 and 20.3.24.
38. F. Syrup, *Hundert Jahre Staatliche Sozialpolitik, 1839–1939* (ed. J. Scheuble, revised O. Neuloh, Stuttgart, 1957), p. 333.
39. ASte, *Verwaltungsbericht*, p. 430.
40. ASte, *Kommunale Mitteilungen*, 1925, p. 191.
41. Ibid.
42. The gradual decrease in the numbers of short-time workers is due to the increase in dismissals in the textile industry.
43. Transition to full-time work in the textile industry.
44. ASte, *Verwaltungsbericht*, p. 430.
45. Michael T. Wermel and Roswitha Urban, *Arbeitslosenfürsorge und Arbeitslosenversicherung in Deutschland* (Munich, 1949), p. 41.
46. StAA, Abg., 10, 1636, *Berechnungsgrundlagen für Kurzarbeiterunterstützung*, 5.3.23.
47. Ibid., 1637, Letter from the Stadtrat, Augsburg to the Min. für Soz. Fürsorge, 28.12.23.
48. Ibid., Stadtrat Augsburg, 4.11.23.
49. Ibid., Letter from the Stadtrat to the Min. für Soz. Fürsorge, 28.12.23.
50. Syrup, *Hundert Jahre*, p. 349.
51. Ibid., p. 348.
52. Ibid., p. 349.
53. StAA, Abg., 10, 1636, Note from the Verband der gemeindlichen Arbeitsnachweise in Bayern to the Landesamt für Arbeitsvermittlung, Munich, 4.12.25.
54. For example, the town of Kempten. StAA, Abg., 10, 1622, Letter from the Arbeitsamt Augsburg, 24.5.23.
55. StAA, Abg., 10, 1622, Note from the Arbeitsamt Augsburg, 8.4.22.

56. Syrup, *Hundert Jahre*, p. 351.
57. StAA, Abg., 10, 1619, Letter from the Arbeitsamt Augsburg, 16.1.19.
58. Ibid., 1621, Letter from the Arbeitsamt Augsburg, 3.7.24.
59. Ibid.
60. StAA, Abg., 10, 1622, Report by the Erwerbslosenfürsorge 17.2.23.
61. Ibid., 1621, Report by the Reichsarbeitsministerium, 28.8.24.
62. Ibid., 1622, Report by the Erwerbslosenfürsorge 17.2.23.
63. Ibid., 1636, Letter from the Augsburg town council to the Min. für. Soz. Fürsorge 5.3.21.
64. Ibid., 1622, Report by the Arbeitsvermittlung, 28.6.23.
65. *Deutsche Sozialpolitik*, p. 172.
66. StAA, Abg., 10, 1619, Letter from the Arbeitsamt, Augsburg, 16.1.19.
67. Ibid., 1621, Letter from the Arbeitsamt, Augsburg, 24.1.22.
68. Ibid.
69. Ibid., Letter from the Arbeitsamt Augsburg, 3.7.24.
70. Ibid., Report by the Stadtbauamt to the Arbeitsamt, 21.1.24.
71. Ibid., 3.7.24,
72. Ibid.
73. Ibid.
74. Cf. for a detailed exposition of the regulations, Syrup *Hundert Jahre*.
75. ASte, *Verwaltungsbericht*, p. 437.
76. The graph is based on figures from ASte, *Kommunale Mitteilungen*, 1925, p. 191.
77. ASte, *Verwaltungsbericht*, p. 437.
78. StAA, Abg., 10, 1638, 7.1.27.
79. The figure is based on figures from: StAA, Nachlass Kleindinst, 7, MS., *Erwerbslosenfürsorge*, p. 6.
80. Ibid., p. 8.
81. Ibid.
82. StAA, Abg., 10, 1637, Letter from Arbeitsamt Augsburg, 22.9.24.
83. Syrup, *Hundert Jahre*, p. 329. See also the rates given in StAA, Abg., 10, 1637, 9.2.25 and 14.12.25.
84. On the controversial calculations see Leibfried, 'Existenzminimum'.
85. StAA, Abg., 10., Report by the Arbeitsamt, 23.12.26.

4 FROM UNEMPLOYMENT INSURANCE TO COMPULSORY LABOUR: The Transformation of the Benefit System in Germany 1927–33

Heidrun Homburg

> Do you know, it's bad enough being a worker but being un-employed is terrible.[1]

I

'Then the unspeakable difficult general situation. The misery. The slide of humanity into the darkness of distress. The repulsive whipping up of political passions.'[2] This diary entry of Käthe Kollwitz from 1932 might be a summary of recent historical depictions of the social and political effects of the Depression.[3] Thus the universality of the crisis became the central issue for later historians as well as for many contemporaries. But some contemporaries, above all those concerned with social policy, social workers in the public welfare sector or welfare agencies and those involved in private welfare organisations, were not content with this general statement. They realised that the economic, social and psychological distress, the 'general misery', was not only a result of the almost natural course taken by the crisis, but also a consequence of the state control exercised as it ran its course. Their attention came to be concentrated on the connection between public distress and the management of the social and economic crisis at the end of the Weimar Republic under the leadership of Chancellors Brüning (from 30 March 1930), Papen (from 1 June 1932) and Schleicher (3 December 1932 to 30 January 1933). One after the other they used extensive emergency decrees to reduce wages and increase taxes, and so cut back the social insurance and public benefit systems in order to balance the budget at a time of rising unemployment and declining state income.

In this chapter I hope to illuminate this general context by looking at public and private welfare agencies.[4] The existence of

73

'decreed distress' — distress produced by government decrees rather than simply by the economic crisis — can be demonstrated using the example of the so-called welfare unemployed. They were deprived of their rights during the course of the Depression, and because the unemployment support system was not reformed they ended up by depending on public assistance. Their fate, repressed after Hitler's seizure of power by the slogan 'the battle for work', and since largely forgotten, forms the central focus of this discussion.[5] While concentrating on this group of unemployed, I shall, however, also touch on the effects of the policy of emergency-decree on both the unemployed and those manual and white-collar workers still in employment. After a short summary of the extent of the general distress in Germany, I shall move on to the unemployed, particularly the long-term unemployed, and the emerging problem of the 'welfare unemployed'. I shall examine the indicators of the creeping and accelerating unemployment crisis and the effects of this on the system of unemployment benefit in operation when Brüning came to power in 1930. Then the chapter will go on to look at the recruitment pattern of the welfare unemployed and their personal histories as claimants in comparison with the two other benefit groups claiming benefit: those receiving unemployment insurance and those on crisis benefit. It will examine the various proposals to solve the problem of the welfare unemployed and the way in which the government established and perpetuated the three classes of unemployment-benefit. The last section will deal with the exceptional position of the welfare unemployed both in terms of labour law and the way in which they were discriminated against and stigmatised. Finally, I shall discuss the short and long-term social and political costs of the multiple differentiation of the workforce which was one of the results of the emergency-decree policy of 1930–3.

II

Almost all employed groups (and pensioners too) were, indirectly or directly, caught up in the whirlpool of the crisis which intensified from the late summer of 1929 until the summer of 1932. At this point the crisis seemed to be coming to an end, but in December 1932–January 1933 it reached a new peak before subsiding decisively in the spring of that year. People were afflicted by many mutually reinforcing processes: more or less long-term unemployment for some; wage and salary losses for those still in employment as a consequence of short-time working; shifts taken out, along with state-ordered reduction of wage-rates; increased taxes and higher social insurance contributions as a result of government action; unemployment brought about by laws of supply and demand in the labour market at a time of surplus labour; state-administered reduction of pensions, of communal benefits for the poor, the disabled and elderly; the reduction of unemployment-related benefits, of crisis benefit and locally administered welfare benefit. Finally, as the crisis continued and the number of unemployed swelled, emergency decree after emergency decree increasingly restricted the number eligible for the miserly rates of public relief. Much damage was done to trade, and many tradespeople and businessmen went bankrupt. Between 1929 and 1933, at the height of the crisis, the economic basis of the existence of many unemployed as well as short-time workers and their families dramatically diminished until, often, even their physical subsistence was endangered.

In the winter of 1930–1 the first Winter Aid Campaign (*Winterhilfsaktion*) was launched by the League for Free Welfare Care (*Liga der freien Wohlfahrtspflege*) and its associate member organisations. They justified this campaign by reference to the 'arithmetic of misery': in January 1931 some 10 million people, or 16 per cent of the population, were dependent on meagre unemployment insurance payments or on continuous public relief. The largest group of benefit recipients consisted of the 4 million unemployed, with their 3.7 million family dependents (7.7 million). At the time of the second Winter Aid Campaign in January 1932, the number of those requiring relief had increased to 12.86 million, about one-fifth of the population. A good 5 million unemployed with 5.5 million family dependents lived on some form of unemployment benefit. These figures had changed little by March 1933.[6] In 1931–2 the crisis had become truly universal. Yet not all groups of the employed and unemployed and their dependents were affected equally. This applies, for example, to the different experiences of manual and white-collar workers. Both absolutely and proportionally, manual workers comprised the largest group of the unemployed, as Table 4.1 indicates.

Table 4.1: Registered Unemployed by Trade, 1930–3

	Total registered unemployed (in 1,000s)	Manual workers (in 1,000s)	%	White-collar workers (in 1,000s)	%	Of every 100 employees[d] Manual workers	White-collar workers
1930[a]	3,075	2,839	92.3	236	7.6		
1931[a]	4,520	4,153	91.9	367	8.1		
1933[a]	4,804	4,236	88.7	541	11.3		
16.6.33[b]	5,055	4,448	88.0	607	12.0		
16.6.33[c]	5,855	4,977	85.0	878	15.0	79.7	20.2

Sources: a. Annual averages, Stat. Jb. V.50,1931, p. 301; V.51, 1932, p. 291; V.53, 1934, p. 207.
b. Registered unemployed according to a special enquiry of the labour exchanges, Stat. Jb. V.53, 1934, p. 312.
c. Unemployed according to the findings of the occupational census, (*Die Arbeitslosenhilfe*, 2, 1935, p. 29.)
d. Occupational census of 16.6.33, (D. Petzina *et al.*, *Sozialgeschichtliches Arbeitsbuch* III, (Munich, 1978), p. 55.)

Many factors account for uneven experience of the crisis: regional variations in the economic structure; the variable susceptibility to the crisis of certain economic sectors and of middling and small businessmen; the different social security provisions for and application of labour law to particular groups of employees; and variations in secondary resources such as occupational ability, savings, house, garden and land ownership, family income and kinship relations, all of which could help in times of unemployment or loss of earnings. The fact that this worker or clerk became unemployed, and that manual workers lost more income on average in the crisis than other groups, can be explained both by 'objective' social and economic structural factors and by 'subjective' factors involved in the cushioning of the loss of accustomed work and loss of income. In each case the extent of the income loss, the duration of unemployment and the diminution of earnings all played an important role for the individual's actual experience of inequality. If one bears these points in mind then other factors come to the foreground. These can be attributed to a considerable extent to the way in which politicians sought to manage the crisis and the effects of the numerous emergency decrees issued between 1930 and 1933. We now turn to these factors as they operated in the case of unemployment policy, particularly the organisation of unemployment benefit.

III

The involuntary unemployment of a million persons in the winter months was already alarming in 1928–9, but this was still nothing very new. Comparable mass unemployment had been experienced as the German economy adapted in a series of crises during the demobilisation at the end of the war, in the final phase of the hyperinflation of 1923, in the first months after the monetary stabilisation in the winter of 1923 and early 1924, and also during the so-called rationalisation crisis from the winter of 1925–6 until the winter of 1926–7.[7] After 1929, however, unemployment reached levels that left all previous experience behind. Yet a concentration on the purely quantitative volume of unemployment also conceals the qualitative aspects of the crisis. Questions about the turnover of the unemployed and the fate of the masses of people enduring long-term unemployment moved to the centre of the debate on the labour market and welfare policy alongside the sheer quantitative dimensions of the problem. For the smaller the turnover of unemployed, the heavier the burden of benefit costs on the public organisations became. Since the great reform of unemployment relief in the Law on Labour Exchanges and Unemployment Insurance (AVAVG) of 16 July 1927, there existed two sources of relief for the unemployed: unemployment insurance (*Arbeitslosenversicherung*) and crisis benefit (*Krisenfürsorge*), both administered by the placement and unemployment insurance authorities (*Reichanstalt für Arbeitsvermittlung und Arbeitslosenversicherung*) and its subsidiary organisations, the labour exchanges.[8] The third source of benefit was the public welfare system.[9] The cost of these three welfare systems was borne by different public sources. Unemployment insurance was financed from the contributions of compulsorily insured employees and their employers in equal parts, and was supplemented by a Reich contribution. Crisis-benefit payments were borne by the state and the local authorities (four-fifths and one-fifth of the costs respectively). The third system of support, so-called welfare support, was financed solely by the local authorities in their role as bearers of the cost of general welfare for the needy. In this way the problem of support entered the realm of politics as the costs were distributed over several sources and were transferred from one to another over time. The three systems were distinguishable by the varying qualifications they demanded for welfare entitlement, the duration and level of benefit payment, as well as the respective marginalisation and stigmatisation of the unemployed recipients. They are outlined here as they stood after Brüning's first emergency decree of 26 July 1930.

Unemployment benefit insurance[10] was a compulsory form of insurance. All manual workers insured against sickness and all insured white-collar workers with an annual income of up to 8,400 RM were under obligation to be insured against unemployment. Higher-earning white-collar workers were insured voluntarily. According to the AVAVG, however, certain occupational groups were not insured. Following the amendment of 12 October 1929 this applied — to name only the most important groups — to those in long-term employment in agricultural and forestry, and those in casual employment involving only up to 24 hours work per week, or those whose income did not exceed 8 RM per week or 35 RM per month. Insurance contributions were levied as a standard percentage supplement to medical insurance contributions. Employers and employees paid half each. In 1927 the rate of contribution amounted to 3 per cent; by December 1929 it was 3½ per cent, 4½ per cent in July 1930 and it finally reached 6½ per cent in October 1930.

Insured employees had a legal claim to insurance-related benefit as long as they were willing and able to work, were involuntarily unemployed and had fulfilled the necessary qualification period. To be eligible for an initial claim a person had to have worked 52 weeks in the last two years; any subsequent claim was dependent on 26 weeks' employment. The benefit had a time limit of 26 weeks (or 39 weeks in exceptional cases of workers over the age of 40). The insured had to register as unemployed with the labour exchange themselves. The benefit was paid only after a waiting period of three, seven or fourteen days depending on the age and family circumstances of the claimant, and the rates were divided into eleven income classes. The decisive factor in the classification of benefit-rate was the average income of the claimant calculated over the last six weeks of employment. The payment was based upon the minimum benefit for single persons (the so-called main benefit) and a limited number of additional payments for dependent family members (see Table 4.2).

Crisis benefit[11] was introduced in 1927 by the AVAVG for times when the labour market was protractedly unfavourable. Its implementation by the Reich Minister of Labour was dependent on the condition of the labour market and could be restricted to certain occupations and/or areas. It was not a system of insurance benefit to which claimants had a legal right once they had fulfilled the necessary qualifications. Nor was it a public welfare payment, but rather fell midway between these two systems. Those eligible were only those insured workers who were willing and able to work, were involuntarily unemployed, and had exhausted the period of payment of insurance benefit. Those who had not fulfilled the qualifying period for an unemployment insurance claim but had worked for at least 13 weeks in the last two years in an insured occupation (so-called short-time claimants) were also eligible. Eligibility for crisis benefit rested on a means test carried out by the labour exchange. According to general guidelines, income from work, capital assets and property and income belonging to other members of the household were taken into account. The level of crisis-benefit payments corresponded to the same income classes as insurance-related benefit (see Table 4.2). Crisis benefit had a time limit of 39 weeks, but could be extended to 52 weeks in cases of hardship when the claimant was over 40 years of age.

Table 4.2: Weekly Rates of Unemployment Benefit and Crisis Benefit, from October 1927 to July 1931[12]

Wage class	Average weekly income (RM)	Minimum benefit for single persons (RM)		Maximum benefit incl. family supp. (RM)	
		unempl. b.	crisis b.	unempl. b.	crisis b.
I	up to 10	6.00	6.00	6.40	6.40
II	10–14	7.80	7.80	9.60	9.60
III	14–18	8.80	8.80	12.00	12.00
IV	18–24	9.87	9.87	15.12	15.12
V	24–30	10.80	10.80	17.55	17.55
VI	30–36	13.20	13.20	21.45	21.45
VII	36–42	14.63	13.20	24.38	21.45
VIII	42–48	15.75	14.63	27.00	24.38
IX	48–54	17.85	14.63	30.60	24.38
X	54–60	19.95	15.75	34.20	27.00
XI	60	22.05	15.75	37.80	27.00

Welfare support[13] was derived from the general public welfare system. The Reich decree on compulsory welfare of 13 February 1924 compelled the local authorities — or to be more precise, the urban and rural welfare agencies — to give full support to 'the needy'. The decree defined 'the needy' as 'whoever cannot provide the necessities of life for himself and his dependents or cannot meet these by his own strength and means and moreover is unable to obtain them from another source, particularly from relatives'.[14] The degree of need was assessed by the welfare department of the local authority, which took into account income from work and property of the claimant, the dependent family members (parents, children, wife) and other members of the household. The 'special character of the situation' was taken into account before payment was made. It was supposed to 'effectively counteract destitution and prevent it becoming permanent'. 'Wherever possible' it was to put the destitute in a position to 'provide the necessities of life for themselves and their dependents.'[15] It had to provide just the 'bare essentials'.[16] It would be paid as long as the recipient was destitute. Claimants of this form of benefit were bound to repay it.[17]

IV

Different groups had various experiences of unemployment depending on the origin, type, duration and rate of benefit received. The recipients of insurance-related benefit were the most homogeneous group in terms of previous occupation and unemployment. Before they registered as unemployed they had to have spent at least 52 weeks in the previous two years in gainful employment. They had not to have been unemployed for more than 26 continuous weeks (or 39 weeks in the case of those over 40) either continuously or with brief periods of intermittent employment. It is more difficult to gain a clear picture of those in receipt of crisis benefits, for they had diverse backgrounds of employment and unemployment. The group included those who had run through their unemployment benefit and had been transferred to crisis benefit, as well as those who had found a job after their unemployment benefit expired but then were dismissed before they had become eligible again for unemployment benefit. Some were employees who had found an insured job but had become unemployed again after 13 weeks. Those in the worst situation were manual and white- collar workers in reciept of crisis benefit who had been continuously unemployed for 1¼ years (1¾ years for the over 40s). Those caught in a benefit cycle (unemployment insurance benefits, crisis benefits, welfare support, etc.) broken by short periods of employment were hardly in a better position. The most heterogeneous group comprised those receiving welfare support, the so-called welfare unemployed. Associated with this label was the idea that they were the 'dregs of the unemployed' or that they were 'work-shy', 'lazy', 'shirkers', 'benefit swindlers', 'asocial' etc. (expressions used by the Nationalist Socialists to promote the concept of 'parasites'). In this situation the welfare unemployed were subject to coercive measures.[18]

These old prejudices had to be revised under pressure of events. In June 1929 towns with over 50,000 inhabitants had a total of 103,000 welfare unemployed. By September of the same year the figure was roughly 133,200; it had risen to some 200,000 by December. In towns with over 25,000 inhabitants the number of people in this category was about 222,500 at the end of December 1929 but swelled to 404,000 in July and 445,000 in August 1930.[19] In view of these figures those claiming welfare support could hardly have consisted entirely of the old and infirm or marginal workers. The great mass of welfare unemployed were able and willing to work, but required support owing to the permanent objective shortage of job opportunities and the mechanics of the benefit system. This view is supported by the age composition of the welfare unemployed, their occupational background and their background as unemployed workers or recipients of benefit.

Workers over 40 could only be found jobs with difficulty, whilst the placement of those over 45 was almost impossible. The welfare unemployed were, however, drawn from all age groups. The over 45s were not grossly over-represented in comparison to their numbers in the other two groups of unemployed. In January 1929, 22.9 per cent of those claiming insurance-related benefits were over 45, and 33.3 per cent of those on crisis benefits. Of a total of 154,899 welfare unemployed who were receiving support in the 90 reporting towns with over 50,000 inhabitants on 15 February 1929, about 35 per cent were between the ages of 45 and 65.[20] Thus, age (being over 45) or personal factors alone hardly explain the swelling number of welfare unemployed. This is also suggested by the distribution of claimants according to their reason for claiming. The picture emerging from an analysis of the welfare unemployed in the 90 towns with over 50,000 inhabitants, and the situation in one particular city, Frankfurt, is shown in Table 4.3.

Table 4.3: The Recruitment of Welfare Unemployed, 1929–30[21]

Key date	Total welfare unemployed	Not insured	%	Insured, but not yet eligible	%	Disqualified from unempl. b.	%	crisis b.	%	Others %
15 Feb. 29										
90 towns	154,899		9.1		31.8		33.4		13.5	11.7
Frankfurt	3,165	189	2.8	1,148	36.3	1,160	36.7	768	24.2	
5 Feb. 30										
Frankfurt	6,750	674	10.0	1,994	29.5	2,161	32.1	1,921	28.4	

The longest uninterrupted period of unemployment for at least nine months had been experienced by those who now received welfare support after exhausting crisis benefit. By 1930 the proportion of the disqualified amongst this group had increased sharply both absolutely and in relative terms compared with the previous year. However, the percentage of over-45s amongst the Frankfurt welfare unemployed decreased from 33 per cent in February 1929 to 27.5 per cent in February 1930. This reduction in the average age of the welfare unemployed together with an absolute increase in the number of permanent claimants shows the widely held assumption that the welfare unemployed were 'downright lazy and shirkers' to be untenable.[22] The reverse was the case.

On the other hand as documented in both surveys, not all the welfare unemployed can be categorised as 'long-term un-employed'. The welfare unemployed were recruited in the first place from former unemployment insurance benefit claimants; in the second place from insured employees who had not fulfilled the qualification period; thirdly from former crisis-benefit claimants, and finally from the heterogeneous group of uninsured workers. This pattern of recruitment and the occupational composition of the welfare unemployed reveals that it was the legal definition of eligibility for unemployment insurance or crisis benefits, and of the duration of payment which drove them to the welfare office.

Surveys of the occupational composition of the welfare un-employed had been carried out since 1929 (see Table 4.4). The results, showing a significant proportion of unskilled workers, are hardly surprising. Unskilled work was precarious, and during the crisis such workers were either dismissed or replaced by better-

Table 4.4: The Occupational Composition of the Welfare Unemployed, 1929–32[23]

Date	15 Feb. 1929	15 Feb. 1929	15 Feb. 1930	23 Dec. 1931	1 Feb. 1931	1 Nov. 1931	31 Dec. 1930	28 Jan. 1931	31 Dec. 1931	28 Jan. 1932	15 Feb. 1929	15 Mar. 1932
	90 Towns	Frankfurt	Frankfurt	Berlin-Kreuzberg	Berlin	Cologne	Reich	Reich	Reich	Reich	Reich	Reich
Total welfare unemployed	154,899	3,165		4,480	117,796	2,421						
Occupational groups (%)												
Assorted wage-earners	over 33	15.8	19.9	48.3	45.3	21.3	41.3	39.0	27.3	10.2	18.7	29.4
Transport	5.7	23.5	23.7	7.5	3.0	1.4			4.6	0.9	5.6	5.5
Domestic service				2.0	10.8	4.3					3.9	
Metalworkers	5.6	3.7	9.0	5.5	7.6	31.5	11.1	11.9	14.2	30.2	15.1	17.0
Woodworkers and carpenters				1.3		8.5			6.7	11.2	5.3	5.2
Building	1.6		3.7	1.3	6.1	2.2	10.2	10.1	10.9	0.5	9.6	14.6
Clothing trades				3.7	5.5	5.6			6.1	1.0	4.6	3.6
Mining, quarrying and excavation	5.1										8.1	10.6
Shop and office workers	8.7	15.7	14.7	12.8	11.0	11.1			4.7	21.9	8.5	7.1

qualified workers. Those in 'minor employment' were exempt from insurance payments, while only a few groups of insured but unskilled workers were eligible for crisis benefit. Thus it was inevitable that they would end up in the ranks of the welfare unemployed. The drama unleashed by the length and severity of the crisis was reflected in two other findings: the relative proportion of casual workers amongst the welfare unemployed contracted, and the percentage of skilled workers, i.e. metal and white collar workers, increased.

For the first time, all local authorities were ordered by the Reich Minister of Labour to produce 'statistics of the welfare unemployed' for the end of August 1930. They were instructed to apply a narrow and precise definition of the concept 'welfare unemployed' as meaning those locally supported claimants whose destitution was exclusively caused by an objective absence of employment, that is, mainly wage-earning workers who were capable of work and had shown their willingess to work by submitting themselves to regular supervision by the labour exchange (see Table 4.4). But the welfare-unemployment figures were not suddenly reduced. From August 1930 to July 1932 the numbers increased from about 453,000 to 2,290,000; after a slight retreat in August 1932 (2,030,000) the increase continued: September 1932 2,046,000, October 1932 2,204,000 until December 1932 when there were 2,407,000 claiming welfare benefit.[24] The official statistics reveal the extent of the participation of the local authorities in unemployment relief as does Table 4.5. In March 1932 welfare-unemployment support, for which the local authorities were responsible, became the most important form of support, taking the place of crisis benefit. During the crisis it became the most important and central benefit system for the unemployed. This reversal provoked only partially successful protests against central welfare policy on the part of the local authorities, and finally led to the transference of the financial burden onto the welfare unemployed themselves. Although the group of 'recognised welfare unemployed' was ever more narrowly defined, its numbers continued to grow. After the welfare-benefit decree of 14 June 1932, effective from 31 August of that year, the welfare unemployed were defined as:

> employees who are willing and able to work and are involuntarily unemployed, who are not over 60 years of age and

Table 4.5: Registered Unemployed by Form of Support (Quarterly Figures and Yearly Averages), 1930–3

	Total unemployed (1000s)	Those on unemployment insurance (1000s)	%	Crisis benefits (1000s)	%	Welfare support (1000s)	%	No support (1000s)	%
1930									
Aug.	2,882	1,507	52.3	441	15.3	453	15.7	482	16.7
Sept.	3,004	1,493	49.7	472	15.7	507	16.9	497	16.5
Dec.	4,384	2,166	49.4	667	15.2	761	17.4	790	18.0
1931									
Mar.	4,744	2,318	48.8	923	19.4	940	19.8	563	11.9
June	3,954	1,412	35.7	941	23.8	1,017	25.7	583	14.7
Sept.	4,355	1,345	30.9	1,139	26.1	1,208	27.7	663	15.2
Dec.	5,668	1,642	29.0	1,506	26.6	1,565	27.6	955	16.8
Annual average	4,520	1,713	37.9	1,045	23.1	1,082	23.9	679	15.0
1932									
Mar.	6,034	1,579	26.2	1,744	28.9	1,944	32.2	766	12.7
June	5,476	940	17.2	1,544	28.2	2,163	39.5	827	15.1
Sept.	5,103	618	12.1	1,231	24.1	2,046	40.1	1,206	23.6
Dec.	5,773	792	13.7	1,281[b]	22.2	2,407	41.7	1,293	22.4
Annual average	5,603	1,086	19.4	1,449	25.9	2,047	36.5	1,019	18.2
1933									
Mar.	5,599	686	12.2	1,479	26.4	2,299	41.1	1,134	20.2
June	4,857	416	8.6	1,310	27.0	1,958	40.3	1,172	24.1
Sept.[a]	3,849	316	8.2	1,109	28.8	1,492	38.8	932	24.2
Dec.	4,059	553	13.6	1,175	28.9	1,411	34.8	918	22.6
Annual average	4,804	531	11.0	1,281	26.7	1,892	39.4	1,101	22.9

a. From 31 July 1933 excludes those employed in labour service.
b. From 28 Nov. 1932 — lifting of time-limit on crisis benefits.

Sources: *RABl* 1932 II, p. 15; *SJB*, vol. 51, 1932, p. 418; vol. 52, 1933, p. 297; vol. 53, 1934, p. 313.

are under the direct control of the labour exchange in their search for work in as much as they receive regular benefit from public welfare funds and such benefit is not petty in comparison with the standard rate of general welfare.[25]

V

Age composition, background, recruitment pattern and finally official registration criteria show the vast majority of the welfare unemployed to have been willing and able to work but involuntarily unemployed. In this way, the dividing line between the systems of unemployment relief and welfare benefit was systematically undermined, although the original intention had been to keep the two separate when crisis benefits had been introduced in 1927. The introduction of this new form of benefit had been a recognition that unemployment insurance was no longer adequate at a time of severe economic depression accompanied by high rates of unemployment, and that 'the unemployed are individuals who can be found a job and who therefore have to stay in touch at all times with the labour exchange, while public welfare support is an institution completely unconnected with this system'.

The introduction of crisis benefit was also deemed necessary because those laid off on account of the slump might rightly regard the 'stringent conditions of welfare relief' (the assessment of the entire assets of the needy and the obligation to repay) 'undeservedly harsh'. A further point in favour of crisis benefits was the fact that the local authority welfare had been stretched beyond its limits by the support of very large numbers of unemployed on welfare support.[26]

In view of such intentions it was inevitable that the crisis benefit system would become the object of a great deal of criticism. From June 1930 the attacks against the 'disastrous' reorganisation of crisis benefit multiplied on the grounds that it had led to a 'landslide' in the number of welfare unemployed and the redistribution of the financial burden in favour of the Reich to the disadvantage of the local authorities.[27] 'In view of the improvement' of the labour market in February/May, 1929, the Reich Minister of Labour in June 1929 decided to restrict the number eligible for crisis benefit and to decrease its duration. The time-limit for crisis benefit (39 weeks, or 52 weeks for those over 40) was raised again, and the extension of crisis benefit to all occupations (including semi- and unskilled factory workers) introduced in February 1929 was rescinded. Basically, from 7 July 1929 onwards, only skilled workers from particular branches of industry and white-collar workers were eligible for crisis benefit. Moreover, the decree of June 1929 excluded all persons under 21 from receipt of these benefits.[28]

The decree established an arbitrary and intolerable threefold division of the unemployed. Many plans and proposals to get rid of 'weaknesses in the system' which pushed those willing and able to work onto local authority welfare and thus created the problem of the welfare unemployed, were addressed to the government.[29] Until the autumn of 1930 they aimed to introduce a clear and systematic division between crisis and welfare benefit. On the one hand, the aim was to organise a unitary system of unemployment insurance and crisis benefits for all those able to work, under the aegis of the Reich and the agencies in charge of job-placement and unemployment insurance. On the other hand, the local welfare authorities were again to be restricted to the care of those no longer able to work or only partly fit to work, as the chances of finding them jobs receded as the pressure on the labour market became more and more intense. In order to carry out the reorganisation quickly the government was urged to extend back-dated crisis benefit to all occupations for the duration of the crisis in the labour market. This urgent reorganisation was not carried out, however. In place of a comprehensive reform the presidential cabinets of the Weimar Republic introduced only 'piecemeal reform'.[30] The first new regulation, and the one with the most serious consequences, was introduced by Brüning's cabinet in October 1930.[31] From the beginning of November the limitation of crisis benefit to certain occupational groups in areas with over 10,000 inhabitants was lifted. On the other hand, serious restrictions were now introduced. As well as the young un-employed (up to the age of 21), agricultural workers and domestic personnel were now generally excluded from crisis benefit. Moreover, it was limited to those whose unemployment insurance benefits had expired. The longest period for which one could receive crisis benefit was reduced from 39 to 32 weeks (52 to 45 in the case of the over 40s.) These restrictions on eligibility for crisis benefit remained in force until the end of the Weimar Republic.

The new regulations of October 1930 appertaining to crisis benefit meant that the longest period of benefit payment from unemployment and crisis benefits combined, amounted to 58 weeks (71 weeks for those over 40 years of age). With effect from 28 November 1932 the limitation of crisis benefit and thereby the disqualification of the unemployed from crisis benefit was abandoned. This inadequate extension and simultaneous truncation of crisis benefit meant that millions of unemployed persons were 'deported' into local authority welfare benefit. So-called short- time claimants were the most affected. Owing to their temporary and unsettled working conditions many of these were unqualified or partly skilled labourers, transport and factory workers, for whom exclusion from crisis benefit meant automatic referral to the local welfare office. The 'long-term unemployed' were also affected after expiry of their unemployment insurance or crisis benefit and they had to resort to the welfare office. Various financial and political considerations of a structural nature (not least the subordination of economic and social policy to foreign policy aims connected with the Allies' extraction of reparation payments) prevented Brüning's cabinet and the subsequent cabinets of Papen and Schleicher from 1930 onwards from recognising the urgent need for a basic reorganisation of crisis benefit and a systematic division between crisis and welfare benefit. Their 'hesitation'[32] was also influenced by the expectation that public bodies would profit from the admittedly inadequate extension of crisis benefit. On the one hand, the pool of recruits was reduced and with it the financial 'risk' of crisis benefit. On the other, it was calculated that the total volume of benefit payments would thereby be reduced. For experience showed that not all the unemployed who were excluded from unemployment and crisis benefits ended up at the welfare office.

These expectations were realised, at least in part. From the autumn of 1930 the main focus of public discussion about the reorganisation of crisis benefit shifted towards arguments about costs and savings, while earlier socio-political arguments in favour of an extension of crisis benefit were increasingly pushed onto the sidelines. Crisis benefit itself now came under attack as being relatively more expensive than the benefits paid by the local welfare authorities. The pressure of the economic crisis, the financial consequences of the government decrees of July to October 1930, and also the latent prejudice against the unemployed casting doubt upon their desire to work, opened the way to a solution at the expense of the unemployed themselves. Welfare associations, the most influential being the German Association for Public and Private Welfare (*Deutscher Verein für öffentliche und private Fürsorge*), as well as the heads of municipal welfare departments and leading community organisations, now sharply attacked the idea of crisis benefit as a generalised system of mass welfare and its hybrid position between insurance and welfare. They demanded that crisis benefits be brought into line with the basic principles and methods of the locally administered welfare in order to put a stop to the intolerable 'waste' of public money. In place of the allocation of benefit-levels according to previous income in line with the income-levels used for unemployment insurance assessment, they advocated the principle of individual welfare, assessment of benefit according to need, and the taking into consideration of specific individual difficulties and dependent family members.

In addition, the German Association for Public and Private Welfare, together with other interest groups, launched even more far-reaching reform proposals, bringing both the 1927 unemployment insurance law *and* work provision into question. Further pressure was put on the government to restore the organisational and social uniformity of the labour market in terms of the provision of work and unemployment benefit. Admittedly, the only unanimity lay in the basic demand for the rejection of the AVAVG and the abolition of unemployment insurance. As far as the rearrangement of unemployment benefit and provision of work was concerned, the chorus of criticism split up into single voices. The proposals ranged from the organisation of welfare benefit under the control of the centralised administration of the labour market, to the total abolition of the 1927 laws and the structure created by the AVAVG. It was argued that such far-reaching changes were justified on the grounds of necessary and possible economies, and of the failure of the labour exchanges to place people in work, to handle the problem of unemployment benefit and to 'weed out' the recipients. The most far-reaching reform model advocated the introduction of a single unemployment benefit scheme of the old style, based on the Poor Law and the provision of welfare by the local authorities. This would mean a return to a decentralised system with responsibility for work provision and unemployment benefit placed back in the hands of community welfare offices. These petitions were more successful than demands for the systematic extension of crisis benefit. In one decree after another from the autumn of 1930 onwards the eligibility qualifications for crisis benefit were brought into line with welfare provision, so that by June 1932 the two systems were virtually identical.

Parallel to these developments radical alterations were made in the unemployment insurance scheme. The founding principles of 1927 — the reciprocity of contributions and payments and the right of insured persons to claim after paying contributions for a prescribed period — were replaced by the welfare principle for several insured groups (wives, young people). The Papen decree of June 1932 extended these measures to all other insured persons.[33] Within the total time period for insured 'care' of 20 weeks, those who had fulfilled the qualifying period were entitled to claim benefit only for the first 6 weeks. Whilst at the end of this period the qualification for further payments was at the discretion of the local authority, the eventual payment was the responsibility of the labour exchanges. The rates were different from previous levels of unemployment and crisis benefits, varying according to local scales and instead of 11 there were now only 5 classes. The emergency decree of 5 June 1931 had already reduced the unemployment and crisis benefit payments by an average of 10–11 per cent. Papen's cabinet introduced further reductions of approximately 23 per cent for unemployment benefits and 10 per cent in crisis benefits. These measures were supplemented by a simultaneous decline in the welfare benefit-levels of a further 15 per cent, having already been reduced by 12 per cent in an emergency decree of June 1931.[34] The Papen decrees blew 'an ice cold wind of social-reactionary tendencies' so that even the normally restrained Catholic welfare association *Caritas* attacked them severely.[35]

These measures damaged the foundations of the AVAVG. Brüning and Papen used their dictatorial powers for radical interventions in the AVAVG, although the facade remained in existence. The tripartite system of benefits that had emerged during the economic crisis of 1929–30, continued to operate. The unemployed were distributed amongst three benefit schemes; two groups of 'labour exchange unemployed', the recipients of insurance-related and crisis benefit, were faced by those supported by local benefit, the welfare unemployed. The Reich government justified the threefold division and the arbitrary separation of crisis and welfare unemployed by reference to the costly crisis benefit system and the necessary discharge of the national debt. The decree of October 1930 had already initiated a levelling-down of the crisis-benefit rates and an assimilation of crisis and welfare-support levels. As a result of subsequent decrees, crisis-benefit payments for single persons and those unemployed with dependent family members fell behind the general welfare rate. In October 1930 the average monthly payment for a single person in income class IV receiving crisis benefit was 39.48 RM. On welfare support it would amount to 40 RM. After the June 1931 decree the average crisis-benefit payment was about 50.50 RM and 50 RM on welfare support. In the last quarter of 1932 the figures were approximately 45 and 44.50 RM respectively.[36] In mid-1931 the Reich government was forced to grant financial aid to the local authorities, who had to fund the ever increasing burden of payments of crisis benefit. The relief of the Reich budget by means of making savings on crisis benefits was countered by additional expenditure in excess of 230 million RM, the amount provided for local authority welfare support in the financial year 1931. In the following financial year (April 1932–March 1933) this state aid increased to 672 million RM and in the following 6 months 425 million RM were accounted for in this way.[37]

The alignment of the two benefit systems in no way alleviated the suffering of the welfare unemployed. They were and remained the only group of permanently unemployed on long-term support who were obliged to repay, often in a humiliating way, the benefits they had received after the government had been forced in October 1931 to revoke the compulsory payment of crisis benefit that had been introduced in June 1931.[38] However, the welfare authorities were ordered not to collect the repayments immediately after the end of unemployment, but were instructed to take account of the after-effects of long-term unemployment.[39] It was not until 22 December 1936 that a 'Law regarding the exemption from compulsory repayment of welfare payments' released the welfare unemployed from repayment of all benefit which they had received before 1 January 1935. With this act, accompanied by corresponding propaganda, the National Socialists put an end to at least one element of the legal segregation of the welfare unemployed.[40]

Apart from having to make repayments, the welfare unemployed were also segregated from the rest of the unemployed in socio-legal terms. The unemployment insurance scheme paid sickness-insurance contributions and the charges for the maintenance of pension rights on behalf of the 'labour exchange unemployed' for the duration of their benefit claim. No corresponding measures applied to the welfare unemployed. Their membership of the health insurance scheme expired as soon as they became unemployed or at the end of their period of receipt of unemployment and crisis benefits. Different systems of sickness care operated in different areas. The welfare unemployed had to accept responsibility for maintenance of pension rights themselves. Even worse was the fact that the welfare unemployed had no legal rights concerning the rate and type of benefit payment, and were dependent on the welfare practice of the individual authority which in turn was influenced by local political alignments.

The welfare unemployed's lack of basic rights was also founded in the fact that the welfare authorities could support those able to work 'in suitable cases by the assignment of appropriate work of a generally useful kind' and they could make 'payments dependent on such work'. From 1924 the welfare unemployed were the only unemployed group for whom compulsory labour was firmly established.[41] The practical operation of the system was subject to local and regional variations. These were present on the one hand in the extent of local authority measures, and on the other in the legal status of those employed by these measures (i.e. whether and to what degree the welfare authorities made efforts to establish a civil contract of employment for them or whether they simply ordered them to work in a 'unilateral administrative act'.[42]) However, both forms of local labour welfare were instituted 'on the basis of a special power relationship' which was justified by reference to the obligations of the welfare authorities to those in their care.[43] If the person in question refused to carry out the assigned work the welfare office could suspend benefit for a time.[44] If he 'persistently' refused work, if he disobeyed the 'legitimate instructions of the competent authority' or demonstrated 'unwillingness to work or obviously uneconomical behaviour', then the welfare agency could restict his benefit payments to the 'bare essentials of life'.[45] Finally, those needing benefit could be accommodated in a 'recognised institution or other work establishment'.[46]

This system of 'welfare through work' can be seen as the legally sanctioned imposition of forced labour imposed by the authorities.

In this way, the welfare unemployed were threatened with labels such as 'work-shy' or 'dregs of the unemployed', and this only contributed to their isolation and remoteness from the labour market. So their willingness and ability to work, the involuntary nature of their unemployment and the undeserved character of their distress were always open to question as long as the welfare unemployed had not undertaken the allotted work according to the stipulations of the welfare authorities. It was not least for this reason that this form of support, which derived ultimately from the old Poor Law, was so controversial. It was only just kept in being as a result of the welfare regulations of 1923–4, which were passed with local authority backing but in the face of considerable SPD and trade union opposition.

As early as 1927 the 40th German Welfare Conference in Hamburg had dealt with the question of 'support through provision of labour (instead of cash benefits)'. This was 'warmly recommended' as a real step forward in the provision of suitable welfare, particularly for those among the destitute who were able to work.[47] Some time elapsed before this recommendation was implemented. With the increasing number of welfare unemployed the discussion about local benefit practice flared up again in 1929–30.

The slogan 'work not welfare'[48] was the concept which underlay the conversion process from financial support to 'support through labour', although this meant different things according to political affiliations and attitudes towards welfare. For the welfare inspector of Lübeck it was simply a matter of judgement and the 'healthy feeling' that anyone would be reluctant to be a 'parasite of the state'.[49] Others, like the mayor of Eckernförde, pointed to 'the terrible moral damage inflicted on individuals and on the nation by the benefit system with no repayments'.[50] Professor Wilhelm Polligkeit, the President of the German Association for Public and Private Welfare, distanced himself from the assumption that the unemployed were simply work-shy. He stressed instead the need for the welfare unemployed to be given work to overcome the oppressive feelings which derived from the hopelessness of finding jobs for themselves.[51] A representative of the Reich League of Municipalities, the main representative of the middling and small towns, emphasised the 'creation of work' as being the most urgent and important measure 'in order that the welfare unemployed can be fed back into the work process' in order to protect them from political radicalism and to test and maintain their enthusiasm for work.[52] The rural administrator in Sprottau hoped for a demonstrative effect on the horde of critics who accused public welfare policy of 'promoting idleness' and 'pampering shirkers' at the expense of the honest working tax payers. 'Compulsory labour', he said, 'has a by no means contemptible, favourable side effect in that it will combat the eternal philistine talk about the idleness of those in receipt of benefit'.[53]

The reasons for the change in public welfare policy from financial support to the allocation of work were diverse and full of contradictions. Motives varied from a fundamental criticism of the Weimar welfare state to the defence of the state's duty to provide welfare for its citizens using the very methods favoured by its critics. They arose from both socio-political and national-economic arguments to the effect that one should preserve 'human capital' in an economic crisis, or that one needed to control the individual and collective psychological processes engendered by the personal and collective experience of long-term unemployment. It was felt that these threatened in the short term to undermine the political system, and in the long run to destroy the economic system as well. Such varied motives were reflected in the various measures undertaken by individual local authorities to implement the new idea of 'work instead of welfare'; and thus in the varied experience of the welfare unemployed in receiving support. Labour welfare measures of this kind had already been adopted by the welfare authorities in the cities by 1928–9. These aimed in a more or less concealed way at placing the welfare unemployed in a situation where social insurance was obligatory for long enough for them to be eligible once more for unemployment insurance or crisis benefits. These procedures presupposed three things. First, the welfare unemployed had to be assigned to 'work that was useful to the community' which, secondly, in terms of type, amount and methods of work undertaken, corresponded to normal economic activity. Thirdly, when benefit ceased, paid work had to take its place and be founded on a private contractual work relationship with all its implications instead of a public one. A public contractual relationship was always present in cases where the support was made dependent on work that did not comply with these conditions.[54]

The employment of the welfare unemployed as 'welfare workers' by the local authorities produced many problems. For one thing, it founded the basis of continual conflict between the local authorities and central government. It also provoked protests from the trade unions and the welfare unemployed themselves, who found short-term work through these schemes. The unions argued that the local authorities were able through these schemes to open a large sector of the local labour market for only temporary employment, thus artificially reducing the supply of permanent jobs and thus producing a separate 'economy of the unemployed'.[55] The welfare unemployed went to the labour courts in an attempt to force the local authorities to offer them regular, permanent employment contracts. These complaints failed, mainly because of the restrictive interpretation of the relationship between welfare authorities and claimants as primarily constituting a power relationship. As far as the trade unions were concerned, their criticisms were progressively undermined by the continued growth of the unemployment figures and the gradual perforation of the employment contract laws by the emergency decrees of Brüning and Papen. The Reich administration was more successful than the trade unions or the unemployed in pushing through its interests against the local authorities. The Reich unemployment insurance office and the labour exchanges accused the local authorities of not being interested in what the workers actually did, but rather of creating a situation in which their 'welfare workers' were once again eligible for insurance-related or crisis benefit. They undertook a number of legislative steps once more to protect themselves against the attempted 'shedding' of the burden by local authorities, and against a renewed influx of claimants from the ranks of the welfare unemployed.[56]

'Labour welfare for the welfare unemployed' was, as the senior municipal councillor of Berlin, Kobrak, speaking about its administration in his own city admitted, an emergency measure. It had been introduced there in the autumn of 1928 in the recognition that there were few other possibilites at their disposal in terms of creation of work for the unemployed (whether jobs specially reserved in the granting of local authority contracts, direction of labour into value-creating emergency work or retraining schemes). These measures were not capable of being expanded indefinitely, and their practical effects became less significant the more the numbers of welfare unemployed increased. 'Compulsory labour' and 'employment on a more or less totally contractual basis in so-called additional labour schemes carrying insurance obligations' were the only remaining possibility. Certainly, as Kobrak and his colleagues in Berlin saw it, compulsory labour was a suitable method of testing willingness to work and 'has proved its worth in the experiences of many towns'. At the same time, however, the Berlin administration rejected making the support of the welfare unemployed dependent on compulsory labour. Compulsory workers were

> subjected to the welfare agency by compulsion and are not equal in the work process in the sense that an employer and employee are. For the maintenance and strengthening of the ability to work, even among fully capable labourers, compulsory labour is not always the right means of labour welfare, because the aim is not simply to test the will to work but also to at least an equal extent to have an effect on the mental attitude of the welfare unemployed.

The emphasis on labour welfare for the welfare unemployed should be placed on the 'provision of supplementary, regular employment with an insurance obligation for an agreed wage rate' — so-called 'welfare labour' — a system within which the occupied welfare unemployed would be 'treated as equals with employees in the free economic market'.[57]

The procurement of suitable jobs was 'relatively easy'.[58] The jobs that were considered were those which afforded only small material costs and where the wage costs were the main source of expenditure. The jobs had to be 'of benefit to the public' and be founded on an employment relationship with insurance obligations. The criteria for 'benefit to the public' were defined by a legal judgement of the Reich Labour Court. They were fulfilled 'if the work is useful to everyone and not indivudal persons or private economic enterprises; or if it is carried out to the general benefit of the public or if it saves public money'.[59] Examples given by Kobrak included the cleaning of parks, playgrounds and cemeteries, reconstruction and cultivation work in public places, statistical investigations, or cataloguing and compiling bibliographies in schools and offices.[60]

But local job-creation efforts were inadequate in view of the ever increasing total of welfare unemployed. Only a small percentage of the welfare unemployed were assigned such work.

In Berlin, for example, in December 1929, some 44,450 unemployed persons on permanent benefit were allocated to the welfare offices in December 1929. By December 1930 this figure had more than doubled (116,150) and by December 1931 it had increased fivefold (220,940). By December 1932 the figure had reached 339,150. The provision of welfare work lagged far behind this rapid increase. In December 1929 a mere 7,617 were employed on 'welfare-labour' schemes; in December 1930 the figure had fallen to 6,500, in December 1931 it had increased but only to 8,536, and by December 1932 it was still no more than 8,897. In the summer months the figures were not substantially higher despite the availability of outdoor work.[61] In October 1930 the total number of welfare unemployed in Germany was 726,242 persons (according to the statistics produced by the district welfare associations). Amongst these only 42,534 or 5.8 per cent were employed as welfare workers. There were 1,030,000 classified as welfare unemployed in March 1931, and 49,542 or 4.8 per cent of these were welfare workers. In July 1931 the proportion reached a highpoint of 8.95 per cent (1,180,000 welfare unemployed), but it declined in subsequent months, managing only 3.6 per cent in February 1932 when a total of 2,039,000 welfare unemployed were registered. The high point of 1932 was reached in the summer but it was less than half that of 1931. In July 1932 only 4.2 per cent of the 2,461,000 welfare unemployed were in 'welfare employment'. The proportion continued to decline: in August–November 4 per cent, and in December 1932 only 3 per cent of 2,407,000 welfare unemployed were on 'welfare work'.[62]

Not all towns, and certainly not all welfare agencies in rural areas, concentrated on the provision of welfare work. It is fair to suppose that in most rural districts labour welfare was administered according to the 'compulsory labour' scheme as the example of one Pomeranian district shows. The regulations of September 1930 regarding 'employment of persons receiving welfare support in compulsory labour schemes' obliged all persons (other than those two-thirds disabled) 'whose public welfare benefit is borne by district D to carry out work that was of service to the public (compulsory labour) in their town of residence or in a neighbouring place'. Benefit was paid while the work was carried out and the rate was dependent on the duration of the work.[63] This extreme variant of welfare schemes was rare in urban districts. In the towns, however, the welfare unemployed were assigned to a mixture of the two systems of labour, and the scheme was used as a means of sifting out and disciplining. Benefit was made dependent on the performance of suitable 'useful work'; then all the welfare unemployed were assigned to compulsory labour, and thirdly, only those welfare unemployed who had proved themselves to be trustworthy were assigned to the welfare-labour schemes.[64] There was no lack of massive and clearly articulated protests by the welfare unemployed against these fixed 'wage' rates and the kind of work allocated. Less common, on the other hand, were cases of collective refusal to work, the absolute rejection of compulsory labour as 'unreasonable' or strikes for higher wages or longer hours to supplement income. Few protests were successful. The welfare agencies reacted to strikes and refusals to work by withdrawing benefit, transferring welfare workers to compulsory-labour schemes or at best with a more careful selection of compulsory labourers from the ranks of the 'work-shy' claimants. Broad sections of the public applauded these measures and were more outraged by the conduct of the welfare unemployed than by the sanctions imposed by the authorities.[65] The judgments of the industrial courts in cases concerning the welfare unemployed highlighted the increasing deprivation of rights of this group under the public welfare system.

Complaints were made to the labour courts about the extensive improper and arbitrary handling of compulsory labour by the local welfare representatives in areas of road construction and repair and office work. But such suits foundered on the wide interpretation of the welfare relationship adopted by the district and Reich labour courts. For instance, in December 1929 the district court of Hamburg dismissed the appeal by welfare unemployed in Lübeck who complained about the appropriate wages for the work done. The statement of the court read as follows:

> There can be no question about the exploitation of those receiving benefit when a state, in order to reduce the expenditure on welfare a little and above all in order to maintain the habit of and willingness to work on the part of the benefit recipients, instead of paying less benefit to the unemployed pays a higher rate for work done, although, on the other hand, this higher benefit rate is not as high as the wages of free workers.[66]

This opinion, which implied that the welfare unemployed were 'work-shy' and regarded long-term mass unemployment as a result of lack of motivation, was echoed in the judgments of the Reich labour court. The protection offered by industrial law was gradually, one might almost say systematically worn down, removing the last obstacle that remained in the way of compulsory state direction of the unemployed, namely the legal rights of employees that had hitherto formed the basis of the distinction between welfare work and compulsory labour. A judgment of the Reich Labour Court of 12 October 1929 which broke with previous jurisdiction, was fundamental in this process.[67] From March 1928 to February 1929 a certain local authority had allocated the painting of two orphanages for eight hours per day as compulsory labour. The regional industrial tribunal decided that this work was not for the common good and as such it could not be recognised as compulsory labour. But the Reich Labour Court in questioning the verdict, prounced that it should be assumed *prima facie* that the employment of claimants constituted a relationship grounded in *public* law. Only if special circumstances could be demonstrated, showing the existence of the 'free will of those concerned', could the relationship be regarded as valid in *private* law, thus enabling the courts 'to accept that a contract of employment is in existence'. The Reich labour court implied therefore that the engagement of the unemployed on work schemes did not involve the exercise of free will. It added that the decision as to whether a job was classified as welfare or compulsory labour lay with the local authority.[68] On the question of the general usefulness of the work the tribunal ruled 'in cases when the work saves the town money . . . it cannot be deduced from that alone that the work serves no general purpose'. 'All jobs of any kind . . . which are on balance useful will always be economically worthwhile and will always somehow and somewhere, save money'. Finally, the most crucial consideration must be 'whether the job was of use to the general public'.[69] With this judgment the Reich Labour Court smoothed the way for the removal of the welfare unemployed from the area where legally enforced industrial norms were in operation.

The more unemployed persons were forced to claim welfare support, the more difficult it became for the most heavily affected towns to employ them in welfare work and to transfer them into jobs with insurance obligations and quasi-regular working conditions. As a result of the intensification of the crisis and the catastrophic collapse of government finance after the bank crisis, it seemed as if the local authorities were beginning to use the legal elbow-room available to allocate extensive compulsory labour in their own interests, especially after the middle of 1931. From the summer of 1930 almost all the barriers blocking the arbitrary handling of compulsory labour had been cleared out of the way. What could have prevented this process? Hardly the protests of the claimants, for the legal right to support did not exist. The arrangement of compulsory labour was financially advantageous in two ways. First, it was up to the discretion of the welfare authority whether any payments were made above and beyond the support-rate. Secondly, the welfare unemployed were still obliged to repay the support they had received even if it had been paid to them for carrying out compulsory labour. The most that a recipient could hope for was that 'his' welfare authority would deduct from the total repayment the 'economic value of the work carried out' as the Prussian Ministry for Public Welfare had recommended in August 1930 in response to enquiries and in agreement with the Reich Labour and Interior Ministries, because of their 'legal doubts' as to the propriety of demanding the repayment of the full sum.[70]

There are no statistics on the numbers of compulsory labourers. They can only be produced by case studies of particular areas. At any rate, the political, economic, industrial and social developments that had taken place in Germany since mid-1930 suggest that the 'hard pressed' local authorities did not need the 'pointer' to the 'reduction of welfare expenditure' given them by Government Councillor Dr Otto Jehle in March 1933: 'Many a local authority can introduce lasting alternatives to its extraordinary welfare expenditure without additional financial burden and through far-sighted handling of the (compulsory) work provisions of the welfare decree.'[71] So compulsory labour in the final years of the Weimar Republic was the first but not the only kind of employment of the unemployed to be taken out of the sphere of labour and contract law and out of the free labour market. The second type of employment was work in the 'voluntary labour service' introduced by Brüning in the summer of 1931.[72] The National Socialists used the same scheme and structures for their job-creation measures. The 'Law on the Reduction of Unemployment' of 1 June 1933 ruled that the employment of the needy on construction projects was not grounded in any contractual relationship of employment in terms of industrial law.[73]

The marginalisation of millions of welfare unemployed had been knowingly taken on board since October 1930 by the cabinets of

Brüning and Schleicher in their regulation of the crisis-benefit system. There was never any lack of penetrating analysis in October 1930 or the following period to point out to the government the grave consequences of its unemployment policy. The arbitrary exclusion of thousands of the unemployed from crisis benefit and the transfer of these persons onto welfare benefit meant, so the critics said, the almost inevitable perpetuation of their unemployment. The repercussions of this policy were spelt out by Ludwig Eiter, expert in the Reich Municipal Federation, as early as 1930.

> The unemployed person himself feels that he is not treated as an equal in the labour market. The employer no longer regards the welfare unemployed as qualified workers, since they have been isolated from the work process for so long. Similar attitudes on the part of the labour exchanges . . . have been pointed out by a large number of local welfare agencies . . . This may be compatible with the insurance principle but cannot be reconciled with the responsibility of the labour exchange to place people in employment.[74]

The urgency and severity of appeals and petitions to the government criticising this 'atomisation of the labour market, the classification of the unemployed into good, medium and bad and the corresponding "mentality" which is produced with every harmful regulation' multiplied as the number of welfare unemployed increased.[75] But the data and evidence produced by the welfare associations were hardly considered, and the cabinets of Brüning and Papen stuck to the limitations on crisis benefit and the threefold division of the unemployed. Their policy accepted the perpetuation of the unemployment of the welfare unemployed and the accompanying processes: their political radicalisation, or more likely their descent into total resignation or apathy, the development of a certain psychological state making them receptive to every kind of putative cure and sometimes expressing itself in violent bursts of hatred.

The presidential cabinets of 1930–3 upheld the policy of atomisation of the labour market, and aimed at the social disciplining of the unemployed by means of the fragmentation and eventually the destruction of the class solidarity that was ultimately founded in the experience of free wage labour. The purposes and aims of this policy were illuminated by the arguments regarding the maintenance of the schism between crisis benefit and public welfare for the welfare unemployed which were advanced by the employers:

> The pressure created by the time limit on unemployment benefit cannot be dispensed with. It is necessary in order to maintain the efforts of the unemployed in finding a job and to increase their preparedness to accept work offered, e.g. the assignment of industrial workers to agriculture.[76] Moreover, it was believed certain that after disqualification from benefits the unemployed themselves increasingly feel the responsibility to provide for themselves and their families. This further strengthens the will to work.[77]

This attempt to isolate the unemployed, to make them morally responsible for their own fate, to draw their attention from the structural crisis of the economic system back onto themselves as individuals was further supported by the unemployment policy of the presidential cabinets. The means by which this was done, as a local politician in Nuremberg pointed out, was to shift the burden on to the local authorities until they were no longer able to bear it:

> The inevitable consequence was a reduction of public welfare payments as well as a reduction in the allocation of the necessities of life and supplementary allowances. The displacement of the financial burden from the Reich meant the burden fell totally onto the unemployed and his family.[78]

Because public welfare denied a person adequate support, he was left only with the hope that he would be able to protect himself and his family from physical ruin with 'alms' from private confessional or free welfare associations or from individuals. The community, the state, the parties and organisations had all failed in a manner that could scarcely have been made clearer to him.

VII

The 'multiplicity and severity' of the economic difficulties facing employees and their families, 'from the most varied directions' 'the extent of emotional unrest', the distress of the unemployed and their families, were frequently described and analysed by the welfare agencies, social workers and social thinkers of the time.[79] In sharp and almost incomprehensible contrast to this are the reports which emphasise the admirable patience, calm and discipline with which the population and the unemployed accepted the drastic government measures taken between mid-1930 and January 1933.[80] As the centuries-old history of poverty shows, need and misery as such have never resulted in rebellion, but rebellion has resulted rather from certain forms of social and political perceptions of poverty. It seems that mechanisms of unemployment policy in the Depression resulted just as much in the suppression of possible direct protests as in political bitterness and impotence. The tense 'calm' observed amongst the unemployed was an effect of the 'policy of atomisation' practised by the government. Mass unemployment, an excessive supply of people prepared to work, made solidarity between those still in work and the unemployed less likely. The emergency decrees prevented any solidarity within the unemployed themselves. This policy of fragmentation and atomisation was described as early as February 1931 by the Prussian Under-Secretary of State and Social Democratic economist, Hans Staudinger:

Hate and jealousy separate those in work and those not. The unemployed envy the employed their jobs . . . At no time has there been such a tension in the labour force as in the last two years. In meetings the workers and the unemployed clash . . . For these groups the capitalist employer is no longer the social enemy but rather the better-situated worker and the married woman worker.[81]

Nevertheless, as subsequent political developments showed, the success of this policy was bought at an extremely high cost, not least for those who had inaugurated it under the umbrella of nationalist and market-economy slogans in the last years of the Weimar Republic.

Notes

This chapter was translated by Lynn Abrams, Richard J. Evans and Dick Geary.

1. 'Bericht eines jugendlichen Arbeitslosen', in, *Arbeiterwohlfahrt*, vol. 6, 1931, p. 117.
2. K. Kollwitz, *Aus meinem Leben* (Berlin, 1957), p. 126.

3. Cf. R. Vierhaus, 'Auswirkungen der Krise um 1930 in Deutschland' in W. Conze and H. Raupach (eds), *Die Staats-und Wirtschaftskrise des Deutschen Reiches 1929–1933* (Stuttgart, 1967), pp. 155–75.

4. The following journals were used: *AH*, vol. 1, 1934; vol. 3, 1936; AWO, vol. 5, 1930–vol. 8, 1933; *Archiv fur Soziale Hygiene und Demographie*, vol. 7, 1932–vol. 8, 1933–4; BWoB, vol. 6, 1930–vol. 9, 1933; *CARITAS*, vol. 36, 1931–vol. 41, 1936; *DZW*. vol. 6, 1930–vol. 12, 1936–7; *FW*, vol. 4, 1929–30–vol. 7, 1932–3; *GZ*, vol. 140, 1930–vol. 42, 1932; *Jugend und Beruf*, vol. 6, 1931–vol. 7, 1932; *ND*, vol. 10, 1929–vol. 17, 1936; *Nachrichtendienst des evangelischen Hauptwohlfahrtsamtes Berlin*, vol. 6, 1929–30–vol. 12, 1935–6; *RABl*, 1927–36; *Sozialberufsarbeit*, vol. 11, 1931–vol. 12, 1932; *SP*, vol. 37, 1928, vol. 45–1936; *VW*, vol. 11, 1930–vol. 13, 1932; ZB, vol. 30, 1930–3; *ZfH*, vol. 33, 1928–vol. 42, 1937; *Zentralblatt fur Gewerbehygiene und Unfallverhütung*, NF, vol. 8, 1931–vol. 13, 1936. The following were also used for the relevant laws, ordinances and decrees: *Reichsgesetzblatt*, Part I (*RGBl* I) and *Reichsarbeitsblatt* official section (*RABl* I).

5. Neither unemployment benefit policy, nor its consequences for the legal and social position of the unemployed and those on welfare support have been systematically researched. The standard works are; L. Preller, *Sozialpolitik in der Weimarer Republik*, (1949, reissued Düsseldorf, 1978.) M. T. Wermel and R. Urban, *Arbeitslosenfürsorge und Arbeitslosenversicherung in Deutschland-Neue Soziale Praxis*, vol. 6, I–III, (Munich, 1949); F. Syrup, *Hundert Jahre Staatliche Sozialpolitik 1839–1939*, (Stuttgart, 1957); the more recent work of F. Niess for an overview up to the 1970s, *Geschichte der Arbeitslosigkeit*, (Koln, 1979); for reports and analyses of the position of the welfare unemployed after Jan. 1933 see *Deutschlandberichten der Sozialdemokratischen Partei Deutschlands, 1934–8*. As the most recent analyses of unemployment and the distribution of the unemployed amongst the benefit systems show, the problem dealt with here is not only of historical interest, cf. *Die Zeit*, no. 49, 2 Dec. 1983; *Süddeutsche Zeitung*, nos. 170–3, 26–8 July 1984. For a more detailed discussion see H. Homburg; 'Vom Arbeitslosen zum Zwangsarbeiter' in *Archiv für Sozialgeschichte*, 25 (1985), 251–98.

6. Cf. *FW*, vol. 5 (1930–1), p. 483; *FW*, vol. 6, (1931–2), pp. 481ff; *DZW*, vol. 10 (1934–5), pp. 164ff.

7. Cf. M. Saitzew (ed), 'Die Arbeitslosigkeit der Gegenwart', *Schriften des Vereins fur Sozialpolitik*, vol. 185/I, (Munich, 1932), pp. 138–74 partic. 151, 1–96; K. Borchardt, 'Wachstum und Wechsellagen 1914–1970' in H. Aubin and W. Zorn (eds), *Handbuch der deutschen Wirtschafts-und Sozialgeschichte*, vol. 2 (Stuttgart, 1976), pp. 685–740.

8. On the AVAVG and crisis benefit: *Das Gesetz über Arbeitsvermittlung und Arbeitslosenversicherung vom 16.7.1927*, (*RGBl*, p. 187), commentary by O. Weigert (Berlin, 1927); F. Spliedt and B. Broecker, *Gesetz über Arbeitsvermittlung und Arbeitslosenversicherung vom 16.7.1927*, edition 4 with addendum regarding the changes to AVAVG by the law of 12.10.1929, (*RGBl* I) p. 153, (Berlin, 1929); on the contributions and payments schemes: paras 142–67 AVAVG.

9. On the Welfare Law after the new regulation of Feb. 1924: H. Muthesius, *Fürsorgerecht* (Berlin, 1928); 'Fürsorgerecht', *Sammlung der wichtigsten einschlägigen Gesetze und Verordnungen des Reichs und der Länder*, (Beck'sche Verlagsbuchhandlung, Munich, 1932), ['Fürsorgerecht']; on the financial-legal instructions, Muthesius, p. 16f, 28.

10. Cf. paras 69–130 AVAVG; Weigert, pp. 203–335; Spliedt and Broecker, *Gesetz*, pp. 127–303; Wermel and Urban, vol. 6/III; Preller, *Sozialpolitik*, pp. 363–76; 422–53.

11. Paras 101, 102 AVAVG; cf. Weigert, pp. 286–91; Spliedt and Broecker, *Gesetz*, pp. 223–33. The Regulation regarding the material rights of crisis benefit in

the *Verordnung über Krisenunterstützung für Arbeitslose vom 28.9.1927* in the draft of the decree of 27.8.1928 (*RGBl* I, p. 373), and the *Erlass uber Personenkreis und Dauer der Krisenunterstützung vom 29.6.1929* (*RABl* 1929, I, p. 161).

12. Spliedt and Broecker, *Gesetz*, pp. 227 and 258.

13. Cf. (*Reichs*) *Verordnung über Fürsorgepflicht vom 13.2.1924* (RFV) (*RGBl* I, p. 100) and *Reichsgrundsätze über Voraussetzung, Art und Mass der öffentlichen Fürsorge vom 4.12.1924*, (RGS) (*RGBl* I, p. 100).

14. Para. 5, RGS.

15. Para. 1; 2, RGS.

16. Para. 5; 6; 10, RGS; para. 6, clause 2 RFV.

17. Para. 15; 25, RFV; Muthesius, pp. 145ff, 165ff.

18. Cf. *SP*, vol. 39, 1930, pp. 243ff; *VW*, vol. 11, 1930, pp. 323ff; *SP*, vol. 39, 1930, pp. 729–34; *ZfH*, vol. 35, 1930, pp. 422–5; *ND*, vol. 11, 1930, pp. 226–9; *AH*, vol. 2, 1935, pp. 321ff; *DZW*, vol. 13, 1937–8, pp. 660ff; Wermel and Urban, vol. 6/III, pp. 70ff.

19. 1929: *ND*, vol. 11, 1930, p. 75f; 1930: *AWO*, vol. 6, 1931, p. 206.

20. *ND*, vol. 12, 1931, pp. 117–20, partic. 118.

21. Frankfurt: *SP*, vol. 39, 1930, pp. 745–50; 90 towns: Helmut Storch, 'Einmalige Erhebung über die laufend unterstützen arbeitsfähigen Erwerbslosen in den Stadten über 50,000 Einwohnern' in *Statistische Vierteljahresberichte des Deutschen*, 2 (1929) pp. 94–125, esp. p. 103. *15.2.1929*, in *SP*, vol. 38, 1929, pp. 860ff.

22. Ibid., p. 862.

23. Decree of 28.8.1930. *Statistik der Wohlfahrtserwerblosen*, in *VW*, vol. 11, 1935, p. 735–8. Table 4.4: 90 towns; Storch, p. 112. Frankfurt Niemeyer; *SP*.vol. 39, 1930, pp. 745–50; Berlin Kreuzberg, 'Die Zusammensetzung der WE in Verwaltungsbezirk Kreuzberg' in *BWoB*, vol. 6, 1930, 93–5; Berlin, Marianne Keiler, 'Einmalige Erhebung uber die laufende unterstützten Wohlfahrtserwerblosen in Berlin', 1 Feb. 1931, in *DZW*, vol. 8, 1931–2. pp. 212–20; Koln, Bernhard Mewes, 'Die ausgesteuerten Arbeitslosen in Köln', (Beliage zum Monatsbericht der Statistischen und Wahlamts der Stadt Köln, Dec. 1931), pp. 1–8; Reich; 'Die berufliche Gliederung der Wohlfahrtserwerblosen', *Reichsarbeitsmarktanzeiger*, no. 7, 9 Apr. 1931. Published statistics on the occupational composition see *ND*, vol. 12, 1931, p. 116; *SJDR*, vol. 49, 1930, p. 325; vol. 52, 1933, p. 300.

24. Concerning this and subsequent statements: *RABl* II, 1932, p. 15; *SJDR*, vol. 51, 1932, pp. 298, 418; vol. 52, 1933, p. 297; vol. 53, 1934. p. 313; vol. 54, 1935, p. 324; vol. 55, 1936, p. 341; cf. Table 5.

25. *Verordnung des Reichspräsidenten vom 14.6.1932* (June NotVO 1932), (*RGBl*, p. 273), section 2, 'Wohlfahrtshilfeverordnung', para 5, clause 2. Criticisms cf.; *SP*, vol. 41, 1932, p. 1069f; *ND*, vol. 13, 1932, p. 221f.

26. Para. 101 AVAVG, 'Entwurf eines Gesetzes uber Arbeitslosenversicherung', RT DS no. 2885 in *Verhandlung des Reichstags* (*VRT*), vol. 413, p. 93.

27. Michel, 'Arbeitslosenversicherungsreform 1930' in *ZfH*, vol. 35, 1930, p. 401–5, 422–5, partic. 402.

28. *2. Bericht der Reichsanstalt*, Jan.–Dec. 1929, *RABl* 1930, appendix to no. 12, pp. 56–60, partic. 56; cf. note 11.

29. Herrnstadt, 'Neuabgrenzung von Versicherung und Fürsorge' in *SP*, vol. 38, 1929, pp. 873–8, partic. 874.

30. Michel, 'Arbeitslosenversicherungsreform 1930' in *ZfH*, vol. 35, 1930, pp. 501–4, partic. 503.

31. Cf. *Verordnung und Erlass über Krisenfürsorge vom 11.10.1930* (effective from 3 Nov. 1930); the same of 23 Oct. 1931 (effective from 9 Nov. 1931; the same of 17 June 1932 (effective from 26 or 27 June 1932); *Erlass über die Unterstützungsdauer in der Krisenfürsorge vom 7.11.1932*, (effective from 28 Nov.

1932). The regulations and decrees are published in the corresponding volumes of the *RGBl* I and *RAB* I.

32. Reich Minister of Labour Stegerwald in the debate on the introduction of the Reich unemployment benefit, 42nd meeting of the Reichstag, 14 Mar. 1931 in *VRT*, vol. 445, p. 1558; cf. the later attitude of the government of 17.3.1932 in RT DS no. 1428, p. 35, VRT, vol. 452, and Preller, *Sozialpolitik*, p. 421–48; Wermel and Urban, vol. 6/III.

33. June NotVO 1932, 1st part, ch. 1: 'Anpassung der Vorschriften über die Arbeitslosenhilfe an die Lage des Arbeitsmarktes': *Verordnung über die Höhe der Arbeitslosenunterstützung und die Durchführung öffentlicher Arbeiten vom 16.6.1932* (*RGBl* I, p. 305), part 1 in effect from 27 June 1932; *Erlass über die Krisenfürsorge vom 17.6.1932* (*RABl* I, 1932, p. 114).

34. On the reductions as a result of the June NotVO 1931: *GZ*, vol. 41, 1931, pp. 640–9 and 679ff; *AWO*, vol. 6, 1931, p. 659; as a result of the June NotVO 1932: *RABl* II, 1932, pp. 225–32; *ZB*, vol. 32, 1932, pp. 169–74.

35. *CARITAS*, vol. 37, 1932, p. 332.

36. *Akten der Reichskanzlei Weimarer Republik. Kabinett Brüning* I u.II, vol. 2, (Boppard 1982), p. 1055; Wittelshöfer, 'Die Neuordnung der Krisenfürsorge' *DZW*, vol. 6, 1930–1, pp. 467–72; *GZ*, vol. 41, 1931, pp. 646–9, 679ff; *DZW*, vol. 9, 1933–4, p. 173. The net monthly benefit expenditure in the unemployment insurance on average amounted to (financial years): 1928 = 78.54 RM, 1929 = 76.87 RM, 1930 = 74.11 RM, 1931 = 60.66 RM, 1932 = 47.50 RM, from Wermel and Urban, vol. 6/III, p. 106.

37. Cf. *ZfH*, vol. 37, 1932, p. 183f; *Kommunales Jahrbuch*, *NF*, vol. 3, 1932, p. 51; Wermel and Urban, vol. 6/III, p. 66; *ZfH*, vol. 38, 1933, p. 471.

38. Para. 101 AVAVG, inserted by the June NotVO 1931, cancelled by the October NotVO 1931 (*RGBl* I, p. 537).

39. Prussian decree of 19 May 1930, in *ND*, vol. 11, 1930, p. 170; Decree of the Reich Labour Ministry of 10 Nov. 1934, in *DZW*, vol. 10, 1934–5, p. 360f.

40. *RGBl* I, p. 1125; cf. *ZfH*, vol. 42, 1937, pp. 17–21; *DZW*, vol. 12, 1936–7, p. 519–27.

41. Para. 19 RFV; para 7 RGS; Muthesius, pp. 67, 162f.

42. *Grundsatzentscheidung des Spruchsenats für die Arbeitslosenversicherung vom 13.3.1933* in *RABl* IV, 1933, p. 80.

43. Diefenbach, 'Der 71 Band der Entscheidungen des Bundesamtes fur Heimatwesen' in *ZfH*, vol. 35, 1930, pp. 33–8, partic. 34.

44. Cf. Herrnstadt, 'Die neuere Rechtsentwicklung in der Arbeitslosenfürsorge' in *ZfH*, vol. 36, 1931, pp. 5–8; Kobrak, 'Die Arbeitsfürsorge für Wohlfahrtserwerblose in Berlin' in *ZfH*, vol. 35, 1930, pp. 179–85.

45. Para. 13 RGS.

46. Para. 20 RFV; cf. 'Preussische Ausführungsverordnung zur Verordnung über Fürsorgepflicht vom 17.4.1931.' paras 21–9 in *Fürsorgerecht*, pp. 177–80.

47. Diefenbach, 'Der 71. Band', p. 34.

48. Kobrak, 'Die Arbeitsfürsorge', p. 179; cf. Wittelshöfer, 'Rezension zu:Th. Marx, Die Rechtsstellung des Fürsorgearbeiters (Karlsruhe 1929)' in *VW*, vol. 11, 1930, p. 840f.

49. *ZfH*, vol. 36, 1931, p. 40.

50. *ZfH*, vol. 37, 1932, p. 177.

51. *ZfH*, vol. 36, 1931, p. 40.

52. L. Eiter, 'Gemeinden und Arbeitslosenversicherung' in *VW*, vol. 11, 1930, pp. 323–30, partic. 330.

53. H. Kranold-Steinhaus, 'Noch einmal "Pflichtarbeit"', in *AWO*, vol. 6, 1931, pp. 100–4, Quotation, pp. 100, 104.

54. Herrnstadt, 'Die neuere Rechtsentwicklung, p. 5; cf. Spliedt and Broecker, *Gesetz*, p. 414; R. Jonas, *Die Grundlagen der wertschaffender Arbeitslosenfürsorge* (Berlin, 1931), p. 12.

55. Cf. B. Broecker, 'Die Beschäftigung von Arbeitslosen' in *AWO*, vol. 7, 1932, pp. 129–38; *CARITAS*, vol. 37, 1932, p. 487ff.

56. Para. 217 AVAVG; cf. Spliedt and Broecker, *Gesetz*, p. 411 and para. 75d AVAVG after the July NotVO 1930.

57. Kobrak, 'Arbeiterfürsorge', p. 181.

58. Ibid.

59. Reich labour court, judgment of 20 Apr. 1932, cited in *ZfH*, vol. 37, 1932, p. 367.

60. Kobrak, 'Arbeiterfürsorge', p. 181.

61. Information from *BWoB*, vols. 6–9, 1930–3.

62. Information from Stt JB. vols 50–2, 1931–3.

63. *ZfH*, vol. 36, 1931, p. 41.

64. Cf. Köln: *AWO*, vol. 6, 1931, pp. 337ff; Hannover: *ZfH*, vol. 35, 1930, p. 373f; 'Wie leben unsere Wohlfahrtserwerblosen', (Hannover, 1932); Hamburg: *ZfH*, vol. 35, 1930, p. 250f; Schleswig: *DZW*, vol. 6, 1930–1, pp,. 472ff; Stettin: *ZfH*, vol. 35, 1930, p. 7f; Lübeck: ibid., p. 87f; Eckernförde: *ZfH*, vol. 37, 1932, p. 177f; Rhein-Main: *DZW*, vol. 6, 1930–1, p. 16f.

65. *ND*, vol. 11, 1930, p. 355. With hindsight the 'Deutscher Verein' complained in 1937 that compulsory labour before 1933 had often been 'ostracised as forced labour'. *ND* vol. 18, 1937, pp. 385ff. Instructive for the attitude of the public: L. Burgert, *Arbeitslos*, (Breslau, 1931); K. Schröder, *Familie Markert*, part 2, (Berlin, 1931); A. Klaus, *Die Hungernden* (Berlin, 1932); H. Rein, *Berlin 1932* (Berlin, 1935).

66. Regional labour court Hamburg, judgment of 28 Dec. 1929, cited in *ZfH*, vol. 35, 1930, p. 88.

67. Cf. Herrnstadt, 'Die neuere Rechtsentwicklung', p. 7; 'Die Arbeiterfürsorge der Gemeinden' in *ZfH*, vol. 38, 1933, pp. 192–6.

68. Cited in *ZfH*, vol. 36, 1931, p. 7; cf. Sommer, 'Der Begriff der Gemeinnützigkeit im para 19 RFV' in ibid., vol. 35, 1930, p. 290.

69. Cited in *ZfH*, vol. 36, 1931, p. 5.

70. Circular letter dated 15 Aug. 1930, cited in *ZfH*, vol. 35, 1930, p. 506f.

71. Jehle, 'Senkung des Fürsorgeaufwandes' in *ZfH*, vol. 38, 1933, pp. 113–16, partic. 113f.

72. Cf. para. 139a AVAVG, introduced by the NotVO 5.6.1931 (*RGBl* I, partic. p. 295).

73. Cf. W. Kraegeloh, 'Arbeitsanweisung als Fürsorgeleistung' in *ZfH*, vol. 38, 1933, pp. 517ff; H. Dersch, 'Rechtsformen des Arbeitsverhältnisses bei der Arbeitsbeschaffung' in *AH*, vol. 2, 1935, pp. 25–9, 45–8.

74. Eiter, 'Gemeinden', p. 325.

75. Michel, 'Arbeitslosenversicherungsreform 1931' in *ZfH*, vol. 36, 1931, pp. 237–41. partic. 237.

76. Cited in *ND*, vol. 11, 1930, p. 185.

77. *Wohlfahrtserwerblose und Gemeinden*, Denkschrift des Deutschen Industrie und Handelstags (Berlin, June 1930) cited in *SP*, vol. 39, 1930, pp. 729–34, partic. 731.

78. Marx, 'Wollen wir helfen', in *FW*, vol. 7, 1931–2, pp. 381–8, partic. 381f.

79. *Familie und Notverordnung*, Denkschrift der Deutschen Liga der freien Wohlfahrtspflege vom Herbst 1932 in *FW*, vol. 7, 1932–3, pp. 333–44, partic. 335.

80. Cf. W. Polligkeit, 'Ein Notprogramm der Wohlfahrtspflege' in *SP*, vol. 40, 1931, pp. 1393–1400; Kobrak, 'Sozialpolitische Bemerkungen zur Notverordnung (vom 14.6.1932)' in *SP*, vol. 41, 1932, pp. 791–5; *ND*, vol. 13, 1932, p. 232; *CARITAS*, vol. 37, 1932, pp. 332–5.

81. H. Staudinger, 'Die Erwerbslosigkeit der Jugendlichen, ihre wirtschaftlichen Folgen und deren Bekämpfung' in *AWO*, vol. 6, 1931, pp. 161–70, partic. 165.

5 FRANKFURT'S WORKING WOMEN: Scapegoats or Winners of the Great Depression?

Helgard Kramer

I

After the First World War there were a number of attempts in Germany to restrict women's work outside the home. They began during the demobilisation of 1918–20. They were continued in the campaigns launched, in reaction to rising unemployment, against married women 'double earners' in 1923–4 and after 1919 (the year of the first state decree on the subject). In 1931 the government ordered that unemployment and crisis benefits should be paid to married women only on proof of hardship. Finally, in 1933 and 1934 there were direct attempts to substitute men for married women in the workforce. In 1931, to be sure, with the decree on the 'removal of double-earning female civil servants', the state's strategy came into conflict with the legal principles of female equality which it had itself established. As a consequence, the state found it difficult to legitimate its own policy in this respect. But with the arrival of Nazi rule in 1933 these legal limitations fell away. Under the Nazis, the catalogue of measures was lengthened by a series of indirect attempts to remove women from the labour market. These included the provision of material and social inducements for them to stay at home and have children. Added to these were the policies followed by the labour exchanges, which now sought to direct women industrial workers and salaried employees back into the shrinking realms of domestic service and rural labour.

These attempts to limit paid female labour constituted a novel and unprecedented experiment in state intervention in the labour market. In the 1920s, the instruments of state direction of labour were still very undeveloped. From 1933 the fascist regime gave these measures a repressive form and a more systematic application. But the purpose for which they were used was no different from that of the democratic Weimar years, nor did it deviate significantly from that of other countries in the same period. Indeed, attempts to limit paid female labour were made, in one form or another, in all industrialised countries in the 1920s and 30s.[1] For this reason it is justifiable to analyse them in the context of a general increase in state intervention in the economy rather than in terms of the specific form of the political system under which they were carried out. Whatever form these policies took, whether they made marriage or the allegedly 'masculine' nature of an occupation the criterion for driving women from the capitalist labour market, in all cases they shared the intention of reducing the real or potential supply of labour. They thus sought the apparent creation of jobs for men, whose higher social priority enabled them to claim the 'right to work' for themselves, and so constituted the initial beginnings of a policy of full employment.

Feminist and Marxist historians have often assumed without question that these campaigns against female labour in the world economic crisis actually did achieve a disproportionate reduction in female participation in the labour market.[2] This view rests on the theory that women formed a particularly flexible part of the reserve army of labour under capitalism, a potential labour force that could be shifted in and out of the labour market at will. In this way, fluctuations in the extent of female labour are seen as dependent on the fluctuations of the economy, and it is claimed that women are always the first to be dismissed in a recession or a crisis. They are thus, it is argued, more affected by unemployment than men. Where this turns out to be unsupported by the statistics, it is still argued that there is a high degree of 'invisible unemployment' among women. This can then be explained in the alleged reluctance of housewives, deterred by state policies or ideological influences, to register themselves as unemployed and so enter the official unemployment statistics. This view makes plausible the conclusion that women were not only among the victims of the labour market in terms of wages, working conditions, promotion prospects and training opportunities — which they undoubtedly were — but were also the first to be dismissed and the least able to find jobs as the general economic situation deteriorated and the unemployment figures rose. Certainly, attempts to limit women's work or stop it altogether were not without success. They were carried through with the support of widely supported arguments. In so far as women were dependent on paid labour, either permanently or temporarily during crises, they were forced to adopt increasingly concealed strategies of representing their interests in this sphere. And they were compelled to tolerate conditions of work that were considerably worse than the forms of 'equal participation' which they had been able to select for themselves in the initial enthusiasm for reform of the Weimar years.

In this contribution, however, I want to argue that state attempts to limit paid female labour are not enough to explain its stagnation. Rather, an important additional factor can be seen in the defensive strategies chosen by women themselves. In part, these strategies ran counter to these state policies; in part, they exploited them in accommodation to the Keynesian 'spirit of the age'. In other words, it is my contention that the fluctuations in the extent of paid female labour during the world economic crisis can be explained, first, by the *active* response of women to their economic situation in domestic work and paid employment outside the home; secondly, by their changing perceptions of the conditions for the reproduction of family life; and finally, by their varying estimations of their own chances of paid employment in specific economic situations and in an ideological climate that was hostile to women's involvement in the labour market.

The evidence on the basis of which I want to argue this case is taken from the regional metropolis and major banking and trading centre of Frankfurt am Main. Here, as elsewhere, the Weimar years brought a sharp increase in paid female labour. The fastest-growing sector of women's employment was in salaried commercial and office work. To be sure, the starting-point for this growth was a relatively low level of female employment before the First World War, compared to that of other large German towns. Frankfurt had a high proportion of men employed in the service sector, in offices and in administration. Their comfortable economic situation meant there was relatively little pressure for their wives and daughters to take on paid employment. Frankfurt's skilled workers, too, were said 'not to let their women-folk go out to work'.[3] Nevertheless, by the 1920s Frankfurt, as a city with a highly developed tertiary sector, had a female labour structure only attained in Germany as a whole in the 1950s, as Table 5.1 indicates. Moreover, the female labour market on the whole developed far more smoothly than the male, even if male and female unemployment rates moved in a roughly similar way to one another. In addition, the crisis in the Frankfurt labour market at the end of the 1920s had a greater effect on male employment than

Table 5.1: Employed and Unemployed in Frankfurt am Main, June 1933

	Employed			Unemployed			Total occupied persons			Women in each group as % of all occupied women
	Total	women	% women	Total	women	% women	Total	women	% women	
Total	201,821	68,971	34.2	68,938	17,509	25.4	270,759	86,480	31.9	–
Salaried	53,344	21,273	39.9	18,610	6,708	36.0	71,954	27,981	38.9	32.4
Commercial and office employees	40,148	16,611	41.4	14,414	5,596	38.8	54,562	22,207	40.7	
Technical employees	13,196	4,667	35.4	4,196	1,112	26.5	17,392	5,779	33.2	
Civil servants	17,863	2,168	12.1	–	–	–	17,863	2,168	12.1	2.6
Domestic servants	13,649	13,532	99.1	2,608	2,532	97.1	16,257	16,064	98.8	18.6
Labourers	71,800	18,797	26.2	47,704	8,267	17.3	119,504	27,064	22.6	31.1
Self-employed and family assistants	45,765	13,196	28.8	0	0	0	45,765	13,196	28.8	15.3

NB: The salaried employees and civil servants (*Angestellte* and *Beamte*) do not include those in leading managerial positions.

Source: *Statistische Übersichten der Stadt Frankfurt am Main 1933/34*, p. 8, *1934/35*, p. 7.

on female. The crisis especially affected typical male industries such as metal-working and engineering (particularly its largest component, automobile and machine-tool manufacture) and building and chemicals. It had less serious consequences for the electrical and the consumer industry where female employment was higher.

If we compare the occupational census of 1925 with that of 1933, it becomes clear that the proportion of occupied women (employed and unemployed) in the female population of working age had fallen from 34 to 32 per cent. Most of the fall must have taken place in the Depression, because the extent of female paid labour actually increased during the 'stable' phase of economic rationalisation between 1926 and 1929. Moreover, the 1933 census underestimated the effect of the Depression on women's employment because it was taken when the economic recovery was already under way. According to the economic version of the 'reserve army' theory, it would follow that the rate of female employment was already 'recovering' in 1933; according to the political version, however, it had not yet sunk to its lowest point since Nazi policies in this area had not yet had a chance to bite. There is another point to take into consideration as well. While the economic crisis may have driven some 'discouraged' women workers from the labour market (and from the unemployment statistics), it may have also have brought in 'added' women workers who had not been in the labour market before, as their husbands and fathers were no longer able to support them because of the loss of their own jobs or the collapse of their businesses.[4] In normal times it was common for bourgeois and better-off working-class men to maintain not only their wives but also their unmarried adult daughters at home — as indeed much of the German social legislation and welfare practice of the time assumed. In a crisis, however, many such women found it necessary to support themselves by going out to work. These were not of course two different groups of women, but two different sets of attitudes, and it was perfectly possible for the same women to change from one to another. Thus, it was clear that political attempts to restrict women's work, such as the demobilisation of 1918–20 and the campaigns against married women workers of 1922–3 and after 1929, could act as a deterrent to women potentially looking for jobs. On the other hand, it was equally clear that the economic dislocations of the early postwar years and the Depression could act as a magnet, drawing women into the labour market because of declining family incomes and high rates of long-term male unemployment. This would have applied to unmarried women who had either not worked before or carried on working longer than they would have done in better times. It would also have applied to married women who would not have sought employment if their family income had been enough to support them. In the Depression, in fact, the income of a large proportion of the population fell below generally accepted subsistence levels.[5] Thus, the crisis increased the general pressures in favour of entering the labour market.

As Figure 5.1 indicates, male unemployment in Frankfurt rose faster than female from the beginning of the Depression onwards. According to the theory of the female reserve army of labour, this could only be explained by the fact that concealed unemployment among women was rising much faster than it was among men. Women would have been removed from the statistics because they did not claim benefits, or were deemed by the authorities to be supported by their husbands or fathers; married women workers would have lost their jobs because of the state decree against 'double earners' introduced in 1929, and the accompanying ideological campaign, and women workers in general would have been among the first to go because of the accepted trade union practice of giving priority to protecting the jobs of married men. At the end of the Depression, in June 1933, however, the proportion of women among unemployed salaried or white-collar workers, at 36 per cent, roughly corresponded to the normal proportion of women in this sector of employment, as Table 5.1 indicates. On the other hand, there were more women among the wage-earners still at work — 26.2 per cent — than among the unemployed wage-earners — 17.3 per cent. This was possible because wage-earners had been affected earlier and more seriously by unemployment than white-collar workers. Thus, many female wage-earners had long since departed into the silent reserve, whilst unemployed women white-collar workers had been without jobs for a shorter period and were still counted in the statistics. It is significant, for instance, that female white-collar workers were over-represented among women claiming unemployment and (later) crisis benefits. Because they had a specialised training and a specific job, it was more difficult for the labour exchanges to take them off their books than the large group of 'general female casual workers'.

Figure 5.1: Persons Registered as Seeking Employment, Frankfurt 1928–33

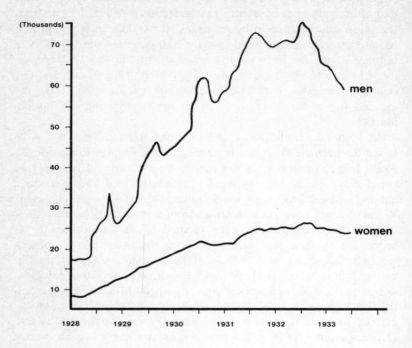

In 1931 the number of jobless claiming crisis benefit for the first time exceeded the number claiming unemployment benefit, thus indicating that most of them had been without work for more than 24 weeks. This preponderance of the long-term unemployed remained until the end of the Depression. This development, however, took place a year later among the female unemployed than among the male, in summer 1932.[6] This could have been because the strictness of the criteria for issuing crisis benefits systematically disadvantaged women claimants. It could also have been because long-term unemployment affected women at a later point in the crisis than men. The decisive question is whether the very great difference between male and female unemployment rates through the 1920s and into the Depression was caused by higher concealed unemployment among women, or whether it reflected a real situation in which women were actually being drawn into the labour market as a consequence of falling family incomes. This question can only really be answered indirectly by looking at the most important female sectors of employment in Frankfurt, and it is to a detailed examination of these sectors, beginning with the unsuccessful attempt to restrict female labour during the demobilisation, that we now turn.

II

In Frankfurt, as elsewhere, the First World War saw a massive increase of women's work in industry and the service sector. Demobilisation at the end of the war saw, correspondingly, the attempt 'to return these workers and those who were not active in industry before the war to their former activities and to exclude them from unemployment benefits', as the city labour office put it in 1920.[7] 'Job creation' for the numerous 'women and girls leaving the munitions industry' was undertaken by a municipal employment centre for clothing-repair workers,[8] which arranged the re-employment of some of the women munitions workers in the clothing industry whence they had come, now that the arms factories were closing down.[9] Women who had been taken in from the countryside for war work in Frankfurt were also to be sent back where they had come from. An attempt was thus being made to reverse the process whereby experienced women workers in the textile and clothing industries had been forcibly 'promoted' into the better-paid munitions and armaments industries, and domestic servants who had had no previous experience of industry had taken their place. This process constituted of course in effect a competition for jobs with male industrial workers.[10] Its reversal was justified with the generally accepted slogan 'a job for every ex-serviceman'. This was no compensation for those women who depended on their earnings or who did not want to go back into service or into the parental home. That women were represented in Frankfurt on the Workers' Council made no difference in this respect, since the council did not deviate from the state's overall aim of giving employment preference to ex-servicemen. The first problem it faced was in fact the closing-down of war production, which employed a large number of women; and unlike in other cities, unemployment benefits for women were no lower than they were for men, although this had been intended in the original German legislation.

According to the Demobilisation Order of January 1919, the jobs to which women were to be returned included not only domestic service but also their 'own' housework. But the city labour office claimed that it was:

> extraordinarily difficult to return these groups to their earlier occupations or to re-educate them for the new circumstances. The re-employment of the male factory workers[11] took place very smoothly as a result of the strength of their organization, the influence of the workers' committees and the short-term notice given to existing occupants . . . Things were much more difficult among the white-collar workers. Here there was often no decisive influence on the part of the salaried employees' committees and it had to be exercised instead by the labour office. Because the employers often refused to co-operate on grounds of economic difficulties, arbitration committees had to be appointed by the labour office in numerous cases.[12]

Thus, against the expectations of the contemporary feminist movement, of the women who had worked in the munitions industry or generally in so-called 'men's jobs' and had enjoyed better pay there, most vacated their posts without protest to make way for the returning ex-servicemen. They could have no interest in enduring the physically exhausting labour conditions in these industries at a time when the work involved in looking after the family was increased by the return of their menfolk and the drastic shortages of food and materials. They were also confronted by well-organised men who insisted on a return to their own 'steady' jobs. Things were very different in the white-collar sector, because women had gladly taken over 'appropriate', that is, clean and physically less demanding, office jobs, and could probably do them better than men who had just spent four years in the trenches. As far as the employees in this sector were concerned, the exchange of male for female labour after the war implied friction and disputes and thus costs — a problem which was also evident in the offices of the city council itself, although it was in theory supposed to take the lead in dismissing women workers as an example to others.

In the main, in fact, ex-servicemen were employed in this sector

without a corresponding dismissal of women workers. State intervention was unable wholly to reverse the feminisation of the
tertiary sector that had taken place during the war, as Table 5.2
indicates, and already by the beginning of 1920 the demand for
female white-collar workers was once more exceeding the
supply.[13] Even the threat of fines for employers who failed to
comply with the decree on the 'release of workplaces' to ex-
servicemen had scarcely any effect. Results were only achieved
when the unemployed men formed salaried employees' committees and made the cause of reducing women's work in favour of
their own. 'Only from this point onwards was there a noticeable
improvement, and numerous cases were referred to the labour
exchange . . . thus making a successful intervention possible', as
the labour exchange reported. Correspondingly, the number of
women registered as unemployed began to increase, although the
demobilisation regulations had intended the removal of a substantial part of the women who had been forced to work during the
war from the labour market altogether. In Frankfurt, the number
of unemployed women reached a peak in June 1919 with 3,389;
already by 1 December 1919, however, this had sunk back to a
mere 225. The labour exchange's attempt to 'lead back' as many
women as possible into domestic service continued to encounter
the greatest obstacles. As far as white-collar workers were concerned, therefore, the labour office tried to divert women into
education with the argument that they needed better qualifications. The demand for trained shorthand typists continued
buoyant, so a small part of the women who were looking for jobs
were placed in a municipal commercial further education college.
Another part was 'hidden' in a domestic economy school. Despite
all this, however, as Table 5.2 shows, state intervention was unable to alter the general trend which the restructuring of women's
work had taken. As the labour exchange concluded,

> There is a crass imbalance in the female labour market between
> the vast number of industrial and white-collar unemployed and
> the high demand for domestic servants. If a balance is to be
> achieved, prejudices on both sides have to be removed. House-
> wives have evinced a distinct disinclination to engage as their
> servants women who have worked in industry, while the latter
> have objected to the wages and working conditions customary
> in domestic service.[14]

Table 5.2: Salaried Employees in Frankfurt am Main, 1907–33

	1907			1918			1925			1933		
	Total	women	% of women	Total	women	% of women	Total	women	% of women	Total	women	% of women
Trade and Transport	6,308	1,407	22.3	2,587	1,090	42.1	42,228	15,228	33.6	25,684	10,734	41.8
Insurance	2,976	627	21.1	–	–	–	2,484	603	24.3	9,693	2,374	24.5
Banking							5,826	1,265	21.7			
Public and private services	–	–		–	–		15,902	5,422	34.1	8,826	5,215	59.1
Industry and manufacture	8,229	1,344	16.3	–	–		25,938	6,531	25.2	18,426	5,023	27.3
Total												
Metalworking	2,878	308	10.7	4,570	1,555	34.0	10,012	2,164	21.6	–	–	–
Electrotechnical	669	83	1.2	–	–		3,137	892	28.5	–	–	–
Chemical				–			4,984	959	21.9	–	–	–

Notes and sources: Städtisches Arbeitsamt, Appendix (1920); *Statistik des Deutschen Reiches 1928*, 94 ff (area of city in 1928); *Statistische Jahresübersichten 1933/34, 7. Erg.-Heft*, 7 f; 1925 public and private services include Abteilung D: Verwaltung, Heerwesen, Kirchen, freie Berufe; und E: Gesundheitswesen, hygienisches Gewerbe, Wohlfahrtspflege; metalworking includes Wirtschaftsgruppe 7 and 8 (*Statistisches Handbuch, 2. Ausg., 4 ff*); chemicals include Wirtschaftsgruppe 9 und 10, (ibid.); insurance figure for 1925 from *Statistik des Deutschen Reiches, Bd. 406, 575* (inner city only); *Statistische Jahresübersichten 1934/35, 8. Erg.-Heft, 44*.

This suggests in turn that the situation of female unemployment varied from sector to sector.

As Figure 5.2 indicates, the short-term cyclical movements of female unemployment were more marked than was the case with male unemployment. This was because the typical women's jobs were more subject to seasonal fluctuations. Domestic servants tended to change jobs either in April or in September, for example, while the clothing industry (the greatest single employer of women) and the shoe industry reached their peaks of production in spring and autumn with corresponding troughs in between. Retailers engaged extra women — and hardly any men — for the summer sales and the Christmas rush, while the demand for women office workers was at its height with stocktaking and the balancing of the books at the end of the year. These seasonal variations also explain the serious differences in the effects of short-time working for men and women throughout the entire period in question (though less so in the Depression). For there was more short-time work in the largely feminised consumer-goods industries such as clothing and textiles, shoes and tobacco, than there was in the male domain of metal-working, engineering and chemicals. The domestic service sector, which still employed 19 per cent of all working women in Frankfurt in the early 1930s,[15] maintained a high level of demand for women workers throughout the 1920s. The servants came into Frankfurt from the surrounding countryside: already in 1907 only 4.1 per cent were born in the town.[16] Wages for Frankfurt's domestic servants were held to be above average in the war years.[17] Nevertheless, the domestic service labour market was in reality little more than a sluice through which women streamed from the countryside into the town, eventually to find work in factories, offices and services of other kinds. The favourable position of women workers in this sector was a result of the high demand for female labour in other sectors, which allowed young women to move on to them with relative ease, and thus improved the position of those women who remained in service by reducing the competition for jobs in this area.

Thus, in the autumn of 1921, for example, for every job on offer in domestic service there were only on average 1.66 women seeking employment there. As the local newspaper, *Frankfurter Nachrichten*, pointed out, 'The demand for servants — as always, especially for the "good-maid-of-all-work" — has remained

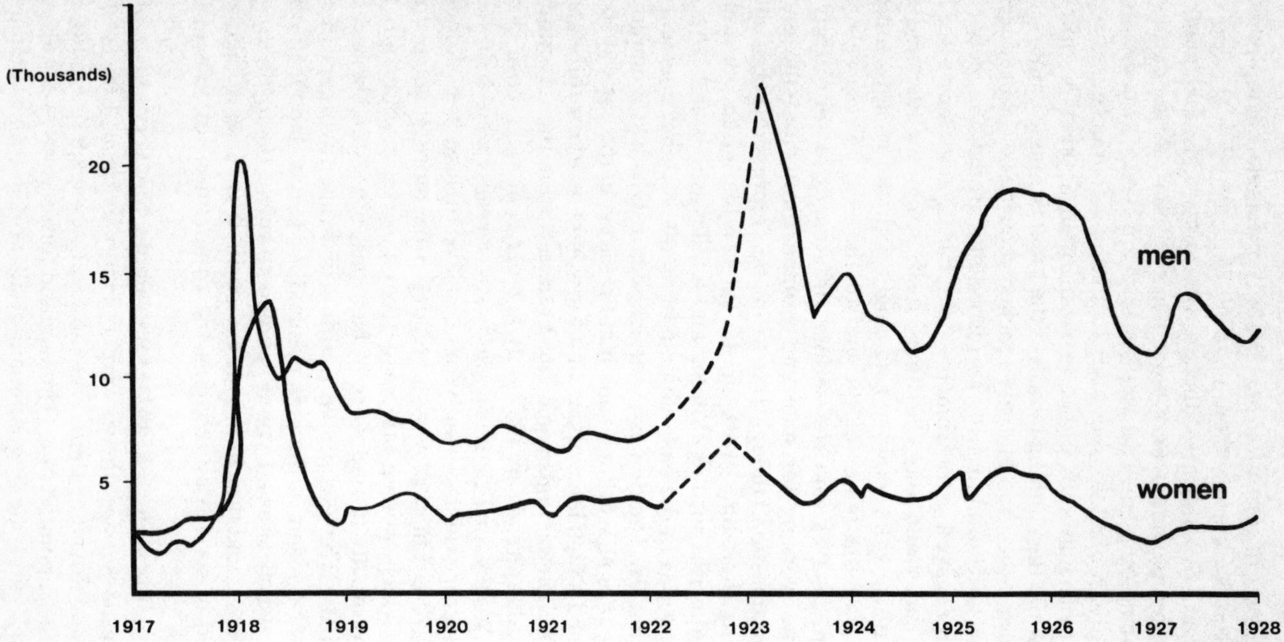

Figure 5.2: Persons Registered as Seeking Employment, Frankfurt 1917–28

buoyant.'[18] The Frankfurt Housewives' Association actually offered to ally itself with the right-wing party in the Frankfurt city council because it had initiated the council's various attempts to send women 'back' into employment within the home. The Housewives' Association pursued the revaluation of middle-class housewives as 'domestic economy managers', and by no means disapproved of attempts to set up domestic economy institutions for training girls in domestic economy and to make such training obligatory for all. For this would have meant that the situation in the domestic service labour market would have changed for the better as far as the housewives were concerned. But 'political' efforts were not enough to bring about this 'change'; it was only brought about, to a certain degree at least, by the economic situation. The broadly favourable situation for domestic servants was again and again temporarily reversed by the fact that the domestic service labour market acted as a port of entry for additional elements into the labour force. In the middle of 1920, it was reported, 'more and more married women are entering domestic service, because their husbands are on short time'.[19] Thus the conventional celibacy of female servants was now abandoned, as a consequence of the long-term demand for cheap domestic labour. Moreover, the introduction of collective wage agreements for domestic servants in 1918, and thus a restructuring of their form and conditions of employment to approximate more closely to the model of regular waged labour, made servants more expensive to employ.[20] The same phenomenon occured at the height of the Depression: married women who received benefits only as dependents of their husbands pushed into the domestic service labour market, where they were able to undercut the wages of the regular labour force of young, single girls. Even then, however, the high demand in this sector of the female labour market kept it more favourable to women than any other. In an economic downturn, the domestic-service labour market thus formed a kind of buffer between the fully developed wage-labour areas of female labour and the unskilled, lower-class housewives who now saw themselves forced to take up some kind of job. During the Depression unmarried working women streamed back into the domestic-service labour market whether under the actual or threatened denial of benefits by the labour offices, or because they could not find any other jobs. Married women also entered it at the same time because it was the easiest way to obtain temporary extra earnings to make up their reduced family income.

The largest individual groups of women workers within the industrial sector were those employed in the garment trade (in needlework for underwear and aprons etc., in ready-made clothes manufacture, in millinery and in tailoring), in shoemaking, in the food and drinks trade and in the growing electrotechnical industry. The garment trade was traditionally the largest industrial employer of women. In the 1920s thousands of needleworkers were employed in small workshops in the inner city or, as was common before the turn of the century, they worked at home. This industry offered the best chances for trained and skilled women workers to become self-employed: in 1925 there were 2,720 self-employed needleworkers and tailoresses in Frankfurt, making 61 per cent of all the female self-employed in the town as a whole.[21] The food and drink trade (tobacco and preserves) and inns and restaurants had also already been strongly feminised before the war. In the shoe industry, on the other hand, the majority of women workers came from the surrounding countryside, while local women workers sought rather to avoid this kind of heavy labour. Women employed in retailing and 'private services' such as laundries and ironing and pressing establishments, which also involved heavy physical work, were also numerically important. Such work, like most heavy female labour in industry, was mostly carried out by women commuting in from the surrounding villages, and was avoided by women from Frankfurt itself.[22] After the war women's employment increased above all in the electrotechnical industry, which counted as part of the metal-working and engineering sector. In this sector as a whole, however, other industries were more important as employers in Frankfurt, and therefore it remained in general a male domain. The larger enterprises in the chemical industry also witnessed a restricted growth of female labour in the production sector. Women — mostly married — were employed here mainly in the packaging department. The metal-working and chemical industries paid higher wages to women than any other industries. Up to 1924 a certain levelling-up of women's wages took place here.[23] In the Farbwerke Hoechst (Hoechst Dye Works), for example, the wage differential between men and women was reduced from about 50 per cent to between 26 and 34 per cent in this period. It remains unclear whether this general tendency in the Weimar period was due to a fall in real wages for men[24] or whether it reflected a genuine improvement in women's wages.

At any rate, it certainly reflected a growing demand for unskilled women workers in the wake of changes in the work process.

Indeed, it was in areas where the introduction of piecework led to women's earnings approaching those of men that male fear of female industrial competition seemed to have a real basis in fact.[25] This had only a limited effect in the metal-working and chemical industries because of the high rate of extra compensation for men in jobs here,[26] but it was more serious in the Frankfurt shoe industry. However, there were few examples of trade union works councils trying to stop the employment of women, once de-mobilisation was over; indeed, there were none at all in Frankfurt's industrial sector: the campaigns of 1922–3 and 1929 against married women 'double earners' were aimed mainly at securing dismissals. So we can reasonably suppose that in the intervening period female employment was able to expand un-hindered by the union factory councils. This does not of course exclude the possibility that they helped create a climate of opinion in some factories that was hostile to the employment of women and that this corresponded to the existing prejudices of the employers. That there was indeed an effective male consensus between unions and employers is suggested by the fact that in periods of high general unemployment most female jobs were lost in male-dominated areas.[27] Thus, Frankfurt employers in building and machine construction reported that it was their practice during the Depression not to employ any new women workers, or even to dismiss them in order to engage male employees[28] in the administration, even in the case of secretaries.[29]

III

Female salaried employees were concentrated in the areas of trade, private services and municipal administration. Women's employment rose fastest in these areas in the 1920s, following the general trend. It also increased in banking, insurance and indus-trial administration, though the increase was slower in the so-called 'male spheres', a phenomenon that was characteristic for the unplanned, 'natural' feminisation of employment in general. More and more women were employed in wholesale and retail trade as well, both on the sales and on the administration side. The majority of Frankfurt's female salaried employees were in fact young saleswomen. If they were organised at all, it was mostly in the Central Union of Salaried Employees (*Zentralverband der Angestellten*), a 'Free' trade union which stood close to the Social Democrats. They were employed in department stores or smaller retailing outlets, and, like the second largest group of female salaried employees, the office workers, they were generally situated on the bottom rung of the hierarchy. Demand for female clerks and shorthand typists, secretaries and book-keepers remained buoyant throughout virtually the whole period from the demobilisation to the Depression.[30] Even at the beginning of the 1920s, however, the trend towards employing ever younger women was already visible, because of the age-related salary scales, which made young shorthand typists into a particularly cheap group of workers, and because of the fast physical burn-out of the labour force through specialisation in typing, a fact which was continually criticised by the trade unions. The generally high demand for women office workers fell in the years of economic stabilisation in which there were frequent dismissals of women salaried employees. Thus, in the economic downturn of 1926 there was a sharp but short-lived contraction in the availability of jobs for commercial employees, 'especially observable in the female job market', as a speaker in the city council observed.[31] This led the right-wing deputies to demand the diversion of girl school-leavers from the commercial colleges into the domestic economy schools so as to stop them 'competing' with male salaried employees for jobs.

In 1925 a decision of the Reich Labour Court in Berlin established the legality of dismissing a female salaried employee if she married. As a consequence of this the Central Union of Salaried Employees thought that many women white-collar workers would not dare to legitimise their relationship with their fiancé or lover because this would cause them to lose their incomes. 'What unworthy and unhappy situations can arise for the woman out of a non-legitimated relationship of this kind!', the union exclaimed.[32] The Court's decision was eventually reversed in 1928. But the official guidelines for unemployment benefit had the same effect of causing women salaried employees to put off getting married until their prospective husband had a good income and a secure job because they made benefits dependent on the family income rather than the income of the individual. Thus, women tended to get married at a later age[33] in this period. In 1927 the reduction in the number of unemployed was then 'proportionately greater in female than in male'.[34] For the end of 1928 a report on the plenary assembly of the Chamber of Industry and Commerce noted that 'most merchants preferred younger people or females in the hiring of salaried employees'.[35]

Shorthand typing work was also the point of entry for women into the male domain of banking. Starting with the demobilisation, employment opportunities expanded here at a tremendous rate until the end of the inflation in 1923, even though the banks made their employees put in up to six hours overtime a day from 1920 onwards — a policy which formed the central issue in trade union conflicts in this area.[36] However, the case of banking also indicates that the feminisation of white-collar work did not proceed wholly without setbacks. With the return to 'business as usual' at the end of the inflation period came not only a reduction of the workforce in Frankfurt's banks by about 60 per cent, but also at the same time a decisive fall in the proportion of women, as Table 5.3 indicates. Female bank workers were most affected by the wave of sackings even though they were here as elsewhere the cheapest to employ.[37] This was probably because the sums of money which the banks had to deal with were drastically reduced by the end of the inflation, and so the less skilled, mostly female extra staff employed to deal with them could be dismissed. Social factors may also have been involved, in the form of status improvement for male salaried employees who had lost prestige through the rapid postwar feminisation of an area which had resisted the encroachments of women into office work longer than most. Some of the dismissed shorthand typists could find jobs elsewhere, for although the demand for bank employees of both sexes fell to a mere twelfth of its former level at the end of the inflation, the demand for shorthand typists in trade and industry increased by 25 per cent at the same time.[38] Nevertheless, the pressure to change from one sector to another also implied that women's career prospects were blocked in banking, where the decisive criterion for salary increments was the presence or absence of a specialised training in the field. This was a requirement which the women who found employment here again in the years of stabilisation were unable to fulfil. Thus, although the period up to 1933 brought once more an increase in the proportion of women working in banks,[39] as Table 5.2 indicates, these were often employed for long periods 'on probation', could be dismissed at a day's notice, and often worked for less than 100 Reichsmarks a month, which was roughly equivalent to the pay of a semi-skilled industrial labourer. These practices suggest a pattern typical for female bank employees in the 1920s: very short-term, underpaid work, and almost exclusively semi-skilled occupations; only the prestige of being 'bank officials' could provide the women concerned with a 'compensation' for their meagre earnings or at least a 'plus-point' on their *curriculum vitae* in the subsequent search for other jobs.

Table 5.3: Bank Employees in Frankfurt am Main, 1923–4

	At 15 Nov. 1923				At 15 Sept. 1924			
	Total	Men	Women	% women	Total	Men	Women	% women
Major banks (9)	7,521	5,153	2,368	31.5	3,538	2,944	594	17.0
Provincial banks (21)	4,577	3,247	1,330	41.0	2,304	1,827	477	14.7
Small banks (54)	1,864	1,364	500	36.7	1,083	844	239	17.6
Total	13,962	9,764	4,198	30.0	6,925	5,615	1,310	19.0

Note: The largest banks were the Deutsche Bank, the Disconto-Gesellschaft, the Dresdner Bank, the Darmstädter und Nationalbank, the Commerz- und Credit-Bank, and the Mitteldeutsche Credit-Bank.

Source: *Bankbeamten-Zeitung*, hrg. vom Deutschen Bankbeamten-Verein, Berlin, XXIX Jg., 31.10.1924, S.135–6.

To give a concrete illustration of the typical situation of women office workers in the 1920s we can turn to the career of a female clerk, as revealed in her personal file in the records of the IG Farben (German Dye Trust). Born in 1899 in Hanau, just outside Frankfurt, M.B. was the daughter of an engine-fitter who probably worked for the Farbwerke Hoechst, since the family lived in company housing. M.B. lived with her parents and graduated from the Elementary School II in Hanau. After leaving school she had private tuition in French, and took a six-month course at a Steinhövel's commercial college. She could type 'perfectly' on an Adler and had mastered 200 shorthand characters. Starting in April 1915 she was employed as an office assistant in the town council of Höchst, but although her contract actually ran until April 1920, she was dismissed in September 1919 while she was working in the food office as a clerk and shorthand typist. 'Her resignation', the file noted, 'took place by mutual agreement to make way for the employment of male assistants returning from the War.' This was a somewhat euphemistic description of the circumstances: women were dismissed during the demobilisation without their agreement having been secured, particularly by state or local authority employers, who generally considered that they had a duty to carry out the demobilisation orders strictly as an example to others. Despite losing her job, however, M.B. quickly found employment as a clerk in the building materials wholesalers company of Karl Pietschmann in Höchst. However, in January 1920 she had to resign in order to look after her mother, who had fallen ill. In May 1920 she joined F. Frankl's South German Transport Insurance Company in Frankfurt as a shorthand typist. Just under a year later, in March 1921, she resigned in order to obtain 'better pay' elsewhere, a move that called down upon her the wrath of her employers, who wrote in the reference requested by the IG Farben that 'her conduct left much to be desired in the last period'.

In April 1921, therefore, M.B. moved to the corn and animal feed merchant Ferdinand Baer in Zeilpalast as a shorthand typist. However, she did not stay long. 'Irregular hours, often up to 10 o'clock at night' led her to resign in July 1921, when she took up a job with the Alliance Insurance Company in Frankfurt. After only a year here, she moved to the Autogen Works in Griesheim, a part of the IG Farben, initially for a probationary period of two months. In March 1923 she was transferred to the technical book-keeping department, where she was occupied with taking dictation for technical correspondence, preparing bills and general accounts work. In September 1924, she was dismissed from this post 'because of general cutbacks in staff', and in November the same year, helped by a positive reference from her former employer, found a job with the Messer Griesheim Company, a producer of chemical gases. From May 1925 to March 1926 she worked for the Naxos Union in Frankfurt, and from July 1926 to January 1928 as a shorthand typist for the Thyssen Rhine Steel Company. In January 1929, prompted by the Association of Women Commercial and Office Employees (*Verband der weiblichen Angestellten*), a conservative trade union that prided itself on supplying employers only with qualified office workers, M.B. applied for a job with the IG Farben. She obtained one as a shorthand typist in the central administration's sales department in Frankfurt in February 1929, as a replacement for a male employee who had 'blotted his copybook'. Her starting salary was 235 Reichsmarks, but it was raised to 255 'because of satisfaction with her performance' in May 1929, although the company was in truth obliged to raise it according to the salaries agreement then in force, since she was now 30 years of age. This broad experience of many kinds of work gave M.B. and women salaried employees like her an intellectual flexibility, a capacity for rapidly adapting herself to new demands, which her male colleagues, who may have had better formal qualifications but whose sphere of work had been restricted to the same department or branch of employment for years on end, generally lacked. Typically, she was not only not paid for the extra qualification of adaptability, but had to gain it at the expense of repeated periods of unemployment between jobs.

When we compare the proportion of women among all salaried employees in 1925 and 1933, it is clear that the feminisation of white-collar work had been increasing in all branches of the economy, as Table 5.2 suggests. Women's 'gains' were greatest in the service sector, especially trade, smallest in the administration departments of industry. Yet the greater proportion of women in these sectors in 1933, as the Depression was coming to an end, does not of itself demonstrate that there had been no replacement of women white-collar workers by men. The degree of 'natural' feminisation might have been greater without political interventions in the female labour market during the crisis.

We may tentatively conclude that there were two periods in which women withdrew from the labour market, one in the demobilisation, when women white-collar workers were less affected than women industrial workers, and a second in the stabilisation of 1924–5, which followed the campaign of 1922–3 against married women 'double earners' and affected both groups of women.[39] External pressures played a role in both periods, which suggests the importance of the so-called 'discouragement effect' which the 'reserve army' theory suggests as the reaction of women to such a crisis. Still, it is clear that such 'discouragement' was only possible where it had a material basis in the continuing provision of support by a husband or father. Those unmarried or widowed women who were dependent on paid labour after the war, protested in public against the 'abuses' of demobilisation and were able to mobilise a fairly broad political and trade union consensus against them. This did not challenge the social criteria of the right to work, which was demanded defensively for women who had no other material basis of subsistence. In both periods there are indications that some women managed to evade the pressure to push them out of their jobs. They included the saleswomen and office workers who directly resisted the political campaign and refused to leave the labour market during demobilisation, and those women who postponed marriage during the stabilisation phase in order to retain their jobs or their claim to unemployment benefits. Finally, there was one group in each period of which it can be said that they voluntarily left the labour market out of family or labour market considerations: female industrial workers during demobilisation, and women workers and salaried employees during stabilisation from 1924 who had put off founding a family up to that point because of the inflation.

As rationalisation began to affect trade administration and banking from 1924 onwards, female employment again increased unhampered. These years saw the formal establishment of female equality in the social insurance system, and the discriminatory practices of the labour offices seemed to be less sharply directed against women than had previously been the case.[40] The demand for 'semi-skilled' women (i.e. women who had completed a short course at a commercial college) led more and more working-class girls to try and move up in the social scale by escaping from the manual labour in industry that would have been their fate under more restrictive economic conditions, when they would scarcely have been able to afford the investment that a training in shorthand typing required. Already in 1933 some 94 per cent of all shorthand typists were women.

IV

Female unemployment among white-collar workers, as indeed generally, followed the same trends as male unemployment both before and after 1929, while remaining consistently lower all the while. An exception could be found in the (quantitatively insignificant) group of the 'salaried office employees' (*Büroangestellten*). Unemployment among women in this group increased sharply from the beginning of 1930 and far exceeded that of the men in the group. For the rest of the Depression women constituted about three-fifths of all the registered unemployed in this category, although the general trend moved in the opposite direction, with the number of female unemployed falling and that of male unemployed rising, for a brief period in 1931 and a rather longer one in 1932.[41]

In general, however, three periods in the development of female unemployment as a whole can be distinguished. From the end of 1929 to the end of 1930 it increased sharply, a trend that is confirmed by the massive increase in the numbers of unemployed women salaried office employees and indeed the proportion of women among all those claiming benefits (*Hauptunterstützungsempfänger*). A second phase began early in 1931 and lasted until the end of 1932, the high point of the crisis. Here, the proportion of women among the employed rose until it reached a point higher than at the beginning of the Depression. This was the case both for female salaried employees and for wage-earning female workers. In the recovery phase, from the beginning of 1933 to 1934, the proportion of women among all people in employment sank once more. Even after two years of Nazi rule, however, it still remained just above its level at the beginning of the Depression, as Table 5.4 indicates.

The decline of female employment in the first phase of the Depression suggests a strategy of reducing marginal labour on the part of capital. The decline would thus have been all the more severe in the areas where women were concentrated in marginal jobs in white-collar work and industry; in other words, in areas which had remained male domains despite an increasing feminisation of the labour force before the Depression. In traditional female-dominated industries and services in which women formed a substantial part of the permanent workforce, they were less affected, or at least not much more severely affected than they were anyway by normal seasonal factors. In addition, these 'women's industries' (consumer-goods production and processing and various services) were less severely affected by the decline in production than was male-dominated heavy industry. To a degree, therefore, the sexual division of labour by sector constituted a barrier against a disproportionate rise in female unemployment.[42] In the second phase, however, from 1931 to 1932, now that marginal labour had been creamed off, there seems to have been a strategy of favouring female employment on the part of capital. A long-term depression, as is well known, has certain advantages for employers on the labour market. It allows

Table 5.4: Women Workers in Firms with Five or More Employees, Frankfurt am Main, 1929–34

	Labourers			Salaried staff			Total	Women	% of women
	Men	Women	% of women	Men	Women	% of women			
1929	111,841			48,842			160,683	43,320	27.0
Middling firms (3,537)	23,328	7,442	24.2	9,608	7,011	42.2			
Large firms (523)	62,320	18,751	23.0	22,107	10,116	31.4			
1930	89,762			40,824			130,586	34,388	26.3
Middling firms (3,390)	19,624	6,500	24.9	8,968	6,463	41.9			
Large firms (465)	49,200	14,438	22.0	18,406	6,987	27.5			
1932	58,268			38,294			96,562	30,446	31.5
Middling firms (2,479)	12,583	4,653	27.0	7,609	5,386	41.5			
Large firms (332)	29,416	11,616	28.0	16,508	8,791	34.7			
1934	72,489			40,523			113,021	32,251	28.5
Middling firms (2,741)	15,817	4,641	22.7	8,085	5,453	40.3			
Large firms (360)	38,208	13,823	26.6	18,651	8,334	30.9			

Source: Calculated from figures given for 31 December each year in *Statistische Jahresübersichten*, 1928/29, p. 111; 1930/31, p. 35–6 1931/32, p. 35; 1933/34, p. 41.

them to undermine security of tenure, to reject wage claims, to weaken the unions and decisively to reduce prospects of permanent employment and promotion. Thus, there is a levelling-down of the position of those who under better economic circumstances would have been able to sell their labour on relatively good terms. And it is precisely these tendencies that give women a strategic advantage on the labour market, because they are generally less qualified, more flexible and less demanding in terms of jobs, and are accustomed to accepting short-term employment.

Thus, Kaethe Gaebel, writing in 1932, characterised the situation in the following words:

> A further recent characteristic development of the labour market must be mentioned: because of the risk and expense of warehousing, seasonal fluctuations are becoming ever more pronounced. Production is increasingly being concentrated within a few months or a few weeks. Apart from this, orders are also coming in at short notice and in bunches. So workers are not even in theory hired on a long-term basis any more, but just for a specific job, and when it is done they are dismissed. A few regular workers are kept or re-hired when the next order arrives, but otherwise employers take whatever the labour market has to offer. They are sure of getting enough people every time, and they feel their responsibility to the workers is satisfied by their contributing to unemployment insurance. Scarcity of resources forces them to act in the most calculating way possible. This system is already spreading to rural labour, where employment contracts are usually so long-term in nature, and to salaried employees. Department stores hire people just for special sales drives, for the 'white week', 'green week', 'household week' etc. and sack them at the end, lawyers even hire shorthand typists by the day, while the court is in session. Words cannot express what the resulting instability means in terms of professional ethos and sense of loyalty to the firm, indeed to work itself. It is shattering to contemplate the loss of internalised work values caused by this system.[43]

It was notorious that the number of white-collar workers of both sexes looking for jobs far exceeded the number of posts offered by labour offices and trade union job lists even in the so-called stabilisation period from 1924. The disparity increased still further in the Depression. Nevertheless, more women's jobs were on offer than men's jobs throughout the Depression. This indicated the employers' preference for cheap female labour and demonstrated a practical resistance on their part to the state campaigns against women's work even while important employers in Frankfurt, such as the IG Farben, were constantly paying lip-service to it.

Regardless of whether the emergency decrees issued during the Depression simply aimed at reducing the burden of unemployment insurance on the empty state coffers, or whether they also tried to create jobs, they still gave support to the strategies of capital described above. Because they brought among other things increased discrimination against women in the area of unemployment insurance,[44] the emergency decrees are generally counted among the various measures directed against women's work.[45] This only applied, however, when the criteria of full-time employment were applied, and these criteria lost their general validity in the Depression. For example, the emergency decree of June 1930, with its implementation regulations of October 1930, removed those working less than 30 hours a week and earning under 45 Reichsmarks a month from unemployment insurance obligations. The emergency decrees of June 1931 and February 1932 made the 'needs test' for receipt of unemployment benefit according to the crisis-welfare regulations more severe and incorporated earnings from 'minor casual labour' into the calculation of benefits; this provision was extended to the earnings of wives of unemployed, married men in the second decree. The dismantling of the welfare state by these emergency decrees forced those women whom they removed from receiving benefits to find employment on the growing 'secondary' labour market. The inclusion in the calculations of 'minor casual labour' by cost-conscious municipal authorities also meant indirect support for the married women whose employment was most strongly opposed ideologically by the state.[46] Every additional lowering of the benefit rate for unemployed women, as undertaken for example by the emergency decree of 1931, must have increased the relative advantage for unemployed women of engaging in any kind of paid employment to compensate for it. The only successful state measures against women's work appeared to be those which offered women financial inducements to leave their jobs.

It remains to ask what influence the organised representatives of male employees had on the extent of female employment. Unfortunately, there is no empirical evidence available on this question as far as Frankfurt is concerned. Successful political pressure of this kind would have had to have been expressed in the preferential treatment of men by unionised councils in the wave of dismissals that took place during the Depression. But the pressure remained merely ideological, because it would have required the preservation of men's jobs to have been successful in practice. In many cases of the temporary or permanent closure of businesses, no distinction was made between the core workforce and marginal labour. And even a male offer to take over worse-paid and insecure female jobs where a firm was dismissing only part of the workforce was bound to fail. The unionised councils generally lost power in the Depression, and this weakened their ability to exert an influence over redundancy policy. The interest of the employers on the other hand clearly lay in cutting labour costs by concentrating on preserving female jobs. Thus, the political pressure to substitute men for women was without effect, because the main problem of the men was how to keep their jobs. In the demobilisation, by contrast, their pressure was 'successful' because — unlike in the Depression — women in some areas had an interest in going back into housework. It was also a question of attitudes. During demobilisation women were supposed to make 'men's jobs' free for men again, whereas in the Depression men would have had to have taken over 'women's jobs' i.e. accept down-grading. This was another reason why men stood a better chance of realising their interests in 1918–20 than they did in 1929–33.

Thus, women were not merely discouraged from working by political and labour market pressures: they also reacted positively to the Depression, rather than simply passively accepting the state's attempts to remove them from the labour market. The Depression did not simply cause an increase in concealed female unemployment but also in hidden female employment. It also once more partially broke down the boundaries between different sectors of the female labour market by forcing women with specialist training to seek and accept jobs below their usual levels of qualifications and pay. Not only housewives but also highly qualified women workers were discouraged from competing on the labour market. They might well have been more prepared to accept downgrading than were men with the same qualifications. Such a consideration was always bound to be present, however, given the continual sexual discrimination to which these women were exposed even in normal times. After a long period of relatively satisfying salaried employment, even without the added pressure of mass unemployment, they may well have regularly opted for marriage and an exclusively domestic life which they might under other circumstances have chosen to avoid. Yet in the crisis a different kind of behaviour seems to have been typical for qualified women salaried employees during the Depression, namely a willingness to be flexible and to reduce their own expectations in terms of job, pay and length of employment. Thus, at the General Assembly of the Association of Women Commercial and Office Employees in September 1931 a speaker exclaimed:

> How much will to life, how much self-sacrificial boldness and courage can be read in the words and writings of our unemployed colleagues! How happily they come to terms with having to take jobs outside their normal sphere of work, with dressmaking, domestic service and the like, so as not to be forced to go on the dole! How faithfully others care for old parents and other members of their family from the meagre pittance that unemployment benefit affords![47]

If it is right that the need of qualified women to maintain their status diminished in the Depression, then it is probably also the case that the behaviour of their male colleagues, who were accustomed to a full-time career, was rather more likely to tend towards 'discouragement', or in other words, dropping out of the employment system altogether, should they be denied benefits. Indeed, long before the economic crisis reached its high point, the male white-collar unions were becoming obsessed with the fear of 'proletarianisation'.[48]

There is further evidence for the argument that as the Depression went on housewives and previously non-earning unmarried women living with their parents began to support their own families by going out to work. As the financial difficulties of the Frankfurt municipal authorities grew, so the benefit rate was cut until it fell below subsistence level. In November 1928 a married unemployed man with no children received a monthly benefit of 100RM. By March 1932 it was only 59.50 RM.[49] This meant a fall of 40 per cent in money terms, and it was not compensated for by any corresponding fall in living costs. These men's wives and daughters were thus driven increasingly to seek casual work of all sorts. Many of them flooded onto the domestic service market, where the situation, with an unemployment rate well below average at 15 per cent, was still better than elsewhere. Labour exchange reports from Frankfurt in 1931 and 1932 recorded increasing numbers of married women doing domestic work. They were, in addition, insured under their husbands, and so the employing household did not have to pay out the social insurance normally due for an unmarried servant.[50] The proportion of living-out servants rose from 14.4 to 24.5 per cent between 1925 and 1933 — a rise caused through the influx of job-seeking housewives, female industrial workers and women white-collar employees into this particular sector of the female labour market.[51]

Two groups of working women who did not appear in the statistics were, therefore, housewives coming onto the labour market and women workers and white-collar employees working below the level for which they were qualified. A third group did appear in the figures for late 1932 — women who kept their jobs or found jobs (again) in the Depression, as Table 5.4 indicates. The clear fall in the marriage rate in Frankfurt in 1930–3 suggests that many women were once more putting off their weddings to save their jobs. The numbers involved may be deduced from the figures of applications for the marriage loans provided by the Nazi regime from June 1933 onwards. A loan was granted on condition that the bride had had a job for at least six months before getting married and did not work for at least three years afterwards. Between June 1933 and December 1934, when the proportion of women among white-collar workers in employment had sunk back to its pre-Depression level, there were 3,543 applications for marriage loans,[52] equivalent to about 10 per cent of the women employed in middling and large firms in the town at the end of 1932. It seems therefore that women took up the financial inducement of the marriage loan in order to realise long-term plans held up by the Depression and now made possible by the recovery, and not because they were simply falling victim to the Nazi glorification of motherhood and family. This hypothesis is confirmed by the fact that the number of applications in the next year, 1935, had fallen by a third.[53] Women's plans for work and family continued to accept the conventional boundaries of marriage, but within these boundaries they put them into action largely independently of pressure from the state. Similarly, the small and relatively privileged group of female civil servants employed on permanent tenured contracts by the municipal administration in Frankfurt had taken advantage of official attempts to reduce the number of working women a few years before. In 1930 the municipality set up a special fund to finance compensation payments for women civil servants on permanent contracts, who were now obliged to resign their posts within six months of getting married. The fund was quickly exhausted, and so it was topped up and extended into the next financial year. It seemed, therefore, that even these highly qualified women who, unlike all others, had securely tenured jobs with long-term prospects, preferred to devote themselves to home life rather than carry the double burden of a career and domestic responsibilities. But they were not prepared to realise this preference, even in an ideological climate hostile to women's paid work outside the home, unless they were given substantial material inducements to do so.

As the economy recovered in 1933–4, the various groups of women who had come onto the labour market during the Depression now left it once more. So the decline of female labour and the fall in the porportion of women among all occupied salaried employees in Frankfurt between the end of 1932 and the end of 1934 was mainly caused not by the effectiveness of political measures directed against women's work, but rather by the behaviour and job planning of women themselves, although their behaviour of course gave the state measures the appearance of actually succeeding.[54] The economic upswing beginning in the middle of 1933 first brought the (overwhelmingly male) core workforce back into their jobs as many middling and large firms resumed production once more after having temporarily stopped it in the Depression, a development readily apparent from the figures provided in Table 5.4. The coincidence of this nascent economic recovery with the severe measures of the Nazis against women's work concealed the essential independence of women's behaviour. Women's own economic calculations — extra earning for their husband or parents, putting off marrying and having children so as to accumulate the financial basis for starting a family, hanging on to their jobs as long as there was no other breadwinner in sight, and a return to housework as soon as all this became unnecessary — were reactions to cyclical developments in the economy at large, but were determined by neither the demands of capital nor the policies of the state.

Notes

This chapter was translated by Richard J. Evans. The research was conducted as part of a larger research project at the Institute for Social Research (Institut für Sozialforschung) Frankfurt, on the history of women's work in Germany since 1870, with C. Eckart and K. Walser.

1. Dörte Winkler, *Frauenarbeit im 'Dritten Reich'* (Hamburg, 1977), pp. 26–8.
2. R. Bridenthal, C. Koonz, 'Beyond *Kinder, Küche, Kirche* — Weimar Women in Politics and Work' in B. A. Caroll (ed.), *Liberating Women's History* (Urbana-Chicago, 1979); cf. also note 45, below.
3. Paul Kuhn, 'Die Lohnregelung durch Tarifverträge unter besonderer Berücksichtigung Mannheimer und Frankfurter Verhältnisse'. (Wirtschafts-u. sozialwiss. Diss., Universität Frankfurt, 1925), p. 121. In 1925 only 5.9 per cent of female salaried employees were married; among female wage labourers the figure was 12.7 per cent, among domestic servants 3.2 per cent (*Statistisches Handbuch der Stadt Frankfurt am Main, 2. Ausgabe, 1928*).
4. Olivia S. Mitchell, 'The Cyclical Responsiveness of Married Females' Labor Supply: Added and Discouraged Worker Effects' in *Industrial Relations Research Association Series* (I.R.R.A.S.) Atlanta, 1979, pp. 251–7; C. Offe, K. Hinrichs, 'Sozialökonomie des Arbeitsmarktes und die Lage "benachteiligter" Gruppen von Arbeitnehmern' in C. Offe (ed.), *Opfer des Arbeitsmarktes* (Neuwied, 1977) pp. 3–62.
5. For example, many workers 'organised' coal, potatoes and other daily necessities and calculated this into their family income. Certain groups such as the male salaried employees in the IG Farben Trust, whose wives of course did not work and had fewer children than average, experienced a rise in real wages in the Depression but may have used it to support more members of their family than usual.
6. Calculated from *Stat. Jahresübersichten, 1927–28*, pp. 94–8; *1928–29*, pp. 28–30; *1930–31*, pp. 28–31; *1931–32*, pp. 28–30; *1932–33*, pp. 31–2; *1933–4*, p. 34; *1934–5*, p. 36.
7. Städtisches Arbeitsamt Frankfurt (ed.), *Die wirtschaftliche Demobilmachung in Frankfurt am Main* (1.11.1918–21.12.1919), (Frankfurt, 1920), p. 29.
8. Städtische Arbeitszentrale für das Bekleidungsinstandsetzungsamt.
9. Cf. the *Jahresbericht der Preussischen Gewerbeaufsicht für 1920*, p. 605, which reports an increase in the number of women working in the consumer-goods industries at the expense of the heavy industrial sector.
10. The factory councils exerted a clear pressure to dismiss married women during the demobilisation and the inflation periods. This was conducted in local and factory-level campaigns against female 'double earners' in 1922–3 (Gisela Losseff-Tillmans, 'Frauenemanzipation und Gewerkschaften (1800–1975)' (sozialwiss. Diss., Ruhr-Universität Bochum, 1975) pp. 380–2.
11. I.e. ex-servicemen.
12. Städtisches Arbeitsamt, *Die wirtschaftliche Demobilmachung*, p. 18–19.
13. Emil Stickler, 'Das Arbeitsamt der Stadt Frankfurt am Main', (Diss., Universität Frankfurt, 1925), p. 19.
14. Städtisches Arbeitsamt, *Die wirtschaftliche Demobilmachung*, p. 8.
15. *Statistik des Deutschen Reiches*, vol. 457/3, 1936, p. 26.
16. Thekla Justus, 'Die weiblichen Hausangestellten in Frankfurt am Main. Ergebnisse einer privaten Erhebung vom Jahre 1920' (Diss., Universität Frankfurt, 1924), p. 38–42; *Statistik des Deutschen Reiches*, vol. 210, 1907, p. 202.
17. Justus, 'Die weiblichen Hausangestellten', p. 87.

18. *Frankfurter Nachrichten*, 14 Oct. 1921, p. 10.

19. *Der Arbeitsmarkt in Hessen und Hessen-Nassau* (ed. by Landesarbeitsamt Hessen), NO. 6, 1920.

20. Thus, there was also a rise in the number of living-out servants, most of whom had their own families. 14.4 per cent of all female domestics were already in this category in 1925 (*Statistik des Deutschen Reiches*, vol. 406, p. 576).

21. Calculated from the *Statistische Jahresübersichten, 1926*, pp. 33–4. The figures apply only to the territory of Frankfurt city before the boundary changes of 1928.

22. Walter Decker, 'Die Tagespendelwanderungen der Berufstätigen nach Frankfurt am Main 1930' (wirtschafts- u. sozialwiss. Diss., Universität Frankfurt, 1931), pp. 24–7.

23. Else Fuldat, 'Die Arbeitsverhältnisse in der chemischen Grossindustrie der Frankfurter Gegend' (Diss., Universität Frankfurt, 1926), pp. 70 and 60.

24. As suggested by Jürgen Kuczynski, *Die Geschichte der Lage der Arbeiter unter dem Kapitalismus — Frauenarbeit* (vol. 18) (Berlin (East), 1963), p. 225.

25. The principles of piecework introduced in the war did not distinguish between men's and women's work and were still valid in the 1920s.

26. Fuldat, 'Die Arbeitsverhältnisse', pp. 30–1; Kuhn, 'Die Lohnregelung', pp. 121–2.

27. Ralph E. Smith, 'Women's Stake in a High Growth Economy in the United States' in Ronnie Steinberg (ed.), *Equal Employment Policy for Women* (Philadelphia, 1980), pp. 350–64.

28. Interview B-3 (employer in building and construction) p. 1.

29. Interview I-1 (employer in machine construction) p. 4.

30. Thus, in the autumn of 1921 there were 4.13 women commercial salaried employees without work for each job on offer, against 13.91 male commercial salaried employees. In office work the figures were 1.33 as against 15.75 (*Frankfurter Nachrichten*, 14 October 1921, p. 5). When the municipal labour office asked girl school-leavers what jobs they wanted, 28 per cent said 'saleswomen' or 'office worker' in 1921; in 1922 it was 38 per cent and in 1923 41 per cent. However, only 20 per cent of the boys in each year wanted to obtain 'commercial employment'. (Calculated from *Die Wirtschaftskurve*, III (August 1923), p. 94).

31. Stadtverordnetenversammlung, StvV 1024, 430. (Stadtarchiv Frankfurt).

32. Zentralverband der Angestellten (ZVA), *Geschäftsbericht 1928*, pp. 144–5.

33. After 1922 once a rather great number of widowed and divorced women had re-married — 13 to 14 per cent each year — the percentage of women's first marriages constituted constantly in between 92 and 93 per cent of all marriages in Frankfurt. In 1924 the total number of marriages reached its lowest point, with the average marriage age of women falling, to rise slightly again until this trend was reversed by the Great Depression from 1930 onwards. During the years 1925 to 1929 the average age of women at marriage varied between 27 and 28 years while the median age of women at marriage rose from 22–24 years in 1925 to 24–26 years at marriage in 1929, staying nearly that high all through the Depression. (Calculated from: *Statistisches Handbuch der Stadt Frankfurt*, p. 78; *Beiträge zur Statistik der Stadt Frankfurt*, Neue Folge, H.13–H.15, H.14, 2–3, 24–5, 46–7, 70–1, 92–3, H.15, 2–3, 26–7; *Tabellarische Übersichten betreffend den Zivilstand der Stadt Frankfurt a.M.*, 1931–35, Frankfurt, 1935, pp. 2–3, 26–7, 50–1).

34. Stadtarchiv Frankfurt: Stadtverordnetenversammlung, StvV 1901.

35. Ibid., StvV 1902.

36. *Jahresberichte der Preussischen Gewerbeaufsicht 1920*, 99; *1921*, 493 and 499; *1923*, 508; *1824*, 508; cf. also *Der deutsche Bankangestellte*, (Orgau des allgemeinen Verbandes der Deutschen Bankangestellten, Berlin) 1921 and the following years.

37. The Frankfurt salaries agreement of 1 January 1924 provided for women's salaries to be 9 per cent below men's at all levels except the highest, where women achieved equal pay after 15 years in the job, so long as they were in 'responsible posts' which required a 'training in banking or equivalent business qualifications'; this affected only a tiny minority of women and so cost the banks next to nothing (*Wirtschaftskurve* no. II, 1924, p. 132; Walter Funk 'Die Lage der kaufmännischen Angestellten im deutschen Bankgewerbe' (phil. Diss., Universität Giessen, 1927), pp. 35–7.

38. Calculated from W. Carlé, 'Der Arbeitsmarkt der Angestellten im Spiegel des Zeitungsinserates', *Die Wirtschaftskurve*, no. I, (1924), pp. 53–5, p. 55. The calculation is based on the classified advertisements in the *Frankfurter Zeitung*, which printed one national and two local editions every day (cf. W. Schivelbusch, *Intellektuellendämmerung* (Frankfurt, 1982) pp. 43–4).

39. The Frankfurt *Wirtschaftskurve* reported from the local trade unions in 1923–4 that 'women members are almost regularly worse affected by unemployment than men, which is attributable to social factors' (Heft II, 1923, 13; Heft III, 1923, 52; Heft I, 1924, 51–2; Heft II, 1924, 133–4).

40. Losseff-Tillmanns, 'Frauenemanzipation', 454–66.

41. The labour exchanges classified only highly qualified individuals as *Büroangestellte*, and it was thus difficult to remove them from the statistics.

42. A further indicator of this development is the fact that while the proportion of women in the industrial unions fell from 21.8 per cent in 1919 to 14.7 per cent in 1929, it hardly fell at all after 1929, and still stood at 14 per cent in 1931 (Kuczynski 1963, p. 249). Kuczynski claims this demonstrates the growing political mobilisation of women, but in reality it says more about the degree of female employment at this level, because people usually left their unions when they lost their jobs.

43. Käthe Gaebel in Bund deutscher Frauenvereine (ed.), *Die deutsche Wirtschaft und das Berufsschicksal der Frau* (Berlin, 1932), pp. 5–6.

44. As Losseff-Tillmanns, 'Frauenemanzipation', remarks, following M. Globig ('Die ökonomische und soziale Lage der Frauen in Deutschland (1933)' in Institut für Marxistische Studien und Forschungen (ed.), *Arbeiterbewegung und Frauenemanzipation 1889–1933* (reprints vol. 3), pp. 139–44; p. 139). 'The emergency decrees were followed by a noticeable decline in the proportion of women among those entitled to benefit, from 42.2 per cent on 15 October 1930 (before the decrees) to 13.9 per cent on 15 October 1932.' This also applies to Frankfurt. However, it is important to note that the reduction took place exclusively through a decline in the proportion of women among those receiving crisis benefits — that means among the long-term unemployed — while their proportion among those receiving unemployment benefit continued to grow.

45. Kuczynski, *Die Geschichte*; Losseff-Tillmanns, 'Frauenemanzipation'; cf. also: Stefan Bajohr, *Die Hälfte der Fabrik. Geschichte der Frauenarbeit in Deutschland 1914–1945* (Marburg, 1979), and Karin Jurcyk, *Frauenarbeit und Frauenrolle. Zum Zusammenhang von Familienpolitik und Frauenerwerbstätigkeit in Deutschland 1918–1933* (Frankfurt, 1976).

46. Magda Puder, 'Die Erwerbstätigkeit der verheirateten Frau' (Diss., Handelshochschule Berlin, 1932) pp. 32–5. Some unforseen effects of the Emergency Decree of 5 September 1932 were analysed by Marguerite Thibert in the following terms: 'In order to encourage the engagement of extra workers, this Decree authorised employers to reduce by 50 per cent all wages paid for the thirty-first to the fortieth hour worked each week, provided that at the same time they engaged additional workers equal to at least 25 per cent of their whole staff. A bonus at the rate of 7.70 Marks per week was also allowed for each extra worker so engaged. As the bonus was the same for both men and women workers, it was obviously definitely to the employers' advantage to engage women wherever possible.' (M.

Thibert, 'The Economic Depression and the Employment of Women', ILR XXVII 14, pp. 443–70; 5, pp. 620–30 (here pp. 469–70).

47. *Die Handels-und Büroangestellte*, Oct. 1931, p. 80.

48. Full benefit was paid only if the receiver sought employment exclusively in a job for which he had been trained and had not attempted to obtain another kind of job in the previous three years. Thus, qualified male white-collar workers had to undergo a long period of unemployment before they could look for other jobs. The labour offices assumed that after this long period, when the period in which these men could claim benefit had run out, they would look for jobs beneath their status; but this was the lowest form of social degradation as far as the Germans were concerned: nothing affected their emotions so profoundly as the unemployed clerk in his pin-stripes and bowler standing on the Kurfürstendamm with the sign 'Will do any job' hung round his neck. It is far more likely that they insisted on getting positions which they regarded as appropriate to their status and so dropped out of the employment system altogether.

49. Franz Fürth, 'Die Entwicklung des Existenzminimums in der Krise', *Die Wirtschaftskurve*, no. II (1932) 138.

50. These payments were stopped by the Nazis in 1934. The unemployment-rate was probably an underestimate: even in normal times many employers managed to avoid paying insurance for their servants.

51. *Statistische Jahresübersichten 1933/34* Ergänzungsheft 8, p. 41; cf. also Gaebel, *Die deutsche Wirtschaft*, p. 7: 'The domestic service labour market is experiencing an excess of supply over demand for the first time in years, although a very large proportion of these women have never done the job before and are thus not always suitable. So there are still plenty of opportunities for really competent people who are willing to adapt to a strange household.'

52. *Statistische Jahresübersichten 1933/34* Ergänzungsheft 8, p. 41; only 2,315 were approved.

53. *Statistische Jahresübersichten 1934/35*, 41.

54. Cf. Mitchell, 'The Cyclical Responsiveness', p. 257.

6 YOUTH UNEMPLOYMENT AND THE STATE: Public Policies towards Unemployed Youth in Hamburg during the World Economic Crisis

Elizabeth Harvey

I

Central and local governments in the world economic crisis that began in 1929 all agreed that the problem of youth unemployment required action. Politicians and officials across the political spectrum put forward a common view of young people as being particularly susceptible to the demoralising effects of unemployment. The experience of work, they argued, played a vital role in socialising young people into industrious habits and into socially and politically responsible behaviour: mass youth unemployment could put the health and morals of an entire generation at risk. There was less unanimity, however, as to what action was necessary or possible. At a time of heavy cutbacks in public expenditure, conflicts over the allocation and use of resources were acute. A wide range of training, recreational and work schemes were discussed and adopted. Although these initiatives were so varied, a distinct trend emerged in public policies towards unemployed youth in the course of the Depression. Overall, the goal of reintegration into production through employment policy gave way to a policy chiefly aimed at subjecting the behaviour of the young unemployed to an increasing range of state and other controls. Measures to deal with unemployed youth were increasingly justified in terms of the need to replace the disciplinary function of the employer, to support the control of the family over its youthful members and to uphold the authority of the state.

This contribution seeks to explain why public policies towards the young unemployed developed as they did. To understand this development, it is important to look at the impact of the measures implemented on a local level and at the reactions of the young people affected by them. For this purpose I have chosen Hamburg as a local case study. The impact of the world economic crisis in Hamburg was severe. The economy of Germany's biggest port town was heavily dependent on foreign trade: this applied both to its large commercial sector and to its biggest industry, shipbuilding. The rate of unemployment in Hamburg was higher than in the Reich as a whole, though comparable to unemployment rates in other big cities. The unemployed as a percentage of the total economically active population in June 1933 was 18.13 in the Reich and 30.02 in Hamburg.[1] Youth unemployment rates were similarly much higher in Hamburg than in the Reich, and higher than the unemployment rate of the overall working population: the unemployment rate among economically active 14–25 year-olds in June 1933 was 19.29 per cent in the Reich and 32.62 per cent in Hamburg.[2] Young people made up a quarter of Hamburg's 136,112 unemployed in July 1932: in that month 21,604 men and 12,470 women under 25 were unemployed in the city of Hamburg.[3] Unemployed youths were not only a more visible problem on the streets of the city than unemployed young women; according to official statistics, they were also worse hit by unemployment than young women. In June 1933, 47.22 per cent of the economically active men aged 20–25 and 26.72 per cent of economically active women in this age group were unemployed.[4]

A regional study also raises the question of the relationship between central and local government policy. From 1930 onwards the Depression, exacerbated by Brüning's deflationary policies, brought a crisis in local government finances, which was used by the Reich government to increase central control at the expense of local government autonomy. Hamburg had been governed since 1925 by a coalition between the SPD, the Democratic Party and the right-wing People's Party. Before the Depression the city government had undertaken large-scale investment in the city's infrastructure, housing and welfare services. In the Depression, with tax revenue decreasing and outlay on financial support for the long-term unemployed growing, the city government came under pressure from the Reich government to balance the city's budget. The Social Democrats in the city government, seeing no alternative, helped to carry through a programme of substantial cuts in public welfare and services in line with central government policy and current economic orthodoxy.[5] As far as policies towards unemployed youth were concerned, the cuts in the city's budget meant that the scope for local initiatives was curtailed: instead, central government funding and guidelines came more and more to determine what was implemented at local level.

II

Unemployed youth fared badly at the hands of the Weimar Republic's welfare system. First, the young unemployed were treated as a special category in need of particular control. Young people were thus the first targets of disciplinary measures to supervise the behaviour of the unemployed, with the withdrawal of benefits as an ultimate sanction in individual cases. Secondly, basic entitlements to benefit were cut across the board: young people as a group were the first to fall through the 'safety net' of the unemployment insurance system.

Before 1927 young people under 18 were entitled to means-tested benefits for 'those without a livelihood' (*Erwerbslosenfürsorge*) only under certain conditions. Under the decree of 16 February 1924, 14–16 year-olds were entitled to benefits only at the discretion of the local authorities and then only on condition that they performed 'compulsory labour' (*Pflichtarbeit*). The 16–18 year-olds could be declared entitled to benefits by the employment authorities of the federal states (*Länder*) according to the state of the labour market, and then on condition either that they performed compulsory labour or that they attended vocational training courses.[6] The national law on unemployment insurance of 16 July 1927 laid down that young workers were in principle entitled to pay contributions and receive insurance benefits. However, formal restrictions were again built in. The unemployed aged under 21 could be required to perform compulsory labour or to attend vocational courses in return for their benefits. A further restriction was applied to apprentices, who were not obliged to pay insurance contributions until six months before the apprenticeship contract expired (which helped to keep apprentices' pay low). No benefits were thus payable if the apprenticeship was terminated before that date. Discrimination against the young unemployed in the insurance system was increased under the Müller government through the exclusion of the under-21s from 'crisis benefits' and, decisively, from 1930 onwards under Brüning: the entitlement of young people to the unemployment insurance benefits was made dependent on a means test applied to the income of the young person's relatives. This regulation initially applied to the under-17s, then only to the under-16s, and finally, under the emergency decree of 5 June 1931, to all under 21. Such provisions rendered void the principle of unemployment insurance where young people were concerned, and paved the way for Papen's policy of dismantling insured benefits generally.[7]

The impact of these increasingly restrictive regulations is shown by the statistics on the number of claimants of insured unemployment benefit in the city of Hamburg. The number of claimants of the standard benefit aged under 21 years fell from a maximum of 5,235 in December 1930 to 2,237 in November 1932. Claimants under 21 as a proportion of total claimants of standard benefit fell from 12.95 per cent in July 1930 to 9.75 per cent in December 1931, rising from July 1932 onwards because of the new restrictions on adults' claims.[8] Access for the young to insured benefit was also restricted by practical difficulties: high unemployment and patterns of fluctuating employment prevented many young workers qualifying for benefit. The mass of unemployed youth was therefore forced to rely on means-tested municipal welfare assistance, the level of which was subject to the discretion of the local authorities. The Hamburg central welfare department fixed minimum rates of benefit for the young, single unemployed living at home at 5 RM per week and for those living away from home at 9 RM per week.[9] The district welfare offices decided whether more than the basic rate was to be paid: some welfare offices added up to 3 RM a week to the minimum rate. The higher rates payable to those living away from home in lodgings offered an opportunity for cutting back on benefits to the young unemployed in the name of 'family self-help'. This was either effected by reducing the level of benefit payable so as to make it impossible for the young person concerned to carry on paying the rent, or by simply cutting off benefit altogether if the unemployed person was under 21 and refused to move back home.[10]

Discrimination against the young unemployed with regard to financial provision was a constant throughout the period in question. It was one of the features of a welfare system which kept the unemployed population divided. Young people allegedly lacked work morale and discipline, hence conditions for claiming benefit were made particularly restrictive in order to prevent 'abuse'. The assumed ties of young people to their families were used to legitimise either lower rates of benefit than those for adults or the denial of benefit altogether. These arguments — that young people were less in need of income than older workers (the reference was usually to fathers of families) but equally if not more in need of occupation — underlay all the policies towards unemployed youth described below.

III

The labour market strategy pursued by the Reich labour ministry and, from 1927, the newly set-up national employment administration (*Reichsanstalt für Arbeitsvermittlung und Arbeitslosenversicherung*), pursued two basic aims: first, to reduce current youth unemployment and to prevent it in future, and secondly, to guarantee the reproduction of the skilled workforce. Here, demographic considerations played a role: the low birthrate during the First World War led to fears in the 1920s that there would be a shortage of skilled workers by the 1930s. The labour ministry and the employment administration focused their efforts primarily on measures designed to improve the chances of individual young workers on the labour market. These ranged from the expansion of vocational training for the unemployed to the redirecting of the unemployed into less overcrowded sectors of the economy such as agriculture and domestic service. In the 'rationalisation crisis' of 1925–6 and in the world economic crisis, the labour ministry (and, from 1927, the employment administration) attempted to carry through a national policy that would co-ordinate the widely varying policies pursued by local authorities for the provision of facilities for unemployed youth, on condition that such measures be undertaken primarily from the point of view of labour market considerations, concentrating particularly on transferring young workers into agricultural work or on obliging them to attend training courses.[11]

The strategy conceived by the labour ministry and employment administration was beset by a variety of problems. On the practical level, the obstacles to a large-scale training and retraining programme were considerable. Financial constraints meant that the courses were short (6–12 weeks, 10–12 hours per week on average) and often inadequately equipped, particularly if the aim was to teach skills applicable in modern industrial production. The main problem was, however, the fact that mass unemployment as it existed from 1929–30 onwards rendered the concept underlying plans for training and retraining illusory. Only with increasing implausibility could unemployment be ascribed to individual deficiencies. The pointlessness of vocational training at a time when skills were no guarantee of a job was hard to dispute. It was well known that, within the under-25 age group, it was the 18–25 year-olds that were worst affected by unemployment, among whom were the large group of newly qualified skilled workers who had been dismissed at the end of their apprenticeships. The realisation of this fact led to the exclusively vocational orientation of the courses funded by the employment administration being revised. A letter from the labour minister to the head of the employment administration in March 1931 summed up this policy shift:

> In view of the limited resources and the large number of the young unemployed, our policy must be to concentrate not so much on giving a relatively small number of young people as thorough a vocational training as possible but on organising the maximum number of young people in the courses, even if only for a short time: to get them off the streets, to maintain their interest in work and their capacity to work, and to protect them from the physical, intellectual and moral dangers of unemployment.[12]

The courses run by the Hamburg labour exchange (*Arbeitsamt*) for the young unemployed in 1926–7[13] were revived in 1930. By the winter of 1931–2, 2,330 young people were attending the labour exchange courses. The development of the Hamburg courses reflected the shift in national policy: the courses became more general, offering rudimentary instruction in various skills combined with general activities, and shorter, to achieve a higher turnover.[14] The labour exchange measures were primarily aimed at the recipients of insured benefit under 21, but they did not succeed in including all such claimants, and reached an even smaller proportion of the total young unemployed. The Hamburg city government therefore sought to supplement these courses by extending the hours of compulsory attendance at vocational school for unemployed 14–17 year-olds.[15]

Attendance at both these types of course was difficult to enforce. The attempt to extend attendance at vocational school encountered the same problem as did the enforcement of vocational school attendance generally: the unskilled, particularly the boys, frequently failed to turn up to the extra courses.[16] The employment office courses also had attendance problems, as the following example shows: Louise B., a 17 year-old unemployed domestic servant, was convicted in August 1931 of fraud. She had forged the signature of the course leader in order to claim her unemployment benefit. The court, however, saw mitigating circumstances in the fact that 'the accused missed the courses in order to look for work. She saw no chance of obtaining work through the employment office.'[17] Although those who considered the courses a waste of time were penalised for non-attendance, the labour exchange itself admitted in March 1932 that attendance was enforced largely for its own sake as a disciplinary measure: 'Although it is still policy that the vocational training measures are primarily intended to serve the purposes of vocational training, the measures do at the same time to a very great degree perform the function of testing willingness to work and preventing illicit work (*Schwarzarbeit*).'[18] The scale and quality of the vocationally oriented courses run by the local labour exchanges and vocational schools could not serve to reduce youth unemployment. Parallel to such courses, therefore, the labour ministry and the employment administration sought other measures to relieve the pressure on the labour market by transferring young workers from the cities onto the land or by delaying their entry into the labour market by raising the school-leaving age.

Transferring 'superfluous' young workers from industrial cities into agricultural work appeared to be a brilliantly simple solution to the political and social problems caused by youth unemployment and by the shortage of agricultural labour. The plans were even seen by some as a step towards a possible re-agrarianisation of Germany. The Reich labour ministry promoted the policy, and after 1926 subsidised the practical efforts of local authorities.[19] In 1930 approximately 18,500 young people under 21 were referred by labour exchanges throughout the Reich to work on the land.[20] Rural unemployment, however, made the plan to relieve a rural labour shortage senseless. By the summer of 1932 the employment administration reversed its policy, and subsidies for retraining urban youth in agricultural work were stopped.[21] The Hamburg labour exchange had never been in favour of the transplanting of the young unemployed onto the land, pointing out that young people were as unwilling to adapt to rural life as farmers were intolerant of city youngsters.[22] Overall, such transfers of labour played a negligible role in combating youth unemployment in Hamburg before 1933.

A different method of relieving the urban labour market for young workers, the extension of obligatory school attendance to nine instead of eight years (raising the school-leaving age to 15), was put forward in the economic crisis as a potential instrument of labour market policy.[23] However, the problem of financing the plan proved to be insuperable, and the ninth school year was introduced in Prussia in February 1931 only on a voluntary basis as an extra year to be spent in the elementary school (*Volksschule*).[24] In Hamburg a compulsory ninth school year had existed from 1919 to 1921. It came in for renewed consideration in 1930 as a response to the Prussian initiative, but was abandoned when Prussia failed to push its plan through. Propaganda efforts were then undertaken to persuade parents to let their children stay on for an extra year. No financial assistance was forthcoming to encourage parents to keep children at school, and the campaign was not a great success: instead of the hoped-for 1,000 that had been expected to stay on at Easter 1932, less than 600 opted to stay on out of 5,572 children due to leave school.[25]

Overall, the various measures making up the labour market strategy of the final Weimar governments towards youth unemployment were limited to attempts to regulate the overall supply of juvenile labour on the one hand, and to improve the individual chances of young workers on the labour market on the other. Efforts to transfer labour back onto the land and to encourage children to stay on at school failed to achieve either of these aims, while the vocational courses increasingly took place in a vacuum as the labour market deteriorated. As a result, attempts to combat the phenomenon of youth unemployment gave way to attempts to combat at most its political and social symptoms. The labour ministry and the employment administration became more and more concerned with ways of simply keeping young people occupied and preventing their 'demoralisation'.

IV

As the Reich government became more concerned with combating the psychological effects of unemployment on the young, its approach to the problem came to converge with that of the youth workers and educationalists who, in local, small-scale, often private, initiatives, were attempting to organise the young unemployed in sports and other recreational activities as a method of keeping young people off the streets. Such social workers and educationalists had criticised the approach of the labour ministry and the national employment administration on the grounds that short courses merely designed to teach vocational skills neither enabled the participants to find a job nor equipped them to cope with long-term unemployment.[26] Instead, they advocated a broader approach in which social work methods could be brought to bear on the task of organising unemployed adolescents. Youth welfare workers saw such schemes as an opportunity to reach hitherto elusive groups of working-class adolescents. As a representative of Catholic youth work in the Rhineland pointed out:

> We are concerned not only with the question of unemployment, but at the same time with the central issue of youth work: the question of the 'unorganised'. From this point of view, all youth workers have a particular interest in the question of the unemployed.[27]

In Hamburg this type of work with unemployed youth was undertaken by the public youth welfare department (*Jugendbehörde*). Its measures were aimed at those who were neither participating in courses run by the labour exchange or the vocational schools, nor were members of youth organisations: in other words, the long-term unemployed, older, 'unorganised' adolescents, both male and female. The organising of such 'problem' adolescents on a voluntary basis (since there was no means of enforcing compulsory participation) was admitted to be the major difficulty. The youth welfare department tried out different methods. Three-week holiday courses were run in the summer of 1931 for unemployed boys, designed to remove the participants 'for at least some time from the oppressive circumstances in which they live' and to 'reach them personally and make them civilised (*kulturfähig*) again', but they were dropped, after only two had taken place, on the grounds that they were too expensive.[28] Day centres for unemployed youth on the Social Democratic Berlin model[29] appeared to be a more economical method of providing supervised activities for unemployed youth over a longer period of time. The Hamburg youth welfare department began to set up day centres in September 1931. Opposition to the scheme was voiced on the one hand by conservative officials in the (general) welfare department on the grounds that cheap canteen meals for young people who could eat at home would 'destroy the family as an economic unit', and on the other by a right-wing member of Hamburg's city government, the Senate, who considered that 'bringing the young unemployed together in a day centre will inevitably cause an increase in delinquency'.[30] Such opposition, coupled with the dwindling financial resources of the city government, hindered the expansion of public facilities: by the end of 1932, the youth welfare department was running only four such centres, while private organisations ranging from the churches to the trade unions were running eleven centres between them, which catered primarily for their own members.[31] A report written in January 1933 for the youth welfare authority by the leader of the largest public day centre (capacity 260 people) gives some impression of the work of the centre.[32] Apart from providing subsidised meals and somewhere to escape the cold, the centre laid on sewing classes for girls and woodwork for boys, sports and entertainments. It attracted about three times as many boys as it did girls, confirming the impression from reports elsewhere that girls were more difficult to organise on such a basis than boys. This was ascribed partly to the pattern of working-class girls' leisure, which took less collective and organised forms than that of boys, and partly to the fact that girls had less leisure overall, tending to be kept at home by their families to do housework. Those young people who did come brought their own ideas of entertainment with them, and this led to disputes with the youth worker. His report on a group of particularly 'rough' streets in a working-class district of the city, the Hammerbrook, noted:

> Since those that come tend to be the sorts of youth that one sees hanging around the Süderstrasse, Idastrasse, Lorenzstrasse, etc. in Hammerbrook it is clear that one cannot reach them if one sticks to the strict 'laws' of the youth movement and of youth work generally. That is why there was from the beginning

a smoking room in the centre and on social evenings not folk dancing but 'modern dance'. These dance evenings (modern) have been a success. The young people initially tried to introduce customs that did not correspond to our purpose, but customs have by now been established of a decidedly superior kind. News of this dance evening got around very quickly, so that on the second evening 160–170 young people suddenly appeared and we were not in a position to control the situation. We have now introduced tickets, so that each youth is invited personally with his girlfriend. We are currently attempting to develop these dance evenings into social evenings with a greater intellectual content.[33]

This extract gives some impression of the problems confronting such social workers, whose sense of cultural mission, inspired by the youth movement, was confronted by a mixture of hostility and grudging compromise on the part of the adolescents whose taste and morals were seen as being in need of 'improvement'. Outright opposition to the day centres in Hamburg was not reported, in contrast to occurrences in Berlin:[34] the youth worker quoted above remarked on the lack of political activism in the centres.

Whether the youth welfare department's efforts to tackle the perennial problem of the 'amorphous underclass of adolescents that refuse to let themselves be organised in any way'[35] would have been more successful given greater resources may appear doubtful: as it was, its experiments were limited by financial and political constraints. The dilemma in which youth welfare administrators found themselves, faced with mounting tasks and dwindling resources, led to their seeking extensive rather than intensive methods of work with the young unemployed, in particular to support plans for work schemes for unemployed youth.

V

The labour market strategy pursued by the final Weimar governments towards youth unemployment neglected the option of increasing demand for young workers through work-creation schemes. The scale of work creation for the unemployed generally in the final years of the Weimar Republic was limited by Brüning's rejection of all major public investment plans to stimulate the economy, while his successors in their short periods of office did little more than lay the basis for Hitler's initial work-creation schemes.[36] The main forms of work creation — the 'emergency labour' schemes (*Notstandsarbeit*) partly financed by the Reich and aimed after 1927 at the recipients of unemployment benefit, and the 'welfare labour' schemes (*Fürsorgearbeit*) financed by the municipalities and aimed at those on municipal welfare assistance, were generally reserved for older, married male workers. Though paid at below normal wage-rates, such work was based on a normal work contract and was subject to the payment of social insurance contributions.[37] Work schemes set up for young people, on the other hand, rarely provided anything approaching normal work at normal wages, nor were they set up with this objective in mind.

Two types of work scheme run in the final years of the Weimar Republic were aimed exclusively or to a large degree at the young, unmarried unemployed: the municipal compulsory labour schemes for those on welfare assistance, and the voluntary labour service. Despite their contrasting origins, these types of work scheme displayed some common features. First, they were intended to reduce the expectations of young workers both regarding the quality of work and regarding pay. Most of the work was unskilled manual labour and could have only a de-skilling effect, running counter to the earlier schemes of the labour ministry to promote the maintenance and improvement of young people's skills. Secondly, the work was defined as being outside the bounds of contract and outside labour law. The remuneration for such work — in the form of a bonus on top of a minimal unemployment benefit or of 'pocket money' — was set at a level below legal wage-rates. A by-product of this policy was that the schemes could function as a source of cheap labour for local authorities. Thirdly, the schemes shared an overall disciplinary function: they aimed to keep the young unemployed off the streets and to control their behaviour.

Of the various forms of municipal work schemes developed in the Weimar Republic, the compulsory labour scheme was the form most commonly foreseen for the young, single unemployed.[38] Municipal interests and the pursuit of efficiency in administering the long-term unemployed led the Hamburg welfare department under its conservative director Martini, who belonged to the German People's Party, to expand work schemes for claimants of welfare assistance during the world economic crisis. While the welfare labour projects (*Fürsorgearbeit*) were a method of enabling the 'deserving' unemployed (above all those on high rates of benefit, such as married men with families) to regain their entitlement to insured unemployment benefit,[39] compulsory labour schemes were to deter claimants and to provide a pretext for stopping benefits to those judged 'undeserving'. The work was officially defined as being performed as a condition for receiving welfare assistance. Those identified by the welfare department as most likely to be defrauding the state and thus as a primary target for obligatory labour were the 'younger, single men, who are supported by the district welfare offices, who have only worked very irregularly and with regard to whom considerable doubts about their willingness to work are justified'.[40] State officials were also convinced that large numbers of fraudulent claims came from unemployed women with 'a certain irregularity of lifestyle'.[41] Obligatory labour was not only imposed on those considered to be abusing the system: it was also part of a strategy for dealing with 'trouble-makers'. Officials tended to label those who dared to complain or protest about the handling of their claims as 'mentally defective and asocial elements' incited by 'political subversives'.[42] Such protests were regarded as all the more suspect when the protesters were young, as the district office in St Pauli reported in December 1929.[43] The welfare department justified its treatment of the unemployed by portraying itself as being in the front line in the battle for law and order. The department worked closely with the police, who sent reinforcements on critical days such as Monday 29 June 1931, the day on which the provisions of the emergency decree of 5 June 1931 came into force, and supplied 61 rubber truncheons from police stocks for the use of the welfare officials who requested them.[44] Fear of the unemployed could lead the welfare department to take steps to relieve the pressure on its district offices and to give the impression that practical initiatives to tackle financial need were being taken. In June 1931 the department spent some time discussing recent disturbances in the welfare offices.

> One district office reports that it has observed that the trouble-makers are mostly youths. The new emergency decree will result in the welfare offices having to deal with a vast new influx of youths: thus the trouble is bound to get worse. Martini remarks that plans have already been made to set up new work sites for young people.[45]

The statistics on the number of young people working as 'compulsory labourers' (*Pflichtarbeiter*) are fragmentary. A breakdown of the workers on individual schemes according to age in December 1932 indicated that young people under 25 made up 40–45 per cent of the workers.[46] There were also sites set aside for young people under 21, where by the summer of 1932 240 young people were occupied. The total number of compulsory labourers reached a maximum of 5,582 in August 1932 before falling to 2,288 in January 1933, so the number of workers under 25 is unlikely to have exceeded 2,500 at the most.[47] This meant that the welfare department's comprehensive plan for disciplining the unemployed was seriously hampered. The argument for the use of coercion — that it enabled the authorities to exert control across the board over the 'hard cases' — was weakened by the fact that the exercise of such control could only be partial and selective. Moreover, the deterrent effect of compulsory labour was undermined as the crisis deepened. In 1929 the district welfare offices in the 'problem' districts of St Pauli and the Altstadt, who were the first to impose obligatory labour on young unemployed claimants, had reported a marked deterrent effect: prospective claimants tended to evade the measure by moving to the district of St Georg.[48] However, by the autumn of 1930 all the district welfare offices reported that the deterrent effect had worn off. Some of the unemployed were even volunteering for work for the sake of the small bonus of 75 pfennigs paid on top of the normal rate of welfare assistance for each of the three days worked per week.

Although fewer people dared to refuse to undertake obligatory labour as the economic crisis worsened, the implementation of the work schemes on the basis of coercion remained a problem. Passive resistance to the work was reflected in extremely low productivity. Active tactics of evasion and opposition emerged more sporadically. Collective opposition was most common on the outdoor sites where unemployed men were made to chop wood, clear scrub, dig drainage ditches and the like, and it was on several such sites in October 1930 that the longest strike occurred. The strike, which lasted from 14 to 29 October, was led by Communist Party activists and co-ordinated with similar action in other cities.[49] It was supported at its height by three-quarters of those employed on the sites concerned — a total of around 470 strikers. The strikers demanded wages at the rate fixed by collective agreement for municipal employees for the days worked on the sites, a demand which was above all in the interests of the unmarried, who were on lower rates of welfare assistance. Of the workers on the site, 80 per cent were unmarried, and it was reported that among those who carried on working, the married men predominated.[50] The Social Democratic and conservative press portrayed the strike as the work of irresponsible youths led by the Communist KPD: 'Communist demagogy has succeeded in driving these mainly youthful unemployed to stop working.'[51] The strike ended in failure, with the strikers losing their entitlement to welfare assistance: Martini reported with satisfaction during the strike that the KPD was supporting those on strike and thereby saving public expenditure on benefits.[52] A second strike wave in June 1932, co-ordinated by the Communist-led committee of the unemployed, was sparked off by the news that the working day for those on the work schemes was to be lengthened. This time the department responded by closing all the men's sites (the women were judged unlikely to strike) for a period of weeks.[53]

Although there were no recorded strikes by women on work schemes, the welfare department nevertheless met with determined opposition on an individual basis from unemployed young women. This was particularly evident in the case of a residential scheme run by the department in premises outside Hamburg owned by the Protestant welfare organisation, the Innere Mission. The scheme was set up in October 1932 with the dual goals of depriving the 'undeserving' of benefits and of retraining the industrious and compliant for domestic service. The girls were required to live in and perform housework, with the occasional free Sunday, in return for board and lodging and 'pocket money' of 50 pfennigs per week instead of their usual welfare assistance. The girls rebelled against this deprivation of liberty: some simply left, and as a result lost their right to benefits, while others sought the support of their parents in finding reasons to leave. The parents generally backed their daughters against the welfare department on the grounds that if their daughters were going to do unpaid housework, they could do it at home.[54] The director of the scheme remarked that the cause of the girls' discontent was a 'feverish desire for the city' and its amusements, for male company and for work in shops or factories rather than as domestic servants.[55] Those who planned the scheme had underestimated the difficulties involved in reducing the expectations of unemployed girls so radically: the scheme succeeded only in stopping a number of girls from receiving further welfare payments.

At the latest by 1932, it was apparent that the conflicting goals pursued by the welfare department in its compulsory labour schemes policy could not be reconciled. A policy of coercing the 'hard cases' and selecting those perceived as 'asocial' for the schemes made for low productivity on the sites and unwelcome adverse publicity. The strikers completely failed in their aim of obtaining the wage-rate they demanded, but they helped to push the welfare department into altering its policy. From the point of view both of projecting a positive image of the city government's welfare provision, and of getting work done for the municipality efficiently, the use of voluntary rather than conscripted labour began to appear a better alternative. Officials in the welfare department were therefore by the autumn of 1932 looking at the voluntary labour service as an alternative model for occupying the unemployed.

VI

The voluntary labour service (*Freiwilliger Arbeitsdienst*) was introduced in June 1931 and was the Brüning government's main initiative to deal with mass youth unemployment.[56] It was run by the national employment administration, which by offering central funding at a time when local municipal initiatives were drying up for lack of funds, could ensure that measures for the young unemployed at local level would increasingly take the form of voluntary labour service schemes. National subsidies under the voluntary labour service were provided for labour-intensive projects of land improvement and the like that were judged to be in the public interest. Regulations regarding work contracts and social insurance payments did not apply to the work, which was to be carried out by young unemployed volunteers. Private organisations were encouraged to become involved in organising voluntary labour service projects. The regulations regarding the eligibility of volunteers for public funding were initially restrictive: recipients of insured unemployment benefit were eligible for Reich funding for a period of 20 weeks; those who would have qualified for insured benefit but were under 21 could be subsidised only at the discretion of the local authorities. These restrictions were lifted by the Papen decree revising and expanding the voluntary labour service in July 1932.

The motives for the introduction of the voluntary labour service were diverse. The scheme was intended as a sop to right-wing advocates of a compulsory labour service, which was regarded by the Brüning cabinet as neither financially nor politically feasible in the short term. It was also envisaged as a cheap alternative to substantial programmes of public works and as such was hoped to have some propaganda effect.[57] Labour-service volunteers would be temporarily removed from the labour market; in the longer term, it was hoped that an elite of volunteers would be recruited for agricultural settlement.[58] Critics of the voluntary labour service argued that it was also designed to force down wage-levels: since work carried out under the scheme could only be of an 'additional' nature which would not otherwise be done, projects would be classified as impossible to undertake under normal circumstances because of high wage costs and therefore only feasible using volunteer labour.[59]

The voluntary labour service also had more general educational and social goals, which became explicit in the decree issued by Papen in July 1932: 'The voluntary labour service gives young Germans an opportunity to perform significant work for the public good on a voluntary basis and at the same time to undergo training for physical, mental and moral fitness.'[60] The recommended form for the schemes was that of the residential camp, which enabled more comprehensive supervision of the urban recruits in an environment which isolated them from the pernicious influence of the city — and incidentally removed from the city a potential source of unrest.[61] This broad ideological and educational purpose was a feature of the voluntary labour service which contrasted with the municipal work schemes, which left those engaged on such schemes to their own devices in their free time — facilitating the organisation of opposition to the schemes.[62] Other features which contrasted with the municipal schemes were the principle of voluntary participation and the role of private organisations. The labour service was set up on a voluntary basis in order to avoid the low productivity and disciplinary problems of work schemes based on coercion.[63] Although its critics argued that most volunteers were motivated by the material reward, however marginal, and were forced by their desperate economic situation to participate,[64] it still remains true that the labour service remained as a rule formally voluntary until 1933 — even taking cases into account where local authorities in late 1932 illegally threatened to withdraw benefits unless unemployed youths 'volunteered'.[65] Regarding the role of private organisations, a national survey conducted in January 1932 showed that public authorities ran less than a third of all projects. The state's aim in relying on private initiative to such an extent was partly financial — the private organisations often used their own assets to subsidise the schemes — and partly political: Brüning's ministers Stegerwald and Treviranus argued that involving the paramilitary organisations in the 'constructive' task of the voluntary labour service would tame them and counteract their 'political dilettantism'.[66]

The voluntary labour service was oriented primarily towards male youth. Girls were marginal to the ambitious plans of right-wing nationalist propagandists who wanted to see the scheme paving the way for the introduction of general labour conscription for all young men, which would act as a 'school for the nation', a substitute army, while at the same time performing the task of reconstituting an agrarian Germany through the reclamation and settlement of agricultural land. This mixture of paramilitary and frontiersman imagery left little room for a role for girls and women except as settlers' wives and helpmates. Although the Brüning government did not fully share the ambitions of such propagandists, its conception of the voluntary labour service was equally focused on schemes for male youth. By July 1932 only 247 of the total of 5,633 projects that had been approved in the Reich as a whole had been for girls.[67] The projects for girls almost invariably consisted of sewing for the 'Winter Aid' programme (*Winterhilfe*), cooking for the unemployed, or laundering and cooking for the men's camps, and they tended not to be residential. The number of girls in the voluntary labour service rose in the autumn of 1932, but they remained a minority of the total participants.[68]

At the beginning of August 1932, 96,067 volunteers were reported to be currently engaged on voluntary labour service projects in the Reich as a whole. When this figure is compared to the 1,456,854 young people under 25 registered unemployed at the end of July 1932, it emerges that 6.59 per cent of the total official unemployed under 25 in the Reich were at this time participating in the voluntary labour service. Participants reached a maximum pre-1933 total of 285,000 at the beginning of December 1932 (no exact youth-unemployment figures are available for that month to compare).[69] In the Hamburg area (including Harburg, Wandsbek and Altona) at the end of August 1932, 4,037 young people from the area were occupied on voluntary labour service schemes, of whom about 1,000 had gone to schemes outside the area. This was equivalent to 7.4 per cent of the 54,302 unemployed under 25 in the Hamburg area in July 1932. However, the proportion was slightly lower in the city of Hamburg, since the figures for the Hamburg area were weighted by the very high number of participants from Altona, where the Social Democratic town government channelled substantial municipal resources into the expansion of the voluntary labour service: in July 1932, about 1,500 young people, 21 per cent of the town's 7,084 unemployed under 25, were occupied on voluntary labour service projects in the town.[70]

A survey of the 121 projects under way in September 1932 in the Hamburg area[71] showed that, while the municipal governments in Altona and Harburg were taking the opportunity of getting routine municipal work done cheaply, the Hamburg authorities were less active in organising their own voluntary labour service projects or promoting those of private organisations. This was partly because Social Democratic officials in Hamburg were initially critical of the voluntary labour service on political grounds, and partly because some conservative officials opposed the development of projects not directly under local public control. Among the private organisations involved in the running of projects, the sports clubs played a major role: the chance of improving club facilities with the aid of public subsidies proved irresistible. Extreme right-wing organisations were less active in Hamburg than elsewhere due to the influence of the trade unions and the strength of the SPD in local government. The Stahlhelm, a right-wing anti-republican ex-servicemen's association, complained that it was being discriminated against by the Hamburg employment office, whose approval was necessary for setting up projects.[72] The political balance of forces among the organisations involved in projects in the Hamburg area lay with Social Democratic organisations, which became involved after the SPD and the federation of free trade unions dropped their opposition to the voluntary labour service at national level in the spring and summer of 1932.[73]

The voluntary labour service schemes run in the Hamburg area did not entirely correspond to the typical schemes portrayed in the press — residential camps for young men undertaking primitive manual work on land improvement. Although the statistics on the Hamburg area for September 1932 confirm that the majority of the projects were schemes to improve the land, only a quarter of these projects were in the form of residential camps, since the volunteers lived at home and travelled each day within the urban area. The tendency for men's projects to predominate was slightly less marked in Hamburg than in the Reich as a whole. In September 1932, 12 per cent of projects in the Hamburg area were for girls, compared with 4 per cent in the Reich as a whole.[74]

Recruiting volunteers does not appear to have been a problem in the Hamburg area up until the winter of 1932–3, except for particularly unpleasant work.[75] Records of two schemes in the Hamburg area confirm that the labour service recruited from among the financially most desperate unemployed. Of 70 youths engaged on the Oejendorf scheme run from October 1932 by the Hamburg welfare department, 34 had been dependent on welfare support and 16 had been receiving no benefits at all. Of the 160 youths, mainly from Hamburg, on a church-run project in Süderdeich in Holstein, 60 had been receiving welfare assistance and 67 no benefits. Only 29 per cent of the Oejendorf volunteers and 21 per cent of the Süderdeich volunteers had been receiving insured benefits.[76] Of the Süderdeich volunteers, 48 per cent had been out of work for more than a year. One of the volunteers on this scheme, a 21 year-old printer from Essen, unemployed since January 1932, wrote for the church authorities organising the camp an account of his 'way to the voluntary labour service':

> After long, planless wanderings from city to city my path took me to the port of Hamburg. But what a disappointment! Here was yet more misery, more unemployment than I had expected and my hopes of getting work here were dashed. What should I do? Without relatives here, I had no desire to become a vagabond and I decided, like so many of my fellow-sufferers, to volunteer for the labour service. It was January 1933.[77]

Some further light is thrown on the organisation of the voluntary labour service and on the recruitment and motivation of volunteers by looking in some detail at one scheme for which relatively detailed records exist.[78] While it cannot be representative, this example may serve to illustrate one type of project and a certain type of participant. The voluntary labour service scheme for girls in Malente (Holstein) was run by the welfare bureau of the Lutheran church in Hamburg from October 1932 to March 1933. The 28 volunteers were accommodated in a church holiday home: the possession of such premises gave the churches a great advantage in setting up projects. The majority of the volunteers were already involved in church youth clubs and girls' circles. The scheme was run on clearly confessional lines, although it served from the point of view of the organisers less to convert the heathen than to consolidate the influence of the church over what was already largely its own flock. As the organiser put it: 'The Church must realise that the voluntary labour service offers opportunities for pastoral care [*seelsorgerische Möglichkeiten*] that no pastor can reckon with today in his parish.'[79] The relatively weak development of the labour service for girls was sometimes ascribed to the low degree to which girls were organised as members of youth groups: projects were most easily set up where, as in this case, there was a basis of already existing clubs and organisations. The work in Malente consisted of sewing and repairing clothes for the church welfare, plus housework and gardening on the premises.The girls received board and lodging and 30 pfennigs pocket money per working day. The list of participants and the essays they wrote while on the scheme (which have to be read with the consideration that they were written for the organisers) provide information about the volunteers. They were aged between 14 and 23. Thirteen had been domestic servants, one a laundress, 3 had been shop assistants and 5 clerical workers; 6 had never been employed. Eight of the 28 had been out of work continuously for over 18 months. Only 4 participants had been receiving welfare benefits; none of the rest had been receiving any sort of support. Five of the girls had fathers who were unemployed. Regarding the girls' motives for volunteering, not all emphasised material need. Some gave as their motives as 'getting away from it all' and 'being with other girls'. Others wrote that the voluntary labour service was an opportunity to learn useful skills, and that employers would be likely to prefer those later on who had been in the voluntary labour service. Several, however, emphasised the poverty in their family as a reason for volunteering.

I am here because I got nothing from the welfare office and my mother cannot support me either, my brother is unemployed and doesn't get anything either, neither of us can live at home because my sister earns so little, and my mother only gets her pension. (Gertrud R., b. 1911)

I wanted to buy a winter coat and my mother said I can't buy you one, my mother hasn't got enough money for food so she can't buy me one, and I can't get any work, so I came to the voluntary labour service and I like it here very much. (Hilde F., b. 1914)

To say this is voluntary . . . for many people it is necessity and if one of you can live away from home and have work, that is a great relief for parents. (Frieda L., b. 1916).

The difficulties of organising the voluntary labour service grew with its expansion, as the resources of the organisations and the public institutions involved were used up. Volunteers were also becoming scarcer by the winter of 1932–3. Private organisations involved in running the voluntary labour service in Hamburg reported in January 1933 that the youth organisations were having problems recruiting for measures currently in progress and that the voluntary labour service in its present form had reached its limits.[80] Disciplinary problems also increased as the labour service grew, and it was these problems as much as financial or economic considerations that were seen in November 1932 by the national director of the voluntary labour service as the chief obstacle to the introduction of compulsory labour service. The Reich government was well aware that conscript armies were liable to subversion.[81]

The voluntary labour service in the Hamburg area also demonstrated that disciplinary problems were not insignificant. Dissatisfaction with the work, the pay, the food and the strict camp regime led to volunteers leaving or to acts of individual protest leading to the expulsion of the person concerned. While the girls' scheme in Malente described above only expelled one participant, the boys' scheme run by the same organisers on the same premises in April –May 1932 began with 31 participants and ended with 15, 5 having left individually of their own accord and one having been expelled: 10 others had left with him out of solidarity.[82] An article on this Malente scheme in the Communist paper *Norddeutscher Echo* attacked not only the inadequate pay for the hard physical work, but also in particular the combination of compulsory Christianity and Nazi indoctrination dominating the camp regime.[83] On the other hand, the organisers of the church-run camps were particularly alert to signs of 'Marxist opposition' among the volunteers, and freely made use of their powers to expel participants — a sanction to which they would have no recourse had compulsory labour service been in force.

Collective action against the voluntary labour service was difficult, though not wholly impossible, to organise. A campaign against the voluntary labour service was waged by the Communist youth organisation.[84] Its tactics at national level were initially to undertake agitation against the voluntary labour service and to boycott schemes already in operation. The emphasis of the campaign was shifted in the summer of 1932 to strike action. In the Hamburg area the campaign made most impact in Altona, where in October 1932 the schemes organised by the town authorities were almost paralysed by a strike which was praised by the KPD as exemplary.[85] The strike, which affected 12 municipally run sites with a total of 1,200 volunteers, was sparked off in mid-October by the announcement by the town authorities that the imposition of new national regulations meant that the pay on voluntary labour service projects had to be cut. The strike failed to reverse this cut and a partial boycott continued. By the beginning of November the SPD newspaper reported that only 250 of the original 1,200 workers were prepared to work for the lower rate.[86] The large number of volunteers in Altona, the fact that the sites were non-residential, and the ease of co-ordination between the different sites helped to make the strike relatively long and well supported. Where volunteers were isolated in the countryside on a residential scheme, without political support from the neighbourhood or the means of making their demands public or their protest visible, the chances of success were even slighter: hence the preference of the Reich government for residential schemes.

A further problem connected with the expansion of the voluntary labour service in the summer and autumn of 1932 was that a growing number of young people was returning after 6 months' 'service' to further unemployment and the street. Schleicher's 'Emergency Programme for German Youth' (*Notwerk der deutschen Jugend*) was aimed particularly at this group of unemployed youth. The measure deserves a brief description as a footnote to this account of policies towards unemployed youth up to the Nazi takeover, although its real development, such as it was, took place after January 1933. The programme was set up, with the aim of attaining maximum propaganda effect, on Christmas Eve 1932. With the aid of subsidies (to be spread as thinly as possible) vocationally oriented courses and recreational activities were to be set up in a minimalised version of the day centres described above. The spectacular numerical expansion of the emergency youth programme — after a month approximately 110,000 participants were reported — was only possible by means of an optical illusion. The organisers of existing courses and activities, including all the courses run by the employment offices, applied for subsidies of up to 25 pfennigs per person per day, and laid on a hot midday meal for participants as prescribed in the regulations of the programme. Such courses were thereby transformed into schemes under the programme and appeared as such in the statistics. Like the voluntary labour service, the emergency programme was heavily dependent on the financial participation of private organisations, and relied on youth organisations running activities for their members with the subsidy contributing to the total cost. Organising the 'unorganised' through the scheme inevitably remained a pious hope on the part of its propagators.[87] In Hamburg the emergency youth programme was set up on 10 Janaury 1933 with the assistance of a grant from the city government, the Senate. The Senate saw the scheme as an opportunity to double the capacity of the municipal soup kitchens with the aid of a national subsidy. Even though the existing courses run by the vocational schools, the employment office and the day centres turned themselves into emergency youth-programme projects, the number of participants failed by February 1933 to reach the target of 15,000, which would have represented half the official number of young unemployed under 25 years of age in Hamburg in January 1933.[88]

VII

'In Hamburg there are no juveniles on the streets who do not have an occupation of any kind.'[89] No employment official during the world economic crisis would have been likely to repeat this claim, originally made in 1927. It was too obvious that none of the policies discussed above had a substantial quantitative impact on youth unemployment. Precise figures are lacking, but an estimate may be made for the summer of 1932: compared with the official youth unemployment figures for July 1932, the vocational courses run by the labour exchange and the compulsory labour schemes each reached about 7 per cent of the city's unemployed under the age of 25, and the voluntary labour service reached about 7 per cent of the young unemployed in the Hamburg area. Taking other miscellaneous measures into account, it still appears that in the summer of 1932 at least three-quarters of Hamburg's young unemployed were not reached by any of the schemes run or subsidised by the public authorities.

How far Hamburg was typical of German cities in the Depression in terms of this failure could only be said with certainty if studies of other cities were available, but it seems that Hamburg was not a notable exception to the rest of the Reich. It is true that the Hamburg city-state government subsidised programmes for unemployed youth less than, for instance, the Social Democratic municipal governments in Berlin and Altona. Nevertheless, what was characteristic of the years in question was that all local governments were reducing their spending on non-statutory measures. This had two consequences: first, an increasing share of the task of supervising the unemployed was taken over by private organisations — this was true of Hamburg as for the Reich as a whole. Secondly, Reich subsidies administered via the local employment offices played an increasing role in determining policy, though particularist and politically motivated opposition to the voluntary labour service within the Hamburg city administration meant that the labour service expanded less quickly there than elsewhere.

Financial constraints and the opposition of the young unemployed themselves limited the effectiveness of the state's attempts to extend control over the young unemployed. The development of 'obligatory labour' schemes in Hamburg demonstrates this. Although the coercive treatment of the young unemployed in Hamburg intensified during the crisis at the hands of the general welfare department, it was only after the political transformation of 1933 that the welfare department could proceed unhindered with its measures to discipline the young unemployed.[90] The policies that came to prevail in the final 12–18 months of the Weimar Republic took into account the difficulties involved in coercion. These initiatives at national level — above all the voluntary labour service, but also the Emergency Programme for German Youth — abandoned the aim of controlling specific categories of unemployed adolescents classified as particularly 'asocial' or 'at risk'. Instead, these schemes set out to organise at low cost on a formally voluntary basis the largest possible number of the young unemployed. Such policies were, apart from anything else, an attempt on the part of the final Weimar governments to achieve a degree of propaganda success. From their impact both at national and local level, it is clear that youth unemployment was one of the problems that these regimes most signally failed even to disguise, still less to solve.

Notes

I would like to thank the F.V.S. Stiftung (Hamburg) and the British Council for their generous assistance, which enabled me to carry out the research for this article.

1. Hertha Siemering, *Deutschlands Jugend in Bevölkerung und Wirtschaft* (Berlin, 1937) p. 253; *Aus Hamburgs Verwaltung und Wirtschaft*, 1935 Nr. 9, p. 174.

2. *Aus Hamburgs Verwaltung und Wirtschaft*, 1935, Nr. 9, p. 174.

3. Sondererhebung über die Zahl der jugendlichen Arbeitslosen im Arbeitsamtsbezirk Hamburg am 30.7.32. StAH, Arbeitsbehörde I, 16.

4. See note 2.

5. Ursula Büttner, *Hamburg in der Staats- und Wirtschaftskrise 1928–31* (Hamburg, 1982) p. 171 ff.

6. Summary of regulations up to 1927 in Ernst Herrnstadt, *Die Lage der arbeitslosen Jugend in Deutschland* (Berlin, 1927), pp. 9–14.

7. Erlass des RAM vom 29.6.29, cf. Ludwig Preller, *Sozialpolitik in der Weimarer Republik* (Düsseldorf, 1978), p. 421; VO zur Behebung finanzieller, wirtschaftlicher und sozialer Notstände vom 26.7.30, *RGBl* 1930, I, Nr. 31, p. 311; VO zur Sicherung von Wirtschaft und Finanzen vom 1.12.30, *RGBl* 1930, I, Nr. 47, p. 517; Zweite VO zur Sicherung von Wirtschaft und Finanzen vom 5.6.31, *RGBl* 1931, I, Nr. 22, p. 279; VO über Massnahmen zur Erhaltung der Arbeitslosenhilfe und der Sozialversicherung sowie zur Erleichterung der Wohlfahrtslasten der Gemeinden vom 14.6.32, *RGBl* 1932, I, Nr. 35, p. 273.

8. Monthly figures in *Arbeitsmarktanzeiger des Landesarbeitsamtes Nordmark, 1930–1933*.

9. On municipal welfare assistance in the Weimar period: Stephan Leibfried, Eckhard Hansen and Michael Heisig, 'Politik mit der Armut. Notizen zu Weimarer Perspektiven anlässlich bundesrepublikanischer Wirklichkeiten', *Prokla*, Jg. 14, 1984, Nr. 3.

10. Survey of districts, October 1930. StAH, Sozialbehörde I, AW 00.12; Niederschrift der Leitersitzung der Wohlfahrtsbehörde, 4.11.31 and 2.12.31. StAH, Sozialbehörde I, VG 24.31 Bd. 3.

11. Rundschreiben des RAM an die obersten Landesbehörden für Erwerbslosenfürsorge, 20.9.26; Erlass des RAM 29.11.26. ZStA Potsdam, RAM, 1500.

12. RAM to RAVAV, 19 Mar. 31. ZStA Merseburg, Ministerium für Handel und Gewerbe, E.I. 1, 80 Bd. 2.

13. Bericht der Behörde für das Arbeitsamt Hamburg, 4.10.26 and 24.1.27. StAH, Senat, Cl. I, Lit. T, Nr. 25, Vol. 17.

14. 'Junge Erwerbslose auf der Schulbank', *Hamburger Echo*, 15 Jan. 32.
15. Protokoll des Senats, 20.2.31. StAH, Arbeitsbehörde I, 153.
16. J. Schult, at meeting of committee on youth unemployment, 9 Mar. 31. StAH, Sozialbehörde I, AW 00.97.
17. Amtsgericht Hamburg, Abteilung 1, Akte 3832 1933.
18. Amtliche Mitteilungen des Arbeitsamtes Hamburg, 19.3.32. StAH, Sozialbehörde I, AW 90.10.
19. Rundschreiben des RAM an die obersten Landesbehörden für Erwerbslosenfürsorge, 20.9.26. StAH, Senat, Cl. I, Lit. T, Nr. 25, Vol. 17.
20. Bernhard Ehmke, 'Die Überführung städtischer Jugendlicher in die Landwirtschaft', *RABl* 1931, T. 2, Nr. 12.
21. Rundschreiben der RAVAV betr: Umschulungsbetriebe Fliegerhorst und Niklasdorf, 29.6.32. StAH, Arbeitsbehörde I, 116.
22. Arbeitsamt Hamburg an die Landherrenschaft, 11.3.30. StAH, Arbeitsbehörde I, 116; Oberschulbehörde an die Behörde für das Arbeitsamt Hamburg, 24.1.31. StAH, Arbeitsbehörde I, 116.
23. Rundschreiben des RAM 22.11.30 and 29.11.30 betr: Verlängerung der Schulpflicht. StAH, Sozialbehörde I, AW 34.01.
24. Erlass des Preussischen Kultursministeriums, 18.2.31. ZStA Potsdam, Reichsministerium für Wissenschaft, Erziehung und Volksbildung, 7544.
25. Merkblatt der Landesschulbehörde Hamburg für die Eltern der Kinder, die Ostern 1932 die Schule verlassen; Vermerk der Oberschulbehörde, 15.3.32. StAH, Oberschulbehörde V, 963.
26. Richtlinien der Vereinigung grossstädtischer Jugendämter für die Fürsorge für jugendliche Arbeitslose, September 1931. StAH, Arbeitsbehörde I, 153.
27. 'Jugendpflege und jugendliche Erwerbslose', *Rheinische Jugend*, Jg. 8, 1930, Nr. 10.
28. Wilhem Hertz to Notstandskommission des Senats, 2.4.31. StAH, Arbeitsbehörde I, 152; Niederschrift über die 4. Sitzung des Ausschusses über Fürsorge für jugendliche Erwerbslose, 4.11.31. StAH, Sozialbehörde I, AW 00.97.
29. Ernst Wauer, 'Die Erwerbslosigkeit der Jugend: Berliner Massnahmen und Erfahrungen', *Berliner Wohlfahrtsblatt*, Jg. 2, 1926, Nr. 7.
30. Kurt Struve (Wohlfahrtsbehörde) memorandum 30.1.31. StAH, Sozialbehörde I, AW 00.77; Senator Hirsch, in Niederschrift über die 2. Sitzung des Ausschusses zur Fürsorge für jugendliche Erwerbslose. StAH, Sozialbehörde I, AW 00.97.
31. Zusammenstellung der Massnahmen, die von den Jugendpflegeorganisationen für erwerbslose Jugendliche getroffen sind, 21.1.33. StAH, Arbeitsbehörde I, 147.
32. H. Hennings, 'Übersicht über die Tätigkeit des Jugend- und Erwerbslosenheims Nagelsweg', 9.1.33. StAH, Arbeitsbehörde I, 147.
33. Ibid.
34. Ernst Wauer, 'Massnahmen der Stadt Berlin für die erwerbslose Jugend', *Berliner Wohlfahrtsblatt*, Jg. 7, 1931, Nr. 5.
35. Käthe Gaebel, 'Berufsfürsorge für erwerbslose Jugendliche', in Deutscher Verein für öffentliche und private Fürsorge (Hg), *Die Verwertung der Arbeitskraft als Problem der Fürsorge* (Karlsruhe, 1927).
36. Helmut Marcon, *Arbeitsbeschaffungspolitik der Regierungen Papen und Schleicher* (Frankfurt/M, 1974).
37. *Notstandsarbeiten*, regulated in AVAVG paras 139–41, cf. Preller, *Sozialpolitik*, p. 367; *Fürsorgearbeit*, regulated in Notverordnung über die Fürsorgepflicht 13.2.24, para 19, Ziffer 1.
38. *Pflichtarbeit* was one of two kinds. It could be imposed on claimants of welfare assistance (NotVO über die Fürsorgepflicht 1924, para. 19, Ziffer 2), or on

claimants of insured unemployment benefit under 21 (AVAVG, para. 91). *Pflichtarbeit* on the basis of AVAVG was not imposed in Hamburg.

39. Senator Neumann, in welfare department discussion 7 Jan. 30. StAH, Sozialbehörde I, AW 00.64; Reg.-rat Marx, memorandum 30.9.30. StAH, Sozialbehörde I, AW 00.62.

40. Oskar Martini, 'Denkschrift über die Belastung der Fürsorge durch Wohlfahrtserwerbslose', 16.11.29. StAH, Sozialbehörde I, AW 00.21.

41. Wohlfahrtsstelle I, report November 1929. StAH, Sozialbehörde I, AW 00.21.

42. Report of discussion between Martini and police president, 18 Jan. 29. StAH, Sozialbehörde I, VG 74.12.

43. Wohlfahrtsstelle IV, report 6 Dec. 29. StAH, Sozialbehörde I, VG 74.12.

44. Chef der Ordnungspolizei to Wohlfahrtsbehörde, 26.6.31; Vermerk, betr: Anfrage des Abgeordneten Westphal im Haushaltsausschuss der Bürgerschaft, Wohlfahrtsbehörde, 17.8.32. StAH, Sozialbehörde I, VG 74.12.

45. Niederschrift über die Leitersitzung der Wohlfahrtsbehörde, 13.6.31. StAH, Sozialbehörde I, VG 74.12.

46. Sozialer Querschnitt der Unterstützungsarbeiterbelegschaften Langenhorn-Süd (15.12.32) und Maienweg (31.12.32). StAH, Sozialbehörde I, AW 60.11. Bd. 1.

47. K. Struve, 'Und wer schafft Arbeit für Männer und Frauen?', *Jugend-und Volkswohl*, Jg. 8, 1932, Nr. 3; Hamburger Staatsrat (Hg), 'Hamburgs Fürsorgewesen im Kampf gegen die Arbeitslosigkeit', *Hamburg im Dritten Reich*, Heft 6, Hamburg 1935, p. 21.

48. Niederschrift über die Leitersitzung der Wohlfahrtsbehörde, 21.10.29. StAH, Sozialbehörde I, AW 00.92.

49. 'Das Verhalten der Fürsorgeverbände bei Streiks von Pflichtarbeitern und Fürsorgearbeitern', *Nachrichtendienst*, Jg. 11, 1930, Nr. 10; Edith Rasche, 'Die Entwicklung des FAD in den Jahren der Weltwirtschaftskrise und der Kampf des KJVD gegen den FAD 1930–1933', Phil. Diss. Dresden 1967, p. 59.

50. The married men with children received a benefit which with the bonus for the days worked was as high as or higher than the wage demanded for the three days worked: cf. 'Ein Bärendienst', *Hamburger Echo*, 16 Oct. 30; Familienstand der Unterstützungsarbeiter, 16.10.30 and Niederschrift der Leitersitzung der Wohlfahrtsbehörde 20.11.30. StAH, Sozialbehörde I, AW 00.93.

51. 'Ein widersinniger Streik', *Hamburger Fremdenblatt*, 17.10.30; 'Die Arbeitsverweigerung der Pflichtarbeiter', *Hamburger Anzeiger*, 17.10.30; 'Ein Bärendienst', see note 50.

52. Martini to Wilhelm Polligkeit, 21 Oct. 30. StAH, Sozialbehörde I, AW 00.93.

53. 'Zwangsarbeit', *Hamburger Volkszeitung*, 1.6.32; 'An alle Pflichtarbeiter Hamburgs' (leaflet of strike committee) June 1932; Becker, report to Martini, 9 June 32. StAH, Sozialbehörde I, AW 00.92.

54. Leiterin E. Ostermann to Wohlfahrtsbehörde, 8 Dec. 32 and report October 1932 on Luise S. and Mariechen H. StAH, Sozialbehörde I, AW 92.10. Bd. 1.

55. E. Ostermann to Wohlfahrtsbehörde 21 Oct. 32. StAH, Sozialbehörde I, AW 92.10, Bd. 1.

56. On FAD generally: H. Köhler, *Arbeitsdienst in Deutschland* (Berlin, 1967); Wolfgang Schlicker, 'Arbeitsdienstbestrebungen des deutschen Monopolkapitalismus in der Weimarer Republik', *Jahrbuch für Wirtschaftsgeschichte*, T. 3, Berlin, 1971; VO über die Förderung des FAD, 23.7.31, *RGBl* 1931, I, Nr. 42.

57. Lehfeldt, 'Die Arbeitsdienstpflicht', *RABl* 1931, T. 2, Nr. 3; Gutachten zur Arbeitslosenfrage. Sonderveröffentlichung des *RABl*, Berlin 1931, T. 2, Abschnitt IV, p. 7f.

58. VO über die Förderung des FAD, 23.7.31, Artikel 18.

59. Ludwig Bregmann, 'Vom FAD' in *Archiv für Soziale Hygiene und Demographie*, 1933, p. 90; Bruno Broeker, 'Arbeit ohne Recht', *Die Arbeit*, 1932, Heft 3.

60. VO über den FAD, 16.7.32. *RGBl* 1932, I, Nr. 45, p. 352.

61. Erlass des Reichskommissars für den FAD vom 7.9.32. Beilage zum Reichs-Arbeitsmarktanzeiger, 7.9.32; Preussisches Ministerium des Innern an die RAVAV, 20.10.32. ZStA Merseburg, Ministerium für Handel und Gewerbe, BB VII, 1, 181.

62. The *Pflichtarbeiter* met in a tavern after work to organise the strike of October 1930. StAH, Sozialbehörde I, AW 00.93.

63. Lehfeldt, 'Die Arbeitsdienstpflicht'.

64. Ernst Schellenberg, *Der FAD auf Grund der bisherigen Erfahrungen* (Berlin, 1932) pp. 12 f.

65. 'Niemand darf gezwungen werden!' *Hamburger Anzeiger*, 16 Jan. 33.

66. Vermerk über Chefbesprechung im Reichsinnenministerium über den FAD, 30.4.31. ZStA Potsdam, Reichsministerium des Innern, 25372.

67. Olga Essig, 'Die weibliche Jugend im FAD' in *Jugend- und Volkswohl*, Jg. 8, 1932, p. 80.

68. Der FAD der weiblichen Jugend, 10.11.32. Beilage zum Reichs-Arbeitsmarktanzeiger, 22.11.32; 'Der FAD für Mädchen' in *Arbeiterwohlfahrt*, Jg. 8, 1933, Heft 6.

69. F. Syrup, 'Der FAD für die männliche deutsche Jugend' *RABl* 1932, T. 2, Nr. 27; youth unemployment figures for July 1932 from Kurt Richter, 'Massnahmen zur Betreuung der erwerbslosen Jugend' in Saarbourg, Richter and Voss, *Handbuch der Jugendpflege*, H. 14 (Eberswalde/Berlin, 1933), p. 18; 'FAD und Werkhalbjahr', *Soziale Praxis*, Jg. 42, 1933, Sp. 269.

70. Hamburg FAD figures for 31.8.32: Struve, Vermerk 20.9.32. StAH, Sozialbehörde I, AW 49.10; Altona FAD figures: 'Der FAD der Stadt Altona', *Hamburger Echo*, 13 July 32; Hamburg and Altona youth unemployment figures for July 1932, see Sondererhebung, note 3.

71. Aufstellung über Massnahmen im FAD im Bezirk des Arbeitsamts Hamburg. Stand von September/Oktober 1932. StAH, Arbeitsbehörde I, 95.

72. Niederschrift über die Sitzung des Heimatwerkes, 29.12.32. KAH, KJA, Heimatwerk.

73. *Hilfe für die erwerbslose Jugend. Die Stellung der Gewerkschaften zum FAD*, Berlin 1932.

74. Aufstellung über Massnahmen im FAD, see note 71; Besprechung des Bezirkskommissars Nordmark mit den Trägern des Dienstes, 29.11.32; Niederschrift über Besprechung über FAD für die weibliche Jugend, 12.9.32. KAH, KJA, 12.

75. G. Donndorf, in Ausschuss zur Förderung der Jugendwohlfahrt, 11.2.32. KAH, KJA, G12; Baubehörde Hamburg to Wohlfahrtsbehörde 11.1.32. StAH, Sozialbehörde I, AW 49.10. Bd. 1.

76. Zusammensetzung der Arbeitsdienstwilligen des FAD Oejendorf. StAH, Sozialbehörde I, AW 49.11; Mitgliederliste, FAD Süderdeich, KAH, KJA, 6.

77. Willy Heisterkamp, 'Der Weg zum Arbeitsdienst' (ms), KAH, KJA, 6.

78. Records of the Malente girls' FAD: KAH, KJA, L2/L4/L6.

79. Donndorf, in Sitzungsbericht über die Zusammenarbeit der evangelischer Dienstträger für FAD in der Nordmark. KAH, KJA, 12.

80. Dulk, in Niederschrift der Mitgliederversammlung des Hematwerkes Gross-Hamburg, 23 Jan. 33. KAH, KJA, Heimatwerk.

81. F. Syrup, 'Der FAD'.

82. KAH, KJA, L2.

83. 'FAD mit Morgenandacht und Bibelspruch', *Norddeutsches Echo*, 1932, Nr. 26, June 1932. KAH, KJA, L2/L4.

84. E. Rasche, 'Die Entwicklung'.

85. Meldungen des Reichskommissars für die Überwachung der öffentlichen Ordnung, 6.1.33, betr: KPD, Kampf gegen Arbeitsdienstpflicht, Bundesarchiv Koblenz, Bestand R134, microfiche 385, frames 35–42a.

86. 'Was die Rechten zerschlug', *Hamburger Echo*, 1 Nov. 32.

87. Erlass des RAM und Erlass der RAVAV betr: Notwerk der deutschen Jugend, 24.12.32. StAH, Arbeitsbehörde I, 120; R. Wiedwald, 'Das Notwerk der deutschen Jugend', *Zentralblatt für Jugendrecht und Jugendwohlfahrt*, Jg. 24, 1932/33, Nr. 10; RAVAV to RAM, 12.2.33. StAH, Arbeitsbehörde I, 120.

88. Auszug aus dem Protokoll des Senats, 9.1.33. StAH, Senat, Cl. VII, Lit. Bd, Nr. 68, Vol. 7, Fasc. 3; Carlberg 'Das Notwerk der deutschen Jugend', *Jugend-und Volkswohl*, Jg. 8, 1933, Nr. 5.

89. Hüffmeier, 'Bericht über die Reichstagung der Arbeiterwohlfahrt in Kiel', 29/30.5.27, Berlin 1927.

90. A. Ebbinghaus, H. Kaupen-Haas and K-H. Roth, *Heilen und Vernichten im Mustergau Hamburg*, (Hamburg, 1984), pp. 10f.

7 THE LOST GENERATION: Youth Unemployment at the End of the Weimar Republic

Detlev Peukert

I

Social historians of Germany have always counted the mass unemployment of the Great Depression among the decisive factors that destabilised the structure of the Weimar Republic, and they have concentrated particularly on the connection between youth unemployment and the growth of a radicalised mass following for the Communists (KPD) and the Nazis (NSDAP). It is surprising, therefore, to find that there are very few detailed historical studies of youth unemployment in the Weimar Republic. We know little about how much there was, nor (apart from a recently published sketch by Dick Geary[1]) about how unemployment was experienced by young people themselves.

It is paradoxical that a phenomenon to which so much historical importance has been attached, should have attracted so little attention from historians. The paradox can be explained to some extent by the difficulties of the sources. It is not easy to obtain even a degree of certainty about the extent of unemployment because there are very few reliable statistics on the subject.[2] It is even more difficult to obtain reliable statistics on unemployment benefits.[3] Nor were young people recorded in trade union statistics unless they had found at least short-term employment after leaving school and so joined a union. And if a quantitative approach to the phenomenon is problematical, then the same is also true of a qualitative approach, the attempt to recapture the behaviour and experience of young unemployed people. The most celebrated study of unemployment in the Great Depression, Jahoda and Lazarsfeld's *Marienthal*, eventually abandoned any attempt to include young people.[4]

In this paper I am not claiming to fill this substantial gap in our knowledge in any complete or thorough way. My aim is only to gather together some of the scattered (but, taken together, in the end far from negligible) sources on the problem of youth

172

unemployment, and to suggest some of the conclusions that can be drawn from them. Beyond this, I also hope to examine the question of whether there were sufficient similarities in young people's experience of unemployment, and in the way of life that it forced on them, for us to be able to speak of those concerned as a distinct generation.

II

In studying the problem of youth unemployment between 1929 and 1933 it would be wrong to confine our attention to these years alone. On the contrary, the special severity of the problem in these years followed on from a combination of three more long-term developments with the more short-term phenomenon of the Great Depression of 1929–33 itself. The first of these more long-term developments was the fact that these years saw the arrival in the labour market of the last age-cohorts born in the fecund years before the First World War. The second consisted in an increase in female employment, particularly in the younger age-groups, which further increased the competition for jobs. Thirdly, even the boom years of 1924–8 were not able to reduce the high base-level of unemployment caused by the damage inflicted on the economy by the 1914–18 war and the rationalisation policies of the 1920s.[5] In this section I shall examine each of these three long-term factors in turn.

First then, demographic developments by themselves were contributing to a particular tension in the labour market in these years. Despite the loss of about 2.4 million war dead and the further consequences of these losses for the birth-rate, the population of Germany (in its postwar boundaries) grew by 8 per cent between 1910 and 1920 and a further 5.2 per cent between 1925 and 1933 alone.[6] The decline of the surplus of births over deaths was already becoming noticeable,[7] but the especially large birth-cohorts of 1900 to 1914 caused a last, absolute and relative increase in the numbers of young people in the Weimar Republic. In 1910 males between 14 and 20 years of age made up 11.9 per cent of the total male population; by 1925 this had grown to 12.9 per cent, before declining with the entry of the smaller birth-cohorts of the war years to 8.1 per cent in 1933. Similarly, the male age-group between 20 and 25 made up 8.7 per cent of the total male population in 1910, 10.1 per cent in 1925 and 9.8 per cent in 1933.[8] In 1925 there were 7.8 million young people aged 14–20; 6.1 million of these had jobs, including 3.3 million of the 3.9 million young males altogether. These developments contributed substantially to the fact that the total number of people in the age-group where one would normally be working (14–65 years of age) was substantially greater in 1925, at 22.9 million or 71.1 per cent of the total population, than in 1910, when it stood at 18.5 million or 63.3 per cent of the total population. Indeed, it was precisely the years up to 1933 that saw this active sector of the population reach an absolute high point at 23.6 million, while in 1933 its proportion in the population had already fallen slightly to 70.5 per cent, thus indicating the future decline.

Whatever the future held, however — and population analysts were already painting gloomy pictures of the approaching 'senility' or even 'death' of the German people — the pressing and distressing experience of the late Weimar Republic was an over-supply of labour through the entry of the last substantial birth-cohorts of the pre-war years into the age-groups of the school-leavers and young people looking for jobs.

With this increase in potential manpower came a noticeable increase in the proportion of employed and unemployed, taken together, in the part of the population theoretically capable of work,[9] from 86.3 per cent in 1895 to 90.5 per cent in 1925 and 93.1 per cent in 1933. The reason for this is above all to be sought in the increase in female labour over this period.[10] These sex-specific changes in the labour market were particularly clear in the younger age-groups. While the young male occupied people (employed and unemployed) between 14 and 20 years of age made up 83.8 per cent of the total resident male population of the same age in Prussia in 1925 and only 82.5 per cent in 1933, the equivalent proportion of females rose from 63.5 to 66.4 per cent in the same period.[11]

At the end of 1927 a publication of the Prussian Welfare Ministry characterised the fluctuations of the youth labour market up to that point in the following terms:[12]

In the war every last ounce of youthful manpower was called up to work and relatively well paid for it. This made the change that took place at the end of the war particularly serious. Now there were for the first time substantial masses of young unemployed people to assist. During the inflation the situation of the labour market was consistently healthy, until the collapse came with the end of the Ruhr struggle in 1923. Now there was once more substantial unemployment among the young. The following year saw considerable improvement once more, but the great world economic crisis which began in the Autumn of 1925 and only began to be overcome in the Spring of 1927, affected the young more than most. During the course of the year 1927 the youth labour market, however, has been in an extraordinarily good situation.

This optimism was shared by the Institute for Research into Economic Trends (*Institut für Konjunkturforschung*). In 1928 the Institute calculated that because of the decline in birth-rate the number of young people would decline by more than a third from 1920 (=100) to 1935 (=60).[13] And indeed, the number of young occupied people (i.e. employed and unemployed aged 14 to 20) in Prussia declined from 3.5 million in 1925 (as against 3.1 million in 1907) to 2.3 million in 1933.[14] But this potential relaxation of the labour market was of course nullified by the Great Depression.

Reliable data on youth unemployment cannot be easily extracted from the census of 1933,[15] so a few tangential figures must suffice to illustrate the problem. In 1926–7 the proportion of young people up to 18 among the recipients of unemployment benefit fluctuated according to the economic situation between 0.7 and 2.7 per cent (about 50,000 people). Most young people did not come into these categories because they were not entitled to receive benefits.[16] On the other hand, the general unemployment census of 2 July 1926 counted 272,137 unemployed between 14 and 21 years of age, or 17 per cent of the total unemployed altogether.[17] This number declined absolutely and relatively in the economic upturn of 1926–7 (to only 9.5 per cent in 1927), but then rose with the Depression until it reached 16.3 per cent of the total unemployed in 1931, or 700,000 young people altogether. Of these, 75 per cent were young males, above all in the age-group 18–21, because they had not found jobs after completing their apprenticeships.[18] Tables 7.1 and 7.2 give some indication of the situation.

The census of 16 June 1933 included population, occupation and workplace data, and took place at a time when Germany had already emerged from the trough of the Depression. Still, it offers,

Table 7.1: Age Structure of Employed and Unemployed in Germany, 1933

Age-group	Employed				Unemployed			
	Male	%	Female	%	Male	%	Female	%
Below 14	64,478	0.4	52,508	0.5	3,113	0.1	1,234	0.1
14–14	382,244	2.4	272,278	2.7	28,720	0.6	17,866	1.6
16–17	550,282	3.4	458,572	4.4	61,904	1.3	39,596	3.5
18–19	822,745	5.1	809,224	7.8	265,526	5.6	107,618	9.4
20–24	2,022,940	12.6	1,854,327	18.0	900,752	19.1	293,222	25.7
25–29	2,093,770	13.0	1,368,125	13.2	883,290	18.8	207,711	18.2
30–39	3,648,768	22.7	2,052,531	19.9	1,172,560	24.9	235,377	20.6
40–49	2,864,865	17.8	1,595,311	15.4	673,996	14.3	140,238	13.3
50–59	2,359,139	14.6	1,189,529	11.5	538,885	11.4	80,487	7.0
60–64	700,844	4.3	351,568	3.4	153,426	3.3	16,626	1.4
65 and over	594,526	3.7	330,482	3.2	30,260	0.6	2,611	0.2
Total	16,104,601	100	10,336,455	100	4,712,432	100	1,142,586	100

Source: Hertha Siemering, *Deutschlands Jugend in Bevölkerung und Wirtschaft*, (Berlin, 1937) p. 287.

Table 7.2: Youth Unemployment in Germany on 30 July 1932

Age-group	Male	%	Female	%	Total
Below 15	10,953	1.1	9,913	2.3	20,866
15–17	68,370	6.6	60,838	14.5	129,208
18–20	339,837	32.8	142,313	33.9	482,150
Below 21	419,160	40.5	213,064	40.7	632,224
21–24	617,536	59.5	207,094	49.3	824,630
Total below 25	1,036,696	100.0	420,158	100.0	1,456,854

Source: Kurt Richter, 'Massnahmen zur Betreuung der erwebslosen Jugend', K. Richter (ed.), *Handbuch der Jugendpflege*, H. 14, (Eberswalde-Berlin, 1933), p. 18.

relatively speaking, the most sharply focused snapshot of the situation during the Depression. In particular, it provides the most exact information for the young because it is based on a survey of the entire resident population.[19] On 16 June 1933 there were in Germany 1,378,002 young male residents aged 14 to 17 inclusive. Of them, 932,526, or 63.3 per cent had jobs, and 90,624, or 11 per cent were unemployed. Of the female resident population of the same age, 732,850, or 54.9 per cent were employed, and 57,462, or 4.3 per cent, unemployed. These figures do not provide any particularly dramatic picture of youth unemployment in the Depression. Obviously this age-group was protected because poorly paid (or even unpaid) apprenticeships had not been destroyed to the extent that fully paid jobs had. Moreover, untrained posts usually reserved for the young and usually paid especially badly, such as messenger boy, page, temporary assistants or day-workers in the transport sector, were apparently not affected.

Yet a completely changed picture emerges when we include the next age-group, and group together those aged 14–19 or 14–24 inclusive. In particular, the figures for the age-group 14–24 make the real problem of youth unemployment dramatically clear. In 1933 the male resident population of Germany aged 14–24 numbered 5,649,818. Of them, 3,778,211, or 66.9 per cent, had jobs, while 1,256,902, or 22.2 per cent, were unemployed. The comparable figures for females were 5,581,491 in total, of whom 3,396,401, or 60.9 per cent, had jobs, while only 458,302, or 8.2 per cent, were unemployed. Three main conclusions about the structure of unemployment in the Depression can be drawn from these figures. First, unemployment among 14–17 year-olds was clearly below average, while unemployment among 18–24 year-olds was not only greater in comparison to the figure for the age-group immediately below, but was also clearly above the general average. If one wanted to be really precise, one would say that the problem was not so much one of youth unemployment as of young adult unemployment. Thus it cut through any sense of continuity or long-term perspective in these people's lives. Just when they had completed their training and had become fully fledged workers at the height of their capacity, they could expect with a considerable degree of certainty to become unemployed. Moreover, these were years in which people normally married and formed their own households, and now there was no material basis for them to do so. From the point of view of a school-leaver at the age of 14 the situation could only get worse, indeed was bound to do so with some statistical certainty the older he or she became. So on top of the actual situation of unemployed people aged 18 came the utterly depressing future expectations that they learned to have from the experience of friends and colleagues who were a few years older.

Apart from these demographic factors, a second development could be observed in the changing sex-structure of the labour market. The lack of future perspectives for young men appeared to them to contrast with the apparently better situation of women of the same age. Here, the statistical data gave the impression of a better situation above all because many young married non-occupied women were not classified as unemployed. However, the proportion of unemployed females in the age-group 14–17 inclusive was also clearly below that of males of the same age, although the percentage of employed females in the total resident female population was comparatively high in this age-group, in fact, only 10 per cent below the comparable figure for males. Evidently the labour of young people of both sexes in the age-group 14–17 inclusive was particularly cheap, and so was preferred in the labour market to that of young adults (aged 18–21 inclusive); beyond this, however, cheap female labour was better placed in the crisis than was male labour.

Thirdly, the extent of unemployment among young males becomes even clearer when we look at the industrial centres which were particularly affected by the crisis. On average, in the nation as a whole, there were 9.7 male unemployed per 100 employed male persons in the age-group 14–17 inclusive, while in the age-group 14–24 inclusive, the comparable figure was 33.3. In Wuppertal, however, the figure for the male age-group 14–24 inclusive was 63.2; in the male age-group 20–24 inclusive, it was as high as 87. In Berlin it was 63.4 for the male age-group 14–24 inclusive, and 82 for males aged 20–24 inclusive. In Oberhausen the corresponding figures were 82.2 and 113, in Gelsenkirchen 84.6 and 113. These figures point to the particularly desperate situation in heavy-industrial regions, and of course the situation was even more desperate in certain specific branches of industry, though to include figures at this level of detail would go beyond the bounds of the present brief survey.

Youth unemployment was thus higher than average in the centres of industry. These were of course the great centres of the labour movement. So the labour movement, both party-political and trade unionist, was confronted in a particularly acute form with the problem of recruitment of the young. After all, up to half of the 20–24-year-olds in these areas were unemployed. That meant that only about half of the younger workers experienced even elementary proletarian socialisation at the work-bench. Only about half of them experienced the problems of wage-labour and the common interests of workers against the employer in a direct way. Already in 1932 the sociologist Theodor Geiger expressed his fears for the young man caught in this situation:

He stands completely outside the traditions of the labour movement. It is not only his youth that separates him from the workers, not only his youthful receptivity to unrestrained slogans and deeds that ignore or evade the real facts; he is separated from the older generation by the fact that he is not even a worker, and so is in no way imbued with the wage-earning worker's mentality in any shape or form, however it might be determined by the circumstances of the time.[20]

One can perhaps soften the rigidity of this judgement by referring to the influence of the proletarian family, the proletarian home or neighbourhood, and proletarian organisational life. The effects of these on this group of young people has still however to be subjected to more precise investigation. In the following section, therefore, I want to discuss the structure of experience of young unemployed people, the relationship of public welfare and youthful self-help, and the place of the experience of unemployment in the longer perspective of their lives beyond 1933.

III

Most young people had already had their first experience of unemployment before leaving school, if only indirectly, as children, brothers or sisters of the unemployed. Even as small children they had already gained a clear picture of what it meant to be unemployed, a picture drawn from the experience of their own family and stamped on their childhood consciousness and behaviour. Ruth Weiland's study of the children of the unemployed in the early 1930s observed in various German *Kindergärten* for example that

> The children's play imitates adult life, and 'going to sign on for the dole' plays a great part in it . . . The children are really terrified when they tear their clothes: 'Papa will beat me, we haven't got the money to buy anything new' . . .
> The children play 'mothers and fathers': 'father' cooks the dinner because he has nothing else to do . . .
> In children's games the father is always at home in a bad mood. A small boy wants to be 'father' because 'then I can scold everyone all the time'.[21]

The children did not simply incorporate playful reflections of family tensions into their games, however, but also included in their understanding of the situation more complex constructions which they had heard from adults:

> In playing 'houses' they build houses 'with lots and lots of rooms' or 'tall factories', so that the unemployed have something to do, or huge dining-rooms, so that the unemployed have something to eat. When some children want to build a machine, however, they are forbidden to do so by the others because 'then my father too will get the sack'.

By the time they themselves left school to look for a job, the children of the unemployed had shared the fate of their own family so long and so intensively that they could write brief biographical sketches with the title, set in December 1932, 'Unemployed', that demonstrated a deep understanding and a differentiated judgement of the subject.[22] Thus, the 14-year-old Hanna S. wrote:

> Father used to joke a lot those days. But after two years the picture changed. One day father came home looking downcast. Mother looked at him and knew at once what had happened. He had lost his job. The first days he was very unhappy. There were no more jokes. I avoided him. This went on for a year. Then he got another job. What joy! I still remember how father put his pay-packet on the table with pride and joy . . . But every year things got worse . . . Now my father has been unemployed for over three years. We used to believe he would get a job again one day, but now even we children have given up all hope.

Another essay, by Margot L., aged 13, was also marked by the negative perspective of the unemployed person's 'career':

> First my father went to sign on for the dole. Later, when the time when he could sign on ran out, he got 'crisis benefit'. He had to collect the money from the welfare. This money was not enough to manage on. I often saw how my mother brooded over the question of clothing and feeding our family of six. We live in a country district which is good for my father. He might otherwise mix with bad company, as so often happens nowadays.

The last remark alludes to a central problem in the experience of unemployment, namely the threat of falling into a social milieu which people normally felt beneath them. Such a fall undermined and destroyed people's self-respect. Social descent became a moral descent in the eyes of the world, often in the eyes of those whom it affected as well. Thus, the children's portrayal of the fate of the unemployed person's family already contains all the main features of the experience of unemployment. Of course, they saw it mainly from the point of view of those who were only indirectly affected, but at the same time they also transferred it to the fate that threatened them in the future as invidivuals, after they left school.

Thus the job choice for school leavers concentrated on one point: to find a position from which one would not be threatened with dismissal after the end of apprenticeship.[23] That is, those who still harboured ambitions for a job wrote in these terms. Others were already bitterly resigned to unemployment and answered the question 'What would you like to be?' with a 'Learn to queue for the dole', 'Go to the University of the Dole'. A youth welfare worker in Bielefeld, Schürmann, perceived a decline in young people's willingness to enter an apprenticeship in 1932:[24]

> The argument that apprenticeship has no point any more because when it is over there are no jobs to be had, is constantly put forward by young people. And given the present situation of the labour market, it is not entirely untrue.

In the autumn of 1931 an investigation was carried out in Hanover among 1,015 young unemployed people between the ages of 18 and 21. In keeping with the overall age-specific distribution of the unemployed, 153 of these were aged 18, 436 were aged 19, and 426 were aged 20.[26] The older they were, the worse their material situation and future prospects became. A large proportion of the 18 year-olds were still receiving unemployment benefit because they had worked before. But the older ones had lost their claim to benefits of a job because of their long unemployment, and if they could not be supported by their families were supported by the welfare services instead. To be sure, there were only 269 unskilled and untrained workers among the 1,015 young people, which also conforms to the general statistical average. But many of them had had to change their apprenticeships once or even more often, or to interrupt it, because their masters had gone bankrupt. After ending their apprenticeships (in 746 cases) only very few could continue for long in their jobs. To be precise, one was able to continue for over 3 years, 19 for 2–3 years, 114 for 1–2 years, 458 for up to a year, and 154 not at all. Of the 269 untrained young people 87 had had work for over 3 years, 77 for 2–3 years, 65 for 1–2 years and 40 for up to a year. Finally, 80 per cent of the trained young people had taken unskilled work in order to earn any money at all, before they finally became unemployed.[26] This was all the more depressing for them since almost all of them had declared that they enjoyed the trade to which they had been apprenticed. Indeed, those who had passed through an apprenticeship clung to the hope that they would one day be able to do the job for which they had been trained and which of course they had originally chosen to do. Thus, their personal identity was caught in a tension between an ideal that was unattainable in the circumstances of the time, and a reality, of looking for stop-gap substitutes, that was dejecting in the extreme.[27]

Resignation came gradually rather than suddenly.[28] When the search for a job turned out to be hopeless, the small group of those who wanted to fill their time with further training or similar occupations crumbled away, leaving a growing number of people whose answer to the question, 'How do you spend your day?' was: 'We go for walks, we play games, we read, we help in the home, we do the garden, we go biking.'[29] With longer-term unemployment, however, even such substitutes for work were abandoned and the day was increasingly spent merely killing time, travelling aimlessly on the electric suburban railway, the S-Bahn (above all on the circle line in Berlin, the Ringbahn, round and round),[30] forming groups of card-players playing endless games in the parks, wandering through the shopping districts and the department stores, looking for sensational amusements in the fairgrounds. At the end of this road was the loss of all experience of time-discipline, not only in terms of the division of the day into different parts, but even for the stages of one's own life. Thus, when young unemployed people in Hanover were interviewed,[31]

the indifference of the young people to their family circumstances was clear even with regard to matters that affected their own person. Thus hardly any of them could answer straightaway when asked in what year they had left school.

Most said, 'Oh, that's such a long time ago'. The question 'how old were you when you left school' elicited the most unbelievable answers. Sometimes they said that they had left school at 11 or 12. This indifference of young people to their personal circumstances, naturally enough, was also experienced again and again with reference to their careers.

In one study, to be sure, that of Tippelmann,[32] optimistic sentiments are more to the fore. This might be due to differences in method[33] or perhaps to the fact that it was carried out a year earlier, in the winter of 1930–1. Nevertheless, its general tendency is the same as in the Hanoverian study. An optimistic mood was particularly noticeable among the younger age-groups, who had not yet been worn down by longer-term unemployment, and among the untrained, whose expectations were not as high as those who had gone through an apprenticeship. This latter group, accustomed to an unstable lifestyle and lacking any specific training, were spared the collapse of any hopes for a specific career. Above all, however, the mood of almost all these young people was inconstant. Many swung from optimism to occasional despondency, from equanimity to irritability and depression. With long-term unemployment came ever more frequently the experience that the 'force of life slowly vanishes and makes room for a desolate inner emptiness'.[34]

Those who neither summoned up the energy to embark on a criminal career (and these were fewer than contemporary lamentations alleged),[35] nor found their identity in a political organisation, experienced their daily life and future life perspective as fragmented and emptied of meaning. Tippelmann summed up these dangers for unemployed youth as follows:[36]

1. Unemployed people easily become egotists. For human society, in denying them companionship acquired through work, also denies them the possibility of experiencing liberation from the ego through service to a greater, impersonal good. Because society compels them, under increasing economic pressure, to defend their most primitive needs against its claims, the need to stay alive compels them to put themselves first.

2. The lives of the young unemployed easily become trivialised. Intruders in human society, they have no possibility of serious

activities or serious human relationships. For the former there is no need; for the latter there is no possibility of coping with the consequences ('But I can't possibly think of marrying!'). It is only logical that trivial amusements dominate their lives, when these are all that the world offers them.

3. The young unemployed easily become unstable. We have already said that the private life of those with jobs is always regulated by consideration of its effect or influence on their career. If this central consideration is absent, as it is from the life of the unemployed, then this by itself means a loosening of the inner structure of their life. The need to do something for its own sake, or to seek company for its own sake, becomes an absolute. From here it is only a step to inner lability and then to complete instability.

4. The young unemployed easily lose the ability to concentrate. This is in part a necessary consequence of the loosening of the inner structure of their character. In part, however, it is also a consequence of the facts outlined in section II above. Where there is nothing that demands the commitment of all one's abilities, there is nothing either that will train one to concentrate them all on one purpose. This lack of discipline extends to thought itself, as teachers of courses for the unemployed experience time and again.

5. The young unemployed easily become enervated after a while. Constant concern with trivialities must in the end lead to boredom. As the desire for new experiences declines so too does the ability to observe and remember. With resignation comes a gradual end to any kind of effort. Indifference and apathy follow and they almost literally 'fall asleep'.

Coming to terms with their fate — unemployment — was difficult for the young above all because they lacked both practical alternatives for the structuring of their lives (both day to day and in the longer term) and any positive models for coming to terms with unemployment in a moral and intellectual sense. Activities such as evening classes, clubs and societies, pastimes and amusements, meeting friends and relations, were normally carried out after work, and in a society whose dominant values were formed by work they were attractive because they formed a contrast to the experience of work. Such activities constituted themselves as separate 'leisure activities' in the first place because they filled the brief hours left over from work.

The young also considered the working day as a normal situation, as an axis around which their life revolved, to such an extent that unemployment appeared to them as a 'stroke of fate' not without a hint of individual failure.[37] Just as the need to work was accepted without question by the young, so too was 'the fatefulness of unemployment'.[38] But this gave no consolation to those whom fate had deprived of a job, not even the assurance that they were innocent in being struck by it. In the judgement of the young, moral decay was the accompaniment of material poverty:[39]

An unemployed person becomes lazy and in the end tries to enrich himself dishonourably. Then he ends up in gaol . . .

I am really sorry for some families where the husband loses his job: the whole family comes down in the world and goes to the dogs . . .

. . . because the longer they are unemployed, the lazier they get, and the more humiliated, because they are always looking at other people who are properly dressed and then they get so irritated that they want that too and become crooks . . . They still want to live! The older ones often don't even want that . . .

In these judgements, the subjective and objective aspects of moral decline come together in a quasi-automatic negative career, from unemployed to poor, lazy, listless, dishonest, criminal, suicidal. Moreover, almost all the young essay-writers, the girls more than the boys, inclined to bigoted value-judgements on the character of the unemployed. These judgements stood in logical contradiction to their simultaneous emphasis on impersonal, fatalistic causes of unemployment, but could apparently coexist with it on a social-psychological level.

The explanation for this phenomenon may lie in the separation of two levels of experience. The causes of unemployment are impersonal, system-bound, they are noted as clichés or formulas. The impact of unemployment, on the other hand, is experienced individually, and how the individual comes to terms with it (or not, as the case may be) is described as the outcome of the individual's character. Situations of deficiency, and deviance from socially expected norms of behaviour, seem to be equated in the judgement of the young.

Young people in the Weimar Republic were unable to come to terms with the experience of unemployment, neither in the perspective of their own lives, nor in constructing their daily activities, nor in fulfilling social expectations, nor in retaining their individual self-respect. Unless, that is, they deviated into areas which, although discriminated against by society, none the less to some extent allowed forms of behaviour and interpretations of reality alternative to those prescribed by industrial values to acquire validity. In the early 1930s, these areas comprised above all the criminal milieu, the youth gangs and the radical political parties. The attempts by philanthropic associations, state and youth welfare to integrate them into society remained without success because they were based on the construction of a world of appearances rather than reality.

IV

The most important 'alternatives' to mere passive surrender to the fate of unemployment which young people were offered in the Great Depression can only be sketched here, for reasons of space. First, publicly financed work, so-called emergency work reached only a small proportion of the unemployed. Improvement work, landscaping parks, building or repairing streets, footpaths, sportsgrounds and so on, were only temporarily available. Because of the hard manual work they involved, the meagre wages paid, and the often poor accommodation and maintenance provided, they were so unpopular that workers could only be persuaded to undertake them when threatened with the withdrawal of their benefits.[40]

Secondly, there were certainly many and varied attempts at providing youth-welfare activities for unemployed in clubs, evening classes or training courses, and the 'founding' of residential work and leisure centres. But until the end of 1932 all these initiatives suffered under the economy measures of the government and reached only a fraction of those affected by unemployment.[41] Whether the symbolically laden 'Emergency Work Programme for German Youth' initiated by a joint declaration of President Hindenburg and Chancellor Schleicher on 24 December 1932 was any more use, is difficult to say;[42] at any rate, the National Socialists could build on this beginning when they came to power the following year. Reports by those involved in residential work centres for young unemployed varied between cautious optimism and profound resignation, because the perspective which they offered was one of boarding-school discipline, progressive ideology and cultural enrichment that completely contradicted the real life-experience and life-perspective of the inmates. The consequences were disciplinary problems, rebellions and expulsion precisely of the young people who were most seriously affected by unemployment and who had least successfully adapted to it.[43]

Thirdly came the camps of the 'Voluntary Labour Service'. Drawing on the traditions of the youth movement, but also on those of the racist movement and the 'Free Corps', these were set up with state aid in 1932 and dealt with a larger proportion of unemployed youth, at least temporarily. But the enthusiastic success stories retailed by youth movement propagandists of the 'true racial community' (*wahre Volksgemeinschaft*) said to have been created in these camps should be treated with caution. In practice, all the usual problems of camp life were there, from poor food and accommodation to the inadequate motivation of those of the youthful 'volunteers' who had been driven into the camp only by the hopelessness of their situation.[44]

Fourthly, although they had been excluded from the core area of proletarian existence, work, the young unemployed none the less sought refuge in the multifarious proletarian milieu of the family, the neighbourhood and the peer-group. These 'attempts at an active connection with the world'[45] ranged from frequently mentioned willingness to help in the home and the family circle to participation in street life, which, in the neighbourhoods where the strongest support for the Left was to be found, characteristically combined social with political activity.[46]

In this context the youth-gangs, the so-called *Wilde Cliquen* (wild cliques) were especially important. They combined the usual leisure activities of young workers with the working-class peer-group relationships which unemployment had prolonged from childhood, into adolescence and adulthood. The members of the cliques possessed an informal group solidarity which gave a structure to their lives. Through the cliques they could themselves provide for their non-working life an alternative meaning and a purpose to that offered by the world of work from which they had been excluded, a meaning and purpose which none the less remained true to the proletarian milieu.[47]

Fifth, a few of the long-term unemployed did of course enter

the criminal underworld. Yet youth crime-rates did not increase
nearly as much during the Great Depression as they had in the
postwar inflation of 1919–23. The press and a part of the youth
welfare services did complain about the growth of 'rowdyism'
among young people in the early 1930s, but in doing so they were
reflecting their own fears more than they were reporting social
reality. Various interpretations were offered for the relative absence of youth criminality in the Depression. On the positive side
it was thought that the youth organisations, including the radical
political groups, had got a grip on young people and disciplined
them, at least as far as their social morality was concerned. On the
negative side, it was suggested that young people had become so
enervated and lacking in initiative with long-term unemployment
that they were no longer able to show any energy at all, even of a
criminal kind. These two explanations are not mutually exclusive
but could apply to different groups of the young unemployed.[48]

Sixth, the attractiveness of radical political organisations, with
their totalising claims on the everyday and long-term existence of
their unemployed members, thus becomes clear:

> Pent-up defences find a way out, and the torture of brooding
> over their fate can release itself in action. Moreover, the young
> are told that this action can effectively shape the world itself.
> Thus they see themselves in contrast to their previous super-
> fluity to the world as people for whom the world hungers, as
> valuable, even perhaps necessary to the world.[49]

Thus, just as youth unemployment, especially in the 18–25 year-
olds (or even up to the age of 30) was growing in the Great
Depression, a clear correlation emerged between the age structure
of the Communist Party and the Nazis and the age structure of the
unemployed. Both parties recruited their most active members,
especially in the paramilitary organisations, from young unemployed men. Both parties differed in important ways in their
programme, their social structure and their political praxis. But
they both had much to offer a generation which was excluded for
several years from the normal social context of the life of a worker.
Above all, they offered a centre of support which structured the
day, provided companionship and transmitted a structure of
meaning to their activities that both recognised their members as
outcasts and made them into an avant-garde.[50] They occupied the

amorphous terrain beyond the closed world of work. Beyond this, they projected apparently meaningful biographical perspectives into a future society that was to differ radically from the system that had ruled up to then.

V

The experience of long-term unemployment in the Depression did not exhaust itself in the shaping of the political constellation of radicalisation and the paramilitary organisations, with their claim to the streets. Its importance went beyond this, as far as the life-history of those it affected was concerned. After all, these were

Young people born in 1910–13, who have experienced little or nothing of what happened in the war, whose first real impressions of life were made in the specious prosperity of the inflation. The first thing they imbibed on their way through life was the intoxication of millions, billions of marks. They never learned the value of money, because it had none. All they saw was their parents' concern to spend their money as quickly as possible because it lost its value with every hour in which they possessed it. They saw how indiscriminately goods were bought, simply in order to avoid the disappearance of money through the purchase of material assets. And as they themselves had left school and entered the world outside as apprentices or untrained workers, they entered a world of economic depression. Their apprenticeship could not be completed because one master after another had to abandon his business. If they were untrained workers they too had to change jobs every other moment for the same reason. At home they saw how father, mother, brothers and sisters became unemployed, or how the only member of the family still working had to support the whole family, even if he or she was the youngest. They grew up in an unhealthy atmosphere of despair or the over-importance of wages. Is it not therefore self-evident that the attitude of these people to work is very different from that of people who have experienced ordered times and circumstances as the first and strongest impressions on their way through life? The latter still have some pride in their work, the former only know it as a miserable fight for the bare necessities of life.[51]

These prewar birth-cohorts born in 1910–13 can with some justification be regarded as a distinct generation.[52] Those born before them experienced the First World War directly, often on the front, while those born later were already fewer in number (with all the consequences that this implied) because of the decline in the birth-rate that took place between 1914 and 1918. If we follow the generation of 1914 beyond the Depression into later life, what do we find? After 1933 they would experience a stay in a camp of the Labour Service, now run by the National Socialists, then, with the introduction of conscription in 1935, military service for three years (those born in 1914–15 were the first to be called up). Next came more military service, this time active, in the war of 1939–45, followed by a more or less substantial period as prisoners of war. Only with the monetary and political stabilisation of 1948 would they begin a 'normal' life with jobs and families of their own, a beginning that most people make between the ages of 14 and 25, but which was denied to them until they reached their mid-30s.

This generational experience helps explain the hectic, compensatory work activity and the compulsive fixation on a conventional bourgeois normality that characterised the years of the so-called 'economic miracle' in the 1950s, when the dominant slogan was 'no experiments'. It also accentuates the reality of the watershed in the social history and experience of the labour force that was constituted by the year 1933. The political socialisation of a whole generation of young unemployed workers took place by necessity not on the shop-floor but in the proletarian milieu outside. The National Socialists then penetrated this milieu through their mass organisations and their police surveillance and destroyed the workers' organisations. So this generation was denied the chance of taking over the experience and the values of the older German labour movement. If they did not spend the whole period from 1933 to 1945 in a paramilitary or military life, then this generation only once experienced the 'normality' of industrial work before the 1950s, and that was in the National Socialist armaments boom of the late 1930s. The generation of 1914, therefore, was not only a lost generation in terms of its own life-history; it also formed the missing link in the discontinuous social history of the German working class.[53]

Notes

This chapter was translated by Richard J. Evans.

1. Dick Geary, 'Jugend, Arbeitslosigkeit und politischer Radikalismus am Ende der Weimarer Republik', *Gewerkschaftliche Monatshefte*, 34 (1983), pp. 304–9.

2. Dietmar Petzina *et al.*, *Sozialgeschichtliches Arbeitsbuch III. Materialien zur Statistik des Deutschen Reiches 1914–1945* (Munich, 1978), esp. pp. 119–21.

3. School-leavers who had never had a job were generally not included in unemployment insurance statistics. Moreover, legal changes in this area in 1927 altered the criteria for inclusion in unemployment statistics, while in June 1931 unemployment benefits were withdrawn from all young people who were still entitled to support from their families. In any case distinctions have to be made between claimants for legally established unemployment insurance benefits, for state 'crisis benefits' and for municipal welfare benefits. See Ernst Riffka, 'Die berufliche Lage der Jugend in der Gegenwart unter besonderer Berücksichtigung der männlichen Jugendlichen im Alter von 14 bis 21 Jahren' in Kurt Richter (ed.), *Handbuch der Jugendpflege* (Heft 1, 1. Teil: Der jugendliche Mensch (Männliche Jugend), Eberswalde-Berlin, 1932) pp. 65–91, here pp. 77–9.

4. Marie Jahoda *et al.*, *Die Arbeitslosen von Marienthal* (1933), 4th edn., (Frankfurt, 1982).

5. See the contributions by Fischer, Petzina, Abelshauser, Köllmann and Reulecke in Hans Mommsen *et al.* (eds), *Industrielles System und politische Entwicklung in der Weimarer Republik* (Düsseldorf, 1974), pp. 26–95.

6. Köllmann in *Industrielles System*, p. 76.

7. The absolute numbers of the birth-cohort sank by 27.5 per cent between 1920 and 1933, according to Köllman *Industrielles System*; see also Klaus Tenfelde, 'Grossstadtjugend in Deutschland vor 1914. Eine historisch-demographische Annäherung', *Vierteljahresschrift für Sozial- und Wirtschaftsgeschichte*, 69 (1982), 182–218.

8. Hertha Siemering, *Deutschlands Jugend in Bevölkerung und Wirtschaft. Eine statistische Untersuchung* (Berlin, 1937), pp. 101–12, pp. 10–12, based on the 1933 census, in *Statistik des Deutschen Reiches*, Bd. 451 (Berlin, 1935). Cf. 1925 figures on pp. 70–1. Here, as elsewhere the continually changing categorisations greatly increase the difficulties of making statistical comparisons.

9. Reulecke, in *Industrielles System*, and for the following remarks on the potential labour force.

10. 95.7 per cent of the potential male labour force was already occupied in 1895, 97.3 per cent in 1933: among women the figure rose from 70 per cent (1895) to 86.3 per cent (1933) (Reulecke, *Industrielles System*).

11. Siemering, *Deutschlands Jugend*, p. 126.

12. Rudolf Herrnstadt, *Die Lage der arbeitslosen Jugend in Deutschland*. (Veröffentlichungen des Preussischen Ministeriums für Volkswohlfahrt aus dem Gebiet der Jugendpflege . . ., Bd. II, Berlin, 1927), p. 5.

13. Quoted in Bernhard Mewes, *Die erwerbstätige Jugend. Eine statistische Untersuchung* (Berlin/Leipzig, 1929), p. 5.

14. Siemering, *Deutschlands Jugen*, p. 126.

15. Cf. note 3, above.

16. Herrnstadt, *Die Lage*, pp. 7, 26–7; on 30 July 1932, some 618,340 of the unemployed under 25 years of age were counted as primary-benefit recipients, but only 78,467 of the 632,224 unemployed who were under 21 fell into this category. See Kurt Richter, 'Massnahmen zur Betreuung der erwerbslosen Jugend' in

Richter (ed.), *Handbuch der Jugendpflege. Heft 14* (Eberswalde-Berlin, 1933), pp. 18–75, here p. 18.

17. Mewes, *Die erwerbstätige Jugend*, p. 10.

18. Riffka, 'Die berufliche Lage', pp. 5–7.

19. This and the following in Siemering, *Deutschlands Jugend*.

20. Theodor Geiger, *Die soziale Schichtung des Deutschen Volkes* (1932) (Stuttgart, 1967), pp. 96–7.

21. Ruth Weiland, *Die Kinder der Arbeitslosen* (Eberswalde-Berlin, 1933), pp. 23–4.

22. Ibid., pp. 40–2.

23. Ibid., p. 55, also for the following quote.

24. Quoted in ibid., p. 56.

25. Reinhold Weisser, 'Hannovers jugendliche Erwerbslose', *Wohlfahrtswoche*, 7 (1932), 67–75, 86–7, 96–7; here 68.

26. Ibid., p. 76.

27. Maria Tippelmann, 'Über die Auswirkung der Arbeitslosigkeit auf Jugendliche', *Freie Wohlfahrtspflege*, 6 (1931), 309–21, 364–77.

28. Ibid.; also for adult unemployed, Jahoda, *et al.*, *Die Arbeitslosen*.

29. Weisser, 'Hannovers jugendliche Erwerbslose', 76; cf. the diaries of young unemployed people in Gertrud Staewen-Ordemann, *Menschen der Unordnung. Die proletarische Wirklichkeit im Arbeitsschicksal der ungelernten Grossstadtjugend* (Berlin, 1933), pp. 90–2.

30. Ibid., p. 92, also for the following.

31. Weisser, 'Hannovers jugendliche Erwerbslose', p. 68.

32. Tippelmann, 'Über die Auswirkung', 310–12.

33. Tippelmann's sample contained 135 persons aged between 15 and 21, including 25 males and 23 females under 18.

34. Ibid., p. 313; cf. also the diaries in Staewen-Ordemann, *Menschen der Unordnung*, pp. 81–3.

35. See Part III, below.

36. Tippelmann, 'Hannovers jugendliche Erwerbslose', p. 376.

37. Ernst Lau and Mathilde Kelchner, 'Die jugendliche Arbeiterschaft und die Arbeitslosigkeit' in Richard Thurnwald (ed.), *Die neue Jugend* (Forschungen zur Völker-psychologie und Soziologie. Bd. IV, Leipzig, 1927), pp. 321–40, esp. pp. 323–5, also for the following. However, this study is based above all in about 400 essays written by Berlin trainees in 1921–2 on the subject 'Pleasure in Work and Unemployment'. This title is not without its problems in relation to the general attitude to work. But the experience of unemployment in the early postwar years was also profound. See Käte Gaebel, 'Die Erwerbslosigkeit der Jugendlichen' in *Deutsche Jugendwohlfahrt. Denkschrift zum Weltkongress für Kinderhilfe* (Red Cross, Geneva, August, 1925), pp. 73–7.

38. Lau and Kelchner, 'Die jungendliche Arbeiterschaft', p. 326.

39. Ibid., p. 328 for the next two quotes: *Die neue Jugend*, p. 86, for the third.

40. This held good for the registration of the young unemployed for rural labour as well. See *Aus der Praxis der Erwerbslosenhilfe an Jugendliche* (Eberswalde-Berlin, 1931); 'Jugendpflegerische Erfassung der erwerbslosen Jugend', *Zentralblatt für Jugendrecht und Jugendwohlfahrt*, 22 (1930), pp. 245–8; *Massnahmen zur Betreuung erwerbsloser Jugendlicher* (Veröffentlichungen des Preussischen Ministeriums für Volkswohlfahrt . . ., vol. XIII, Berlin 1930); Herrnstadt, *Die Lage*; Richter, *Handbuch*; Joachim Bartz and Dagmar Mor, 'Der Weg in die Jugendzwangsarbeit. Massnahmen gegen Jugendarbeitslosigkeit zwischen 1925 und 1935' in G. Lenhardt (ed.), *Der hilflose Sozialstaat. Jugendarbeitslosigkeit und Politik* (Frankfurt, 1979), pp. 28–94; F. Gräsing, 'Erziehungsarbeit am erwerbslosen Jugendlichen' in *Erfahrungen der Jungen*

43. Very critical, and sensitive to the world of experience of young people: Albert Lamm, *Betrogene Jugend, Aus einem Erwerbslosenheim* (Berlin, 1932); cf. note 40 above, and Walter Friedländer, 'Zur Frage der Tagesheime für jugendliche Erwerbslose', *Zentralblatt für Jugendrecht und Jugendwohlfahrt*, 23 (1931), 322–4; Erna Magnus, *Werkheime für erwerbslose Jugendliche* (Berlin, 1927).

44. Wolfgang Benz, 'Vom Freiwilligen Arbeitsdienst zur Arbeitsdienstpflicht', *Vierteljahreshefte für Zeitgeschichte*, 16 (1968), 317–46; Hellmut Lessing and Manfred Liebel, 'Jungen vor dem Faschismus. Proletarische Jugendcliquen und Arbeitsdienst am Ende der Weimarer Republik' in Johannes Beck *et al.* (eds), *Terror und Hoffnung in Deutschland 1933–1945* (Reinbek, 1980), pp. 391–420; and the literature cited above, Note 40.

45. Tippelmann, 'Über die Auswirkung', 367.

46. Staewen-Ordemann, *Menschen*, p. 89, describes the tent 'cities' of mostly young unemployed outside Berlin. Cf. also the film 'Kuhle Wampe' and the novel by Walter Schönstedt, *Kämpfende Jugend* (1932, repr. Berlin, 1972).

47. Eve Rosenhaft, 'Organising the "Lumpenproletariat": Cliques and Communists in Berlin during the Weimar Republic' in Richard J. Evans (ed.), *The German Working Class 1888–1933* (London, 1982), pp. 174–219; Helmut Lessing and Manfred Liebel, *Wilde Cliquen* (Bensheim, 1981).

48. Reining, 'Arbeitslosigkeit und Jugendkriminalität', *Westfälische Wohlfahrtspflege*, Nr. 1/2 (1933), 14–15; 'Die Entwicklung der Kriminalität der Jugendlichen in den Jahren 1930–1933', *Zentralblatt für Jugendrecht und Jugendwohlfahrt*, 26 (1934), 124–128; Elsa von Liszt: 'Die Kriminalität der Jugendlichen in Berlin in den Jahren 1928, 1929 und 1930', *Zeitschrift für die gesamte Strafrechtswissenschaft*, 52 (1932), 250–71; Heinz Jacoby, 'Die Kriminalität der Jugendlichen in den Jahren 1930 und 1931', ibid. (1934–5), 85–117; Justus Ehrhardt, 'Die Kriminalität der Jugendlichen in den Jahren 1932 und 1933', ibid., 665–91.

49. Tippelmann, 'Über die Auswirkung', 369.

50. Eve Rosenhaft, 'Working-class Life and Working-class Politics: Communists, Nazis and the State in the Battle for the Streets, Berlin 1928–1932' in Richard Bessel and E. J. Feuchtwanger (eds), *Social Change and Political Development in Weimar Germany* (London, 1981), pp. 207–40; Christoph Schmidt, 'Zu den Motiven "alter Kämpfer" in der NSDAP' in Detlev Peukert und Jürgen Reulecke (eds), *Die Reihen fast geschlossen* (Wuppertal, 1981), pp. 21–43.

51. Weisser, 'Hannovers jugendliche Erwerbslose', p. 67.

52. Wilhelm Fliedner, 'Die junge Generation im Volke' in *Die Lebenswelt der Jugend in der Gegenwart* (Berlin, 1928), pp. 7–26.

53. Cf. the contribution by Alexander v. Plato, Ulricht Herbert and Michael Zimmermann to Lutz Niethammer (ed.), *"Die Jahre weiss man nicht, wo man die hinsetzen soll". Faschismuserfahrungen im Ruhrgebiet* (Berlin/Bonn, 1983); Detlev Peukert, *Volksgenossen und Gemeinschaftsfremde. Anpassung, Ausmerze und Aufbegehren unter dem Nationalsozialismus* (Cologne, 1982).

8 THE UNEMPLOYED IN THE NEIGHBOURHOOD: Social Dislocation and Political Mobilisation in Germany 1929–33

Eve Rosenhaft

I

It is impossible now to think of the problem of unemployment in Germany during the 1930s without seeing the Depression as a prelude to, indeed a precondition for the Nazi takeover of power. At the national level, the curve of the vote for the National Socialist Party (NSDAP) followed closely the changing rates of unemployment, rising spectacularly from 1928 to September 1930, the first autumn of full-scale economic crisis, peaking in the summer of 1932 and falling off in November 1932, as the national economic indicators were showing signs of recovery. This is not to say that the unemployed themselves were won to the Nazi Party in significant numbers; regional voting patterns show relatively low Nazi votes in the areas of greatest concentration of unemployment. The relationship was clearly a less immediate and more complicated one than that, and may best be approached by considering the Depression as a period of general political mobilisation and radicalisation in which the fact of mass unemployment had an influence on the perceptions of social groups extending beyond the ranks of jobless themselves. That the Nazi gains were a consequence of general mobilisation was made clear by the fact that voter participation rose in the Depression elections; that they were part of a pattern of general radicalisation was demonstrated by the concurrent electoral success of the party of the extreme Left, the Communist Party (KPD). Indeed, the Communists continued to gain votes into the winter of 1932, when the National Socialist vote was already declining.

The attractiveness of the radical parties was not only registered periodically at the polls. It manifested itself more alarmingly in urban areas as demonstrative action in the streets and neighbourhoods. This ranged from individual gestures like the wearing of party badges, through painting slogans, distributing or selling party literature, and taking part in organised marches and demonstrations. And some sections of the population experienced a polarisation of daily life in the wake of the general mobilisation, which led to disputes between family members, factional activity within local clubs and associations and at its most extreme to murderous violence between members of the 'paramilitary' sections of the respective parties. What made the political enmities all the more perilous, especially as far as the situation of the working class was concerned, was that they divided the labour movement, setting Communists against Social Democrats as well as against Nazis. National Socialism as a movement was thus an important element in the general crisis of political, economic and social life apparently precipitated by the Depression. It was also its principal beneficiary, of course, and once in power the Nazi regime claimed to have produced a solution to the crisis.

As far as the problem of unemployment is concerned, that 'solution' was more gradual than is sometimes acknowledged. In Berlin, for example, the end of 1932 saw the unemployment figures at their worst, with over 600,000, or about 28 per cent of the workforce, out of work. But at the end of 1933, nearly a quarter of the city's working population was still registered as unemployed, and the government measures of the next two years, including the introduction of compulsory labour service for young people and the reintroduction of conscription, succeeded only in reducing this figure by just over one half.[1] In the realm of social relations, recent research has suggested that the National Socialist system operated to solidify the divisions opened up by the Depression crisis, rather than to overcome them in the way that the official myth of restored national community or *Volksgemeinschaft* implied.[2] At the same time, the fact that many Germans remember the Third Reich as above all a period of economic stability prefiguring the 'economic miracle' of the 1950s,[3] and the recurrence in popular memory of the vision of the unemployed disappearing from the streets in 1933, suggest at least the depths of the trauma that Germany's working population suffered during the Depression.

The most familiar interpretation of popular politics during the Depression has always been that it was in the nature of that trauma that it acted on suffering individuals to radicalise them. The combination of economic and political disturbance was not unexpected at the time. Each of the social upheavals that Germany suffered in the wake of the First World War was greeted with warnings that the physical and moral distress of the population could breed only opposition and instability. In each period particular anxiety was expressed about children and youths, their prospects blighted in turn by wartime shortages, the catastrophic inflation and mass unemployment of 1923, rationalisation and the structural unemployment that accompanied it in the late 1920s, and finally Depression. It was here, of course, that the notionally irrevocable effects of material distress on the formation of character were most often cited. Political unpredictability was implicit, too, in the observation of the sociologist Theodor Geiger that structural unemployment compounded by the Depression tended to generate a sort of class of the unemployed, which differed from the traditional classes in being 'socio-economically without location'.[4] And assumptions that this psuedo-class was characterised by a reactive and inarticulate 'radicalism' with a dynamic of its own that ignored all the familiar patterns of political allegiance, appeared to be borne out by the high visibility of its members among the activists of the extreme Right and Left, as well as by a pattern of exchanges of membership between the two 'extremist' parties. Historians have periodically returned to the argument that mass unemployment created a new class of victims. Most recently it has been deployed to account for the presence of a significant number of young workers among the stormtroopers of the Nazi SA, on the premise that political violence of the kind that was the hallmark of the SA is the characteristic mode of the rebel without a cause.[5]

Broadly psychological arguments are hardly irrelevant to the discussion of the political consequences of unemployment. The virulence of political conflicts as they arose during the Depression demands explanation in terms not only of people's search for solutions to acute material need but also of their acting out of collective anxieties. But the psychological effects of unemployment depend on the social, political and institutional circumstances in which the unemployed live. This is borne out, for example, by the observed differences between parts of Austria and Germany during the Depression[6] or even between different sections of the unemployed in contemporary Britain. Nor are the consequences of unemployment, psychological or economic, confined to the unemployed themselves. This chapter will explore some ways in which joblessness and poverty contributed to the situation of general crisis in which Nazism took root alongside other forms of militant politics, focusing less on unemployment's impact on individuals than on its consequences for the network of social relations and institutions through which everyday life was carried on in urban neighbourhoods.

II

The erosion of more or less established institutions is a familiar theme in the historiography of Weimar's closing years. It is a commonplace, for example, that German democracy did not suddenly disappear with the advent of Hitler's chancellorship. By 1933 democratic and representative institutions at the national level were already seriously weakened. Since the summer of 1932 the republic had been governed by chancellors who relied on the confidence of the President, Hindenburg, rather than that of the Reichstag, and who executed their policies in the form of presidential decrees rather than parliamentary legislation. In spite of the increasingly frequent and bitterly fought elections, the Reichstag's role during the Depression was reduced to that of a Greek chorus. This development was a direct consequence of the worsening economic situation, which brought to a crisis existing conflicts between the representatives of labour and industry over the distribution of power and wealth in the new 'social state'. In 1930 the republic's last Social Democratic chancellor had been forced to resign over the question of how the unemployment insurance fund, whose resources were already in danger of being stretched to the limit, should be financed. Government by decree recommended itself as the only effective means of enforcing the unpopular deflationary policies of his successors, Heinrich Brüning and Franz von Papen.[7]

This is the crassest and the best-known example of how the strains placed by mass unemployment on public institutions prepared the way for the triumph of fascism. Recent German research has reminded us of the extent to which the same mechanisms were operating to undermine local government and administration. Not only did the municipalities suffer a cash crisis in the natural course of things, as more and more of the unemployed exhausted their insurance entitlement and moved onto the locally administered and funded welfare rolls. They were also limited in the exercise of their traditional autonomy by the provisions of central-government decrees which were designed to ensure that the policies of stringency being pursued nationally could not be frustrated by local initiatives. It may be an unnecessary exaggeration to describe this erosion of representative institutions as a process of 'fascisation',[8] but it is none the less clear that the abandonment of pluralist participatory democracy which was the central aim of Nazism and one of its earliest achievements was already under way in the years before Hitler's takeover, and that the movement was accelerated, if not caused, by the social cost of mass unemployment. This process had its counterpart in the lives of localities and neighbourhoods. There, poverty and idleness worked to corrode not only the private certainties and psychological defences of individuals but also the fabric of informal institutions that bound people together and the material that insulated conflicting groups from one another.

The accounts that follow are intended to show that unemployment had a disrupting and (in every sense) dislocating effect, as well as an irritant one. They should also illustrate in various ways some of the circumstances that conditioned the political consequences of unemployment and dictated that it would result in mobilisation and polarisation rather than either confused resignation or solidaristic class action. Among these was the persistence in government and administration of an attitude which stigmatised as 'political agitation' any articulation of a conflict of interests between the state and citizens or between different groups of citizens represented within official institutions. In its most virulent form, this was a relic of German conservatism; it had served to legitimise the demand for subordination that buttressed Wilhelmine autocracy and had weathered the revolution to reappear in influential quarters of the civil service and the political parties. It was in this form that the criminalisation of 'politics' was most common during the last years of the republic, dominated as they were by outspokenly anti-parliamentarian chancellors and by the figure of Hindenburg, who took very seriously the view that the president of the republic had to play the role of an *ersatz* Kaiser, representing the interests of the nation as something 'above politics'.[9] The result was that the authorities saw 'political agitation', inherently subversive, behind every form of popular self-organisation and action, particularly as it affected public institutions. To some extent, then, conservative administrators created for themselves a chimera of general radicalisation out of people's struggles to remedy their grievances within the system.

In another form, the desire to avoid social conflict, or to make it obsolete, contributed to its politicisation under the Weimar Republic. The 'social state' involved more than unemployment insurance, which was in fact introduced relatively late in the life of the republic.[10] It also implied the regulation by law of conflicts between opposing interests, beginning with the establishment of national machinery for collective bargaining, and the introduction of elements of representative self-government and local control into such institutions of social policy as schools and labour exchanges. Inspired to some extent by a Social Democratic vision of class compromise, these measures did not so much reduce conflict as shift its focus; once relations between members of different classes were institutionalised within the state apparatus, they became an object of open political conflict, at the lowest as well as the highest level of social negotiation. And since the functions of the welfare state were always associated directly or indirectly with the distribution of material resources, economic crisis implied a sharpening of political sensibilities.

In justification of their fears of political subversion, the authorities could, of course, point to a further peculiarity of the German political scene: the existence of political parties whose express aim and practice it was to translate economic discontent into active and articulate political opposition, namely the Communist Party and the National Socialist Party. Communism and Nazism each posed a real and independent challenge to the *status quo*, not least because both were able to draw on traditions of influence and support outside the unemployed section of the population. The Communist movement was nourished by the historical militancy of skilled industrial workers, the Nazis by the discontent of the urban and rural lower middle classes. Neither was exclusively a movement of the unemployed, although the Communist leaders often expressed fears that their party might become one; but when people who had lost their jobs sought an answer to their problems in political organisation (and this reflex itself had a long tradition, especially in the working class), or when those still in work looked for a champion of normality against the different kinds of threat posed by unemployment and the unemployed, the organisations were ready and waiting.

The association between radicalisation and action in the neighbourhoods, as distinct from, say, workplace action, was most obviously a result of the economic circumstances. The unemployed were the targets of radical agitation, and the neighbourhoods were where the unemployed were to be found. But the KPD and the NSDAP themselves helped to shape the political struggle by defining it as a fight for control of the streets and neighbourhoods. They chose to adopt the neighbourhood as the arena for their kind of politics. That choice was backed up by the more or less explicit construction of competing ideologies or images of the working-class neighbourhood, a process in which the labour movement enjoyed the strength of tradition while the National Socialists had the advantage of novelty and consistency. In the political mythology of Berlin, for example, the 'redness' of the city's working-class districts had been unquestioned since the end of the nineteenth century, and by the 1920s 'red' meant solidly and self-consciously proletarian as well as actively Socialist or Communist. This image of the proletarian neighbourhood almost certainly originated in pre-war Social Democratic observations about voting patterns rather than in any coherent view of the texture of social life or expectations about collective action. The principal source and theatre of solidarity in the German socialist tradition was, of course, the factory; the Social Democratic and, in the Weimar years, the Communist leadership remained deeply suspicious of those realms of life which were subject neither to the discipline of the workplace nor to that of the Party and its organisations.

While the assertion of the solidity of the working-class neighbourhood in itself constituted an effective moment of solidarity and confidence (for labour movement activists, at least) into the 1920s, there was always an element of ambivalence in socialist rhetoric.[11] National Socialism challenged that confidence, and exploited its ambivalence, with its own assertion that it was time to break the hegemony of socialism in the neighbourhoods. The context of SA terror against the labour movement in Berlin was a counter-ideology of reconquest. Like most aspects of Nazi rhetoric, this was explicitly anti-socialist, though ostensibly not anti-proletarian — hence the marching song of the SA in the Kreuzberg district: 'Workers' quarter, deepest need/ but faith in Germany is not dead/ Red Southeast, the underworld/ every week a new victim/ But Hitler's banner is planted firm/ in the heart of Moscow's murder-nest.'[12] Where the old myth asserted that the whole neighbourhood was proletarian and that every proletarian was a socialist, the new one also addressed the neighbourhood as proletarian, but insisted that the socialists were a minority who terrorised their neighbours. As the brownshirts established bases in 'red' areas from which they organised provocative demonstrations and physical attacks on Social Democratic and Communist targets, they could present themselves simultaneously as champions of the underdogs and restorers of order in a world where criminals and terrorists had unaccountably been given the upper hand. This was a vision of the neighbourhood that by its nature drew people's attention to conflicts and antagonisms between individuals and groups of residents. It was thus ideally adapted to a situation in which differences of perception and interest that were always implicit in the life of the neighbourhood were made acute by changing material conditions. I have already suggested that the Depression represented such a situation.

Although neighbourhood life is most easily characterised, for industrial workers, as that complex of activities that goes on outside the workplace, its patterns and rhythms are none the less crucially determined by the fact of work and its proceeds. This is true even of those relations whose significance is peculiar to the neighbourhood: the division of labour within households between the sexes and the generations, the differential use of private and public space which arises out of that division. In the case of groups of people it's also associated with expectations about lifestyle and respectability, the uneasy mutual dependence of tradespeople and their customers, landlords and their tenants, the direct intervention of agents of the state (police and welfare) in daily life, and the forms of sociability through which leisure time is organised. In European industrial societies these have all been premised on the full-time employment away from home of men as well as on the more or less steady income gained from work. Large-scale unemployment therefore implies the disruption of familiar patterns and the reversal of expectations at once more intimate and more wide-ranging than individual poverty.

III

In urban neighbourhoods social relations are reflected, among other things, in the way space is occupied and used. This applies both to relations between individuals and to those between groups. Within households, an individual's sense of self and of his or her social role is closely associated with a sense of place.[13] This reflects in turn the division of labour between the sexes and the generations. In the case of factory workers in industrial cities, work removed men from the context of the neighbourhood for long stretches even when the factory was not far from home. For housewives and homeworkers, on the other hand, the neighbourhood was the workplace, where productive and household activity interpenetrated, and this was also more true of women factory workers than of men. We can locate in the notorious conditions of overcrowding in Berlin the material basis for a shared need to appropriate space outside the home. The classic tenement flat, with its one or at most two small rooms and an 'alcove', would have been constricting for one or two people; for even a small family, the pressure of four walls must have been intense. On holidays and during the summer months, whole families would seek relief in the local parks or further afield. Vacation colonies, or 'tent cities' of the kind depicted in the film *Kuhle Wampe*, began to appear in Berlin's green belt before the First World War, as did the more permanent institution of allotment gardens. By the 1930s there were around 140,000 allotments in Berlin; each had its shed, many of which were used as summer residences by slum-dwellers.[14]

These colonies probably represented the furthest extent of mobility for a whole family within the city. During the working week individuals had to find their own solutions to the problem of overcrowding. Photographs of working-class interiors in which the family members appear reading or working show very clearly how difficult it was to find a comfortable place to sit, let alone withdraw into any kind of privacy.[15] In effect, the only action one could take to relieve the situation was to get out. But the escape itself had specific social dimensions; what was gained by getting out was not so much privacy as wider publicity and more specialised forms of sociability. Men, youths and children stayed out as long as possible, the adult men fleeing to the local bar and the younger ones, depending on their age, to the company and entertainments offered by the courtyards, streets and parks, and by the 1920s dance-halls and cinemas. Women and girls remained tied to the home by their work, paid and unpaid, although in the Weimar years teenage girls were not only going out to work more often than their mothers had, but demanding and to some extent exercising the freedom to spend their free time, like their brothers, outside the house.[16]

There is clearly more involved in this division between men's and women's activities than relative freedom to escape cramped conditions. For one thing, the flat became relatively less cramped as various family members cleared out, and this was an advantage which the woman who had to stay there could achieve by driving them away. Testimony about men's visits to the bar includes statements that husbands went there to get away from nagging wives as well as the observation (from a woman) that they fled the house when they were bored, 'to meet people' — other men.[17]

On the other hand, women would presumably leave the house when they could in order to shop (the usual alternative being to send a child) and would have to go into the public parts of the tenement to use the communal toilet, dispose of rubbish, and so on. Moreover, any mother of small children needed to be aware of the general area in which her own and neighbours' children were playing, and the need to keep an eye on them as well as her preoccupation with household tasks would tie her within a fairly limited range of home even when she did manage to escape the dark and damp of the flat to stand in the open and chat with neighbours. Different members of the household thus maintained their own preserves. In Berlin, where tenements were characteristically built around a series of inner courtyards, the rear court, or *Hinterhof*, close to home, easily surveyed from an upstairs window, and entered only by neighbours or those having business with them, was the preserve of women and children, while men and youths ranged more widely — to the front of the tenement and the immediately adjacent streets in their visits to taverns, and even further afield in club or political activities.[18]

Similarly, expectations about the allocation of space, the use of shared space and behaviour in public are intrinsic to relations between families and social groups in neighbourhoods. Until very recently, historical accounts of the Berlin working class tended to suggest that relationships between near neighbours were characteristically conflict-ridden. In this, they followed a labour movement tradition. It is surprising how rarely the myth of mutual aid was invoked in German socialist and working-class accounts of slum life, although it does appear in the observations of outsiders. Indeed, it was implicit in the labour movement's male- and workplace-centred ideology that forms of solidarity generated outside the workplace should be ignored or devalued. It was probably for this reason among others that gossip, the constant and reciprocal exchange of news through which informal networks were maintained among neighbourhood women, was persistently presented in the socialist press as a source of bitterness between neighbours. But socialist observers were not alone in their perception of the ambivalence of relationships in the neighbourhood. The House Rules (*Hausordnungen*), posted in the entry of every tenement, which banned children from playing in stairwells and courtyards, regulated the use of public areas of the buildings and fixed the hours during which music might be played, presented themselves as rules for neighbourliness, and the Tenants Protection Act of 1923, one of the most important pieces of Weimar social legislation, made failure to observe one's responsibilities to one's fellow-tenants grounds for eviction. Press reports and criminal statistics suggest that conflicts arose among neighbours with some regularity and that they could be pursued tenaciously, by means ranging from verbal disputes through physical violence to litigation (rates of which remained surprisingly high even in a period of general crisis like the First World War).[19]

Clearly too, there existed more diffuse antagonisms between different working-class groups within neighbourhoods. These are perhaps most sharply reflected in fears of crime and violence. Local perceptions of 'criminal areas', expressed as a reluctance to walk down certain streets, are symptomatic of a view that some local residents are not only different from the speaker but inherently menacing by virtue of their way of life. A woman reminiscing about life in Kreuzberg remembers her attitude to a particular street, notorious during the 1930s as both a depressed area and a Communist stronghold:

> Well, I always said, I wouldn't go into the Nostizstrasse for love nor money. Well, if they'd offered me a flat there, if they'd made me a gift of it, I wouldn't have taken it . . . It was all so antisocial: They had the toilets one flight up or down, and there was never just one family on a corridor, but always two or three . . . I don't know about fights, I just had an aversion to the Nostizstrasse, just like I took against the Naunystrasse [now a street with a high concentration of immigrant Turks]. I was supposed to get a flat there, but I said I'm not moving in there, they can do what they like! It was really the poorest of the poor that lived there.[20]

Even in the 1920s ethnic difference was an element in conflicts between near neighbours. In 1927, for example, some residents of a street in the Wedding district appealed collectively for police intervention against others who were thought to be the source of a series of street robberies. The crime wave, in a notoriously depressed street, was blamed on gypsies who had recently taken up residence.[21]

Ethnic conflicts are a special case of the way in which antagonisms between lower-class groups in the neighbourhood tend to be articulated in terms of lifestyle and public behaviour. Such antagonisms imply a sense of moral and physical threat, in contrast to conflicts in the industrial sphere, which typically crystallise around issues of competition and threats to livelihood. The same sense of threat is implicit in the notion of respectability. Respectable behaviour involves not only maintaining a clear distinction between public, private and work, but also avoiding appearances in public except in circumstances where one is unequivocally going about one's business or in highly structured situations like the Sunday promenade or organised processions. The breaching of these distinctions may be perceived as threatening in itself; those who habitually spend their time in public places open themselves to the suspicion of unreliability and even criminality.[22]

There was also a spatial dimension to the relations between different classes in the neighbourhood. In spite of the myth of 'red Berlin', Berlin's neighbourhoods were not closed to 'outsiders', either physically or socially. Except for the proletarian areas of the old city centre, all its working-class districts were punctuated by public squares and criss-crossed by broad streets and boulevards. These high streets carried traffic into and through the neighbourhoods; along them were located the larger shops and restaurants, places of entertainment and meeting-halls. They both served as boundaries and destinations for the members of the immediate community, which revolved around the corner and cellar bars and shops located in the side streets, and brought the 'outside world' into close proximity with it. As early as the 1890s, the streets were patrolled regularly by police officers who were officially expected to live in their own precincts and who operated out of local precinct-houses. These policemen had extensive powers to enquire into and interfere with the lives of families in the neighbourhood themselves and also acted as the executive arm of other authorities like the school board and welfare board.[23]

Socially, the working-class districts at large were not homogeneous, although by the 1920s there clearly existed solidly proletarian enclaves like the Nostizstrasse. Indeed, the original design of the Berlin tenement house, or *Mietskaserne*, represents an early attempt at social engineering through the manipulation of space. The classic tenement was intended to provide housing for tenants of all classes, each building forming a social microcosm whose members would learn mutual tolerance and practise mutual aid; in this plan, the rear courts, cellar dwellings and upper stories were reserved for the poorest, while the more prosperous classes would occupy the largest and best situated flats at the front.[24] The ideal remained at best partially realised in most areas, but there was always a significant presence within the population of people who earned their living other than as wage-workers, and the association between social position and location persisted in various forms. Apart from the small masters, for example in the metal-trades, who continued to carry on their businesses in some of the courtyards well into the 1930s (a feature particularly of Kreuzberg), the most important non-proletarian members of the community were publicans, shopkeepers, and landlords and their agents. The relationship of working-class residents with these members of the petty bourgeoisie constitutes a form of class relation that was specific to the neighbourhood and that was the more ambivalent by virtue of being largely bounded by it; publicans and small traders ordinarily lived in or near their premises, while many tenements were owned by individuals who themselves lived and/or traded in the house or had commercial premises in the nearest high street. The role of small businessman implied at the very least accommodation with working-class customers, and the solidarity of publicans with their patrons was legendary.

At the same time, the stability of that accommodation depended on the continued ability of local residents to pay for the goods and services on offer. And the internal economy of the local petty bourgeoisie did not necessarily take account of the sentiments of the working class. Owning a tenement was the characteristic adjunct, or happy ending, to a life in trade; shopkeepers and publicans aspired to the status of landlord, in spite of the fact that landlords were probably the most uniformly disliked group in working-class neighbourhoods.[25] There were also sources of conflict in the stratification within the lower middle class, since some could be tenants as well as competitors of others. Thus, in an incident in the Prenzlauer Berg district in 1932, local Communists attempted to organise the boycott of a bakery because the manageress' husband had threatened to evict another baker whose shop he owned.[26] Here, we see not only the collapse of any notional trade solidarity accompanying a crisis in landlord-tenant relations, but also evidence of how social rifts followed geographical fault-lines; the landlord's wife traded in a main thoroughfare, his tenant in a side street.

IV

The case of the Prenzlauer Berg bakers may stand as our first illustration of how political confrontation during the Depression frequently arose out of the disruptive and transforming effect that mass unemployment had on relationships and activities specific to the neighbourhood. The background to that case was provided by the inability of the local population to spend enough to support all the small businesses that normally operated in the neighbourhood. This is the most obvious way in which generalised poverty opened up divisions between the members of different social groups, whose respective interests could then be taken up by the activist parties.

The poverty itself was real enough. The system of un-employment insurance was only two years old at the onset of the economic crisis, and although it had been introduced in the first place as a response to the emergence of significant structural unemployment, it was not designed — no insurance system is — to alleviate large-scale, long-term joblessness. By the end of 1932 more than one half of Berlin's unemployed had exhausted any insurance entitlement they might have had and were entirely dependent on municipal relief. Moreover, the government's de-flationary policies included a series of cuts in benefits. By De-cember 1932 an unemployed man with two children under 6 and a non-working wife who had been earning the relatively high wage of 50 RM a week when in work could officially expect to receive 19.80 RM weekly from the state insurance fund (six months earlier the system would have allowed him 35 RM) or about 17 RM per week in municipal welfare benefits. The average rent on the cheapest (and smallest) flat would absorb about one-third of that, and what was left after other fixed expenses had been met would probably provide a diet, consisting largely of bread and potatoes, that was just adequate in calories.[27] Public health physicians who watched for signs of malnutrition in the school population noted that, as had been the case during the First World War, the children were the last members of most families to suffer the effects of shortage, but by 1932 reports like that from a Berlin kindergarten were becoming commonplace: 'Most of the children live on what they get at the centre. Many a mother asks us to give her child something to eat right away, because she has no bread for the child.'[28]

Impoverishment had serious consequences not only for relations between the classes, but also for the operation of informal institutions and structures of association within the neighbourhood. In some cases the two problems were combined. Many traditional leisure-time pursuits required the spending of money, the most obvious of these being the visit to the local bar. Between 1929 and 1933 the *per capita* consumption of beer in Germany fell by 43 per cent.[29] When people couldn't afford to buy drink, the social function of the tavern was undermined; it threatened to lose its role as hangout and meeting-place for local men. The fall-off in trade was also a threat to the livelihood of the publicans. In a letter of February 1932 to Berlin's Social Democratic daily newspaper a publican estimated that three-quarters of his colleagues were 'ruined or on the verge of ruin' as a result of higher taxes and lower takings since 1929; he himself could no longer afford to keep up his membership in the health insurance fund and the trade union.[30] Bars closed, changed hands, or looked, if they could, for new customers. The effect of this was to erode the relationship of mutual tolerance and dependence between the different classes represented by publican and customer, and in this case the change was as disturbing to customers as to publican. In the attitude of a local population to 'its' bar the cash basis of the relationship is overlaid by the rhetoric of hospitality (the 'host' and his 'guests') on one side and the customary right of occupation that free entry and regular visits seem to establish on the other. The shock of withdrawal or the wariness of a newcomer in the key social role of publican is thus particularly acute. The shock rapidly took on an explicitly political edge in Berlin's working-class neighbourhoods, where taverns were traditionally the centres of party as well as social life; the SA, looking for bases in Communist-dominated neighbourhoods, made calculated approaches to publicans who were new in the area or obviously suffering, and won their acquiescence with the promise of guaranteed sales. Once these bars were occupied by the SA, they became the targets of violent attacks by local Communist (and some Social Democratic) activists, and it is not difficult to see in that violence the expression both of economic powerlessness and of rage at the reversal of deep-seated expectations about the structure of daily life.[31]

Other kinds of transformation occurred as a whole generation became preoccupied with raising money. Again, it was the situation of the children, mercenary beyond their years and before their time, that excited official comment. The youngest children revealed a clear understanding of the intricacies of the welfare system, their games and conversation peppered with the officialese of the labour exchange and welfare bureau. When offered the opportunity to fantasise about their wants, the children of the unemployed were found not only to be preternaturally sober and realistic, but also to have taken upon themselves a degree of responsibility unknown among their better-off peers: 'My father has been unemployed for three years. He says I'm not going to get anything for Christmas. I'm going to go out hawking again this year. Whatever I earn my mother will spend on food.'[32] The inhabitants of the proletarian provinces were reduced in some cases to a state approaching the pre-industrial subsistence economy. Thuringian villages which had been entirely dependent on local craft industries presented an appalling picture of moral and physical ruin, while unemployed families in the Silesian coalfields turned to sinking their own pits alongside the closed mines.[33] In the metropolis there was still sufficient economic activity to make it possible to scrape an income on the margins.

Children, young people and able-bodied men and women joined the once ubiquitous war veterans in selling whatever they could get their hands on to supplement their state benefits. There were errands to be run for the businesses that were still operating; one Berlin mother boasted that her 11 year-old son had five after-school jobs as a messenger. In a working-class population that was both relatively literate and highly politicised, selling and delivering newspapers had always provided part-time work, delivery being particularly a job for housewives and pensioners. In the mid-1920s a man could earn just enough to live on as a street-trader selling books and papers, and the fall-off in sales of newspapers during the Depression did not deter the unemployed from trying their luck. Young people could also earn payment, though often only in kind, by performing such household services as baby-sitting, house-cleaning, or carpet-beating. Singing and playing instruments in the streets and courtyards was another way of raising money. And the centres of Berlin's nightlife were plagued by gangs of children offering to 'look after' people's cars — an offer which, then as now, implied the threat of vandalism to those vehicles whose drivers refused.[34] In short, there were very few things that could not be turned into money with energy and determination, and very little that was not tried.

In this scramble for cash the means of entertainment and recreation were turned into a source of income. Social life thereby took on a new edge, as informal institutions ranging from the family to the drinking club either dissolved or acquired new functions. The hiking and social clubs that young people formed to occupy their free time from work or school could become vagrant or criminal communities as their members were forced out of job and home. Spending their days in the open and their nights in cellars and abandoned buildings, stealing or scavenging, tolerated or feared by the local population, and constantly risking confrontation with the police, these gangs lived out a process by which the realms of leisure and recreation were infected with the urgency of a new kind of physical necessity. When both the KPD and the SA attempted, with some success, to recruit these gangs to the political fight, they were acknowledging that they had taken on a role in the lives of their members comparable to that played by the workplace under normal conditions.[35]

The scramble for cash in the neighbourhoods had broader political implications in that it took the unemployed to the margins not only of the economy, but also of the law. The temptations to theft were very great in the conditions of the Depression; although the incidence of reported crime did not increase as dramatically after 1929 as it had during the crises of 1914–18 and 1923, the number of arrests for crimes against property edged up, with thefts multiplying at a greater rate than other offences (about 24 per cent between 1929 and 1932). In 1931 there was a rash of thefts of food in Berlin which the police described as plundering and the Communists encouraged as 'proletarian shopping trips'.[36] (Of course, 'plundering', bullying and boycotts directed against shopkeepers were not purely economic acts; they both reflected and engendered social resentments.) And the handling of stolen goods was an offence that any unlicensed trader could unwittingly commit. But many of the activities of the unemployed fell into a grey area, where the offer of 'protection' merged into extortion and the sale of small, cheap items like shoelaces was not easily distinguished from illegal begging. Moreover, those who were receiving state benefits were legally prohibited from earning additional income. A political order that demanded that citizens break the rules in order to survive jeopardised its own legitimacy, while at the level of more immediate experience people's daily encounters with the agents of the police and bureaucracy became increasingly hostile.[37]

V

Within the official institutions that operated in the neighbourhood too, explicitly political confrontations could emerge from conflicts precipitated by generalised poverty. The atmosphere in the labour exchanges, where so many of the unemployed spent so much of their time to so little purpose, was notoriously tense.[38] But conflicts arose even in those establishments which were only indirectly involved in the distribution of resources. This can be illustrated by an incident from the annals of one of the characteristic institutions of the Weimar Republic, namely the secular schools. These were public schools established by local authorities under the provisions of Article 146 of the Weimar Constitution, which allowed for schools of different *Weltanschauungen* to be set up at the request of local parents. They aimed to provide a progressive, comprehensive education to working-class children. Unlike the majority of state schools, they were non-confessional. Berlin's secular schools tended to attract young, left-wing teachers, and for a few years the schools of the Neukölln district in particular, where the Social Democratic educational reformer Kurt Löwenstein was schools councillor, were well known as centres of educational and cultural experimentation. A further reform of the 1920s was the participation of parents in the governing of schools through an elected parents' advisory council.[39] The combination of the political stance implicit in the schools' commitment to social change with practices that brought any partisan divisions existing within the neighbourhood into the school through the parents, indeed the very programmatic openness of the new schools to the world outside, made them particularly sensitive to agitation by the political parties. The development of these genuinely popular local institutions illustrates with particular poignancy the fragile political economy of 'Weimar culture'.

In 1930 the schools in Neukölln were already coming under 'attack' from two sides: Communists organised school strikes in protest against cuts in the education budget, while the right-ward swing in national politics drew attention to conservative doubts about their aims and practices. When the Social-Democrat-led Prussian government demanded action against political sub-version in the schools, the mostly Social Democratic heads of the secular schools were quite clearly torn in their responses between loyalty to the new order, their own desire to see social change carried further, and concern for the smooth running of the in-stitution. One reply read in part:

> recent events have shown that staff, parents and pupils of this school wish to follow their own path, guided solely by the desire to protect the school and its development from disturbance and internal disruption . . . For Primary School 41/42 there is not the slightest reason to attempt to counter any existing political subversion of the schoolchildren . . . It may be added that the members of the teaching staff represent a political view that is opposed to the tendencies of the class state and sees the organisation of a people's state [*Volksstaat*] as its aim. Certain differences of emphasis within this attitude cannot in this case be regarded as particularly divisive . . . The teachers . . . have before them almost without exception children from the work-ing class, in whose homes political issues of the day are dis-cussed more frequently than in those of the propertied classes . . . The school can and must give time to the objective discussion of political events of the day . . . Only by referring to current problems is it possible to show children whose pre-judices are already fixed that an organic evolution of the demo-cratic idea is sensible, that it is utopian to imagine that a sudden revolution could produce an immediate improvement in the economic circumstances of our country.[40]

By the end of 1932 the teachers at Primary School 41/42, who had prided themselves on their good relationship with both parents and pupils, had become convinced that it was impossible to con-tinue working with the parents' council. The council was by now riven by party-political divisions, and a 'significant minority' of the parents, led by the representatives of a 'Unity List of Working Parents' consistently adopted a posture of open hostility to the school. Since the activities of this Unity List continued until February 1933, it is safe to assume that they represented right-wing opinion in the neighbourhood.[41] At Primary School 16, in the heart of one of the more depressed sections of Neukölln, the head's response to the decree on political subversion suggested that the threat came from the Left. This school had been one of those most affected by the school strike, and the head was at pains to explain that only about 10 per cent of the parents had supported it; the children of the others, he claimed, had been kept away from classes by bullies from the Communist Young Spartacus League. Other records from this school show that its operation was characterised by mistrust of the parents' council and an overwhelming desire to maintain peace within the institution by avoiding political debate.[42]

If Primary Schools 41/42 and 16 and the kinds of opposition they inspired seem to represent two extremes of political posture within the broadly liberal and progressive institution of the secular school, then the situation in Primary School 53/54 represents an intermediate position. There, differences of perspective among teachers and between teachers and parents were negotiated with some success during the years of relative normality, but came to a head under the conditions of the Depression. The minutes of staff meetings for 1927 include the record of a remarkable meeting between teachers and parents' representatives, at which some of the parents complained that the academic education of the children was being neglected:

The class was behind in arithmetic and German . . . Too much time was spent on gardening, too much playing around, e.g. aquarium, catching spiders in the schoolyard. Discipline was not all it might be. [One of the teachers] points out the nature of our profession, that we work with living material, with people. In that situation misunderstandings could arise between parents and teachers [. . . Another teacher] emphasises the thought which guides us all: What helps the children on? We wait. We let things develop. This kind of approach requires a lot of patience on the part of the parents. [One of the parents] notes with satisfaction that this meeting has proceeded more calmly than the last one. He too is satisfied with the teachers' declarations. [A teacher] points to Marx's great idea of renewal. These ideas must be carried over into the realm of education . . . Socialist culture is based not on performance but on attitude.[43]

A few weeks later the situation was reversed when one of the more radical teachers joined one of the parents in accusing the man who had since replaced him as head of abandoning the 'proletarian' character of the secular school. Too much time was now spent, he argued, on practising correct pronunciation and on humanistic 'life-studies', and not enough attention had been given to articulating the school's political vision and long-term programme. That there was already a party-political element to the tensions within the school became apparent when it was pointed out that the Social Democratic parents' representatives were satisfied with the work of the school and had invited the teachers to join them in publicising it; the new head welcomed the invitation and 'expressed the hope that the KPD would issue the same invitation'.[44]

In the years that followed the school weathered accusations of pro-Soviet bias in the teaching of some of the staff and a Communist leafleting campaign against the head.[45] But in September 1932 a minor scandal developed over the conduct of a class trip. Early in the summer, the children in Class 3B had been taken by their teacher on an outing to Rauen, a village some 50 kilometres southeast of Berlin. During the several days they spent in the country, the children had been provided with three meals a day by the youth hostel there. But after their return, some of the parents complained to the head teacher that the meals — mostly in the form of soups and stews — had not been sufficient to keep the children from feeling hungry most of the time. Questioning of the pupils and their parents revealed that the teacher had said that the trip would have to be cut short if they had any more meal breaks. He had also used the cutting of rations as a disciplinary measure, and had personally censored postcards that children sent home asking their parents to meet them at the station with food. On one occasion three of the children were found begging cakecrumbs in a bakery. In the depths of the Depression, when social workers were regularly reporting that parents relied on the schools and other state institutions to provide their children with meals that they could not afford themselves, such complaints demanded to be taken at their face value.

Certainly, it was the conclusion of the headteacher and other teachers involved in the investigation of the case that 'things had not gone as they should' on that trip. But the teacher against whom the complaints were directed was unrepentant. Costs had to be kept to a minimum, he argued, and the food was of the kind and amount provided in reform schools; it therefore could not have been inadequate. (The force of this complacent and tactless analogy cannot have been lost on the parents.) He objected both to the parents' breach of protocol in complaining directly to the headteacher and to the discussion of the case with the children themselves, and used the occasion of a meeting with the parents to air his doubts about the parents' council. In the staff meeting that followed he described the events as a 'witch-hunt' and declared that he 'had no interest in working in a political school'. Shortly thereafter he applied to be transferred to a well-to-do district which had no secular schools.[46] Given the school's history, it is not difficult to guess at the party affiliations of the protagonists in this incident, but it is characteristic of the crisis that there is no suggestion here of partisan calculation or agitation. The politics, that is, the articulation of an opposition of interests as a political one, arose very rapidly out of circumstances which might otherwise have been regarded as relatively trivial but which the scarcity of financial and physical reserves transformed into a matter of urgency.

Of course, it was also characteristic of the national mood of 1932 that 'political' should have been used as an accusation when applied to a public institution. The last entry in the minute-book of Primary School 53/54, dating from April 1933, records the first meeting attended by the new head; expressing the hope of 'fruitful collaboration' with the staff, he explained, 'A change of head had taken place because the national government felt that this was the only way in which young people could be assured of a national education'. Dissolution of the secular schools and reintroduction of strict discipline and traditional authority were among the measures of which the National Socialist local authorities boasted after 1933; they were presented publicly as measures to improve the productivity of the education system.[47]

VI

If certain legal and institutional measures characteristic of the Weimar Republic opened up the schools to political conflict, the same was true of that most fraught of neighbourhood relationships, the one between landlord and tenant. During the First World War rents had been fixed by law, and during the 1920s the rents on prewar housing stock continued to be held at a relatively low level. At the same time, tenants were granted new rights to security of tenure, their relations with landlords and with one another regulated by statute. Although these measures were intended in the first instance to mitigate the social effects of the postwar economic problems, they were widely read as an attempt to shift the burden of inflation from the working class onto the lower middle class. Middle-class opposition organised itself very early into a householders' movement, which spearheaded the principal interest-group party of the 1920s, the Economics Party (National Party of the German Middle Class). This party in turn was in a state of dissolution by 1930, losing much of its support to the more extreme parties of 'national opposition', including the Nazi Party.[48] It was therefore to be expected that when unemployed tenants were unable to pay their rents and landlords called on the power of the law to enforce their claims the resulting conflict would be absorbed into party-political rivalries.

In 1932 there was a wave of rent strikes in Berlin. These were very largely organised by the Communist Party, but they originated in the inability of large numbers of tenants to pay very high rents for very inadequate housing in the wake of a new series of cuts in living standards decreed by the Papen government. (It was one of the paradoxes of the rent-control system that the worst housing was often the most expensive, since flats that had been converted since the war were not covered by legislation, and conversion could entail the breaking-up of larger flats through the construction of cheap partitions without any increased provision of plumbing or cooking facilities.) The conflict was fuelled by the readiness of landlords to evict tenants, even under the most pitiful circumstances, and it became part of the popular image of the tenants' movement that neighbours would mobilise to prevent evictions or to carry furniture back into the house once the bailiffs had left. In their public pronouncements, both Communist and Nazi Parties in Berlin expressed their solidarity with the tenants (while carefully avoiding any expression of sympathy for each other), but the terms in which they presented the movement are strikingly different, and characteristic of the two competing visions of community. The Communist Party approved and encouraged the formation of special 'Tenants' Defence Troops'. Its press complained that the SA had called the police on rent-strikers. The Communists addressed their propaganda to a proletarian public undifferentiated by sex, age or material circumstances, to 'workers', and at the same time made it very clear who the enemy was, by naming the landlord and describing his visit to the building. The tendency of the National Socialist accounts, by contrast, was to depoliticise the scene: an SA-man lifts one end of an evicted tenant's table, a Communist the other, in a moment of spontaneous neighbourliness accompanied by the singing of a current popular hit. The landlord is invisible, an anonymous 'profiteer' somewhere in the background, and the bailiff appears only as a 'representative of the System', mildly comic in his consternation. Where it was impossible for tactical reasons to blame the 'reds' for trouble in the neighbourhood, the National Socialists managed to interpret social conflict in such a way as to avoid blaming anybody.[49]

Rent strikes and resistance to eviction were ways of responding to the dislocation of individuals and the rupturing of communities that the Depression brought with it. Poverty alone forced many families to seek alternative housing, with the result, among others, that vacation camps and allotment gardens ceased to be places of recreation and were transformed into permanent settlements. At the end of 1933, 44,000 allotments were being used as year-round habitations. The inappropriateness of the garden sheds for full-time habitation was made up for by their cheapness, but there were other disadvantages. Observers who noted that children who lived on allotments clearly benefited from the availability of fresh vegetables, also commented on the isolation that families suffered at the fringes of the city. This is contradicted to some extent by evidence that the allotment-dwellers developed a sense of community and local identity themselves, an extension of relationships developed during earlier seasonal visits or perhaps based on the fact that some of them had been near neighbours in town. Indeed, many of the allotments came to be known as Communist strongholds and provided the cover for underground activity well after the Nazi takeover, with the result that the National Socialist regime made a particular point of 'cleaning up' the allotment gardens and restoring them to their proper functions.[50] But the general impression left by accounts of people's housing situation in the 1930s is that the Depression set people in motion, or at any rate interrupted familiar patterns of mobility and stability.

Mobility was not always voluntary, particularly when it resulted not from poverty itself but from administrative diktat. The Second Emergency Decree of the Brüning government, issued with effect from June 1931, raised the minimum age for receipt of unemployment insurance from 16 to 21, regardless of one's work history. The only exceptions to this were young people living away from their families. Since many families depended on all their adult members receiving some kind of income, and very few could afford to support a non-earning adult, the effect of this decree was to drive more young unemployed away from home. If they could not afford lodgings on their own (a young single person with a history of low-paid jobs could expect to receive between 5 and 10 RM per week in unemployment benefit) they could register at a fictitious address and then either sleep rough or spend their days on the street and their nights at home and risk being caught out.[51] To be sure, most young people spent as much of their free time as possible on the streets even under normal conditions of overcrowding, but this enforced vagabondage was something new. Moreover, it was in the nature of the way the Depression was managed that while young people were forced out onto the streets their mobility within the city was often limited. The system of making weekly appearances at the labour exchange in order to maintain one's entitlement to benefit, ostensibly to demonstrate that one was available for work that was never there, was regarded by the young unemployed as an illegitimate and artificial restriction on their freedom of movement, and they took considerable risks to avoid the tedious queues. It was a common complaint that unemployed people could not afford to travel even to find work, much less to entertain themselves. If they did travel on the underground or suburban railway, men were liable to be stopped by inspectors from the labour exchange who suspected them of being engaged in illegal work. And by the 1930s any group of young people hanging about on the street could expect to be moved on by the police.[52] The fugitive existence to which some young people were reduced in the Depression thus prefigured the situation of many working-class militants in the early months of the Nazi regime, when it was safer to disappear from the neighbourhoods where they were known, to sever their ties for the duration or find ways of meeting old neighbours and political associates away from home.[53]

VII

Finally, we may return to the idea of the neighbourhood as a realm in which social relations are expressed in and reinforced by the organisation and use of space. While some people, especially young men, were cut adrift from their familiar haunts, others experienced a radical invasion of space, private and public, which posed a material if unspoken threat to their sense of social order. The most notorious example of this is the disruption caused to family life by the presence of unemployed men in the household. Survey after survey attested to the difficulty that unemployed men had in fitting in to a household routine premised on their absence. They very rarely contributed to the tasks of housekeeping. One observer (a woman) commented: 'Often — and this is particularly true of the lower middle class — the wife attempts to spare her husband the housework, out of a mistaken sense of the [appropriateness of] different forms of work.'[54] Such attitudes persisted in spite of the fact that the shortage of money increased the housewife's workload. The strains of the situation were such that children were found more and more often identifying with their mothers in a common front against the man who took on the role of an intruder, alien not only by his presence but by virtue of the character changes wrought by idleness and psychological depression.[55]

Ultimate poverty also laid the household open to intrusion by total strangers. Municipal welfare support was paid on the basis of need, and need was determined and entitlement confirmed by regular home visits from officers of the welfare board. While these were familiar events in the lives of the tens of thousands of Berliners who relied on the city for their whole income or income supplements before the Depression, by 1932 new recipients were joining them at the rate of some 10,000 per month. For many of them, the indignity of inspection was a new experience. Reporting on the working day of a typical social worker, Berlin's Social Democratic daily remarked in 1931, 'She is about as popular as the bailiff, and even though, unlike him, she takes nothing away but may actually bring them something, people don't expect much good of her.'[56]

That extension of the working-class home, the courtyard, also appears to have been more than usually crowded during the Depression. Politics retreated into the *Hinterhof* with the men.

Always the place where gossip went on, the courtyard became in the late 1920s the neighbourhood newsroom and stage for political theatre — literally and figuratively — and the worse conditions became the more attention the *Hinterhof* received. Agitprop troupes performed; individuals and propaganda squads of the working-class organisations made political announcements, summoned their members and denounced their enemies. Teams of foreign journalists and groups of radical doctors pursued their investigations. Where tenants were organised, the notices of tenants' committees hung in the courtyard. And in a rent strike of 1932 in Kreuzberg, large notices and arrows invited the public into the Hinterhof to sign a protest letter, tour the building, and examine jars of vermin collected in the flats.[57] It is difficult to assess what the reaction of the courtyard's traditional *habitués* might have been to this invasion. The Communists hoped by taking politics into the neighbourhood to politicise women and youth, and the reports of the Communist press on the Kreuzberg rent strike show women and children as the activists in delegations to the landlord (whom they presented with bags of sweepings from the building). But in the long term the efforts of the labour movement to mobilise women were notoriously unsuccessful.[58] In those residents who were not already incorporated into a political culture, this invasion of the *Hinterhof* might just as well have inspired anxiety and resentment.

Even those members of the household who fled the home and courtyard could not go far, with the result that the Depression witnessed an overcrowding of public areas within the neighbourhood. This was not an entirely new problem; every period of social strain in Berlin seems to have brought with it a series of conflicts over people's right to 'hang out'. In the summer of 1923 the police cleared a parade-ground in Prenzlauer Berg of 'rabble of both sexes' whose 'shameless carryings-on' (thus the press report) had been annoying local residents. Forty-two people were taken into custody. A year later, residents of Wedding, driven by the June heat into the largest local park, refused to move on and even formed themselves into troops to stone the police when challenged. And during 1927, citizens' complaints led to night-time roundups in the Friedrichshain, another large park in Berlin's proletarian east, which resulted on one occasion in 76 arrests of both men and women.[59] The police actions of the 1920s coincided with periods when unemployment threw large numbers of people into the street and with moments of heightened public awareness of crime, especially juvenile delinquency. They reflect not only the ambivalence of relations between police and residents in working-class neighbourhoods, but also the way in which relations among residents are informed by expectations about public behaviour. During the Depression these conditions, conditions under which respectability was most embattled in the neighbourhood, were reproduced in a particularly intense form. Idle men and youths were everywhere. They spilled out of the overburdened labour exchanges and blocked the surrounding streets. They dozed in the public libraries. They loitered on streetcorners and occupied the local parks, driving children from their playgrounds and pensioners from their reserved benches by their very presence: 'Old people pass timidly by these seats which they have used for years. Park keepers dare not intervene.'[60] In this vignette, taken from a 'dole chronicle' published in 1932, the sense of threat is apparent, although the author makes no mention of overt violence.

The disruption that unemployment caused to expectations about mobility and use of space, both for the unemployed and for their neighbours, is among the least tangible and least easily demonstrated consequences of the Depression. It is considerably easier to chart the radicalising effects of general impoverishment within institutions that were already more or less political in character, or to show that material need led to conflicts over resources which were subject to being articulated in terms of radical politics. What suggests that this aspect of collective psychology is worth investigating is the fact that both the radical parties defined their fight in the neighbourhood as a fight for the streets, a struggle to recover control of space which had been lost. Using a shared vocabulary of territorial conquest they addressed different patterns of expectation with different notions of terror. They thereby gave voice to the anxieties of sections of the neighbourhood population who had divergent political experiences as well as potentially conflicting economic interests. The Communists appealed to people with a long experience of legal and physical coercion from above, and urged them to fight off the intrusions of an alien and brutal SA as they had resisted and would again resist the 'terror' of the police. The National Socialists, on the other hand, associated themselves with the forces of order, while promising at the same time to be radically effective in a way that the existing police could not. This was implicit in the association made between the activities of political opponents and actual criminal violence in Nazi propaganda about 'red terror' (as indeed in Social Democratic and Communist counterblasts at the 'brown murder gangs'). But the National Socialist notion of terror was at once more diffuse and more comprehensive. In *Mein Kampf*, in which Hitler justifies the active terrorism of the Nazi movement as a response to Socialist terror, he cites two incidents in prewar Vienna which shaped his political vision. His first taste of 'terror' comes when he is chased off a building site by fellow-workers because he has refused to join a trade union and objected to the political tone of their conversation. But the conviction that Social Democracy is essentially terroristic appears to crystallise out of his experience of watching a Socialist demonstration, at a time when he was inclined to feel that 'the mass of those who could no longer be counted among their own *Volk* was growing into a menacing army', and shortly thereafter reading a Socialist paper for the first time.[61] For the prototypical National Socialist, then, Marxist terror consists not simply in physical violence, but pre-eminently in the stating of demands and the organised appropriation of the street.

This notion of the terror of the mass must have had some resonance for people whose social or political existence was denied by the myth of red Berlin — for non-Socialist workers and members of the lower middle class — particularly when the links that bound them to their Socialist and working-class neighbours had been eroded and they felt themselves surrounded by alien and threatening presences. There are echoes of anxieties about control of space in people's memories about the time of the Nazi takeover. The daughter of a woman who worked in an SA soup kitchen in Kreuzberg before 1933 offers as her most vivid recollection of the depression years: 'When we went down to the Mariannenplatz, the whole square was full of men, of all ages; they were playing cards.' — while another says of the first phase of the Nazi regime: 'There was no more hardship, no more suffering; it was so safe on the streets that they [you?] kept the front doors unlocked, well there was no crime, not like before.'[62] The more or less sudden disappearance of men from the street recurs in the memory of the apolitical as a sign of the new regime's success in creating work, although such disappearances, in so far as they actually occurred, were in the first instance the result of Nazi actions against their political opponents or of the cosmetic introduction of draconian measures against 'scroungers'.[63] The conviction that the Nazis made not only the streets but even people's homes safer (implicitly by bringing an end to poverty though conceivably by other means which escape the speaker's recollection), so much a cliché of our own day, has a pedigree in repeated complaints about urban crime in the 1920s. But the recollection cited above is curious in that it comes from the daughter of an academic. Her reference to unlocked doors cannot therefore bespeak a nostalgia for some ideal-typical working-class community based on shared poverty and mutual respect; assuming it refers to her experience at all, the point of reference seems rather to be the sort of *Volksgemeinschaft* in which people once again knew their place.

Notes

1. *StJB*, 1933, p. 105; 'Berlin 1929–1945. Eine Chronik' in *'Wer sich nicht erinnern will . . .' Kiezgeschichte Berlin 1933* (Berlin, VAS in der Elefanten Press, 1983).

2. E.g. E. Brücker, 'Gemeinnützige "Siedlung Lindenhof" E.G.m.b.H. — Berlin-Schöneberg' in Berliner Geschichtswerkstatt e.V. (eds), *Projekt: Spurensicherung. Alltag und Widerstand im Berlin der 30er Jahre* (Berlin, Elefanten Press, 1983), pp. 9–49; and in a rather different context: G. K. Wagner and G. Wilke, 'Dorfleben im Dritten Reich: Körle in Hessen' in D. Peukert and J. Reulecke (eds), *Die Reihen fast geschlossen* (Wuppertal, Hammer, 1981), pp. 85–106, esp. p. 93.

3. U. Herbert, '"Die guten und die schlechten Zeiten". Überlegungen zur diachronen Analyse lebensgeschichtlicher Interviews' in L. Niethammer (ed.), *'Die Jahre weiss man nicht, wo man sie heute hinsetzen soll.' Faschismuserfahrungen im Ruhrgebiet* (Berlin/Bonn, Dietz, 1983), pp. 233–66.

4. T. Geiger, *Die soziale Schichtung des deutschen Volkes* (Stuttgart, F. Enke, 1932), p. 97.

5. C. Fischer, *Stormtroopers* (London, Allen & Unwin, 1983), esp. Chapters 6 and 7.

6. Cf. M. Jahoda *et al.*, *Marienthal. The Sociography of an Unemployed Community* (1933, reprint London, Tavistock, 1974), p. 41.

7. W. Jochmann, 'Brünings Deflationspolitik und der Untergang der Weimarer Republik' in D. Stegmann *et al.* (eds), *Industrielle Gesellschaft und politisches System* (Bonn/Berlin, Neue Gesellschaft, 1978), pp. 97–112. Cf. W. Conze, 'Die politischen Entscheidungen in Deutschland' in W. Conze and H. Raupach (eds), *Die Staats- und Wirtschaftskrise des deutschen Reiches 1929/33* (Stuttgart, Klett, 1967), pp. 176–252.

8. See V. Wunderlich, *Arbeiterbewegung und Selbstverwaltung. KPD und Kommunalpolitik in der Weimarer Republik* (Wuppertal, Hammer, 1980).

9. Cf. T. Eschenburg, 'The Role of Personality in the Crisis of the Weimar Republic: Hindenburg, Brüning, Groener, Schleicher' in H. Holborn (ed.), *Republic to Reich. The Making of the Nazi Revolution* (New York, Vintage, 1973), pp. 3–50.

10. L. Preller, *Sozialpolitik in der Weimarer Republik* (1949, reprint Kronberg/Düsseldorf, Athenäum/Droste, 1978); A. Gladen, 'Probleme staatlicher Sozialpolitik in der Weimarer Republik' in H. Mommsen *et al.* (eds), *Industrielles System und politische Entwicklung in der Weimarer Republik* (Düsseldorf, Droste, 1974), pp. 248–58.

11. For further discussions of political conflict in working-class neighbourhoods and socialist attitudes to it, see E. Rosenhaft, 'Working-class Life and Working-class Politics' in R. Bessel and E. J. Feuchtwanger (eds), *Social Change and Political Development in Weimar Germany* (London, Croom Helm, 1981); E. Rosenhaft, *Beating the Fascists?* (Cambridge, Cambridge University Press, 1983); A. P. McElligott, '"Das Abruzzenviertel". Arbeiter in Altona 1918–1932' in A. Herzig *et al.* (eds), *Arbeiter in Hamburg* (Hamburg, Erziehung und Wissenschaft, 1983), pp. 493–507; A. McElligott, 'Street Politics in Hamburg 1932–33', *History Workshop Journal* 16 (1983), 83–90.

12. J. K. von Engelbrechten, *Eine braune Armee entsteht. Die Geschichte der Berlin-Brandenburger SA* (Berlin/Munich, Franz Eher, 1937), p. 103.

13. Anthropological and historical studies discussing the social organisation of space include R. Reiter, 'Men and Women in the South of France' in R. Reiter (ed.), *Toward an Anthropology of Women* (New York, Monthly Review, 1975), pp. 252–82; E. Ross, 'Survival Networks: Women's Neighbourhood Sharing in London Before World War One', *History Workshop Journal* 15 (1983), 4–27; T. Kaplan, 'Female consciousness and collective action: the case of Barcelona 1910–1918', *Signs* VII (1982), 545–66.

14. E. Mahler, 'Kleingarten' in *Berlin und seine Bauten. Teil X. Gartenwesen* (Berlin, Ernst & Sohn, 1972), pp. 218ff; A. Schorr, 'Die Wohnlauben in Berlin', *Der Gemeindetag* XXIX (1935), 793; *Vorwärts* 25 July 1927 (E), 4 May 1924 (M).

15. Cf. the collection of photographs taken for the Ortskrankenkasse für den Gewerbebetrieb der Kaufleute, Handlesleute und Apotheker, published as G. Asmus (ed.), *Hinterhof, Keller und Mansarde* (Reinbek, Rowohlt, 1982).

16. R. Dinse, *Das Freizeitleben der Grosstadtjugend* (Berlin-Eberswalde, R. Müller, [1932]) pp. 62ff, 75ff; L. Franzen-Hellersberg, *Die jugendliche Arbeiterin* (Tübingen, Mohr, 1932), p. 47; H. Schönrock, *Wir kamen gerade so hin. Meine Kindheit und Jugend in Berlin-Moabit* (Berlin, Dirk Nishen Verlag in Kreuzberg, 1983), pp. 20ff; *Kreuzberg 1933. Ein Bezirk erinnert sich* (Berlin, Kunstamt Kreuzberg, 1983), pp. 17ff; R. Bridenthal, 'Beyond "Kinder, Küche, Kirche": Weimar Women at Work', *Central European History* VI (1973), 148–66; J. Wickham, 'Working-class Movement and Working-class Life: Frankfurt am Main during the Weimar Republic', *Social History* VIII (1983), 315–43, esp. 332ff.

17. Cited by R. Beier, 'Leben in der Mietskaserne', in Asmus, *Hinterhof*, p. 251; cf. J. S. Roberts, 'Wirtshaus und Politik in der deutschen Arbeiterbewegung' in G. Huck (ed.), *Sozialgeschichte der Freizeit* (Wuppertal, Hammer, 1980), pp. 124–6. On the social function of nagging, cf. Ross, 'Survival Networks'; Reiter, 'Men and Women'; S. Harding, 'Women and Words in a Spanish Village' in Reiter, *Toward*, pp. 283–308.

18. Children often played in the streets, since games were officially banned from the courtyards in most tenements; cf. Schönrock, *Wir kamen*, pp. 7f. But the drawings of Heinrich Zille, which are among the best evidence we have for the character of slum life in Berlin, suggest that in the summer at least the prohibition was taken lightly. Cf. also the photographs reproduced in W. Römer, *Kinder auf der Strasse Berlin 1904–1932* (Berlin, Dirk Nishen Verlag in Kreuzberg, 1983).

19. One observer who saw poverty and lack of privacy as the sources of a kind of 'Communism' was, of course, Paul Göhre, *Drei Monate Fabrikarbeiter und Handwerksbursche* (1891, reprint Gütersloh, Mohn, 1978), p. 39. Cf. O. Rühle, *Kultur- und Sittengeschichte des Proletariats* (1930, reprint Berlin, VSA, 1978); *Vorwärts* 20 Mar. 1914 (M), 24 July 1923 (E), 12 Oct. 1927 (M), 2 June 1931 (E). For

historians' views, cf. L. Niethammer and F. Brüggemeier, 'Wie wohnten Arbeiter im Kaiserreich?', *Archiv für Sozialgeschichte* XVI (1976), 61–134; H.-h. Liang, 'Lower-class Immigrants in Wilhelmine Berlin', *Central European History* III (1970), 94–111; A. Lüdtke, 'The Historiography of Everyday Life' in R. Samuel and G. Stedman Jones (eds), *Culture, Ideology and Politics* (London, Routledge & Kegan Paul, 1982), pp. 38–54. Tenants Protection Act: *Reichsgesetzblatt* I (1923), p. 353. M. Liepmann, *Krieg und Kriminalität in Deutschland* (Stuttgart, Deutsche Verlags-Anstalt, 1930), pp. 41f.

20. *Kreuzberg 1933*, p. 22. Of course, such perceptions are by no means universal, even among non-residents of 'criminal areas' and their political consequences are not always predictable; cf. J. White, 'Campbell Bunk: a Lumpen Community in London Between the Wars', *History Workshop Journal* 8 (1979), pp. 1–49 and letters in issue 10, pp. 24f.

21. *Vorwärts* 18 Dec. 1927 (M).

22. An example of a clash over public lifestyles is provided by an incident in Zurich in 1896, when members of the native population rioted against immigrant Italians; objections to the newcomers included 'noisy games in and near their bars, blocking the streets and pavements, wearing dirty clothes on Sundays, catching birds, etc.': H. Rathgeb, *Der Ordnungsdiensteinsatz der Schweizer Armee anlässlich des Italiener-Krawalls im Jahre 1896 in Zürich* (Frankfurt a.M./Bern, Lang, 1977), p. 19.

23. *3. Verwaltungsbericht des Königlichen Polizeipräsidiums von Berlin* (Berlin, 1902), pp. 482, 503; cf. H.-h. Liang, *The Berlin Police Force under the Weimar Republic* (Berkeley, University of California Press, 1970).

24. W. Hegemann, *Das steinerne Berlin* (1930, reprint Berlin, Vieweg, 1979), pp. 232f. See Schönrock, *Wir kamen*, p. 16, for an example of this principle in action in the 1920s; in this case, the landlord's agent, who lived in the building, was a recognised Nazi.

25. *Berliner Adressbuch* gives the names and addresses of landlords and their agents. G. Dehn, *Die alte Zeit, die vorigen Jahre. Lebenserinnerungen* (Berlin, Kaiser, 1962), pp. 167f. On the role of small-scale producers with National Socialist sympathies as employers in the neighbourhood, see *Kreuzberg 1933*, p. 92; cf. McElligott, 'Abruzzenviertel', p. 497.

26. Documents on the case in LAB1n, Rep. 58 [Files of the state prosector], vol. 1496. Cf. McElligott, 'Street Politics', p. 89.

27. *StJB* 1933, pp. 106f, 194; H. R. Knickerbocker, *Germany — Fascist or Soviet?* (London, John Lane, 1932), pp. 19ff. Fifty RM per week would have been a low to middling wage for a metal-worker in Berlin in 1928, but by 1932 wage-levels had been eroded through the combination of an overcrowded labour market and government-enforced pay cuts; cf. *SJDR* 1931, pp. 269ff.

28. R. Weiland, *Die Kinder der Arbeitslosen* (Berlin, R. Müller, 1933), p. 18.

29. *SJDR* 1933, p. 319. Cf. Knickerbocker, *Germany*, pp. 13ff; *Kreuzberg 1933*, p. 47.

30. *Vorwärts* 26 Feb. 1932 (M).

31. For details of such incidents, see Rosenhaft, *Beating*, pp. 117ff.

32. Weiland, *Kinder*, pp. 39f; cf. p. 23.

33. A. Stenbock-Fermor, *Deutschland von unten. Reise durch die proletarische Provinz* (Stuttgart, J. Engelhorn, 1931), pp. 87ff; L. Machtan, 'Die "Elendsschächte" in Oberschlesien' in H.-G. Haupt (ed.), *Selbstverwaltung und Arbeiterbewegung* (Frankfurt a.M., Europäische Verlagsanstalt, 1982), pp. 141–56.

34. Weiland, *Kinder*, pp. 25ff; *Kreuzberg 1933*, p. 61; H. Benenowski, *Nicht nur für die Vergangenheit. Streitbare Jugend in Berlin um 1930* (Berlin, Dirk Nishen Verlag in Kreuzberg, 1983), pp. 16, 19; Police report on an itinerant bookseller, May 1927, LAB1n 58/364, I, p. 109.

35. Cf. H. Lessing and M. Liebel, *Wilde Cliquen* (Bensheim, päd-extra, 1981);

E. Rosenhaft, 'Organising the "Lumpenproletariat"' in R. J. Evans (ed.), *The German Working Class 1888–1933* (London, Croom Helm, 1982), pp. 174–219.

36. *StJB* 1931: p. 249, 1932: p. 199, 1933: p. 211, 1934: p. 259; Rosenhaft, *Beating*, pp. 53f; *Kreuzberg 1933*, pp. 64f.

37. Cf. Weiland, *Kinder*, pp. 51f.

38. See Rosenhaft, 'Working-class Life', pp. 219f.

39. J. Nydahl (ed.), *Das Berliner Schulwesen* (Berlin-Friedenau, Deutscher Kommunal-Verlag, 1928), pp. 46ff. For first-hand accounts and photographs of schooling in Neukölln during the Weimar Republic, see 'Widerstand in Neukölln' in *"Wer sich nicht erinnern will . . ."*, pp. 15–26.

40. 41./42. Volksschule to Schulrat des Bezirks Neukölln-Ost, 15 May 1930, LAB1n Rep. 214 (files of Bezirksamt Neukölln), no. 525, vol. 31, insert.

41. Staff meetings of 18 Aug. 1932, 16 Jan. and 20 Feb. 1933, LAB1n 214/525, vol. 32, pp. 76f, 91, 94.

42. Report on staff meeting, 5 May 1930, LAB1n 214/525, vol. 28, insert; staff meetings of 28 Aug. 1931, 12 December 1932, 11 Jan., 21 Feb., 13 Mar. 1933, LAB1n 214/525, vol. 29, pp. 7, 44–52.

43. Staff and parents' meeting of 23 Sept. 1927, LAB1n 214/525, vol. 33, pp. 78ff (reverse pagination).

44. Staff meeting of 13 Dec. 1927, LAB1n 214/525, vol. 33, pp. 59ff.

45. Staff meetings of 28 June 1928, 23 Oct. 1930, LAB1n 214/525, vol. 33, p. 6, vol. 34, p. 6.

46. Staff meetings of 5 and 26 Sept. 1932, LAB1n 214/525, vol. 34, pp. 83ff.

47. Staff meeting of 29 Apr. 1933, LAB1n 214/525, vol. 34, insert. *Bericht des Bezirksbürgermeisters fur Neukölln, 1.4.1932–31.3.1936* (Berlin, 1936), pp. 66ff.

48. M. Schumacher, 'Hausbesitz, Mittelstand und Wirtschaftspartei in der Weimarer Republik' in Mommsen, *Industrielles System*, pp. 823–36.

49. Press reports on the rent strikes are reprinted in R. Nitsche (ed.), *Häuserkampfe 1872/1920/1945/1982* (Berlin, Transit, 1981), pp. 163ff. Cf. *Kreuzberg 1933*, pp. 19f; Rosenhaft, *Beating*, p. 54.

50. Schorr, 'Die Wohnlauben', pp. 793ff; Weiland, *Kinder*, p. 7; Stenbock-Fermor, *Deutschland*, pp. 149ff. Cf. R. Klages, 'Proletarische Fluchtburgen und letzte Widerstandsorte? Die Zeltstädte und Laubenkolonien in Berlin' in *Projekt: Spurensicherung*, pp. 117–36.

51. E. Haffner, *Jugend auf der Landstrasse Berlin* (Berlin, Cassirer, [1932]); *StJB* 1933, p. 106; Schönrock, *Wir kamen*, p. 23; *Kreuzberg 1933*, pp. 60f.

52. *Kreuzberg 1933*, pp. 61ff; Schönrock, *Wir kamen*, p. 26.

53. Examples in Benenowski, *Nicht nur*, p. 25; Schönrock, *Wir kamen*, p. 28; Klages, 'Proletarische Fluchtburgen', pp. 125f; *Kreuzberg 1933*, p. 164.

54. Weiland, *Kinder*, p. 8; cf. Franzen-Hellersberg, *Die jugendliche Arbeiterin*, p. 49.

55. Weiland, *Kinder*, pp. 44f.

56. *Vorwärts* 15 Jan. 1931 (M); *StJB* 1933, p. 105; Weiland, *Kinder*, p. 43.

57. Nitsche, *Häuserkampfe*, pp. 169f. *Kreuzberg 1933*, pp. 19, 23, 40; Stenbock-Fermor, *Deutschland*, pp. 153f; Rosenhaft, *Beating*, pp. 120, 146.

58. S. Kontos, *Die Partei kämpft wie ein Mann. Frauenpolitik der KPD in der Weimarer Republik* (Basel/Frankfurt a.M., Stromfeld/Roter Stern, 1979).

59. *Deutsche Allgemeine Zeitung* 14 July 1923 (M); *Vorwärts* 20 June 1924, 12 Sept. and 10 Oct. 1927 (all E).

60. B. N. Haken, *Stempelchronik. 261 Arbeitslosenschicksale* (Hamburg, Hanseatische Verlagsanstalt, [1932]), p. 26. Cf. photographs in Stenbock-Fermor, *Deutschland*, opposite p.129; *Kreuzberg 1933*, pp. 13, 62. On the increase in the number of visitors to libraries, and the sharp decline after the Nazi takeover, see the reports of the district administrations, e.g. *Bericht des Bezirksbürgermeisters des Verwaltungsbezirks Horst Wessel, 1.4.1928–31.3.1936* (Berlin, 1937), p. 36; F.

Hansen, '"Die Scheidung der Geister" oder: "Frontarbeit am Leser". Die Volksbüchereien in Schöneberg' in 'Leben in Schöneberg/Friedenau 1933–45', *"Wer sich nicht erinnern will . . ."; Kreuzberg 1933*, pp. 123ff.

61. A Hitler, *Mein Kampf* (Munich, Franz Eher, 1935), pp. 42ff.

62. Cited in I. Wittmann, '"Und es ging ja auch voran, nicht wahr?" Spurensicherung durch Interviews' in *Projekt: Spurensicherung*, pp. 51, 53.

63. The latter included the narrower interpretation of terms of eligibility in order to discourage applications to official agencies and the enforced removal of some welfare recipients and vagrant youths to workhouses or detention centres: Reports of social services departments, Neukölln, 1938, LAB1n 214/794, vol. VI; *Bericht über die Tätigkeit des Landeswohlfahrts- und -jugendamt, 30.1.1933–31.12.1934* (Berlin, 1936), pp. 70f; *Bericht der Bezirksverwaltung Charlottenburg, 1.4.1932–31.3.1936* (Berlin, 1936), pp. 48f.

9 MOBILISING THE UNEMPLOYED: The KPD and the Unemployed Workers' Movement in Hamburg-Altona during the Weimar Republic

Anthony McElligott

I

Did unemployment lead to political mobilisation or political apathy? The best-known contemporary accounts of the phenomenon in Weimar Germany tend to suggest the latter. For instance, in the film *Hunger in Waldenburg* (1929), the nearest we find to any form of protest by the unemployed miners who lived in the village of the title, is an individual and isolated outburst by one man (who had only recently moved there in the hope of finding work) against an exploiting landlord. This protest is undertaken after a long series of fateful events which have brought him down into the social abyss. After a short physical struggle with the landlord, the man is thrown down the stairs of the tenement and dies. The film is heavy with symbol and conveys not only the bleakness and poverty of unemployment but also the futility of protest. While the film gives glimpses of working-class solidarity, this is inward-looking and has absolutely no political overtones. Similarly, the celebrated field-study of the effects of unemployment on working-class family and community life carried out in the industrial village of Marienthal in Austria in 1932–3 also makes no mention of any collective protest by the villagers against their lot.[1] This does not mean that there was none. It merely reflects middle-class perceptions of how unemployment impinged upon the 'lower orders'. Both the film and the study were executed in a period of political and social disequilibrium when the public sphere was characterised by a charged political atmosphere in which violence came to play an increasingly important part. More recently, two historical studies of Communist efforts to mobilise the unemployed in this period have also reached negative conclusions on mobilisation.[2] The German Communist Party (KPD) is convicted of merely succumbing to its own rhetoric, while the unemployed are said to have been unreceptive to Communist appeals for action because of the subjective factors of social isolation, hunger, psychological and physical demoralisation, all of which led to political apathy. Instead, it is argued, a collective political answer to the problem of unemployment was replaced by individualist strategies of survival on a daily basis.

There is certainly a grain of truth in these arguments. A large body of work exists from the period, and much of this work studies the social effects of unemployment and underlines some of the points made above.[3] And while some interesting observations are made on the proclivity of youth not to accept its lot without a murmur,[4] little is said in these earlier studies about the relationship between the unemployed and the political parties. In the case of the KPD, its relations with the unemployed were far from unproblematic. A closer look at this relationship, using local evidence, soon begins to reveal a more complex picture than the one conventionally presented. The unemployed were not simply addressees of political messages from above: they also had their own independent world of popular politics. They frequently forced the KPD to adjust its tactics to messages coming from below. Nor were the unemployed steeped in apathy. It appears that the longer the period of unemployment was, the more likely 'hunger' was to translate itself into 'anger'. Mobilisation was linked to specific conditions relating to specific junctures in the experience of unemployment.

The first evidence of the political mobilisation of the unemployed can be found in 1918–19, when Committees of the Unemployed sprang up around the country as part of the revolutionary Workers' Council Movement. The members of the Committees not only saw themselves as defenders of those out of work, but also formed a conscious element in the political organs of the Revolution. The fortunes of the Committees were thus closely tied to those of the Revolution. When the tide began to turn against the revolutionary forces, the unemployed movement also found itself being pushed onto the defensive. Its political bite became feebler as unemployment decreased after the postwar adjustment. A number of internal factors added to the inability of the Committees to form an effective national movement, in spite of two national congresses held in Berlin and Hamburg in April and August 1919. The Committees of the Unemployed, conceived as a national movement, were supposed to be based on popular democracy, but the volatile nature of politics on the Left in the early phases of the revolution meant that struggles between the Communists, the Independent Socialists (USPD) and (to a lesser degree) the Social Democrats, militated against the formation of a concerted movement. From 1920 a third force, in the shape of anarcho-syndicalist trade unionists who subsequently formed the so-called radical 'Communist Workers' Party' further added to the schism within the movement. It remained, therefore, a conglomeration of disparate parts which could defend local interests but failed to perform consistently at the level of the Reich.[5] This was to be its fate in later years too, despite Communist efforts to implement firm control at a national level.

The two congresses of the unemployed held in 1919 did succeed in laying down the basis for a national structure. Committees at local level were to be co-ordinated by regional bodies which ultimately led up to a national co-ordinating body in Berlin. The idea was that of a federal organisation, allowing for flexibility in responding to regional variations in the nature and dimensions of unemployment. After this organisation fell into oblivion, however, no further attempt was made to form a national movement until the KPD began to resuscitate the Committees of the Unemployed in late 1924. The KPD used the blueprint of 1919, but gave it a stronger centralising accent. The Central Committee of the Party, situated in Berlin, sent out circulars in July 1924 urging the mobilisation of the unemployed in committees firmly in the hands of 'trustworthy comrades'. These cadres, numbering between five and seven for each committee, were instructed to organise demonstrations and meetings of the unemployed and to ensure a close contact between the jobless and those still employed in factories and workshops. Furthermore, the leaders of the unions and the Social Democratic Party (SPD) were to be provoked into taking an unequivocal stand on the issue of unemployment, which the Communists felt they had so far neglected to do.[6]

The KPD, and indeed the international Communist movement in general, had always recognised the radical potential of the unemployed during periods of capitalist crisis. Accordingly, they were keen to secure control over this revolutionary reserve. But unemployment in itself was not necessarily the primary concern. This was summed up by a leading Communist, Fritz Heckert, during the Congress of 1920 which sealed the amalgamation of the left wing of the Independent Socialists with the KPD. He stated that the Communist Party's 'aim is to save the working folk by saving the unemployed; and the unemployed must be saved in order that the other workers do not go under'.[7] The 'redemption' of the working class was to take place by harnessing the energies of the unemployed in support of the struggles of those still in work. In this way it was also hoped that a division within the working class as a politically conscious force would be prevented. This strategy was neatly illustrated during a wave of unofficial strikes in March 1924, when the unemployed were expected by the Party

> to form large raiding parties and to stop the strike-breakers on their way to the factories or to expel them from the same. Moreover, the unemployed have to make sure that at labour-exchanges or signing-on offices or any other places where the unemployed gather, strike-breakers cannot be hired or found . . . Furthermore, the unemployed, if they are active enough, can certainly arrange a large demonstration in order to provide moral support for the strikers whether or not these join the demonstrations. In this way the unemployed would have shown their solidarity and have done their duty.[8]

However, the unemployed proved unwilling to be used by the Party for purposes which had little bearing on their immediate situation.[9] This posture must have alarmed the Party leadership. Yet integration of the unemployed into the struggles of their working colleagues, and the consequent subordination of their interests to that of the class struggle, remained the leitmotif of KPD strategy for the rest of the 1920s. This created an area of continuing tension between leadership and large sections of the grass-roots.

At the Tenth Party Congress in 1925 the KPD leadership ordered a restructuring of party organisation in which neighbourhood cells were to be replaced by factory cells as the nucleus for party activity,[10] in order to underpin the Party's policy of unconditional support of strikes and the fight against rationalisation in the middle years of Weimar. Although this was seen as an indirect means of combating unemployment, the attempt to get unemployed workers to enrol in factory-based organisations remained thwarted because of disputes between those in work and those out of work. Moreover, even in 1925 a majority of KPD members were unemployed.[11] They soon began to resist the new policy. In addition, there was a flow of complaints from the leadership about local functionaries' failure to show a satisfactory level of interest in the unemployed in their areas.[12]

It took the changed climate of late Weimar, with unemployment numbering 4.5 million by 1931 (or 21.9 per cent of the employable population) and rising steadily, to force a change of track. In August 1931 the International Revolutionary Union, part of the Communist International, held a special conference in Prague on world unemployment, with delegates from the USA, Britain and France as well as from Germany. At this meeting the unemployment question was placed at the forefront of the international Communist struggle. As a result of the congress, the tactics of the KPD's trade union organisation, the *Revolutionäre Gewerkschaftsopposition* (RGO) were partly revised, along lines first suggested by the Reich Committee of the Unemployed in 1928. The unemployed were now to be accorded a degree of autonomy and allowed to create their own cells alongside the factory cells rather than continue to be subordinated to the latter. However, it was intimated that the RGO should still keep at least an indirect control over the unemployed movement.[13] Thus, the Party moved away from its original concentration on the point of production as the locus of organisation. This shift was further accentuated by the creation, in 1932, of the Antifascist Action (*Antifa*). This was both part of a so-called 'United Front' of labour, and an attempt to organise the working class in the neighbourhoods in the face of Nazi incursions. The *Antifa* was organised around street cells which were explicitly intended to mobilise 'housewives and the unemployed'.[14] The shift away from the factories, indeed, had already begun in practice if not in theory in 1928 as the economy went into the Depression.[15] At that time the Reich Committee of the Unemployed began to exhort Party functionaries to make a more concerted drive to mobilise the unemployed, as part of a broader thrust against the state. 'Each struggle in forcing through demands', it was said, 'reveals the class face of the state and arouses in the unemployed the need to take up the struggle' against it.[16] From May 1928, when there was an SPD-led coalition government in Germany, such a slogan could be applied to lend weight to the KPD's depiction of the Social Democrats as 'Social Fascists' or class collaborators.

Increasingly, through the medium of organisation, the subjective interests of the unemployed became conflated with the objective interests of the KPD:

The conquest of the mass of the unemployed is not only a question of putting forward the right solutions. Above all it is a question of creating an organisation for their ordinary struggles for the demands of the day. Only through systematic daily work, through successful leadership of their partial struggles, can we secure and consolidate our influence among them.[17]

This was to be achieved within a comprehensive organisational scheme which even included the independent registration by the Party of all the unemployed in the working-class neighbourhoods, although in the event this later aim was never carried out. Regional Committees were to be called into life at mass meetings; if this was not possible then meetings at the dole offices would do. From the regional bodies an inner working caucus was to be established which had the task of carrying out daily organisational work. At the mass meetings, in addition to the regional committees, an intermediary (*Vertrauensmann* or V-man) was to be elected for every 50 to 100 unemployed persons. These V-men were of the utmost importance, for they were the last link in the Party's chain to the base. Members of the Committees did not necessarily have to be *bona fide* Party cardholders, but they had at least to be 'revolutionary' workers. Once established, the regional committees had to organise local meetings at the dole offices and labour exchanges, where smaller committees were then to be established. From these, further caucus groups had to be formed. These in turn sent out intermediaries to establish contact with the mass of the unemployed. In the case of those who had exhausted all claims to benefits under the unemployment scheme, the local committees sent out their intermediaries into the soup kitchens, day-centres and warm rooms, in fact wherever the unemployed were to be found. And for those unemployed who stayed at home there was also an intermediary or committee either in the tenement block or in the street.[18]

In 1918–19 the initiative for organisation had come from below, moving upwards within the organisation. But now the whole structure was conceived in terms of an exercise of Party authority from 'top to bottom'. Information was to be passed on via numerous meetings and local newspapers or flysheets specifically directed at the unemployed. The party also provided legal advice centres, where any unemployed person could go and receive help free of charge. The KPD thus hoped to politicise all aspects of everyday life on the dole. A revived class consciousness would be channelled into a broad unified movement against the prevailing order. The KPD's drive to organise the unemployed was part of a general policy of trying to control both the political and social spheres of the working class; for with the coming of the economic Depression and the consequent alienation of a vast section of the working class from formal labour organisations, this extension of political mobilisation from the workplace to the street and the dole queue was now the only way of keeping the working-class movement intact on a coherent basis and of lending it weight in national politics.[19]

The KPD has been rapped on the knuckles by some historians for the primacy it accorded to organisation. This organisational fetishism is said to have resulted in the KPD's ultimate failure to mobilise the unemployed.[20] The Party's inclination towards a policy of exercising tight control certainly stemmed from the 'bolshevisation' which it had undergone in the middle years of the Weimar Republic.[21] But it was also conditioned by the Party's general experience of Weimar politics. It was not a phenomenon peculiar to the Communists. The Social Democratic Party was no less centralist in its organisation. The discipline demanded from a tough organisational structure was therefore not new to the un-employed. Any worker with experience of the German labour movement would already have been initiated into this type of hierarchy. There are no grounds for supposing that this was otherwise among those unemployed workers attracted to the Communist movement.[22] The real problem of discipline appears to have been much greater among younger workers who never had a job and so lacked this experience. The Communists, the Social Democrats and the Nazis all experienced problems of maintaining discipline in their paramilitary organisations, where mostly younger members predominated.

Until now studies of the KPD and the unemployed have con-centrated on the failure of the Communists to mobilise the jobless. At a national level the movement did experience difficulties and was ultimately brought to a halt by the events of January 1933. But regionally the story was rather different. Given the hostile climate of Weimar political society towards any form of worker organisation operating beyond the consensus of support for re-publican institutions, the local and regional successes of the KPD in mobilising the unemployed surely deserve closer attention. It was the ability of the KPD to place itself at the forefront of local and regional protest and to lend it the appearance of national cohesion, which was the real test of Party influence among the unemployed; and as we shall see in the rest of this chapter, it was indeed able to do this on occasion, admittedly at the price of making concessions to pressures coming from below.

II

The area chosen for the study is the conurbation of Hamburg-Altona, in North Germany. The city of Altona had a population of just over 240,000 in 1933, making it the twelfth largest in Prussia. Over 70 per cent of its population was dependent on industry and crafts and the trade and transport sectors of the regional economy. Around 45 per cent of the population was classified as working class.[23] Because of Hamburg's position as a world trading port much of the manufacturing industry situated in the two towns was export oriented.[24] Thus, the downturn of 1926 and the Depression of 1929 onwards hit both Hamburg and Altona hard, as Table 9.1 indicates. Both in the middle and the later years of Weimar, building and metal-workers joined with dockers and sailors in forming a large proportion of those standing in the dole queues. White-collar workers from the tertiary sector associated with the harbour and export trades also had a precarious hold on their jobs in these circumstances. In July and August 1925 it was expected that 50 per cent of clerical workers would forfeit their jobs. When trade contracted in the second half of 1928 and continued to shrink thereafter, white-collar employees again found themselves on the streets in large numbers.[25]

In terms of Communist Party organisation, Hamburg-Altona formed part of the region known as the Wasserkante, which included the rural hinterland of Schleswig-Holstein as well as that province's industrial centres in the Hamburg-Altona conurbation. In terms of KPD membership the Wasserkante, with its head office in Hamburg, was the third largest district (*Bezirk*) until 1928. By 1931 it had slipped to fourth place, though without any absolute decline in its proportional share of membership, which stood at 8.9 per cent of the total Party membership in Germany as a whole. Within the Wasserkante, the KPD was strongest in Hamburg and Altona, and was also strong in industrial towns and

Table 9.1: Unemployment in Hamburg and Altona 1919–33

Year	Hamburg		Altona	
	Unemployed persons	As % of working population	Unemployed persons	As % of working population
1919	-		-	
1920	-		-	
1921	-		6,957	7.3
1922	-		8,560	9.0
1923	-		9,265	9.8
1924	23,887	4.0	1,722	1.8
1925	40,087	6.8	6,592	6.9
1926	50,146	8.5	6,107	6.4
1927	49,624	8.4	4,653	3.9
1928	50,162	8.5	9,712	8.2
1929	60,483	10.3	10,649	9.0
1930	94,343	16.0	17,680	15.0
1931	139,803	23.8	26,088	22.1
1932	164,359	28.0	31,295	27.2
1933	-		25,730	21.8

Note: Hamburg's working population in 1925 was 586,407, Altona's was 94,337; from 1927 the Altona figures are measured against the 1933 level of 117,690 since in that year (1927) the city and its population was expanded.
Sources: StAH 424–44/307, 424–27/G7/G8; StaBr 4,65 IV4e Vol. 4–7; Stat.JB Altona (1928), p. 279; U. Büttner, *Staats- und Wirtschaftskrise*, p. 681.

ports such as Neumünster, Rendsburg, Kiel, Büdelsdorf and Brunsbüttelkoog. Yet the attempt to mobilise and control the unemployed workers' movement during the middle years of Weimar ultimately failed on the Wasserkante too. A regional Committee of the Unemployed was set up in October 1926.[26] The Committee, which comprised fourteen persons, was dominated by activists from Hamburg and Altona. Only three members represented Kiel, Harburg and Pinneberg. The Chairman of the Committee was Anton Becker, a tried and trusted Party member. Becker was also a member of the Hamburg Parliament, and edited the KPD newspaper aimed specifically at the unemployed — *Der Arbeitslose* (The Unemployed Man). The fourteen-man Committee met once a month and elected a smaller committee of five to carry out the week-to-week business.

The timing of the Committee's birth augured ill for its survival. It came at a time when regional membership was suffering a decline, also partly reflected in national trends,[27] and unemployment was receding as the economy picked up. The first conference, called to coincide with the election of the Committee,

Figure 9.1: Unemployment in Hamburg and Altona (absolute figures) 1919–33

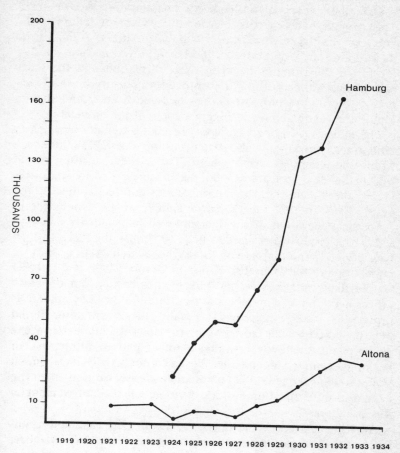

was to have been accompanied by a mass demonstration, but barely 400 turned up and of these 207 were delegates to the Conference.[28] Reports from the end of 1926 and from early 1927 suggest that the Party was having increasing difficulty in attracting the unemployed, with a consequent adverse effect on the size and success of the factory cells.[29] During the same period the Party was afflicted with serious internal squabbles related to an ideological power struggle between the Moscow-supported leadership of Ernst Thälmann and the 'Ultra-Leftists' around Ruth Fischer and the Hamburg teacher and leading functionary Hugo Urbahns.[30] This struggle had repercussions for the success or failure of the Party among the unemployed in Hamburg-Altona. Almost from its inception the Committee of the Unemployed became another forum for the disputes between the radical Left and the Party orthodoxy. Although the Party enjoyed a majority on the Committee (there were only two representatives of the 'Left') it soon found itself alienated from the grass-roots of the movement.

The radicals, described as *Parteilose*, that is free of formal party affiliations, quickly set about neutralising the KPD's influence within the unemployed movement. Their leader, a fitter named Martin Sander, was able to do this since he had sympathisers among the V-men (i.e. the intermediaries in direct contact to the base). Early in 1927 the *Parteilose* split from the Committee to form the Federation of the Unemployed, a decidedly anarcho-syndicalist group which harked back to similar tendencies which had played a major role in both the March Action 1921 in Hamburg and the Hamburg Rising in October 1923.[31] The KPD was further weakened by the police raids carried out on Party offices and homes of members at this time, followed by numerous prosecutions and convictions.[32] Finally, the elimination, at the behest of Stalin and Zinoviev, of the radical Left in the Party alienated the unemployed and resulted in a period of isolation and declining membership on the Wasserkante.[33] Only later on, in 1929–33, did changed KPD tactics and a greater caution about the composition of the Committees bring about an improvement in this situation.[34]

As the economic crisis deepened from 1929, unemployment levels began to rise more steeply than ever before. Under the pressure of long-term unemployment, the burden of financial relief for long-term unemployment began to move from central government to the hard-pressed local authorities.[35] It was under these radically different circumstances that the KPD re-addressed itself to the question of mobilising the unemployed, undeterred by its failures of the mid-1920s. According to the Communists' analysis, the period of capitalist stability had now come to an end. The inner contradictions of the capitalist system would lead to its rapid disintegration in the economic, political and social spheres. To escape the impending crisis the bourgeoisie would have to seek a means by which to survive the crisis intact. Fascism would be its choice. Indeed, Brüning's authoritarian style of government was already seen by the Communist leadership as fascist. The working-class response to this should be to resist and then to transform this resistance into a revolutionary class struggle. Even in a period of rising unemployment, the Communist leadership insisted that strikes should be the central tactic in this struggle.[36] In so doing, the KPD was maintaining the orthodoxy that the struggle should take place in the factories. Yet in reality the focus of the struggle had shifted from the factories to the neighbourhoods as a consequence of unemployment and as a result of growing Nazi challenges to the KPD and SPD in working-class districts.

The KPD leadership placed great emphasis on the work of the Revolutionary Trade Union Opposition (RGO), an organisation that enjoyed little more than a few scattered regional successes.[37] In the Hamburg-Altona region the RGO had little impact on the factories. By 1929 the KPD had already lost most of its influence among the workers on the docks, although it had traditionally been strong there.[38] In the metal-workshops on dry land, reformist trade unionists held sway. Whereas areas such as the Ruhr, Berlin, the lower Rhine and Saxony could boast RGO membership levels from between 22,000 to over 56,000 in 1931,[39] the Wasserkante had a mere 2,800 members, or barely 2 per cent of the national membership of the RGO. By August 1932 membership had trebled to just over 9,000,[40] but it had been growing no faster than the national membership, so the proportion remained at 2 per cent. The RGO was also supposed to organise the unemployed, although from 1931 the Committees were allowed a further degree of autonomy. The degree of success of the RGO in gaining a foothold in the factories was not overestimated by the Party leadership,[41] and yet from mid-1932 the numbers of unemployed workers joining the union declined in favour of persons in work. By October, new members in work — generally adult men — formed around 90 per cent. This was an obvious contradiction to the profile of the Party, which was composed overwhelmingly of unemployed persons;[42] nor could a slight improvement in the regional economy taking place at the time really account for this. A possible explanation might be that those workers designated as 'employed' were in fact unemployed workers involved in municipal emergency work schemes.

The system of organising the unemployed was not a simple or

Figure 9.2: Employed and Unemployed Members of the KPD-Wasserkante 1930–2

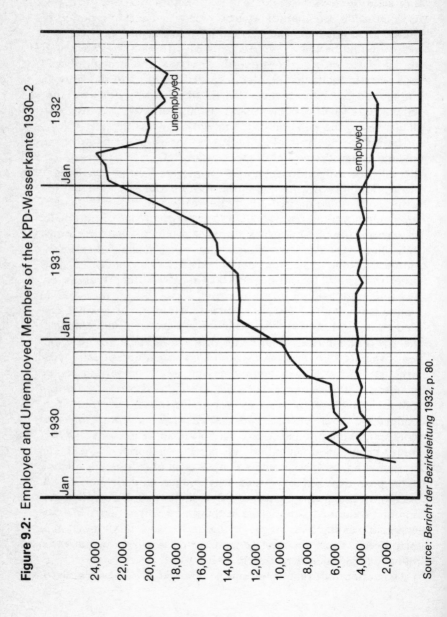

Source: *Bericht der Bezirksleitung 1932*, p. 80.

even congruous one. First, the unemployed were to be involved in the Committees of the Unemployed; then they were expected to be affiliated to the RGO, thus presenting the image of unionised and factory-related workers; and finally, unemployed persons were expected to be involved in either the Fighting League Against Fascism (*Kampfbund gegen den Faschismus*) or the Antifascist Action (*Antifa*).[43] There was a considerable amount of overlap in the duties of each of these organisations, and an unemployed person could find him or herself in two or three of these simultaneously. This also makes it rather complicated for the student of these organisations or of the unemployed in ascertaining precisely who was organised in what. A reasonably safe indicator is the street cell, for it was here that the unemployed person was most likely to be found; but again the street cell could just as easily be the nucleus of a *Kampfbund* or *Antifa* group. According to a report of the Central Committee of the Party in Berlin the *Kampfbund/Antifa* on the Wasserkante totalled about 13,000 members organised in 260 formations by summer 1932. This was about 12 per cent of the national total from the end of 1931. Nearly half of the Wasserkante membership was concentrated in the Hamburg-Altona-Wandsbek conurbation. The report stated that around 70 per cent of the *Kampfbund/Antifa* were unemployed.[44]

The Committees of the Unemployed also mushroomed on the Wasserkante during the final years of Weimar. In the course of 1930 their number rose from 43 to 59, and by 1932 there were 70. This compared to a total of 1,300 as a whole in mid-1931. However, they reached only a fraction of the unemployed, in Hamburg as on a national level, where the Committees contained a mere 80,000 out of the 5 million to 6 million registered unemployed in these years.[45] Yet it would be rash to assume that the organisations mentioned above had little influence beyond the regular Party constituency.[46] Their appeal could be considerable. The activities of the Committee of the Unemployed in Altona covered a wide range of actions, even including the protection of families of unemployed persons from evictions and possibly also the organisation of strikes of unemployed workers drafted onto municipal job-schemes.[47]

Moreover, Party membership in general nearly trebled between 1928 and 1932, as Table 9.2 shows. This was due almost exclusively to the recruitment of the unemployed. In 1930 the unemployed

Table 9.2: KPD Membership in the Wasserkante, 1919–33

Year	Wasserkante	Reich
1919	11,450	106,656
1920	2,250	78,715
1921	40,000	359,000
1922	23,263	224,689
1923	22,714	294,230
1924	-	-
1925	8,000	122,755
1926	-	
1927	13,525	124,729
1928	10,000	130,000
1929	10,680	116,735
1930	12,081	126,000
1931	19,978	246,513
1932	28,016	252,000
1933	-	-

Sources: Weber, *Wandlung*, p. 364; *Bericht der Bezirksleitung* (1932), pp. 69–71, 79–80.

probably formed barely 20 per cent of Party membership on the Wasserkante. By 1931 this figure had risen to just over 30 per cent. But by 1932 over 90 per cent of the KPD's members on the Wasserkante were unemployed,[48] as Table 9.3 illustrates.

Table 9.3: Percentage of KPD Members Employed, Wasserkante 1930–2

	December 1930[a]	December 1931	April 1932	July 1932	October 1932
Wasserkante	42.3[a]	19.1	13.8	13.7	9.2
Hamburg	36.1	18.0	14.4	13.2	12.3
Altona	26.5	14.5	9.0	7.0	8.75
Wandsbek	19.5	7.0	9.0	10.0	7.0

Note: a = March

Among *new entrants* to the Party in October 1932, 73.3 per cent on the Wasserkante were unemployed. In Hamburg the figure was 62.4 per cent; and in Altona 69.2 per cent. Some individual areas of Hamburg, such as St Pauli, Neustadt, Altstadt, Barmbeck-Centre and Hammerbrook, had percentages of between 72 and 77; while a major city as Kiel had 84.4 per cent and Hamburg-Wandsbek 88.8 per cent.[49] The trend for the Wasserkante more or less corresponded to the profile of the party at national level,[50] and it was also reflected in the growing contrast between the role

of the factory cells and the street cells which catered for unemployed men and housewives in the party. While the number of
KPD factory cells on the Wasserkante grew from 145 in January
1931 to 192 in August 1932, showing a growth rate of 32 per cent,
the number of street cells grew in the same period from 320 to 641,
that is, at a rate of 100 per cent.[51]

Overall, the KPD made more and more inroads into the ranks
of the unemployed. Increasingly, it became the 'Party of the
Unemployed'. So whereas in earlier days the leadership had ex-
horted the unemployed to assist the employed factory workers in
strike actions, now the message was partially reversed:

> The following measures are necessary in order to eliminate the
> weak contacts between the unemployed workers' movement
> and factory workforces:
> a) Every action, even the smallest partial action of the un-
> employed, must find its echo in the factory. All Party Comrades
> and red factory committee members are duty-bound at work
> and at meetings of the employees to raise the issue of the
> unemployed's struggles and to initiate a common organisation
> of the struggle as well as the mutual representation of factory
> workers and unemployed.
> b) . . .
> c) The demands of the unemployed are to be taken into con-
> sideration in all of the fighting demands of the factory workers,
> so as to create the political basis for the mobilisation of the
> unemployed for strike support.[52]

The sentiment of the message could not disguise the fact that the
KPD leadership was issuing directives into a void. By the end of
the Weimar Republic, trade unions in general had lost all their
muscle-power and were in fact showing a sharp decline in
membership levels.[53] But still more pertinent was the fact that the
RGO (and it was this union that the Party leadership was
addressing) was itself composed mostly of unemployed workers.

III

A great deal of party energy was also directed to spheres other
than factory and union work. The KPD leadership knew that it

had to address itself to the streets and neighbourhoods as well as to the factories. For the KPD leadership it was important to combine factory-oriented, parliamentary and street politics in order to display the Party's strength and bring pressure to bear on the Weimar state. In an announcement on 'National Unemployed Workers' Day', set for 10 September 1930, the Reich Committee of the Unemployed declared how: 'In broad columns, we unemployed public-relief-scheme workers, factory workers, women and youth will conquer the streets. The streets belong to us!'[54] This emphasis on 'visibility' was nothing new to politics during the Weimar Republic. Public holidays, such as Constitution Day (11 August) and May Day were usually celebrated with mass displays of strength in the form of meetings and marches. Lesser occasions also led to similar manifestations. Organising the unemployed in a similar way diverged from the usual association of unemployment and the public sphere. Normally, unemployment is kept hidden, tucked away in the proletarian neighbourhoods or at the dole offices and labour exchanges — also often in more obscure locations. Polite society is thus spared the spectacle of troops of the jobless roaming the streets — especially in the bourgeois quarters of a city.[55] The KPD recognised the importance of politicising unemployment by bringing it into Weimar society centre stage. Here, most probably, lay the basis for the continuing and increasing success which the KPD enjoyed among working-class people in the final few years of the Weimar Republic. While the Social Democrats succumbed to political paralysis before the onslaught of conservative forces,[56] the KPD, by virtue of its political ideology, saw this precisely as the moment for increased activity and mobilisation.

Thus, in early January 1928 the Reich Committee of the Unemployed was quick to comment on the Decree of 20 December, dealing among other things with seasonal labourers, whose period of waiting between being fired and starting receipt of benefits was now extended. The Committee noted that protests had been made at a local level and took the view that these should be intensified.

We must also point out . . . that the administrative council of the Reich Ministry, which issued the Decree, contains leading functionaries of the Social Democratic Free Unions, who are all for this measure. The Social Democratic press is not carrying out any struggle, instead it is appealing to the reactionary labour exchanges. In contrast our slogan is 'don't beg, but fight the Decree and Law, and fight for the payment of benefits'.[57]

The KPD sought to carry out its strategy by organising marches and demonstrations, and by petitioning local authorities. This would draw the issues at hand to the public's attention. By the same token, it would show the unemployed what the policies and attitudes of the different political parties and groupings in local town councils really were. More importantly, the Party sought to prevent the unemployed from becoming atomised and morally and politically apathetic, by drawing them into the parliamentary sphere of politics through extra-parliamentary activity. This strategy was made explicit in the words of the Reich Committee in Berlin:

> It will only be meaningful to draw up demands and run campaigns when the parliamentary activity is successfully linked to extraparliamentary action, and when these campaigns are brought to a successful conclusion. Naturally, it will not always be possible to achieve these demands, but the unemployed will see that although the requests could be granted, the municipal representatives of each political party will reject them. Only in this way is an intensification of the movement possible . . . The unemployed should always be informed of their entitlement to benefits and how large these should be; and which political parties turn down their demands.[58]

This strategy for the mobilisation of the unemployed applied not only to seasonal workers in 1928, but also to all unemployed workers as the economic situation of the country deteriorated from 1929. At the same time the KPD was able to exploit the situation to its advantage, not least because so often it was Social Democrats in municipal governments who had the task of implementing unfavourable policies towards the unemployed both before and after September 1930.

The KPD thus sought to combine its presence from within local town councils and city parliaments with pressure from outside these institutions. The basis for the marriage of parliamentary politics with street politics was laid down in guidelines issued by the Party Central Committee in 1924.[59] This was an important part of Communist political strategy since the party generally had minority representation in many of the municipal councils throughout the country.[60] Of course, pressure could be extended only when unemployment was high, so this policy did not meet with much success until the end of Weimar. The format of such politics was to call a rally, followed by a march on the town hall timed to coincide with a meeting of the town council. There would be a public address, usually decrying the iniquities of the capitalist order, the cuts in benefits and the collusion of the Free Trade Unions and the Social Democrats in the attacks on the welfare of the working class. This part of the political theatre done with, the crowd would select a delegation which would present demands to the council meeting in progress. Thus the KPD sought to obtain *de facto* recognition of the Committees of the Unemployed by the municipal authorities through the use of demonstration and delegates.

Demands in Altona and Hamburg differed little from those made elsewhere in the Reich.[61] Reductions in the price of, or conversely, increases in the subsidies for heating and cooking materials such as oil, coal and wood, or in the cost of gas and electricity, were called for on numerous occasions. The cancellation of rent arrears or the lowering of rents for unemployed families, and the cessation of evictions, were also demanded. During the winter of 1931–32 the KPD parliamentary group in Altona demanded that a winter bonus of 60 RM should be paid to each unemployed head of household, plus a further 20 RM for each dependent, and the free distribution of warm clothing, and hot meals or food at either a minimum cost or free of charge. Again, in June 1932 the KPD representatives in the city parliament of Altona took up the case of about 90 unemployed families who were no longer able to pay rent for accommodation in the city and who had built huts on plots of land rented from the city for growing vegetables or as weekend retreats in so-called *Schrebergärten*. All these people complained of not being able to make ends meet on the low benefits they were receiving, and requested the city to abrogate their rents for the duration of their unemployment.[62] The KPD's espousal of these demands was obviously functionalist and agitational. It was a propagandistic tactic to expose the bankruptcy of the bourgeois state and its hostility to the working class. For all that, it was directed at the real needs of the unemployed, and the granting of these demands would certainly have brought about a real improvement in their situation.

The labour movement in Germany had always enjoyed a

tradition of visibility, and it derived much outward strength and confidence through displaying itself in public as a disciplined force. The marches and demonstrations of the KPD and the unemployed recalled this tradition. And in a city such as Altona or its larger neighbour Hamburg, with strong labour traditions, a march of any reasonable size was bound to be impressive, especially given the close spatial dimensions of the inner-city areas. The marches had all the trappings of a traditional labour parade: a colourful arrangement of flags, music, singing and drums. The songs stemmed from the traditional repertoire of the German labour movement, with the *Internationale* taking pride of place. For the KPD, such displays also satisfied the need of the Party leadership to reassure itself that here it was at the head of the labour movement. They showed that it was contesting or even assuming the traditional claim of the SPD to lead the German working class as an organised force. Coming out onto the streets also served another, more subjective need: the desire of many of the unemployed to resist marginalisation and find a way of social and political reintegration into labour politics. For many unemployed people, the participation in a large, colourful spectacle had a positive function, because it offered a moment of convergence with the traditions of the working class normally associated with the factory and the union. The march was a form of organisation, of being organised. The KPD was providing for the unemployed a sort of surrogate labour movement. On the level of psychology and morale, it also constituted a moment of rediscovered pride in the individual and the collective. This must have been a welcome alternative to being ignored by the mainstream labour movement and a useful antidote to the humiliation and intimidation of life on the dole. These marches enjoyed a patchy success in the middle years of Weimar,[63] but after a lull in activity until late 1929, they became almost daily occurrences.[64] Their importance lay not so much in the level of participation — although this was obviously an important factor — but in the effect they had on the public eye and on working-class morale, especially in a period when the working class as an organised force was being steadily beaten back.

IV

Relations between the police and certain sections of the working class had always been strained throughout the Weimar Republic. Violence had often occurred between them, even before the bloody

May Day demonstration of 1929. During the referendum on the confiscation of the property of the German royal and princely families in 1926, rioting broke out in parts of Hamburg and Altona after police moved in to disperse demonstrating crowds. Several police officers were injured and weapons were used to break up the crowds.[65] In August 1927 during nationwide demonstrations against the execution in the United States of two immigrant trade unionists, the Italians Sacco and Vancetti, rioting again broke out in Hamburg's *Gängeviertel* — tumbledown, disorderly and strongly Communist areas in the inner city — after a police agent had been spotted in the crowd and identified as an *agent provocateur*. He was stabbed to death, whereupon the crowd is reported to have sent up a cheer.[66] And at the end of the Weimar Republic, any violence or brutality exercised by the police against members of the working class in working-class districts was reciprocated on certain occasions by the people of those areas.[67]

The police were often at a disadvantage on such occasions: they did not have the same knowledge of the streets and alleyways as did their opponents, and so the latter were usually able to gain the upper hand.[68] These incidents were usually centred around specific issues. But they also constituted general challenges to public authority in which the collective actions of the crowd can be interpreted as the 'crystallization of community identity . . . against a society felt to be dominating and oppressive'.[69] For many of Germany's inner-city areas, similar to the *Gängeviertel* in Hamburg or the Altstadt-St Pauli area of Altona, traditional hostilities were accentuated in the crisis. With the imposition of mass unemployment upon already precariously balanced local economic and social structures, public order now became seriously endangered as the work-based fulcrum of local community life and social relations disintegrated. The police authorities were not interested in understanding the structural problems of the poorer quarters of the big city, where most violence occurred. Instead, they identified these areas as bases of Communist support and sympathy, and correspondingly called for a more offensive style of policing.[70]

In December 1929 the chairman of the Reich Committee of the Unemployed, Johannes Schröter, issued a call for hunger marches to take place throughout Germany on 19 December. The march was to be the first test of the recently reactivated Committees of the Unemployed. But the day ended in violence and death. In Berlin and Chemnitz fighting broke out after police forces moved in to disperse the marchers. One person was shot dead in Berlin and four others killed in Chemnitz.[71] Similar clashes with the authorities took place in other parts of the country. A hunger march in Altona in December 1929 ended in violence,[72] as did another march organised by the Altona KPD in January 1930. This latter demonstration had been prohibited after the Christmas disturbances, but the KPD and its followers ignored the ban. When police officers again tried to break up the 800 strong crowd, they found themselves at a hopeless disadvantage on their opponents' home territory. Amid a 'general howling, whistling and shouting', the young demonstrators waded into the police showing a 'particular enthusiasm for attack' as they did so.[73] Three months later, in March, a similar pattern of events was repeated, when police clashed with a smaller crowd of 500. The officers, making free use of their truncheons, were met with cries of 'Beat the dogs down!'[74] In the early stages of the Depression, a remarkable feature of these clashes between the police and the unemployed in Altona was the relative absence of really severe casualties, in contrast not only to other parts of the Reich but also to the pattern of police violence against demonstrators in the early years of the Weimar Republic in Hamburg-Altona itself.[75] But all this was to change in the course of the next two years. By 1932 the scene was set for a regular massacre of civilians by the police in the notorious 'Bloody Sunday' incident that took place in Altona in July.

The KPD's attitude towards violence was ambivalent. Naturally, it believed that the overthrow of the capitalist system would be violent; but it is arguable that it did not see itself as the initiator of violence. The Party had its own paramilitary organisations, notably from 1924 the Red Front-Fighters' League. But these were for the defence of the proletariat. They were not expected to engage in wanton or undisciplined violence.[76] The idea of direct action had been unpopular within Communist orthodoxy ever since the days of Lenin's tirade against the 'infantile disorder' of Communist leftism. The bad experiences of the early 1920s were still fresh in the Party's memory. Not only had the Party been weakened by the secession of an 'actionist' group to form the very radical 'Communist Workers' Party',[77] but many of the Party's leading men had been either killed, maimed or imprisoned in the October rising of 1923 and the Party had even been banned for a period.[78] These experiences made the Party leadership acutely aware of the negative consequences of violence which was associated with the party while not necessarily being under its control or direction. Even the police authorities recognised that the Communists disapproved of undirected violence, although they thought that the KPD was certainly ready to exploit it.[79]

So one should be careful of apportioning to the KPD the sole responsibility for outbreaks of violence when they occurred.[80] In the disturbances arising from unemployed marches the KPD was treading a thin line between breaching legality on the one hand and losing its credibility among the working class on the other. It was clear that an activist minority of the unemployed would respond to state violence by reciprocating within the same grid of reference, but the appeals of the KPD to this group were founded in the belief that it could channel and direct this energy in a broader political front. The Communists indeed thought they would eventually be banned, but they did not want working-class energies to be dissipated or the unemployed activists to become disillusioned after suffering unnecessary defeats at the hands of the police. On the other hand, the Party did not want to alienate the activists either.

A good example of how the KPD tried to solve this dilemma occurred in the hunger march of 20 December 1929, held the day after similar demonstrations elsewhere in the Reich. The march began with a meeting in the spacious rooms of the 'Café Fatherland', situated in the busy thoroughfare of Altona's Grosse Bergstrasse. The meeting started at 1.50 p.m. and was over in just two hours.[81] Although the KPD only managed to fill the rooms with about 400 people, another thousand gathered outside. Party officials in the meeting complained of the 'lukewarm' response to their appeals. Towards the end of the meeting the doors were shut in order to prevent people leaving early. Representatives of the reactivated Committee of the Unemployed in Altona, and of the regional Committee of the Wasserkante, addressed the meeting in strong terms. A representative from the Regional Committee, a man called Eichhorn, sketched a heroic picture of the progress of the RGO and went on to analyse fascism in Germany and Italy and the role of the Free Trade Unions and the SPD in terms of the Party's theory of Social Fascism. He spoke in sharp tones against the role of the police. They were under Social Democratic control, yet they had broken up meetings and other gatherings of the unemployed. Eichhorn paid particular attention to a case which had only occurred the same day at the dockers and seamen's labour exchange at the Kohlhöfen in the Hamburg *Gängeviertel*. Eichhorn was followed by Heinrich Stahmer, a Communist member of Hamburg's city parliament (*Bürgerschaft*). He told the meeting how the KPD was persevering on the political plane in order to achieve an improvement in the lot of the unemployed and the working class. Real headway in this direction, he said, could only be arrived at in Hamburg and Altona through a coalition of the SDP and KPD, but the SPD was unwilling to work towards this aim. Instead, according to the KPD, it preferred to do business with the Democratic Party (*Deutsche Demokratische Partei* or *Staatspartei*). Thus, it betrayed the working class and played into the hands of the Right. Stahmer concluded by declaring that 'the present-day government, and also the Hamburg parliament, do not give a thought to improving the lot of the unemployed'. But, he assured his listeners, 'the KPD will do everything in its power to wring benefits for you!' In return, the unemployed were expected to give their support to the Party by joining the Committee of the Unemployed and by following the calls to mobilise on the streets.

The mood of the meeting became clear when one speaker read out a letter from the Altona city council, headed by the Social Democrat Max Brauer, containing a refusal to grant requests for the amelioration of the condition of unemployed families. This was met with cynical laughter from the audience. Another speaker let it be known that there had been isolated clashes between police and unemployed individuals or small groups in the city. The meeting then passed a resolution and agreed on the motion of the organisers to take it to the Hamburg City Hall, where it would be presented by a delegation of five. The meeting was wound up with several more speeches laying particular stress on the actions of the police against the unemployed at the Kohlhöfen. 'We can chase this government to the devil!', came the response: 'The KPD won't let itself be deprived of its right to the streets!' The leaders of the meeting then called for good behaviour once the march began. The slogan 'Discipline! Hamburg's workers on the streets!' was greeted with the response 'Red Front!', repeated three times. The audience then rehearsed the following refrain, 'What do the unemployed have? Hunger! What do the unemployed want? Work! Give us bread, or we'll strike you dead!' Although the organisers of the demonstration thus appeared to have arranged a 'spontaneous' march to Hamburg, the Committee of the Unemployed had in fact previously obtained permission from the police to march to Hamburg along an agreed route. Trouble began when some marchers deviated from the official route and tried to penetrate some other streets. Here they were met by contingents of police. Clashes ensued. Four police officers who tried to usher this part of the crowd back onto the proper route were badly beaten with sticks and stones. One officer received knife wounds. Attempts to make arrests were thwarted by the other participants. Meanwhile, a group of demonstrators corresponding roughly in size to the meeting (400) and under the stewardship of the KPD-dominated Committee, gathered quietly in a neighbouring square, Gählersplatz. As soon as the authorities had the situation more or less under control, a spokesman for this 'official' group asked for permission to proceed to Hamburg. The spokesman stated explicitly that the Committee's intention was to abide strictly within the terms of the law. Thus, he carefully distanced the KPD from the violence which had just occurred between the police and the larger group of unemployed protesters.[82]

This event revealed the dual nature of the unemployed workers' movement during the Weimar Republic. On the one hand, there was the Party-defined and imposed movement, with its emphasis on organisational discipline; its main aim was to be effective within an accepted code of political action. On the other hand, there existed a more spontaneous movement which jelled into form to confront the state only at particular moments, then dissolving until the next confrontation. The basis for action of this latter movement lay in the social conditions of the unemployed community. It responded to Communist appeals to action only where it appeared that the KPD was acting as a radical party. In terms of political tradition, this movement had much in common with the early anarcho-syndicalists who had dominated grass-roots unemployed politics in the middle years of the Weimar Republic. Thus, while the KPD espoused the case of the unemployed in order to transform it into a political cause, it would be wrong to assume that the unemployed were the mere tools of the KPD.[83] It would be equally wrong to assume that where the KPD failed to establish itself as master of the unemployed this was due to the latter's psychological and physical demoralisation. The appeal to the unemployed began to pay off in electoral terms too. The KPD vote began to increase, mostly at the expense of the Social Democrats, most spectacularly in precisely those areas where unemployment was heaviest.[84] Thus, among the mass of the unemployed, political mobilisation against unemployment and government unemployment policy manifested itself in ways other than continuous demonstrations on the streets.[85]

The final crisis of capitalism did not, of course, transpire, either worldwide or in Germany. It was solved by various means. The bourgeoisie in Germany opted for Hitler while the Left crumbled and its leadership was sent packing with its tail between its legs. Despite the coalition of interests and forces, the unemployed workers' movement did not succeed in its KPD-defined aim: the redemption of the working class in the face of the class offensive. The united front of employed and unemployed workers did not crystallise in a solid form. The clash of interests between those in work and those out of work was partly sanctioned by members of the Free Trade Unions, while even within the ranks of the unemployed there were differences of interest related to the various stages in the experience of being unemployed. New claimants were less likely to revolt in the queues than longer-term unemployed.[86] Although the Party knew that some sections of the unemployed were not lacking in punch,[87] it also knew that what was needed in order to stay the rounds and to finally put the opponent out of action was a co-ordination of the multiplicity of small punches to create a knock-out blow, and this simply never happened.[88] In the early summer of 1932, when the KPD was increasing in political power, the Reich Committee of the Unemployed complained of the inability of regional and local officers to put themselves at the head of actions undertaken by unemployed workers:

> If during the months of May and June, in nearly all the regions of Germany, great and even in most cases successful actions against the Benefits Robbery by the municipal authorities have been carried out, the fact remains that we have not managed, on the basis of these struggles, to activate an even broader movement against the Papen Decree.[89]

This may have been due partly to the burdens of organising election campaigns, but if the example of the 1929 Hunger March in Altona is anything to go by, then it would seem that the real problem lay elsewhere. The KPD leadership could not match its revolutionary rhetoric with revolutionary action. Its stress on party and organisational discipline, although appropriate to a Leninist vanguard Party, meant that any actions which threatened the integrity of its perfection were to be disowned. At once fascinated with and appalled by the radicalism of unemployed activists, it went along with them but stopped short of ultimately leading them. Had it done so, then perhaps the winter of 1932–33 might have seen a different political outcome to that of January and beyond.

Notes

I am indebted to Christa Hempel-Küter (Hamburg) with whom I wrote an earlier paper on unemployment in Altona (1918–23). I have benefited greatly from our discussions. Alan Kramer and Eve Rosenhaft read critiques of an earlier draft of the present article. To them my thanks. My thanks also to Dr Stig Förster, of the German Historical Institute, and Jim Knight for their hospitality. Finally, I am very grateful to Richard Evans whose interest in this chapter went beyond normal editorial duty.

    1. M. Jahoda, P. F. Lazarsfeld, H. Zeisel, *Die Arbeitslosen von Marienthal* (Leipzig, 1933; reprint, Edition Suhrkamp, Frankfurt 1982), *Hunger in Waldenburg*, film directed by Piel Jutzi, 1929.

    2. Rose-Marie Huber-Koller, 'Die kommunistische Erwerbslosenbewegung in der Endphase der Weimarer Republik', *Gesellschaft. Beiträge zur Marxschen Theorie*, 10 (Frankfurt, 1977), pp. 107–9; Hildegard Caspar, 'Die Politik der RGO. Dargestellt am Beispiel der Arbeitslosenpolitik in Hamburg', *Deutsche Arbeiterbewegung vor dem Faschismus* (Argument-Sonderband AS 74, Berlin, 1981) pp. 77–9.

    3. *The Economic Depression and Public Health* (Memorandum prepared by the Health Section of the League of Nations. Document A II.1 1932). Reprinted in *Quarterly Bulletin of the Health Organisation* vol. 1/3 Sept. 1932; a summary of the memorandum is in *ILR* vol. 26/6 Dec. 1932 841–6. For an overview of Household Budgets in this period, see 'Recent Family Budget Enquiries', *ILR* XXXVIII/5 Nov. 1933.

    4. Jean Rosner, 'An Enquiry into the Life of Unemployed Workers in Poland' in *ILR* vol. 27/3 Mar. 1933 (= Reports and Enquiries) 378–92.

    5. Klaus Dettmer, 'Arbeitslose in Berlin. Zur politischen Geschichte der Arbeitslosenbewegung zwischen 1918–1923' (PhD Berlin, 1977), pp. 84, 96: StA Br.4, 65 IV 4d vol. 2 Report Nr. 143, 22 Aug. 1919 and weekly report Nr. 2, 25 Aug. 1919. StA Br. 4, 65 IIE 1a3 vol. 1, Bl. 153, Auszug aus dem Bericht des RK Nr. 28, 15 Feb. 1921 (Abtlg. NB Nr. 432. Geheim!).

    6. Staatsarchiv (StA) Bremen 4, 65 IIE 1a3 vol. 1, Bl. 119 Abschrift, Zentrale der KPD Abtlg. Gewerkschaften, Betriebsräte, Berlin 18 July 1924 (Rundschreiben an die Bezirks- und Unterbezirksleitungen betr. Betriebsschliessungen, Kurzarbeit und Erwerbslosenbewegung); cf. *Hamburger Echo* 248, 9 Sep. 1924: 'Schamloser Missbrauch der Arbeitslosen für bolschewistische Zwecke'.

    7. Fritz Heckert on the unemployment question in *Bericht über die Verhandlungen des Vereinigungstages der USPD (Linke) und der KPD (Spartakusbund)* vom 4. bis 7. *Dezember 1920 in Berlin* (Berlin, 1921).

    8. StAB, 4, 65 IIE, 1a vol. 1. Bl. 109 Abschrift (Anlage 7) 'An alle Erwerbslosenräte und Funktionäre', 21 March 1924; Eva-Cornelia Schöck, *Arbeitslosigkeit und Rationalisierung. Die Lage der Arbeiter und die kommunistische Gewerkschaftspolitik 1920–28* (Frankfurt and New York, 1977), pp. 99–110, pp. 133–40.

9. StA Bremen, 4, 65 IIE 1a3 vol. 1, Bl. 109.

10. 'Resolution zur Arbeit der Kommunisten in den Freien Gewerkschaften (1925)' printed in Hermann Weber (ed.), *Der deutsche Kommunismus, Dokumente 1915–1945* (Cologne, 1963), pp. 174–6; *Der Parteiarbeiter* (1926), Heft 2, Sonderbeilage (in Landesarchiv (LA) Schleswig, 309/22718); Schöck, *Arbeitslosigkeit*, pp. 56–65.

11. Schöck, *Arbeitslosigkeit*, pp. 57–9.

12. Huber-Koller, 'Die Kommunistische Erwerbslosenbewegung', passim.

13. StA Bremen 4, 65 IIE 1a3 vol. 2, Abschrift 1A. AN. 2164 d.SH 2/5.9.31 (report on the Prague Conference); Cf. *Die Rote Fahne*, Nr. 173, 9 Sep. 1931. 'Kampf aufgaben der revolutionären Erwerbslosenbewegung'; Caspar, 'Die Politik der RGO', p. 76; W. Ulbricht, 'Mobilmachung der Erwerbslosen!', *Die Rote Fahne*, Nr. 169, 4 Sep. 1931, LA Schleswig 309/23056, RK u.d. ö.O. 159/28 II, 11 January 1928, Betr. Rundschreiben des Ausschusses (des Reichskomitees der Erwerbslosen). For the RGO, Freya Eisner, *Das Verhältnis der KPD zu den Gewerkschaften in der Weimarer Republik* (Cologne/Frankfurt-Main, 1977). *Die Generallinie. Rundschreiben des Zentralkomitees der KPD an die Bezirke 1929–1933* (Quellen zur Geschichte des Parlamentarismus und der politischen Parteien, Band 6, Dritte Reihe: Die Weimarer Republik), edited and introduced by Hermann Weber (Düsseldorf, 1981), Document 56, pp. 426–32, here, 427–8.

14. Erika Kucklich and Elfriede Liening, 'Die antifaschistische Aktion. Ihre Rolle im Kampf um die Abwehr der faschistischen Gefahr im Jahre 1932 und ihr Platz in der Strategie und Taktik der KPD, mit einem Dokumentenanhang' in *BzG*. 4 Apr. 1962, pp. 872–97. Anthony McElligott, '". . . und so kam es zu einer schweren Schlägerei". Strassenanschlachten in Altona und Hamburg am Ende der Weimarer Republik' in Maike Bruns *et al.*, '*Hier war doch alles nicht so schlimm'. Wie die Nazis im Hamburg den Alltag eroberten* (Hamburg, 1984), pp. 58–85, here pp. 67–75; Eve Rosenhaft, *Beating the Fascists? The German Communists and Political Violence 1929–1933* (Cambridge, 1983), pp. 97–9; *Die Generallinie* Document 64, 26, 5 1932 Rundtelefonat des ZK zur Antifaschistischen Aktion; *Bericht der Bezirksleitung Wasserkante an den Bezirksparteitag vom 2.–4. Dezember 1932*, p. 71.

15. Karl Hardach, *Wirtschaftsgeschichte Deutschlands im 20. Jahrhundert* (Göttingen, 1976), p. 50. D. Petzina, *Die deutsche Wirtschaft in der Zwischenkriegszeit* (Wiesbaden, 1977), pp. 96–107.

16. LA Schleswig 309/23056, RK u.d.ö.O. 159/28 II, 11 Jan. 1928, Betr. Rundschreiben des Ausschusses (des Reichskomitees der Erwerbslosen).

17. StA Bremen 4, 65 IIE 1a3 vol. 2, Abschrift 1A. 2164 d. SH 2/25.9.31; cf. *Kommunistische Internationale*, Heft 33/34, 15 Sep. 1931: 'An die Spitze der Millionenarmee der Erwerbslosen'.

18. LA Schleswig 309/23056, RK u.d.ö.O. 159/28 II 11 Jan. 1928. Betr. Rundschreiben des Ausschusses (des Reichskomitees der Erwerbslosen).

19. Peter Grottian and Rolf Paasch, 'Arbeitslose: Von der gesellschaftlichen Randgruppe zum politischen Faktor? Einige Hypothesen zur zukünftigen Entwicklung der Interessenvertretung von Arbeitslosen' in W. Bonss and R. G. Heinze, *Arbeitslosigkeit in der Arbeitsgesellschaft* (Frankfurt, 1984), pp. 331–48.

20. See Ben Fowkes, *Communism in Germany under the Weimar Republic* (London, 1984), pp. 158–9, 183–9. Huber-Koller, 'Die Kommunistische Erwerbslosenbewegung', pp. 105–6, 109 and passim.

21. See Hermann Weber, *Die Wandlung des deutschen Kommunismus. Die Stalinisierung der KPD in der Weimarer Republik* (gekürzte Studienausgabe, Frankfurt, 1969), Part 1; more recently in English, Fowkes, *Communism in Germany*.

22. According to a party survey in 1928 just over 60 per cent of those asked (88.8 per cent of membership) stated they were in Trade Unions; 31.6 per cent of these

for more than ten years and 12.5 per cent over 25 years. Furthermore, 38.9 per cent of the sample had been previously organised in the SPD; of which 23 per cent from between 10 and 20 years (StA Bremen, 4, 65 4e Bd 6 Report 2. 16.6.1928, p. 6).

23. *Statistisches Jahrbuch deutscher Städte*, 30. Jahrgang 1935 (results of the 1933 Census), p. 370.

24. U. Büttner, *Hamburg in der Staats- und Wirtschaftskrise 1928–1931* (Hamburg, 1982), Chapter 2; LA Schleswig 301/5038: letter of the Verein der Industrie und des Grosshandels von Altona-Ottensen to the Preussische Staatsministerium in Berlin, 6 January 1926; ibid., Industrie- und Handelskammer zu Altona, B.Nr. 2738/26, to the Oberpräsidenten der Provinz Schleswig-Holstein, 12 May 1926; ibid., Denkschrift of the Verein der Industrie- und des Grosshandels von Altona-Ottensen, 2 Apr. 1926.

25. StA Bremen, 4, 65 4e vol. 5, Reports for July and August 1925; ibid. vol. 7, Reports for October 1928, Büttner, *Hamburg in der Staats-*, p. 114.

26. StA Bremen 4, 65 IV 4e. vol. 5 reports for October, passim.

27. Weber, *Die Wandlung*, pp. 362–3.

28. StA Bremen, 4, 65 IVe Bd. 5 Report 9–14, 12.10.26.

29. Ibid., 4.65 IVe Bd.6 Report 3, 19.3.1927, p. 8.

30. Ossip K. Flechtheim, *Die KPD in der Weimarer Republik* (Frankfurt, 1969, second edition, 1976), pp. 228–31; Weber, *Die Wandlung*, Part 1, chapter 3 and p. 128; Fowkes, *Communism in Germany*, pp. 140–1; StA Bremen IVe, Bd.6, Report 2, 19.2.1927.

31. Wolf D. Hund, 'Der Aufstand der KPD 1923' in *Hamburg-Studien, Jahrbuch für Sozialökonomie und Gesellschaftstheorie* (Opladen, 1983), pp. 32–61; Angelika Voss, Ursula Büttner, Hermann Weber, *Vom Hamburger Aufstand zur politischen Isolierung. Kommunistische Politik 1923–1933* in *Hamburg und im Deutschen Reich* (Hamburg, 1985), pp. 9–54; V.I. Lenin, '"Left-Wing" Communism — An Infantile Disorder', Part V in *Selected Works* (Moscow, 1968), pp. 526–31.

32. Gotthard Jasper, *Der Schutz der Republik* (Tübingen, 1963); idem, 'Justiz und Politik in der Weimarer Republik' in *Vierteljahreshefte für Zeitgeschichte*, 30. Jahrgang, 1982, Heft 2, pp. 167–205; LA Schleswig 309/222722, Der Polizeipräsident, Altona P2 904/25 to Oberpräs. in Kiel, 8 Aug. 1925; LA Schleswig 309/22695, Pol.Präs. Altona P2 779/23 to Oberpräs. in Kiel, 7 Dec. 1923, betrifft politische und wirtschaftliche Bewegungen in Gross-Hamburg, Bl. 14.

33. LA Schleswig 309/22572, Pol.Präs. Altona P2 090/24 to the Regierungspräsident Schleswig, betr. Betätigung der kommunistischen Partei innerhalb des Bezirks Wasserkante, Bl. 56 (report from Oct.), and Bl. 211 (report from Dec.); LA Schleswig 309/222754, Pol.Präs. Altona P2 836/24 to Regierungs-Präs. Schleswig, 13 Sep. 1924 (Bl. 218), betr. Betätigung der kommun. Partei; Weber, *Die Wandlung*, p. 371. In 1923 membership on the Wasserkante stood at 22,714; by the end of 1925 it had sunk to a miserable 8,000.

34. StA Bremen, 4, 65 IIE 1a3 Bd.2, Rundschreiben (Abschrift) des Bezirksausschusses der Erwerbslosen Ruhrgebiet (betrifft Kampfcongress, Dortmund 31.8.1930). Cf. *Die Generallinie*, Dokument 56, 16.12.1931, Anweisungen der Orgabteilung über die Aufgaben in der Erwerbslosenbewegung, p. 428.

35. See above, p. 5–9.

36. Ernst Thälmann, *Der Revolutionäre Ausweg und die KPD. Rede auf der Plenartagung des Zentralkomitees der kommunistischen Partei Deutschlands am 19. February 1932 in Berlin* (Kleine Bücherei des Marxismus-Leninismus, Band 4, Frankfurt, 1971), p. 54.

37. Eisner, *Das Verhältnis*, pp. 222–47; Siegfried Bahne, 'Die Kommunistische Partei Deutschlands' in E. Matthias and R. Morsey (eds), *Das Ende der Parteien 1933. Darstellungen und Dokumente* (1960, reprint Düsseldorf, 1979), p. 664; Weber, *Die Wandlung*, p. 368.

38. Also due to a rise in unemployment among its supporters: StA Bremen 4 IV 4e vol. 7, Reports Nr. 1, 3 Mar. 1929, and Nr. 2, 7 June 1929; Büttner, *Hamburg*, pp. 47–8.

39. RGO membership in Berlin: 56,000; Ruhr: 36,900; Lower Rhine: 23,600; Saxony: 22,200 (Bahne, 'Die Kommunistische Partei Deutschlands').

40. Caspar, 'Die Politik', p. 56.

41. Thälmann, *Der Revolutionäre Ausweg*, p. 56.

42. *Bericht der Bezirksleitung*, p. 29a, b, p. 80. ibid., pp. 67–8.

43. StA Bremen, 4, 65 IV 4e Bd. 7, Report from 29 Nov. 1930, p. 12. IML/ZPA, St.3/622 Bl. 2, Bl. 55/56 (Kampfbund in Altona).

44. *Die Antifaschistische Aktion. Dokumentation und Chronik Mai 1932 bis Januar 1933* (edited and with an introduction by Heinz-Karl and Erika Kucklich with the assistance of Elfriede Förster and Käthe Haferkorn, East Berlin, 1965), Dok. 78, p. 230, p. 260; Weber, *Die Wandlung*, p. 365.

45. *Bericht der Bezirksleitung* (1930), p. 53; *Bericht der Bezirksleitung* (1932), p. 60; StA Bremen, 4, 65 IIE 1a3 vol. 2, *Nachrichtenstelle* R.D.I. 1AN 2164 d/3.11 (3 Nov. 1931), p. 2.

46. Bahne, 'Die Kommunistische Partei Deutschlands', p. 668.

47. Ibid., p. 57: 'In Altona for a long time it was not possible to carry out an eviction, even under a heavy police guard.' For strikes, see LA Schleswig 301/4578, Bedeutungsvolle Lohnbewegungen 1924–1929, Ed.I, Report of 6 April 1927 (1 January–31 March 1927), Nr.1; Special issue of *Hamburger Volkszeitung* (1 Nov. 1932) in IML/ZPA v DF VIII 14OU.

48. *Bericht der Bezirksleitung Wasserkante an den Bezirksparteitag*, 1930 (Hamburg), p. 13. *Bericht der Bezirksleitung* (1932), pp. 70–1, p. 80.

49. Ibid., pp. 67–8; cf. the slightly varying figures for the Bezirk Wasserkante in Ursula Büttner 'Politik und Entwicklung der KPD in Hamburg 1924–1933' in Voss *et al.*, *Vom Hamburger Aufstand*, p. 102.

50. Bahne, 'Die Kommunistische Partei Deutschlands', p.661.

51. *Bericht der Bezirksleitung* (1932), p. 64, p. 71.

52. *Die Generallinie* Dokument 55, 22.10.1931, Anweisungen des Sekretariats "über Streiks und Erwerblosenaktionen", part V, Streikkämpfe und Erwerbslosenbewegung, pp, 421–2.

53. Casper, 'Die Politik', p. 57. D. Petzina, W. Abelhauser, A. Faust (eds), *Sozialgeschichtliches Arbeitsbuch III. Materialien zur Statistik des Deutschen Reichs 1914–1945* (Munich, 1978), p. 111.

54. StA Bremen, 4, 65 IIE 1a3 vol. 2, 'Aufruf zum Reichs-Erwerbslosentag 10. September (1930)'.

55. This has been captured in a number of novels written at the time portraying, sometimes from first-hand experience, the problems of life as an unemployed outcast, for instance: Otto Nagel, *Die Weisse Taube oder das nasse Dreieck (Kleine Arbeiterbibliothek)*, n.d., or Bruno Nelissen Haken, *Der Fall Bundhund* (Jena, 1930).

56. Hans Mommsen, 'Die Sozialdemokratie in der Defensive: Der Immobilismus der SPD und der Aufstieg des Nationalsozialismus', in idem. (ed.), *Sozialdemokratie zwischen Klassenbewegung und Volkspartei* (Frankfurt, 1974), pp. 106–33. Hagen Schulze, 'Die SPD und der Staat von Weimar' in M. Stürmer, *Die Weimarer Republik* pp. 272–86; Ludwig Preller, *Sozialpolitik in der Weimarer Republik* (Düsseldorf, 1978), p. 419.

57. LA Schleswig 309/23056, RK u.d.ö.O. 159/28 II, 11 Jan. 1928 (Rundschreiben des Reichsausschusses), p. 6. The reference is to Reichsanstalt für Arbeitsvermittlung und Arbeitslosenversicherung (Reich Ministry for Labour Exchange and Unemployment Insurance), Preller, *Sozialpolitik*, pp. 372, 419.

58. LA Schleswig 309/23056, Rundschreiben des Reichsausschusses, p. 5.

59. Instruktion für die neugewählten kommunistischen Gemeindevertreter (Hrsg. ZK der KPD, Abteilung Kommunalpolitik), 2 May 1924; Kommunistische

Gemeindepolitik. Richtlinien u.Erläuterungen, Hrsg. v. ZK der KPD (Berlin 25.9.1925); Instruktion für neugewählte Gemeindevertreter, 1924 (Manuscript im IML). See in general: Beatrix Herlemann, *Kommunalpolitik der KPD im Ruhrgebiet 1924–1933* (Wuppertal, 1977), pp. 11–25.

60. The Ruhr was, however, an exception (Herlemann *Kommunalpolitik*, p. 29); Volker Wunderlich, *Arbeiterbewegung und Selbstverwaltung. KPD und Kommunalpolitik in der Weimarer Republik mit dem Beispiel Solingen* (Wuppertal, 1980), esp. p. 40.

61. J. Wollenberg, L. Heer-Kleinert, M. Muser and D. Pfliegendorfer, *Von der Krise zum Faschismus. Bremer Arbeiterbewegung 1929–1933* (Frankfurt/Main 1983), pp. 113–17, here p. 115; Rolf Schwarz, 'Rendsburg and Büdelsdorf. Lokale Aktivitäten der Arbeiterparteien SPD und KPD' in E. Hoffmann and P. Wulf (eds), '*Wir bauen das Reich.*' *Aufstieg und erste Herrschaftsjahre des Nationalsozialismus in Schleswig-Holstein* (Neumünster, 1983), p. 160.

62. StA Hamburg 424–2 IVE 48, Drucksache 1590 (KPD-Fraktion) 30 January 1932; ibid. Drucksache Nr. 1622, 11 June 1932; cf. ibid., Drucksache Nr. 1660 (KPD-Fraktion), 26 Oct./18 Nov. 1932, protest against the Magistrat's Decision to cut the level of benefits to the hutters (Laubenkolonisten). See also the demand for better conditions in the waiting rooms of the dole offices of the hutters of Stellingermoor (ibid., Drucksache Nr. 1662, 19 Nov. 1932).

63. LA Schleswig, 309/22722, Pol. Präs. Altona 1a to the Oberpräs. in Kiel, 12 June 1926, betr. Erwerbslosendemonstration; LA Schleswig 309/22718. Pol. Präs. Altona P2 254/26 to the Oberpräs. in Kiel, 27 Mar. 1926, betr. Ablauf des Erwerbslosendemonstrationstages; StA Bremen 4, 65 IV 4e vol. 5, Report Nr. 8, 4 Sep. 1926, p. 10.

64. LA Schleswig, 309/22669 and 22996.

65. LA Schleswig 309/22739, Pol. Präs. P 2 84/26, 29 Jan. 1926 to Oberpräs. in Kiel, betr. Zusammenstoss zwischen Kommunisten und Polizeibeamten.

66. LA Schleswig 309/22781, Lks. Flensburg 1A 397/27 to Reg.Präs., betr. RFB-Versammlung 29.8.27, p. 77; StA Hamburg, Polizeibehörde I-697, Kommunistische Unruhen und Hinrichtungen 1927/28.

67. LA Schleswig 309/22721, Übersicht über politische Ausschreitungen im Polizei-Bezirk Altona-Wandsbek (statistical compilation of disturbances). For a selective account of violence between police, political parties etc., see Wolfgang Kopitzsch, 'Politische Gewalttaten in Schleswig-Holstein in der Endphase der Weimarer Republik' in Erich Hoffmann and Peter Wulf (eds), '*Wir bauen das Reich*', p. 25.

68. See note 66 above; A. P. McElligott, 'Das "Abruzzenviertel". Arbeiter in Altona 1918–1932' in Arno Herzig *et al.* (eds), *Arbeiter in Hamburg. Unterschichten. Arbeiter und Arbeiterbewegung seit dem ausgehenden 18. Jahrhundert* (Hamburg, 1983), p. 501.

69. Lewis A. Coser, *Continuities in the Study of Social Conflict* (New York Free Press, 1967), p. 104; A. P. McElligott, 'Petty Complaints. Plunder and Police in Altona 1917–1920. Towards an Interpretation of Community and Conflict' (MS., 1985).

70. StA Bremen, 4, 65 II Z 11, 1528/249, Denkschrift über Kampfvorbereitungen radikaler Organisationen 1.11.1932–1.12.1933, pp. 39–51. The KPD found their strongest support in such areas: cf. Büttner, *Wirtschaftskrise*, pp. 439–40; p. 666.

71. StA Bremen 4, 65 IIE 1a3 vol. 1, Erwerbslosenkrawalle zu Weihnachten 1929 im Reich.

72. LA Schleswig 309/22996. Nachweisung (Versammlungstätigkeit), Report for December; LA Schleswig 309/23056, Pol.Präs. Altona 1A. 21.12.29 to Oberpräs. in Kiel betr. Erwerbslosendemonstration (including reports for Kiel and Itzehoe);

StA Bremen 4, 65 IV 4e vol. 7, Report Nr. 4, 31.12.29; LA Schleswig 309/23056; Pol.Präs. Altona 1A, 21.12.29 to Oberpräs. in Kiel, pp. 1–2.

73. LA Schleswig, 309/23056, Pol. Präs. Altona 1A, 21.12.29 to Oberpräs. in Kiel, report of police authorities from 24 Jan. 1930; LA Schleswig 309/22996, compilation for Jan.

74. LA Schleswig 309/22996, compilation for Mar. LA Schleswig 309/23056, Report from 7 Mar. 1930 (the clash occurred on 5 March).

75. *Die Generallinie*, p. 93, footnote 30; H.-J. Zimmermann, *Wählerverhalten und Sozialstruktur im Kreis Herzogtum Lauenburg 1918–1933* (Neumüster, 1977), p. 107; A. P. McElligott, 'Petty Complaints', parts 2 and 3.

76. Helmut Gast, 'Die proletarischen Hundertschaften als Organe der Einheitsfront im Jahre 1923' in *Zeitschrift für Geschichtswissenschaft* iv (1956), pp. 439–65; Eve Rosenhaft, 'Die KPD der Weimarer Republik und das Problem des Terrors in der "Dritten Periode", 1929–1933' in W. J. Mommsen and G. Hirschfeld (eds), *Sozialprotest, Gewalt, Terror, Gewaltanwendung durch politische und gesellschaftliche Randgruppen im 19. und 20. Jahrhundert* (Stuttgart, 1982), pp. 405–8.

77. Heckert, *Bericht*, pp. 136–7. Hans Manfred Bock, *Syndikalismus und Linksradikalismus von 1918–1923. Zur Geschichte und Soziologie der Freien Arbeiter-Union Deutschlands und der Kommunistischen Arbeiter Partei Deutschlands* (Meisenheim/Glan, 1969).

78. Fowkes, *Communism in Germany*, pp. 104–9; Larry Peterson, 'A Social Analysis of KPD Supporters: The Hamburg Insurrectionaries of October 1923', *International Review of Social History* XXVIII (1983), 2, 200–39, here pp. 203–9.

79. StA Bremen 4/65 II ZII, 15828/249 Denkschrift, p. 40.

80. Wilhelm Hoegner, *Die Verratene Republik* (Munich, 1958); Ursula Büttner, 'Das Ende der Weimarer Republic und der Aufstieg des Nationalsozialismus in Hamburg' in Ursula Büttner, Werner Jochmann, *Hamburg auf dem Weg ins Dritte Reich, Entwicklungsjahre 1931–1933* (Hamburg, 1983), pp. 30–1; Harald Focke and Hartmut Hohlbein, *Stationen auf dem Weg zur Macht. Von der Weimarer Republik zum NS-System. Die Jahre 1932/33 in Deutschland* (Hamburg, 1982), pp. 38–41.

81. LA Schleswig 309/23056 Pol. Präs. Altona LA 21.12.29. The following is based on this report.

82. For source references, see notes 72 and 81, above. At the first meeting of the revitalised Committee of the Unemployed on 12 Dec., about 270 are reported to have attended (LA Schleswig 309/22996, Nachweisung, report for Dec.). Cf. LA Schleswig 309/23056, Pol. Präs. Altona 1A to Oberpräs. in Kiel, 14 Dec. 1929).

83. Büttner, *Wirtschaftskrise*, p. 293ff (at the same time Büttner underplays the influence and mobilisation chances of the KPD). An excellent study of the British Unemployed Workers' movement (NUWM) is to be found in Ralph Hayburn, 'The National Unemployed Workers' Movement, 1921–1936, A Reappraisal', *International Review of Social History* XXVIII (1983), 3, 279–95; Hayburn emphasises the degree of autonomy of the local councils in the day-to-day running of the movement.

84. *Vorwärts und nicht vergessen. Arbeiterkultur in Hamburg um 1930. Materialien zur Geschichte der Weimarer Republik* (Hamburg, 1982). For a refutation of the myth that Hitler owed his electoral success to the unemployed, see Jürgen Falter, 'Arbeitslosigkeit und Nationalsozialismus. Eine empirische Analyse des Beitrages der Massenarbeitslosigkeit zu den Wahlerfolgen der NSDAP 1932 und 1933', *Kölner Zeitschrift für Soziologie und Sozialpsychologie*, Jg. 35, 1983, 525, 544; Cf.Wilhelm P. Burklin und Jürgen Wiegand, 'Arbeitslosigkeit und Wahlverhalten' in Bonss and Heinze, *Arbeitslosigkeit*, pp. 273–97.

85. A. P. McElligott, 'Strassenschlachten', p. 67ff; StA Bremen 4, 65 IIE 1a3 Bd. 2, Auszug aus dem Lagebericht der M.D.l. 22.12.30 Nr. 33040/11 II, Hungermarsch der KPD am 27.11 bzw. 3.12.1930, p. 7.

86. StA Bremen 4, 65 IIE 1a3 vol. 3, Abschrift, Nachrichtenstelle d. Reichsministeriums d. Innern, IAN 2164, 28.1. (1932); ibid., Report from Aug. 1932; ibid., vol. 4, Report from 7 Nov. Rosner, 'An Enquiry', 384.

87. StA Bremen, 4, 65 IIE 1a3 vol. 2. Abschrift of the Resolution der Prager Conferenz über die Erwerbslosenfrage; ibid., vol. 3, report of the *Nachrichtenstelle* RM. d.I., IAN 2164 d 3./6.1.1932.

88. W. Ulbricht, 'Mobilmachung der Erwerbslosen!'

89. This referred to the protest movement which sprang up against the Papen Decree of 14 June 1932 which foresaw further dismantling in benefits for the unemployed. (see pp. 9 above).

10 UNEMPLOYMENT AND WORKING-CLASS SOLIDARITY: The German Experience 1929–33

Dick Geary

I

'It cannot be reiterated too often that unemployment is not an active state; its keynote is boredom — a continuous sense of boredom. This boredom was invariably accompanied by a disbelief which gave rise to cynicism.'[1] Thus reported a British enquiry into the condition of the unemployed between 1936 and 1939. Its findings were to a large extent in agreement with those of Jahoda and Lazarsfeld, who conducted a detailed investigation of unemployment in Austrian Marienthal in the mid-1920s. They concluded that its major impact was to be found in increasing apathy, privatisation, and in some cases, despair. Contrary to what one might have expected, participation in social as well as political organisations and activities actually declined quite dramatically.[2] And as contemporary British experience of mass unemployment further testifies, that sad phenomenon in no way necessarily correlates with political radicalism.

At first sight one would hardly imagine that such conclusions could also apply to the Weimar Republic in the depths of the Depression. The years of the international economic crisis also witnessed an increase in electoral participation, which exceeded 80 per cent of the electorate in the Reichstag elections of July 1932. Moreover, the major beneficiaries of this increased participation, and at least to a certain extent of the votes of the army of unemployed, appeared to be the parties on the extremes of the political spectrum, namely the German Communist Party (KPD) on the Left, and the National Socialist Party (NSDAP) on the Right. The KPD attracted almost 5 million votes in 1930, 5,370,000 votes in July 1932, and 5,985,000 votes in the November elections of the same year; whilst the NSDAP became the largest single party in the Reichstag in this period. The KPD, which was already recruiting heavily from the unemployed in the so-called 'stabilisation crisis' of 1924 after the end of the 'Great Inflation', found that over 80 per cent of its membership was out of work at the end of 1932, and in some party branches only 8 per cent of members had jobs.[3] Unemployment was disproportionately concentrated in the larger industrial towns, and it was precisely here that the Communists gained their greatest electoral successes. It was true, for example, in the huge conurbations of the Ruhr, like the mining town of Herne, where the KPD was the largest single party throughout the Depression and where almost half the total population was dependent on some form of public welfare in mid-1932.[4] From 1928 the 'Committees of the Unemployed' here were also controlled by members of the KPD.[5]

Even Ernst Thälmann, the KPD leader, had to admit, however, that the KPD was not the only beneficiary of mass unemployment, and that the Nazis also made gains from the dole queues.[6] The giant German Metalworkers' Union and the socialist General German Trade Union Federation both heard reports at their conferences that some sections of the unemployed working class were deserting to the Nazis or the SA (the Nazi paramilitary organisation of stormtroopers); and it was the case that many young Nazi Party members were jobless.[7] According to a KPD source, the Nazi Party and other 'fascist' organisations had some success amongst the unemployed in the Ruhr as early as 1927.[8] In an interview with a local newspaper, the *Rheinisch-Westfälische Zeitung*, Hitler himself claimed, five years later, that no fewer than 300,000 of the 400,000 members of the SA and the SS were out of work.[9] Later, in the memoirs of Carl Severing, Prussian Minister of the Interior until the Papen coup of July 1932, we find the report of the Director of the Workers' Academy (*Arbeiterakademie*) in Frankfurt, who claimed that

the working class split into two fundamentally distinct parts, those in employment and the unemployed . . . Standing in the dole queues were young people, some of whom had been without work since their apprenticeships, who had never enjoyed the benefits of a trade union education and who were therefore amenable to radical slogans of all kinds . . . It was there that the National Socialists broke into the ranks of the proletariat, whose unemployed had become atomised and had often sunken back to the level of the lumpenproletariat.[10]

Nazi success among the unemployed was not only stated by contemporary witnesses but has been confirmed by subsequent historical investigation. It has been claimed that there exists a correlation between the rise in unemployment and the increase in the size of the Nazi vote.[11] Research in East Germany has conceded that youths in Chemnitz who had been unemployed for several years sometimes found their way into the SA;[12] whilst the British historian Conan Fischer has claimed that a significant level of Nazi support and SA membership was recruited from the unemployed working class.[13] In the Ruhr town of Bochum 40 per cent of the SA were jobless, according to recent research.[14] It is also true that some of the greatest electoral successes of the NSDAP were registered in areas of high unemployment, such as Chemnitz-Zwickau in Saxony and the shoe-making town of Pirmasens in the Palatinate. In the latter town no fewer than 10,000 of a total labour force of 24,000 were unemployed and a further 10,000 on short time as early as 1929.[15]

That some of the unemployed found their way into the NSDAP or, more likely, the SA, therefore, would seem to be beyond question. We should none the less be on our guard in positing too close a relationship between the *working-class* unemployed and Nazi Party membership. Amongst the working class in general we know that the Nazi factory-cell organisation (the *National-sozialistische Betriebszellenorganisation*) did badly in factory-council elections, and that in those elections it had more success with white-collar workers than their manual colleagues. Within the Nazi Party the manual working class was under-represented; whilst at elections, the larger the town and the more proletarian the quarter, the lower the percentage of the vote mobilised by the Nazis. Many of the large towns which did provide significant numbers of Nazi voters, such as Hanover and Brunswick, had a disproportionately large sector of white-collar workers. Many workers in the Nazi Party and the SA came from non-proletarian backgrounds: it should be remembered that the Depression was characterised by high levels of *downward* social mobility. Many of the apprentices who turned to the Nazis or the SA worked in small craft shops and came from middle-class backgrounds. Some young 'workers' in the SA 'homes' had in fact come from the countryside.[16]

This does not mean that the Nazis did not win the support of a significant number of industrial workers; but, as in the case of some of the groups already mentioned, they were rarely workers rooted in working-class communities with union or socialist traditions. Deserters from the Social Democratic Party (and for that matter Catholic workers who had previously voted for the Centre Party) were much more likely to switch their allegiance to the Communists, especially when they lived in large towns. Where the Nazis were much more successful was amongst rural labourers, domestic workers, workers in the industrial *villages* of Saxony (rather than the factories of Chemnitz itself); also amongst workers who had previously given their support to the German Nationalist People's Party, and *new* workers recruited from nationalist labour exchanges in the Depression.[17] This explains why the combined vote of the Social Democrats and the Communists remained more or less constant at precisely the same time as electoral support for the National Socialists surged dramatically.[18] And it suggests that unemployment, which again was disproportionately concentrated in the urban, industrial working class, did not lead to mass desertion from *organised* labour. (This is also reflected in the ability of the Free Trade Unions to retain 4 million members, despite the fact that roughly half of those members were without jobs in early 1932.)

It may be that the unemployed worker, and especially the *young* unemployed worker, deviated from this behaviour, as suggested by the contemporary reports cited earlier. Certainly, the Depression forced many to seek work away from home. Others left home to be eligible for unemployment relief, especially after the second emergency decree of the Brüning government came into force in mid-1931. The so-called 'Voluntary Labour Service' took many youngsters out of the family household and into what were effectively sealed camps.[19] In this situation patterns of socialisation, in which the father found employment for his son, introduced him to the union and took him to the socialist club or pub, could easily break down, especially where the father himself was out of work. There is much evidence of increased tension in the households of the unemployed in this period; and violence within the family was not unknown.[20] Having said this, party-membership analysis and a breakdown of voting behaviour by residential community still suggests that the young, manual, urban unemployed were more likely to opt for the Left than for the Nazis, and other evidence confirms this. Unemployment in Germany was disproportionately concentrated in towns with a population of over 100,000 inhabitants, i.e. precisely where the Nazis performed poorly at elections, as was pointed out some time ago.[21] More detailed analysis of the Reichstag elections in the

Ruhr, an area more solidly urban and working class in its social composition than most, and again an area with high levels of unemployment, confirms this picture. In Bochum the Nazi vote was lowest in the almost monolithically working-class mining colonies where unemployment was high, its share of the vote sometimes falling below 20 per cent, and at its highest in the merchant community of the town centre, and above all in the semi-rural area of Bochum-Stiepel, where the Nazis won 42.63 per cent of the vote (32 per cent in Bochum as a whole, almost 38 per cent in the whole Reich).[22] In Herne the picture was even clearer, with Nazi support overwhelmingly and *increasingly* concentrated in the central shopping area of the Bahnhofstrasse. Around the pits of 'Constantine the Great' and especially 'Teutoburgia', which had been closed down as early as 1925 and where levels of joblessness were frighteningly high, the Nazis won only 12 per cent of the vote, whereas the Communists attracted nearly 70 per cent![23] In these cases it is clear that the Communists were strongest where the Nazis were weakest, and vice versa; which may also explain why the KPD directed so much of its hostility at its real competitor for votes, the SPD. The recent work of Thomas Childers and Jürgen Falter makes this even clearer, for they establish a strong positive correlation between unemployment and KPD support, but a *negative* correlation between unemployment and Nazi-voting.[24]

II

What the above suggests is that unemployment did not lead *organised* labour to desert to the Nazis, even if it did provide the latter with support from the previously unorganised, non-factory working class, which, it should none the less be admitted, was numerically very significant.[25] Thus, the combined SPD/KPD vote remained constant, and in the November elections of 1932 actually outstripped that of the Nazi Party. This obviously raises the hackneyed question of why such a numerically powerful labour movement failed to halt the arguably resistable rise to power of its most vicious enemy. The question usually elicits an equally hackneyed answer, which seeks an explanation in terms of the isolation of the labour movement from the rest of German society, which seems to this author undeniably true; the failures of the leadership of both the SPD and the KPD (though especially the latter) to recognise the seriousness of the Nazi threat, equally undeniable; and above all the internecine warfare between Communists and Social Democrats, which split the labour movement and reduced it to a fatal impotence. Again this last point is beyond dispute, though I have argued elsewhere that the errors and omissions of the Social Democrats are often overlooked and those of the Communists sometimes exaggerated or misunderstood.[26] What the hackneyed answer normally ignores, however, is the *social* as well as the political basis of the split within the German labour movement and of its impotence, namely mass and long-term unemployment.

It is the contention of this paper that the German working class and its organisations became increasingly fractured and atomised in the course of the Depression in two different senses: increasingly, the split between Communists and Social Democrats hardened into a split between different sections of the German working class, whilst in their daily lives in the home and the factory various pressures independent of politics made solidaristic action less and less likely. In a sense this second point is perhaps more crucial than the first. The period between 1918 and 1923 witnessed a fairly broad-based radicalism among skilled and unskilled, employed and unemployed. The divisions between Communists, Left Communists, Syndicalists, Independent Socialists and Majority Social Democrats were still fluid at the time and still far from clear-cut. Industrial and political militancy still went hand in hand, and an attempted coup in 1920 on the part of the German Right had been frustrated by joint working-class action in the mines and factories. But by 1932 the picture had changed completely.[27]

So, at least, believed many contemporaries. Observing the working class of the Depression years, the Austro-Marxist Max Adler commented,

> The working class has been burst asunder. By its loss of unity and striking power . . . the German working class has dug its own grave instead of being the gravedigger of capitalism . . . [The origin of this] is the differentiation within the proletariat . . . which has existed for decades at the upper levels, but has also become especially marked at the lower levels since the world crisis and its long-term unemployment.[28]

Friedrich Stampfer, editor of the Social Democratic newspaper *Vorwärts*, commented in January 1933 that 'trenches' were being dug within the working class as evidenced by a comparison of the skilled membership of the SPD with unemployed Communist demonstrators.[29] Another member of the SPD executive made much the same point at a party meeting in 1932;[30] whilst the Social Democratic economist and Prussian Under-Secretary of State Hans Staudinger spoke of an unprecedented split within the labour movement, of hatred and envy between the employed and unemployed, and of conflicts between the two.[31]

This social division at the very base of the German working class in the Weimar Republic was reflected in the fact that the Social Democrats and the Communists became increasingly different in their social composition and their constituencies increasingly distinct. The fluidity of the immediate postwar period disappeared as the KPD became a party almost all of whose members were unemployed. The Social Democratic Party, on the other hand, had a membership that was disproportionately *employed* in 1932 (something like 30 per cent of Social Democrats were unemployed then), when one considers that male unemployment was officially registered at around 40 per cent, that the membership of the Party was predominantly male and that it traditionally recruited from trades with even higher levels of unemployment, such as the skilled building trades and metalwork.[32] This Communist success amongst the unemployed was facilitated by the development of neighbourhood as well as factory strategies, all the more necessary as few of its members remained at work: rent strikes, the prevention of evictions, committees of the unemployed, action involving women and children.[33] On the other hand, the SPD initially welcomed 'rationalisation' in the factories — arguably a major cause of redundancies,[34] tolerated the deflationary economic policies of the Brüning governments between 1930 and 1932, and rejected the ambitious work-creation programme of the ADGB.[35] All this hardly made the SPD seem responsive to the needs of the jobless.

Level of unemployment was not the only factor that distinguished one socialist party membership from the other. According to KPD sources, the members of the SPD were increasingly recruited from the better-off sections of the German working class;[36] and whilst the KPD became increasingly unskilled in its social composition in this period, the SPD, though still predominantly recruiting from the manual working class, extended its social base amongst white-collar workers and the non-working wives of relatively affluent workers.[37] Whereas the KPD membership was 90 per cent manual working class in this period, resembling the composition of the SPD on the eve of the First World War in its manual if not skilled definition, the SPD was only 73 per cent proletarian in 1926, and 60 per cent in 1930.[38] Increasingly, the division between Communists and Social Democrats was also a division between younger and older workers. In 1927, 64.5 per cent of KPD members were under 40 years of age and 31.8 per cent under 30. As the Depression deepened, the Party's membership became even younger. On the other hand, fewer than 45 per cent of the SPD membership were under 40 three years later, and only 8 per cent under 25. In the Reichstag the Social Democratic delegation was in fact the oldest of all the political parties.[39] Indeed, there is a substantial amount of evidence that the leadership of both the SPD and the Free Trade Unions found it increasingly difficult to communicate with the younger generation, as Willy Brandt, amongst many others, has testified.[40]

Patterns of recruitment also increasingly hardened along residential lines, with the SPD doing well in some of the newer areas of municipal housing, whilst Communist support tended to concentrate to a certain extent in the inner-city slums and areas characterised by relatively high levels of criminality. In fact, a kind of residential segregation seems to have taken place in at least some working-class communities in this period, though it was far from total.[41] This development also coincided with and partly explains what appears to have been an increasing gulf between the 'rough' and the 'respectable' working class, which again overlapped to some extent with that between KPD and SPD constituencies. The militants of the SPD distanced themselves from various aspects not only of crime but even of popular culture in this period, criticising smoking, drinking and the frequenting of cinemas and dance-halls, making a clear distinction between themselves and the 'Lumpen'.[42] Their refusal to condone working-class crime, even in circumstances of poverty, was even more likely to alienate former or potential supporters, as some recent research has suggested, especially as the KPD was prepared to advocate 'proletarian shopping-trips' and looting as forms of class conflict and self-help.[43] And this issue was far from peripheral in the circumstances of the Depression, when youth criminality seems to have escalated.

It is true that one must be cautious in positing too close a relationship between a rise in crime and the increasing number of the unemployed, as Detlev Peukert has pointed out.[44] A similar cautionary note has also been sounded on many occasions in the British case.[45] It is also true that the development of youth criminality in Germany had roots that stretched back to the period before the First World War, and that the number of indictable offences, which had peaked between the end of the war and 1923, actually fell in the years 1929 to 1933 in the Reich as a whole.[46] However, at least some contemporaries did detect a connection between youth criminality and unemployment, pointing out that the crimes which increased by the greatest extent were those associated with economic hardship, namely crimes against property.[47] Moreover, if one looks at those towns with especially high levels of unemployment, such as Herne, the picture changes: for there crimes committed by the under-18s tripled between 1929 and 1931.[48] In this context the SPD's respectable self-image may well have been a handicap rather than an advantage, especially as the gains made by organised labour in the Weimar Republic, such as the welfare institutions and the works councils, also landed the SPD and the Free Trade Unions with some of the responsibility for the maintenance of factory and communal discipline.[49] Institutions of welfare could easily be seen by those on the receiving end as increasingly remote organs of supervision.[50] The fact that the police forces of the major cities, which had to supervise the dole queues and often broke up demonstrations of the unemployed with no small degree of violence, were sometimes under the control of Social Democratic police chiefs, most notoriously in Berlin, hardly helped the cause of the SPD amongst those without work;[51] and it is scarcely surprising in this context that the KPD had more success at the dole queues than in the factories.[52]

This increasing gulf between the social composition of the SPD and the KPD perhaps goes some way towards explaining their mutual incomprehension, and suggests that co-operation between the two parties was hindered by far more than the suicidal myopia of the two leaderships. But of itself it does not demonstrate that any form of co-operation was impossible. Some Communists did complain about the KPD's perpetual vilification of SPD leaders. In some places the local branches of the two parties co-operated, especially in 1932; whilst the Communist and Social Democratic youth organisations joined together in anti-fascist initiatives.[53]

The bourgeois press also claimed that members of the SPD's 'Iron Front' were to be seen marching arm in arm with KPD members; and at the same time the SPD leadership was presented with demands for joint action from several groups of party militants, amongst whom the Communists' demand for a 'unity front' (*Einheitsfront*) was reputedly popular.[54]

Perhaps even more significantly for daily behaviour and working-class solidarity was the fact that trade union leaders claimed their members would willingly work short time to create jobs for the unemployed;[55] and there are a few examples of firms in which the labour force agreed to a reduction of hours to this end.[56] Examples of the unemployed picketing pits and works in which the labour force was on strike were far from uncommon,[57] though as we shall see, this could cause problems; whilst the KPD claimed that few of the unemployed acted as strike-breakers.[58] This last point has been repeated by later research, both in the German Democratic Republic[59] and in the Federal Republic.[60]

These examples of collaboration must be treated with extreme caution, however. Between 1929 and 1933 it is the absence of strike action which is most apparent: at their peak the strikes of 1932, when the economy began to move out of its trough, involved only 129,000 workers, compared to over 2 million in 1919. The lack of success of these later strikes and the fact that few were long lived obviously obviated the employers' need to bring in blacklegs on many occasions.[61] Furthermore, some of the literature which claims co-operation between the employed and the unemployed subsequently proceeds to give examples of strike-breaking, or at least of strikes being called off because of the threat of strike-breaking,[62] especially where blacklegs were not local. Reports to employers' organisations also suggest that workers were not keen to reduce hours to create more jobs for their less fortunate and jobless colleagues,[63] whilst union organisations complained about the high levels of overtime still worked when so many were without work.[64] On several occasions attempts by union officials to persuade men to vote in favour of the *Krümpersystem* (a system to share jobs more widely and one that thus meant more short time) were not supported by pit assemblies.[65]

In addition to the contemporary observations concerning conflict between the employed and unemployed cited earlier,[66] the employed rarely participated in demonstrations initiated by the unemployed, and the local branches of the trade unions tended to shun the Committee of the Unemployed.[67] And, the unions had little success in trying to mobilise the unemployed.[68] The organ of the German Metalworkers' Union received numerous letters complaining of the failure of those still in employment to sympathise with their jobless brothers.[69] In Hochlamarck, a mining district near Recklinghausen, tension between the employed and the unemployed certainly existed;[70] and the failure of the two to act together has been noted more generally.[71] Catholic newspapers of the time also warned their working-class readers against strike action, precisely on the grounds that an army of unemployed would take their jobs.[72] The absence of solidaristic feeling was also increasing by petty criminality *within* working-class communities in the Depression;[73] and possibly testified by the fact that trade union officials often received a hostile reception from the crowds at labour exchanges.[74]

An analysis of strikes initiated by the KPD in this period, in which the employed and the unemployed are supposed to have co-operated, is also instructive. After the bloodshed of May Day 1929 in Berlin, the KPD called for strikes throughout Germany. In the mines of the Ruhr, where Communist support was large — at least to judge from the works-council elections — and where one might have therefore expected a fair degree of solidarity, this strike-call was only partially successful: although some pits were closed, only a minority of the labour force participated in strike action,[75] which might well explain why the KPD had to extend its activities beyond the shop-floor and coalface into the settlements, and why the large numbers of miners who continued to work normally were subjected to the threat or use of violence.[76] However, the strike-call of 1929 was far more successful than some later Communist initiatives, as in the winter of 1930–31, when the Communist trade union organisation, the RGO, called further strikes against wage cuts.[77] Although strike participation was initially in the order of 50 per cent in the Mülheim-Hamborn-Duisburg district, an area long noted for its industrial and political militancy, only 14 pits of 54 around Recklinghausen, that is between 10 and 12 per cent of the labour force, joined the action. Around Bochum the participation rate fell to a miserable 0.5 per cent of the labour force; whilst in the Westphalian pits as a whole it stood at 4–5 per cent, despite the fact that the KPD was far from weak there.[78] It was precisely in this situation that the KPD was forced to strengthen its activities outside the pits and in the mining colonies, amongst women, children and the unemployed,[79] that it used outsiders as pickets,[80] and that it resorted to violence and sabotage.[81] This had a number of consequences.

First, the use of outsiders and of violence alienated many of those still working, even though some of these would have voted for the RGO in works-council elections and for the KPD in elections to the Reichstag. It was not infrequent for 'agitators' (*Hetzer*) to be reported to the management by other workers.[82] It further meant that some of those involved in picketing incidents at the pithead were not from the pit involved, were often unemployed, and in some cases had not even been miners,[83] all of which tended to alienate the working locals, even where their voting behaviour expressed hostility to the Weimar system. In fact, a KPD group in Bochum involved in the organisation of these strikes included not a single *employed* miner.[84] Furthermore, involvement in 'wild' strikes, and especially involvement in their initiation, led to instant dismissal, thus reinforcing the unemployed status of Communist Party membership and robbing it of some of its best cadres.[85] In this context it is hardly surprising that even Communist members of works and pit councils failed to follow the strike call on occasion.[86] This failure of KPD strike initiatives was not restricted to the Ruhr, nor to the strikes of the winter of 1930–1. The strikes of October 1931 and January 1932 mobilised even fewer workers, whilst massive RGO initiatives for strike action in mid-1931 failed to bring a single plant in the metal-working industry to a halt, despite the fact that this was another sector in which KPD support was substantial.[87]

III

That German workers should prove all too human in such a crisis may be tragic, but it is not difficult to understand. The sheer scale of mass and long-term unemployment at the nadir of the Depression acted as a real deterrent to action. (Significantly the RGO's rather miserable strike record began to improve with the up-turn of the economy in the second half of 1932.)[88] Calculations of the 'real' number of the unemployed vary wildly, but even the official figure of the *registered* unemployed was gruesome enough, standing at over 6 million, approximately a third of the active labour force. Some sources suggest that a further half-million needs to be added to this, some a further million; and one calculation reaches the staggering figure of over 9 million unemployed.[89] Significantly, the unemployment rates were even higher for trades which had traditionally formed the backbone of working-class militancy, standing at 60 per cent for metal-work and a dreadful 90 per cent for the building trades in the winter of 1931–2. Of trade union members almost a half were without work, and another quarter on short time in this period.[90] The consequences of this were massive.

Those out of work were obviously deprived of all industrial muscle; and this again explains why a party such as the KPD, the overwhelming majority of whose members were unemployed, had to develop new strategies based upon the neighbourhood rather than the place of work. On the other hand, those in jobs showed every sign of being glad to have them,[91] a further factor which produced the decline in strike action in the Depression. In fact, long-term unemployment, which in some cases went back to the stabilisation crisis and the industrial rationalisation of the mid-1920s, had intimidated some sections of the labour force even before the onset of the world economic crisis of 1929. As early as 1927 a French team sent to investigate conditions in the Ruhr mines reported that all levels of management believed they had recovered 'control' in the pits.[92] Mining officials spoke of resignation in the pits around Dortmund in January 1929,[93] and said of the workforce of the Herne pits in December 1930 that it had no desire to follow the KPD's attempts to exploit the wage negotiations of that time to bring work to a standstill.[94] On 8 June 1931 the *raporteur* of the Committee for Trade and Industry of the Prussian *Landtag*, or parliament, claimed to have received a letter from the chairman of a works council which spoke of such fear among the workers of his mine that no-one dared say a word of criticism for fear of being left on the street.[95] In some pits the fear of unemployment led men to sign away their holiday entitlement;[96] whilst both management and unions reported that miners were afraid to take sick leave for fear of dismissal.[97] On occasion, management adopted policies which exacerbated this insecurity. A miner from Hochlamarck claimed that one pit took on 31 school-leavers when it was known there were only 30 vacancies, the result being that the lads then lived permanently under the 'sword of Damocles'; 'one thing', he added, 'was especially obvious then: the lads were afraid of unemployment, and that was exploited', especially as methods of payment were introduced which set one individual miner in competition with another. 'No-one wanted to lose his job . . . and you were always afraid that you'd get the sack. The threat was always there.'[98]

The way in which unemployment exacerbated divisions within the working class can also be seen in the actions of unions and miners to save the pit in their locality rather than one elsewhere, a situation that led the reactionary mine-owners' journal to report gleefully that the 'artificial' solidarity of the class and union had been replaced by the 'natural' solidarity of the individual pit.[99] Workers in some pits were reluctant to take on colleagues from redundant mines, fearing that their own jobs would be jeopardised.[100] Such sad divisions were further reflected in the many cases brought before the labour courts, often with the backing of the *Betriebsrat* or union, in which, in complaining about their dismissal, many miners who were married, had children or were relatively old, demanded that they be reinstated, and that younger, single workers should be the first to be shown the door.[101] However morally justifiable in terms of 'hardship' such a position was, it was hardly a testimony to solidarity, and obviously alienated younger workers from their older colleagues and the official union organisations, especially where the plaintiffs actually mentioned their younger and still-employed colleagues *by name*![102]

The fact that unemployment was distributed unevenly both in terms of region and industrial sector may further have divided the German working class, as may have a system of unemployment relief aimed at atomising the recipients.[103] Certainly, there is evidence that some workers resented the markedly different treatment of other groups of the unemployed.[104] Even in daily life in the mining colonies, those supposedly most solidaristic of all working-class communities, the residents observed an increase in thefts (miner from miner).[105] In short, a host of factors, many of them generated by mass unemployment, militated against common action. If anything, these mundane pressures were more important in undermining successful resistance against Nazism than the political divisions at the summit of the organised labour movements. For the events of 1920 had demonstrated that political divisions within the working class could be eclipsed, at least temporarily, where the economic and social basis for solidaristic action still remained in the factory and the neighbourhood.

IV

So far we have seen that although unemployment served to divide and weaken the labour movement, it did lead to an apparent radicalisation in the form of increased votes for the KPD; but it was far from clear whether those voting for the KPD would also participate in various forms of more direct *action*. As we have seen, some KPD militants failed to follow the party's strike-calls. The KPD's attempt to mobilise the unemployed, it is now recognised, met with far less success than has often been imagined, and failed to worry some police-chiefs.[106] Significantly, the greatest and most effective demonstrations of the unemployed were in 1923, a year of mass unemployment, but one which followed almost three years of full employment. The strikes and actions of the unemployed were again more successful in 1929 than in 1931. The paramilitary actions of Communist street-fighters became increasingly professionalised, rather than spontaneous confrontations of large numbers of the jobless. Thus, the quality if not the quantity of KPD support was problematical, as in the pit strikes. Indeed, the Party was well aware of the problem, deploring the poor attendance at shop-floor meetings as well as the massive turnover of its membership, which reached something in the order of 80 per cent per annum in the Depression. We may perhaps speculate that whilst a hard core of militants became increasingly active, many of the unemployed dropped out of the political scene altogether, except for the occasional trip to the polling station, which in many cases meant the pub.

That resignation became the hallmark of the young unemployed has already been suggested by Detlev Peukert, even if that resignation came only gradually. Increasingly, the young came to kill time rather than engage in any kind of activity.[107] Though some youths found their way into the Communist youth organisation, many others responded to unemployment with a helpless resignation and an increased concern for their own individual selves.[108] Again, it may be speculated that this atomisation of the working class, especially its younger members, is precisely what facilitated the subsequent Nazi control of labour. It was not only in Marienthal but also in Weimar Germany that attendance at the social activities of the organised labour movement declined. The unemployed left the sports clubs and libraries.[109] In short, many of the consequences of unemployment recorded in Britain and Austria were also to be observed in Weimar Germany, despite a radical political culture. For underneath radical politics there persisted the divisive and stupefying effects of long-term unemployment. The quotation with which this paper began, therefore, is not without relevance to the history of German labour in the Depression. For as a newspaper report in Herne remarked, the consequences of unemployment, especially amongst the young, were often a loss of the sense of time, a retreat from social life and into one's self, a disinclination to overcome difficulties, and a lack of confidence.[110]

Notes

1. Quoted in Stephen Constantine, *Unemployment in Britain Between the Wars* (London, 1980), p. 97.

2. M. Jahoda, P.F. Lazasrsfeld and H. Zeisel, *Die Arbeitslosen von Marienthal* (Bonn, 1960; orig. 1933).

3. Ossip K. Flechtheim, *Die KPD in der Weimarer Republik* (Frankfurt, 1971), pp. 318–21; Beatrix Herlemann, *Kommunalpolitik der KPD im Ruhrgebiet 1924–1933* (Wuppertal, 1977), pp. 90–107; Sigfried Bahne, *Die KPD und das Ende der Weimarer Republik* (Frankfurt, 1976), p. 16.

4. Helga Reiners and Hermann Meyerhoff, *Auf dem Wege zur Grosstadt* (Herne, 1953), p. 37.

5. Herlemann, *Kommunalpolitik*, p. 176.

6. Ernst Thälmann, *Der revolutionäre Ausweg* (Berlin, 1932), p. 58.

7. *DMV Verbandstag . . . Dortmund 1932*, p. 132; Eberhand Heupel, *Reformismus und Krise* (Frankfurt, 1981), p. 80; Wolfgang Uellenberg, *Die Auseinandersetzungen der Jugendorganisationen mit dem Nationalsozialismus* (Cologne, 1981), p. 5.

8. Roter Frontkämpfer Bund (Gauführung Ruhrgebiet), *Faschismus im Ruhrgebiet* (Essen, 1927), p. 8.

9. *Rheinisch-Westfälische Zeitung*, 15 Apr. 1932.

10. Carl Severing, *Mein Lebensweg*, vol. 2 (Cologne, 1950), p. 357.

11. *Die Arbeit* (1930), 9, 658; Albin Gladen, *Geschichte der Sozialpolitik in Deutschland* (Wiesbaden, 1974), pp. 103–4.

12. Gerhard Uhlmann, *Der Kampf der Chemnitzer Werktätigen unter Führung der KPD gegen die Errichtung der faschistischen Diktatur* (PhD, Leipzig, 1966), pp. 65–6.

13. Conan Fischer, *Stormtroopers* (London, 1983); idem, 'The Occupational Background of the SA's Rank and File Membership' in Peter D. Stachura (ed.), *The Shaping of the Nazi State* (London, 1978); idem, 'Nazis and Communists: Class Enemies or Class Brothers?', *European History Quarterly* (1985) no. 3.

14. Johannes Volker Wagner, *Hakenkreuz Über Bochum* (Bochum, 1983), p. 87.

15. 'Niederschrift, Vollversammlung der Industrie- und Handelskammer Bochum 18 March 1929', p. 20; in Bergbauarchiv Bochum 32/3898.

16. For a summary of material on the relationship between workers and Nazis see Dick Geary, 'Nazis and Workers: a Response to Conan Fischer', *European History Quarterly* 15 (1985), no. 3, 453–64.

17. Ibid.

18. Ibid.

19. For a discussion of the specific problems of youth unemployment see

Chapter 7. See also Dick Geary, 'Jugend, Arbeitslosigkeit und politischer Radikalismus am Ende der Weimarer Republik', *Gewerkschaftliche Monatshefte* May (1983), 304–9.

20. Karl Heinz Jahnke *et al.*, *Geschichte der deutschen Arbeiterjugendbewegung* (Dortmund, 1973), p. 395.

21. Ross McKibbin, 'The Myth of the Unemployed', *Australian Journal of Politics* XV (1969), 2, 25–40.

22. Wagner, *Hakenkreuz*, pp. 57–9.

23. Herne Stadtarchiv V/38, 39, 42, 50, 63, 64, 66.

24. Thomas Childers, *The Nazi Voter* (Chapel Hill, 1983), esp. pp. 184–5, 243, 253, 256; Jürgen Falter, 'Wer verhalf der NSDAP zum Sieg', *Aus Politik und Zeitgeschichte* 14 July 1979, 3–21; idem, 'Wählerbewegungen zur NSDAP 1924–1933' in Otto Büsch (ed.), *Wählerbewegungen in der europäischen Geschichte* (Berlin, 1980), pp. 159–202.

25. Geary, 'Nazis and Workers'.

26. Dick Geary, 'The Failure of German Labour in the Weimar Republic' in Isidor Wallimann and Michael Dobkowski, *Toward the Holocaust* (Westport, Connecticut, 1983), pp. 177–96.

27. The fluidity of political allegiance in the early period is stressed in Dick Geary, 'Radicalism and the German Worker' in Richard J. Evans (ed.), *Politics and Society in Wilhelmine Germany* (London, 1979), pp. 267–86. The contrast between the broad base of radicalism in the early period and later developments is also a central theme in James Wickham, *The Working Class Movement in Frankfurt am Main during the Weimar Republic* (PhD, University of Sussex, 1979).

28. Quoted in Tom Bottomore and Patrick Goode, *Austro-Marxism* (Oxford, 1978), p. 221.

29. *Vorwärts* 26 Jan. 1933.

30. Künstler, quoted in Hagen Schulze (ed.), *Anpassung oder Widerstand* (Bonn, 1975), p. 52.

31. See above p. 103.

32. See note 3 above.

33. See reports of strikes and demonstrations in 1931 in Bergbauarchiv 32/4290; *Herner Anzeiger* 2/3/4/5/6 Jan. 1931. Communist strategy in the neighbourhoods has been discussed in several articles by Eve Rosenhaft, including Chapter 8 in this volume, and especially in her book, *Beating the Fascists*? (Cambridge, 1983).

34. *DMV Verbandstag . . . Bremen 1926*, p. 151f; *Verbandstag des Holzarbeiterverbandes . . . Stuttgart 1925*, p. 227 and 308; Gunna Stollberg, *Die Rationalisierungsdebatte 1908–1933* (Frankfurt, 1981); Eva Cornelia Schöck, *Arbeitslosigkeit und Rationalisierung* (Frankfurt, 1977).

35. For an account of this job-creation scheme and the SPD's reaction to it see Michael Schneider, *Das Arbeitsbeschaffungsprogramm des ADGB* (Bonn, 1975).

36. This calculation was made on the basis of SPD accounts of membership dues: *Betrieb und Gewerkschaft*, 15 Sep. 1929, p. 500.

37. Bahne, *Die KPD*, p. 15f; Richard N. Hunt, *German Social Democracy in the Weimar Republic* (Chicago, 1970), p. 133; W. L. Guttsmann, *The German Social Democratic Party* (London, 1981), p. 121ff; *Die Arbeit* (1930), 10, 654–9.

38. S. Miller and H. Potthoff, *Kleine Geschichte der SPD* (Bonn, 1983), p. 115.

39. Ibid., p. 120; Hunt, *German Social Democracy*, p. 107; Bahne, *Die KPD*, p. 15.

40. K. Rauschwalbe, *Geschichte der lippischen Sozialdemokratie* (Bielefeld, 1979), p. 232; Geary, 'Jugend', p. 1.

41. On the increasing division between the rough and the respectable working class see the articles in Richard J. Evans (ed.), *The German Working Class* (London, 1982), especially the article by Stephen Bajohr, 'Illegitimacy and the Working Class', pp. 142–73. Bullock suggests that few workers could afford to

move into the new council houses in Berlin, whilst in the case of Frankfurt James Wickham argues that council houses creamed off a section of the labour aristocracy from the older working class districts (I am indebted to Wickham's unpublished paper, 'Working-class Culture in the Weimar Republic' for this information). Similar points appear in Hermann Hipp, 'Wohnungen für Arbeiter' in Arno Herzig *et al.*, *Arbeiter in Hamburg* (Hamburg, 1983), pp. 471–81.

42. Wickham, *Frankfurt*, p. 181f; Dieter Langewiesche, 'Politik, Gesellschaft, Kultur', *Archiv für Sozialgeschichte* XII (1982), 390f.

43. See, for example, Michael Grüttner's account of pilfering in the Hamburg docks 'Working-class Crime' in Richard Evans (ed.), *German Working Class*, pp. 54–79.

44. See above p. 183.

45. Constantine, *Unemployment*, pp. 41–2.

46. Jürgen Reulecke, 'Bürgerliche Sozialreformer und Arbeiterjugend im Kaiserreich', *Archiv für Sozialgeschichte* XII (1982), 299–329; Klaus Tenfelde, 'Grossstadtjugend in Deutschland vor 1914', *Vierteljahresheft für Sozial- und Wirtschaftsgeschichte* 69 (1982), 2, 182–218. The number of indictable offences per 100,000 inhabitants of the age of criminal responsibility developed as follows:

1923	1693 (total)	1082 (youth)
1928	1188	536
1929	1191	517
1930	1187	566
1931	1125	561
1932	1125	623

(Source: D. Petzina *et al.*, *Sozialgeschichtliches Arbeitsbuch* vol. III (Munich, 1978), p. 137). For an account that is sceptical of the relationship between economic change and criminality in the Weimar Republic, see Eric A. Johnson, 'Socio-economic Aspects of the Delinquency Rate in Imperial Germany', *Journal of Social History* 13 (1980), 3, 384–402.

47. *Arbeiterwohlfahrt* (1932), 2, 60; ibid. 5, 133f; Verwaltungsbericht der Stadt Herne 1932/3, p. 37: Herne Stadtarchiv Bestand V.

48. Verwaltungsberichte der Stadt Herne 1929–33. It is true that the figure drops in 1932. However, contemporaries attributed this to the fact that people had simply ceased to report petty crimes: Verwaltungsbericht der Stadt Herne 1932/3, p. 37.

49. See Dick Geary, 'Welfare Legislation, Labour Law and Working-class Radicalism in the Weimar Republic' in Douglas Hay *et al.*, *Law, Labour and Crime in Historical Perspective* (forthcoming).

50. Ibid. Tests of need incensed some of the unemployed attempting to collect dole money, as did impolite and hostile treatment at the hands of the officials of the labour exchanges, and the requirement to perform various kinds of low or non-paid labour to receive benefits: *Die Arbeit* (1932), 7, 415–24; ibid. (1931) 5, 328; *Arbeiterwohlfahrt* (1933) 1, 2f; ibid. 2, 323.

51. Interestingly the 'social-fascist' line of the KPD only became widespread after May Day 1929, when the Social Democratic Police President, Zörgiebel, attempted first to ban, then to prevent a demonstration. In the subsequent fighting, which spread over several days, there were many fatalities.

52. Flechtheim, *Die KPD*, pp. 318–21; *Protokoll . . .4. Kongresses der Roten Gewerkschafts-Internationale . . . Moskau . . . 1928*, p. 97; Eve Rosenhaft, 'Working Class Life and Working Class Politics' in Richard Bessel and E. J. Feuchtwanger (eds), *Social Change and Political Development in the Weimar Republic* (London, 1981), pp. 207–40.

53. Bahne, *Die KPD*, pp. 24–6; Detlev Peukert, *Ruhrarbeiter gegen den Faschismus* (Frankfurt, 1976), pp. 17–19, 21 and 28; Henryk Skrzypczak,

'*Kanzlerwechsel und Einheitsfront*', *IWK* 18 (1982), 4, 482–99; Jahnke, *Geschichte*, p. 448; Reiner Tossstorff, 'Einheitsfront' in Wolfgang Luthardt (ed.), *Sozialdemokratische Arbeiterbewegung in der Weimarer Republik* (Frankfurt, 1978), pp. 209–10.

54. As note 52. Also *Kölnische Zeitung* 22 July 1932; *Rheinisch-Westfälische Zeitung* 19 July 1932.

55. Minutes of a meeting in the Reicharbeitsministerium 29.4.1932: in Bergbauarchiv 15/1081.

56. This was the case, for example, at the Howaldtwerke in Kiel and Hamburg: *Metallarbeiterzeitung* (1931), 49, 65 and 92.

57. As in the case of the Berlin metal-workers' strike in 1930: *Betrieb und Gewerkschaft* November (1930), 444.

58. Thälmann, *Der revolutionäre Ausweg*, p. 57; *Protokoll . . . Roten Gewerkschaftsinternationale 1928*, p. 323f.

59. Uhlmann, *Der Kampf*, p. 90.

60. Ludwig Eiber, *Arbeiter unter NS-Herrschaft* (Munich, 1979), p. 31.

61. Petzina, *Sozialgeschichtliches Arbeitsbuch*, p. 114.

62. Eiber, *Arbeiter*, p. 34; Uhlmann, *Der Kampf*, p. 84.

63. See, for example, a report of the VDAV on 22 Apr. 1931, *Bergbauarchiv* 15/184.

64. See, for example, *Protokoll . . . 13en Kongresses der Gerwerkschaften Deutschlands* (Berlin, 1928), p. 87.

65. Bergbau-Verein to Chancellor 4 Apr. 1932, p. 4f, in Bergbauarchiv 15/1081; *Wanne-Eickler Zeitung* 4 Feb. 1932; *Herner Anzeiger*, 5 and 12 Feb. 1932.

66. See above p. 266f.

67. *Rotengewerkschaftsinternationale 1928*, 323f.

68. *Metallarbeiterzeitung*, August 1931, 103.

69. Ibid., 95–101, 237–61.

70. Michael Zimmermann, '"Ein schwer zu bearbeitendes Pflaster"' in Detlev Peukert/Jürgen Reulecke, *Die Reihen fast geschlossen* (Wuppertal, 1981), pp. 70–1.

71. Andreas Dorpalen in *VfZg* 31 (1983), 1, 102.

72. Quoted in *Betrieb und Gewerkschaft* 15 July 1930.

73. Zimmermann, '"Einschwer"' pp. 71–2.

74. Freya Eisner, *Das Verhältnis der KPD zu den Gewerkschaften in der Weimarer Republik* (Frankfurt, 1976), pp. 188–9.

75. See the detailed breakdown by mining officials in StA Münster Bergamt Herne A8–141; and Bergbauarchiv 32/4290.

76. See note 32 above.

77. This emerges from that breakdown (note 74) and from the figures for 1931 cited below.

78. See note 32.

79. Ibid.; also note 51.

80. Police report of a meeting of KDP functionaries in Essen 26 Jan. 1932: StA Münster. Bestand Polizeipräsidium Bochum, Nachrichtenstelle 50.

81. As note 51; also note 79.

82. Bergbauarchiv 32/4290 (report from pit Wilhelmine Viktoria 1/4, 2 Jan. 1931); ibid. 4/620, report in the Communist newspaper *Ruhrecho* 14 Feb. 1930.

83. *O-Dienst Nachrichten* 16 Jan. 1931, 7, in Bergbauarchiv 32/4290.

84. *Herner Anzeiger* 6 Jan. 1931.

85. Many examples in Bergbauarchiv 32/4290. For the mass sackings of KPD members see *Herner Anzeiger* 21 July 1932; *Betrieb und Gewerkschaft* 15 Nov. 1929, 616; ibid., 30 Jan. 1931, 20; Dorpalen, *VfZq*, 86.

86. See the example of the Wilhelmine Viktoria pit, 20 Jan. 1931, in Bergbauarchiv 32/4290. In this case even the management reported that the Communist official concerned dared not risk strike action, as he had a wife and several children to support.

87. Josef Wietz, *KPD-Politik in der Krise* (Frankfurt, 1976), p. 436 and pp. 553–5.

88. Ibid., p. 555.

89. *Die Arbeit* (1932) 9, 310; Ludwig Preller, *Sozialpolitik in der Weimarer Republik* (Düsseldorf, 1978), p. 165f; Frank Niess, *Geschichte der Arbeitslosigkeit* (Cologne, 1979), p. 40; Helmut Drüke *et al.*, *Spaltung der Arbeiterklasse und Faschismus* (Hamburg, 1980), p. 95.

90. Preller, *Sozialpolitik*, p. 165f; *Protokoll des 15. ausserordentlichen Kongresses der Gewerkschaften Deutschlands* (Berlin, 1932), p. 26.

91. Letter of the Berg- und Hüttenmännischer Verein zu Wetzlar to Fachgruppe Bergbau 21 Nov. 1931, in Bergbauarchiv 15/137.

92. Report of the Reichskohlenrat 30 May 1929, in Bergbauarchiv 32/3861.

93. Stimmungsbericht of Oberbergamt Dortmund 14 Jan. 1929: StA Münster Oberbergamt Dortmund 1871.

94. Stimmungsbericht of Bergamt Herne 23 Dec. 1930: StA Münster Bergamt Herne A8–141.

95. Bergbauarchiv 15/137.

96. *Bergbauindustrie* 16 Apr. 1932.

97. Committee for Trade and Industry of the Prussian *Landtag* 8 June 1932, in Bergbauarchiv 15/137.

98. Quoted in Zimmermann, "'Ein schwer'" p. 118f.

99. *Deutsche Bergwerkszeitung* 15 Jan. 1933.

100. Cases of 21 June 1930 and 6 Aug. 1930, in Bergbauarchiv 15/142.

101. See the many cases reported in Bergbauarchiv 15/142.

102. As in a case that came before the Landesarbeitsgericht in Essen 3 Oct. 1928, in Bergbauarchiv 15/140.

103. See Chapter 4.

104. *Die Arbeit* (1931), 5, 328.

105. Zimmermann, "'Ein schwer'" pp. 71–2.

106. Dorpalen, *DfZg*, 87.

107. See above p. 182ff.

108. Jahnke, *Geschichte*, p. 395.

109. Wickham's dissertation makes this pointed contrast between the vitality of socialist cultural organisations in the earlier period and their decline in the Depression.

110. *Herner Anzeiger* 5 Aug. 1932.

11 THE THIRD REICH AND THE UNEMPLOYED: National Socialist Work-creation Schemes in Hamburg 1933–4

Birgit Wulff

I

The work-creation schemes undertaken during the Third Reich are still frequently regarded as models of progressive policy which caused a decisive recovery of the German economy. Yet in reality this view is only one of many legends still current in the discussion of the Nazi regime. When the Hitler government came into office at the beginning of 1933, it was able to make use of a whole range of already existing official plans for reducing unemployment, and indeed it could call on a good deal of practical experience within the state apparatus in their practical application. It did, of course, set new quantitative standards in the work-creation area by making the immense sum of one billion Reichsmarks available for public works. Moreover, we should not underestimate the psychological effects of the Nazis' work-creation policy on the attitudes of German investors. The Nazis took the question of overcoming the Depression out of the narrow confines of economic policy and created through their intensive propaganda campaigns an atmosphere of trust that helped the recovery considerably. We have to ask, therefore, whether the economic up-turn already in evidence at the end of 1932 might have occurred with the same intensity even without the policies of the Nazi regime from January 1933 onwards.

Work-creation schemes had already been carried out at a local level at the beginning of the 1920s. They began with the un-employed welfare system set up in 1920. After the founding of a central authority in 1927 (the *Reichsanstalt für Arbeitsvermittlung und Arbeitslosenversicherung*), these measures were also financed from central government funds and from unemployment insurance payments. During the Depression, labour-market policy became a central problem of economic and financial policy for every German government. But the Weimar parties were too reluctant

281

to abandon their belief in the self-healing properties of the economy to engage immediately in an active work-creation policy financed by deficit spending. During Brüning's Chancellorship the discussion of such a programme intensified as opposition to the government's deflationary policies grew. Reform-oriented economic theorists, trade unionists and central-government civil servants demanded state intervention in the economy and criticised the passivity of the government. They drew up plans for an active programme of recovery and provided details of how it might be financed. Much of this work provided a basis for subsequent work-creation schemes. Although they never got beyond the planning stage under Brüning, they were in fact put into effect under Papen's government, which provided 302 million RM for road building, agricultural improvement and suburban housing estates. But the scheme's main emphasis was on a system of tax vouchers, which could be used not only to secure tax rebates but also as compensation for providing the jobless with employment. It was Schleicher who first made help for the hard-hit provinces and communes a high priority with his work-creation programme and his nomination of Gereke as Commissioner for Work Creation (*Arbeitsbeschaffungskommissar*). Five hundred million RM were made available to the provinces and localities for their own projects under Gereke's work creation scheme.

The Hitler government was able to build on these existing programmes in two ways. It could put their well-oiled machinery for formulating and effecting work-creation schemes to practical use, and it could present the resulting decline in unemployment as all its own work. When they came into office the Nazis had no plans of their own for combating unemployment. Even in the March elections of 1933 Hitler avoided committing himself to a particular economic programme. His desire to leave his options open in this respect was particularly obvious in his public speeches, in which he spoke only in the vaguest terms of (for example) 'the rescue of the German worker by a massive and comprehensive assault on unemployment'.[1] For tactical political reasons he left the details of this 'assault' entirely vague. Only after the March elections were plans for work creation actually drawn up in the Ministries of Labour and Finance, and they were discussed in the Cabinet only after the destruction of the trade unions, the dissolution of the political parties, the dismantling of employees' rights, the emasculation of the Reichstag and the central take-over of the provinces. The Nazis needed the monopoly of political power which these measures gave them before they could put their economic policies into effect, for these involved from the very beginning the linkage of work creation with rearmament: the view that these policies took place in two stages — first work creation, then rearmament — is untenable. A final important precondition for the realisation of work-creation plans was the replacement of Luther by Schacht as head of the Reichsbank, which provided the guarantee that the bank would take a flexible line in monetary policy.

The First Law on the Reduction of Unemployment was not in fact published until 1 June 1933.[2] The Nazis allowed four months for carrying out the 'Immediate Programme' of the Schleicher government, and they increased its funding by 100 million RM. However, this increased financial support did nothing to help the suffering communes and municipalities, because it all went towards projects at the national level, among which rearmament was pre-eminent.[3] Together with these measures, the Second Law on the Reduction of Unemployment, passed on 21 September 1933, the Law on the Autobahns (27 June 1933) and the work-creation schemes of the railways and the post office constituted the effective core of the Nazi work-creation programme, which was completed by a variety of state inducements to investors and measures taken to reduce the labour supply. State work-creation schemes are usually held to achieve their effects by an initial 'pump-priming' in the form of direct measures such as road construction, agricultural improvement, flood protection and other public works designed to improve the infrastructure, and a range of indirect measures to encourage investment, such as tax reductions and allowances, state loans, and the provision of a political and psychological context in which capital will be willing to invest. The idea is not simply to create temporary employment to alleviate hardship, but to stimulate the economy so that as the 'pump-priming' measures fall away, economic recovery will lead to the re-absorption of the unemployed into the labour force of its own accord.

The Schleicher 'Immediate Programme', now backed by 600 million RM, of which about 400 million went to the provinces and communes in the form of work-creation loans, involved only direct measures, above all road construction, agricultural improvement, the extension and improvement of municipal works and public utilities.[4] The so-called First Reinhardt Programme[5] provided a billion RM for repairs and extensions to provincial and communal buildings, for the renovation, division and conversion of houses and farms, for suburban housing estates, for flood-control schemes on rivers, for civil-engineering projects and so on. The Law on the Reduction of Unemployment, of which the Reinhardt Programme was a part, contained among other things measures to encourage marriages and the return of women to the home, and a fund for 'national' work, or projects other than those explicitly included in the work creation scheme. On 21 September 1933 the Second Law on the Reduction of Unemployment, also referred to as the Second Reinhardt Programme, came into effect. It sought above all to encourage private building and construction work by indirect means to counter the coming seasonal rise in unemployment during the winter.

The work-creation projects proposed by the provinces and communes were required not only to correspond to the measures proposed in the overall programme but also had to be 'valuable' in themselves. They were also required to be supplementary to normal work, that is, the provinces and communes had to prove that they could not have undertaken them without the work-creation scheme. The scheme was financed in three main ways. First, through the so-called preparatory financing on the basis of bills of exchange provided for the Papen, Schleicher and First Reinhardt Programme, i.e. by central state guarantees, by tax vouchers, and by treasury work bonds respectively. The central government promised to redeem these bills of exchange, which were eligible for rediscount, within a fixed period. Secondly, provision was made for the schemes from the state budget, above all via the central authority (*Reichsanstalt*) founded in 1927. Finally, and especially in the Second Reinhardt Programme, state credits were guaranteed only when the borrowers provided some of the funding themselves.[6] Up to the end of 1934 the total expenditure on work creation in the broadest sense, including marriage loans, the costs of the voluntary-labour project, the post office and railway employment scheme, the autobahns and so on, came to around 5 billion RM.[7]

II

In order to judge the effectiveness of these schemes, it is best to turn to a detailed examination of the situation at local level. The example chosen for the present study is the seaport and federal state (*Land*) of Hamburg, Germany's second-largest city. If by the end of January 1933 the Depression was past its worst in Germany as a whole, this was not yet apparent in Hamburg. The seaport's dependence on the world economy, which was still deep in the Depression, prevented it benefiting from the recovery now taking place inland. Shipping, trade and shipbuilding dominated Hamburg's economy, and a large part of the city's other industries were dependent on these branches since they were based on the processing of imported foreign raw materials or of German materials for export. Shipbuilding, shipping, trade and other industries dependent on the world market were the life-blood of Hamburg. They employed around 70 per cent of the people who worked in the city. These sectors of the economy were all disproportionately affected by the collapse of international trade caused by the world economic Depression. At the beginning of 1933 there were 145 German and foreign ships with 568,693 tonnes capacity[8] lying empty in the city's harbour. Altogether 303 German ships with 917,818 tonnes capacity were affected by the crisis, or 26.5 per cent of the total tonnage of the German merchant navy.[9] This over-capacity obviously affected shipbuilding as well. In 1932 only 15 ships were launched in Germany, with 80,799 tonnes capacity, the lowest figure since 1893.[10] The turnover of seaborne goods in Hamburg's harbour fell from 28.6 million tonnes in 1929 to 19.6 million in 1933.[11]

At the end of January 1933 there were 145,509 unemployed or 167,207 people seeking jobs in Hamburg's urban territory, which had a total population of 1.2 million.[12] The unemployment-rate (measured as the percentage of unemployed among occupied persons, excluding the self-employed) thus stood at about 30 per cent in Hamburg compared with 22 per cent in the Reich as a whole. If one takes the whole area dependent on the Hamburg labour market, including Altona, Harburg-Wilhelmsburg and Wandsbek, the number of unemployed in March 1933 reached 206,208.[13] Hamburg had a particularly high proportion of long-term unemployed (the so-called 'welfare unemployed'). The welfare unemployed were supported by the municipal and communal authorities once the period in which they were entitled to claim unemployment and crisis benefit from the *Reichsanstalt* had expired.[14] Sixty-six per cent of the unemployed in Hamburg's urban territory fell into this category at the end of February 1933, a higher proportion than in any comparable large German town. (In comparison to the other federal provinces, Hamburg came fourth after Prussia, Bavaria and Saxony.) The largest groups among the 167,207 people looking for jobs in Hamburg at the end of 1933 were those in industries dependent on the world market:[15] 34,024 in transport, 31,273 unskilled labourers, 20,365 in the metal industry, 15,387 salaried commercial and office employees, 10,748 building workers. 'Transport', which included regular dockers as well as drivers and others, accounted for 20 per cent of the unemployed in Hamburg compared with a Reich average of 6 per cent. A large proportion of the unskilled workers would also have found employment as casual labourers in the docks when times were good. However, as trade collapsed, so the average number of dockers employed on the state quays fell from 5,947 a day in 1929 to 1,710 a day in 1932.[16] The shipbuilding workers who had lost their jobs because of the collapse of their industry, provided a major part of the unemployed in the metal industry. The great Hamburg shipbuilding firm of Blohm & Voss cut its workforce from 10,553 at the end of 1929 to 2,449 at the end of 1932.[17] Unemployment among white-collar workers was a further reflection of the world trade crisis. Workers in shipping and trade also predominated among the long-term unemployed.

The financial position of the jobless can be gauged from the fact that an unemployed person without a family but with an independent household received 5.10 RM per week in unemployment benefit at the end of 1932, while the standard rate of public poor relief, also by no means generous, stood at 9 RM a week at the same time.[18] The standard rate of the public welfare for a family of 2 adults and 5 children stood at 132 RM a month at the beginning of 1933, while the basic cost of living for such a family was officially calculated at the same time as 153.61 RM a month.[19] These figures explain why many of those who received unemployment and crisis benefit — in March 1933 almost 30 per cent — were at the same time receiving support from the welfare authorities. Because benefits were below subsistence level, the unemployed were to a large extent dependent on welfare payments or black market labour, concealed from the authorities, for the basic necessities of life. Rent alone often accounted for a third to a half of the benefit payments, so that recipients were forced to cut back drastically on expenditure on food, clothing, heating and so on. Malnutrition, physical weakness and susceptibility to disease were the physical consequences of long-term unemployment. The medical authorities began to register an increase in the frequency of rickets and tuberculosis.[20] Welfare payments which were borne by the municipalities, placed an increasing burden on the finances of Hamburg's government, the Senate. In 1932 they accounted for some 30 per cent of the city's expenditure.[21] Against this background of the increasing impoverishment of broad strata of the population and in view of the growing problems of the city finances, it is not surprising that the new Nazi Senate which took over the reins of power in Hamburg early in 1933 expected effective assistance from the work-creation schemes soon announced by the national government in Berlin.

Hamburg received loans from the central government corresponding to its unemployment rate, and in addition was granted some 6.3 million RM up to the end of the financial year 1934 for the Immediate Programme originally conceived under Schleicher.[22] The Senate had submitted a list of suggestions for urgent projects costing about 18.9 million RM, including building schemes (especially schools and hospitals) costing 3.9 million, steet, bridge and sewer renovation costing 5.4 million and repairs and extensions to the harbour costing 8.5 million.[23] However, the Reich authorities approved only a portion of the street, bridge and harbour construction and rejected the rest. As part of the first Reinhardt Programme the Hamburg Senate was loaned 1.37 million RM up to 31 March 1935.[24] A further 5.4 million went to the Hamburg Mortgage Loan Fund (*Hamburgische Beleihungskasse für Hypotheken*)[25] and was placed at the disposal of houseowners. Even when taken together, these sums were quite insufficient to support any really major projects. They were used mainly for repair and renovation, particularly in the harbour, where the budgetary deficits of the preceding years had caused the postponement of important reconstruction work. A list of projects that were to be undertaken in the following winter, should the Reinhardt Programme be extended, contained schemes with a total estimated cost in excess of 85.6 million RM.[26] It is not possible to specify exactly what sums were transferred to Hamburg under the Second Reinhardt Programme, but this programme was mainly devoted to the support of private building projects, and a rough comparison between applications and grants shows that the Hamburg authorities' wishes were not met here either.

The resources of the first Reinhardt Programme were already exhausted by the end of 1933. The Reich authorities refused to approve any new projects, indeed they could not even guarantee the remaining financial support for projects already begun. Extensions of the deadline for the completion of such projects were not accompanied by the granting of more funds, but took place because there was simply not enough money to complete them. The so-called 'battle of labour' (*Arbeitsschlacht*), which had been announced with a huge propaganda campaign as the main weapon of the two Reinhardt Programmes, had also run out of money by the end of 1933. The Reich Ministry of Labour rejected further applications for support and informed the Hamburg authorities that although job creation was to continue in a sense as part of the planned rearmament measures, it was impossible 'for obvious reasons' to make any propaganda capital out of this fact.[27] It was not only lack of money that prevented all the projected work-creation projects in Hamburg from being carried out, and thus hindered the reduction of the city's unemployment figures. The implementation clauses of the laws to reduce unemployment also restricted the ways in which the loans offered could be used. These clauses created particular problems for Hamburg. The loans were handled by the Society for Public Works Ltd (*Gesellschaft für öffentliche Arbeiten* AG) among other credit institutions, and the Society required half the money to be spent on renovation and extension works and a quarter each on municipal services and civil engineering.[28] These requirements might have been appropriate for some places, but for Hamburg they were not. Although civil-engineering projects were favoured by the Reich because they were labour-intensive and required little capital investment, the possibilities for carrying them out were very limited in the heavily built-up area of the city-state. The Hamburg Senate wanted to carry out repairs and renovations to its valuable public building stock, after having had to put them off for years because of budgetary deficits. The rules, however, forbade the loans being used for such projects on grounds of cost. Yet at this time there were 15,000 unemployed building construction workers in Hamburg and only 4,000 or 5,000 unemployed civil-engineering workers in the city.[29] Moreover, building work would have suited the structure of the Hamburg construction industry better, because it could be carried out by small firms of the kind that predominated in the city, while civil-engineering projects generally needed to be carried out by large organisations. Despite all this, however, the Society for Public Works turned down the Senate's request for increasing the sums available for building-renovation work, just as it turned down all the other amendments to the scheme requested by the provinces.

III

In December 1934 the unemployment rate in Hamburg still stood at 17.5 per cent.[30] The number of registered unemployed fell from 145,509 in January 1933 to 129,659 in January 1934 and 98,223 by December 1934, a fall of 32.5 per cent overall. The number of those seeking employment fell over the same period from 167,207 to 151,622 and finally 111,872, a fall of 33.1 per cent. As for the long-term jobless, the welfare unemployed, their number fell from 97,372 in January 1933 to 80,009 in January 1934 and 56,437 in December 1934, a fall of 42.04 per cent overall.[31] As these figures indicate, Hamburg lagged behind the general development in the Reich as a whole, where the number of unemployed had fallen by 56.7 per cent over the same period. On 30 September 1934 Hamburg's unemployment rate was the highest of any comparable large city, higher even than Berlin and Leipzig, where it was also above average. From January 1933 to December 1934 the average fall in the number of welfare unemployed in cities with over half a million inhabitants was 60.4 per cent, while in Hamburg it was only 42.4 per cent.[32] Still, it remains clear that the number of welfare unemployed fell faster than the number of unemployed in general. This was partly because they were given preferential treatment by the work-creation schemes, but it was also because the basis for calculating the welfare-unemployment statistics was changed. Thus, for example, people temporarily employed on relief work, young people sent to work on the land, and unemployed persons enrolled in the Voluntary Labour Service were no longer included in the figures. However, the decisive change lay in the fact that the statistics no longer took information supplied by the welfare authorities into account but were now based exclusively on the figures provided by the labour exchanges.

In Hamburg these two sets of figures diverged from one another quite substantially. For example, the labour exchange registered 54,079 welfare unemployed on 31 March 1934, while the welfare office registered no fewer than 59,528.[33]

Unemployment also fell unevenly across different parts of the job spectrum. In Hamburg the number of unemployed in the transport sector fell by 57.2 per cent between January 1933 and December 1934, in the metal industry it fell by 50.4 per cent and in the building trade by 45.6 per cent. However, among unskilled labourers in the city the unemployment-rate fell by only 6.4 per cent, and among salaried commercial and office employees the fall, at 2.4 per cent, was even less.[34] An actual increase in unemployment in the trade and shipping sector, which was particularly affected by Nazi 'autarky' policies, was avoided only by slowing the pace of work and by job-sharing. In September 1933 the Chamber of Commerce issued a call for the strict observation of the 40-hour week. Eight hundred firms, including above all those suffering from the system of quotas for raw materials and foreign exchange, followed this advice and reduced working hours for their employees.[35] In the absence of any compensatory average rise in working time the reduction from 48 to 40 hours a week in these businesses meant a cut in income of 16 per cent for their employees. As far as the white-collar workers were concerned, the labour market situation of male employees was improved by the introduction of interest-free marriage loans for working women. The terms of the loans required women to leave their job and not to take up another position until they had paid back the whole sum, which could be as much as 1,000 RM. These state inducements were backed up by cash payments from private firms. Thus, the Reemtsma cigarette company paid its female employees a reward of 600 RM if they gave up their jobs on marriage.[36] The sharp decline of unemployment in the metal industry was due on the other hand to arms contracts for the Hamburg shipyards, and soon enough a shortage of skilled labour in some departments of the shipbuilding firms was becoming noticeable.

Those who profited directly from work-creation programmes included above all building workers and lorry drivers, who could find work in the state construction schemes without difficulty. Unskilled workers obtained employment mainly on street repair and civil-engineering projects, but the latter were only possible to a limited extent in Hamburg, as we have seen. Detailed figures for the number of unemployed people directly provided with jobs by the work-creation programme are only available for the principal recipients of benefits (*Hauptunterstützungsempfänger*) paid by the labour office. According to these figures the provision of relief work[37] reached its peak as far as 1933 was concerned in October, when 845 unemployed people (1.2 per cent) were occupied in this manner. The number increased to a maximum of 2,933 in 1934 (3.8 per cent). In 1934 as a whole, some 14,443 unemployed people in this category were provided with relief work, or 13.3 per cent of the unemployed in Hamburg: in the Reich as a whole the percentage was 56.6 per cent.[38] The difference between the two figures indicates the lack of success of the work-creation schemes in the seaport. That the fault lay with the kind of work that had to be provided is suggested by the fact that such massive dyke construction and agricultural improvement schemes were possible under the job-creation programme in Schleswig-Holstein, that those who found work in the huge earthworks which the schemes involved included a number of the unemployed from Hamburg.[39] In some parts of Schleswig-Holstein and Mecklenburg the unemployment-rate fell by 83.3 per cent to 92.7 per cent between the beginning of 1933 and the beginning of 1935; the average decline for the Reich as a whole was 71.5 per cent and for Hamburg only 47.9 per cent.[40] Relief-work schemes outside the city became increasingly important for Hamburg's unemployed as time went on, since the possibilities for providing them with jobs inside the city boundaries were soon exhausted. For example, in June 1937 the number of Hamburg workers employed in relief-work projects outside the city was 2,300; inside the city the figure was a mere 468.[41] Against the background of this limited effect of work-creation schemes on the Hamburg labour market, a ban was placed on immigration to Hamburg from other parts of Germany at the end of 1934. The effect of this measure was little more than psychological, however, since immigration to Hamburg had declined more sharply since 1932 than to any other German city.[42]

The lack of success of the programme to reduce Hamburg's unemployment can also be read from the labour exchange's reports. In general, the labour exchange was able to provide only short-term employment for the jobless: permanent jobs predominated only in August–October 1933.[43] These months saw the campaign against married women 'double earners' and the measures to encourage single women to get married and leave their jobs. Since the number of jobs on offer remained constant throughout these three months, it seems reasonable to conclude that the predominance of permanent jobs among those provided by the labour office in this period was mainly a consequence of the removal of women from such posts, as well as of various work-sharing arrangements such as the introduction of the 40-hour week.[44] The fact that young people were going into the Voluntary Labour Service and the 'Work Comradeships' (*Arbeitskamerad-schaften*)[45] also facilitated the employment of older jobless men. Unmarried men and young people under 25 were massively pressured to enter the Voluntary Labour Service or to accept an 'auxiliary placement' on the land. The welfare offices threatened them with a tougher examination of their eligibility or in some cases even with forced labour or incarceration if they refused.[46] The Senate tried to counter the dangerous political consequences that continued economic stagnation threatened to bring with it by adopting precisely aimed measures to reduce the possibility of unrest and integrate the workers into the 'national community'. One such measure for example was the creation of a 'Harbour Assistance Agency' at the end of 1934. The intention of this institution was to guarantee assistance to all permanently or casually employed dockers living in the Hamburg economic region: because of the decline in trade, hours of work had been drastically reduced and the possibilities of earning had been considerably diminished. From its introduction on 2 November 1934 up to the end of the year, the Agency issued 9,216 assistance payments totalling 54,000 RM. The Agency continued to expand the scope of its operations and in 1935 it was issuing up to 15,000 payments a month,[47] financed mostly from central government, with the rest coming from the welfare offices situated in the Hamburg region.

Finally, the relative failure of work-creation schemes in Hamburg can also be gauged from the fact that they led to no observable increase in purchasing power, even among those who obtained employment in this way. The people who found work under the Immediate Programme, and the Reinhardt Programmes were not allowed to work more than 40 hours a week and were paid according to customary local wage-scales for unskilled labourers.[48] Many of those employed on relief work schemes were thus earning little more than was provided by welfare support: the wages of men with families indeed often fell below the benefit rate. The figures for July 1933 are shown in Table 11.1.

Table 11.1 Unskilled Weekly Wages and Benefits in Hamburg, July 1933[49]

	Weekly wage (0.70 RM hourly)		Weekly welfare benefits
	Gross	Net	
Unmarried	28.00	23.70	11.82
Married	28.40	25.10	18.12
with one child	28.80	25.40	22.83
with two children	29.20	25.80	25.23
with three children	29.60	25.80	27.63
with four children	30.00	26.20	30.03

The situation for those employed in relief-work schemes outside Hamburg, where wages were customarily lower, was even more precarious. Moreover, those who worked too far from the city to be able to go home every day also had to bear the cost of keeping two homes going at the same time. They thus still relied on income from the welfare services, and complained about their continued reliance on assistance despite the fact that they were now employed. Finally, in addition, it should be borne in mind that between January 1933 and December 1934, at a time when wages were fixed and hours of work restricted, the cost of living rose by some 6 per cent.[50]

The work-creation schemes thus gave rise to a stratum of 'second-class workers' who were unable to support their families despite working a 40-hour week. Because the wages of these formerly unemployed people were barely above welfare support level, or even below it, the increase in purchasing power that would have been an important stimulant of economic recovery failed to materialise. The job-creation programmes had no real effect in increasing long-term employment because the continued stagnation of the economy prevented employment for more than the maximum of 26 weeks provided for. These programmes thus had no more effect in Hamburg than those of earlier years, which had been merely therapeutic. The distinction between relief work and work creation thus virtually disappeared since the programmes brought no longer-term possibilities of employment in the city.[51] As a consequence, the savings in welfare benefit payments which the Senate hoped to achieve by these programmes fell far short of expectations, and Hamburg became the only state in the Reich which was unable to meet its credit repayments. For although expenditure on welfare benefits fell by 18 per cent in

1933–4, the central government in Berlin cut its support grant for these benefits by 38 per cent over the same period.[52] In addition, Hamburg also had to meet out of its own budget the costs of amortisation and interests on these loans plus the subsequent costs arising out of the projects begun under the work-creation schemes. Hamburg's requests for an extension of the repayment time were unsuccessful, and the granting of new loans was made dependent on the city's meeting its current obligations. This created a vicious circle from which the Senate could only be rescued by action from the Reich. But the Reich had its own financial difficulties and declared that it was not in a position to do anything. So the Senate was compelled to make swingeing cuts in its budget in order to carry out further work-creation projects, and it made them among other things by freezing appointments and declaring redundancies in the public services.[53] On the other hand, the Reich Savings Commissioner's suggestion that the basic rate of welfare benefits be cut by 20 per cent and that welfare payments in kind be reduced was not put into effect.[54] These suggestions had been justified by the claim that Hamburg had a 'more than averagely expensive standard of living' which had led to the city having the highest benefit-rates in the country alongside Stuttgart. The Reich Governor (*Reichsstatthalter*) of Hamburg, Kaufmann, protested very strongly against these proposals because of the political effect he feared they would have. They were only introduced in 1938 as part of a general restructuring of the unemployment insurance system, when the labour market had become much easier.

IV

Kaufmann's protests were part of a series of attempts on Hamburg's part to convince the Reich authorities of the desolate situation of the local economy and the limited ability of the work-creation schemes to improve it. Kaufmann also complained to the Reich Finance Minister in August 1934 that the difficult position of Hamburg's economy was being ignored in Berlin. Hamburg, he declared, was still a deprived area and had not participated in the general economic recovery of the Reich as a whole. The only way to help the city was to stimulate the trade that was its life-blood. The work-creation schemes could not function as pump-priming for Hamburg's economy because they were specially designed for the economic structure of areas further inland.[55] Kaufmann noted that a recent referendum[56] had produced a higher proportion of 'no' votes in Hamburg than anywhere else, and pointed to the political danger threatened by the continued high unemployment in the city. The situation of the workforce, he said, was 'in the highest degree unfavourable' and 'everything' must be done in order to improve it. This view was shared by local businessmen, whose 'liberal attitudes' were frequently registered in the situation reports of the secret police.[57] They thought that import and export trade should be stimulated and complementary industries brought to the city in order to revive the economy.[58] The Senate also shared these views. Yet they were difficult to put into effect. Industrial overheads (wages, rent etc.) in Hamburg were far above the Reich average, and strategic considerations on the part of the Reich authorities also stood in the way of attracting industries to the city. Thus, the Reich Ministry of Air virtually forbade new buildings in the city by describing them to enquiring firms as undesirable because Hamburg was already an intensely built-up area and this would pose problems in the event of an air attack.[59]

The Senate also tried to get lucrative armaments contracts for Hamburg firms by taking them before the *Reichsausgleichstelle*, the authority responsible for orders for the Party, the Labour Front, the Labour Service, the Reich Naval Ordnance Department and the Army Weapons Office.[60] Yet it usually only succeeded in obtaining large orders for the shipyards: the small, undercapitalised specialist producers, who were such an important part of the Hamburg economy and who were not always able to raise the necessary investment capital, were seldom considered. Trade, the life-blood of the local economy, far from being stimulated by the economic policies of the Reich government, was further damaged by the restrictions placed on imports and exports.[61] The position of the German seaports had worsened considerably since January 1933, when Hitler became Chancellor.[62] Not only were Nazi policies increasingly oriented towards the internal economy, with all the implications that this carried with it, but external trade was also increasingly oriented towards the southeast European states. These new trading links hardly affected Hamburg and Bremen, which were situated in the northwest. The boycott actions of some Western industrial states in reaction to the German persecution of the Jews can also be considered to have had an effect. The necessary restructuring of

Hamburg's economy and its reorientation towards the internal German market could be achieved only with massive protection from the Reich government. In this situation the Senate and Reich Governor Kaufmann turned directly to the government, which met on 2 November 1934 to discuss the city's problems.[63]

Right at the beginning of the meeting Hitler gave Kaufmann the floor. The Reich Governor emphasised above all the exposed position of the city in relations with foreign countries. This meant that foreigners who came to Hamburg immediately concluded that the city's economic situation was representative of that of the Reich as a whole. It also meant that Hamburg was especially exposed to foreign propaganda, and its population was thus permanently at risk. If the economic situation did not change, they had to reckon before long with a sharp contraction of import and export industries and thus of shipping. The only substitute for foreign trade was the relocation of new industries to Hamburg. Yet this was difficult because of the city's high wage-levels, and there was no question at the moment of lowering wages, since the cost of living was continually on the increase. As far as the political situation in Hamburg was concerned. Kaufmann reported that the desolate state of the labour market and the poor economic position of trade was causing disaffection among the city's merchants, industrialists and financiers and was leading them to reject government policy. Burgomaster Krogmann took a similar line when it was his turn to address the meeting. Foreign trade had declined by a third since January 1933, he said, and this was not least because of the economic policies pursued by the Reich. The competitive position of Hamburg in relation to Rotterdam and Antwerp was worsening all the time. Hamburg must be given more armaments contracts as compensation for the decline in foreign trade. As far as the work-creation measures and the general financial situation was concerned, the Burgomaster remarked that grand projects such as the extension of the underground railway system or the construction of a sewage-treatment works could not be undertaken because the Society for Public Works had opposed the necessary funding. The Reich government was refusing credits and had told the Senate to take care of the matter itself while also insisting on the fulfilment of obligations already undertaken. This however could no longer be done. More than a third of the annual budget of the city-state was taken up with welfare payments, and this could be reduced only if unemployment was reduced. Only then, concluded the Burgomaster, could the Senate begin to put the city's finances in order.

In his reply, Hitler underlined the political danger to Hamburg from the influence of foreign propaganda. He understood the city's wish for an equivalent to replace its stagnant foreign trade. Military experts had told him there were no strategic objections to the relocation of new industries into Hamburg. Hitler then spoke strongly in favour of the modernisation of the merchant navy, which, he said, had to remain competitive in relation to foreign fleets. However, he declared the Senate had to come to terms with the fact that the economic conditions under which Hamburg had become a city of millions were no longer operative. He suggested that more state contracts should be given to Hamburg in order to solve the problem of the high wage-rates in the city, for a wage cut was politically unthinkable. The Reich Finance Minister added a few words on the problem of the work-creation loans. Not only Hamburg, he said, but also the Reich, was going round in circles on this issue. As soon as he tried to plug one gap in the Reich budget, another one immediately opened up elsewhere. So he had no positive proposals to make for the necessary ordering of the city's finances. At this point, Hitler intervened once more:

> I have an idea that it might be better to give a decisive stimulus to one sector of Hamburg's economy through really large-scale measures, in order to pull others along with it. If we, for example, put a sum, let us say 30–50 million, into the modernization of the harbour pool, so as to make it as competitive as possible, do you not think that other branches of the economy would be carried along with it? We must seriously consider whether it is better to make individual smaller contributions for various purposes at various points, or to invest really big sums in Hamburg's own natural basis of life, thus for example in shipping.

With the instruction that these problems should be discussed in detailed sessions, Hitler then closed the meeting.

With his 'idea', Hitler had fundamentally contradicted the existing conception of work-creation programmes. In opposition to the 'watering-can principle' of widely distributed measures of equal volume and importance, which had characterised the basic conception of the schemes so far undertaken, he now pleaded — appropriately for the Hamburg situation — for a regionally concentrated support according to the specific demands of the local situation. But this 'idea' did not in fact lead to any restructuring of the support system. No new work-creation schemes were announced in 1934. For one thing, there was work available in existing schemes still in operation. For another, the replacement of Hugenberg by Schmitt in the Reich Economics Ministry led to an increased preference for indirect measures. In view of the continuing high unemployment, however, the Reich government felt compelled to continue the work-creation schemes through 1935. However, the money it offered went only to the regions specifically hard-hit by unemployment, the so-called special relief areas. The support consisted of loans and grants from the Reichsanstalt. This did at least give the affected regions a stronger say in how the money was used, but it did not conform to Hitler's idea of a massive shot in the arm for one particular sector.

Hamburg counted officially as a special relief area until 1939, and continued to get regular support loans and grants until the beginning of 1938.[64] Yet it was not the continuation of the work-creation schemes but above all the increase in armaments production that led to a steady decline in the city's unemployment, even though trade continued to stagnate and currency and raw material supplies became ever more problematical. It was armaments contracts, above all for warships, that made the shipbuilding sector the source of the long-awaited stimulating effect on the other sectors of the local economy. The other side of this recovery lay in interventions in the structure of unemployment insurance and a sharper operation of the eligibility criteria. These led again to a drastic reduction in the unemployment figures. The Reich authorities wanted to reduce the social-welfare budget at all levels and to alleviate the acute labour shortage in some branches of the armaments industry. So it pressed for a rigorous 'combing-out' of the unemployment statistics. So-called 'life histories' of unemployed individuals were compiled, which provided the basis for deciding whether they were 'usable' members of the 'national community' and if they could be described as eugenically 'valuable'. Compulsory and welfare labour schemes were no longer urged to alleviate the physical and psychological effects of a long period of unemployment, but were applied as instruments to measure whether those concerned were willing and able to work or not. They came to assume the character of forced labour. Those among the unemployed who did not pass these various tests were officially stamped as 'anti-social' and were removed from the unemployment welfare system. Thus laundered, the unemployment figure in Hamburg fell to a mere 10,000 by early 1939. Finally, the Decree on Unemployment Assistance of 5 September 1939 passed the support of all those unemployed judged capable of work to the Reichsanstalt. Henceforth, the local welfare authorities remained responsible only for supporting those incapable of working. This meant the abandonment of the existing three-tiered structure of unemployment insurance.[65] At the same time, these measures of selection and compulsion effectively destroyed the basic principle of insurance in the unemployment support system, a principle which had been one of the major achievements of the Weimar Republic in the field of social policy.[66]

Notes

This chapter was translated by Richard J. Evans.

1. Schulthess, *Europäischer Geschichtskalender*, N.F., Jahrgang 1933, p. 36.
2. Gesetz zur Verminderung der Arbeitslosigkeit, 1.6.1933, *RGBl* I, 1933, pp. 323–5. The law had five sections, on job creation, tax exemption for substitute jobs, voluntary contributions to encourage national work, the return of women to the home, and encouragement of marriages. The first section was often called the Reinhardt Programme after the main author of the section, the State Secretary in the Reich Finance Ministry, Fritz Reinhardt.
3. Deutsche Gesellschaft für öffentliche Arbeiten to Reichsfinanzministerium, 6.11.1933, in BA Koblenz R2/18656.
4. *RGBl* I, 1933, pp. 11–13.; see also BA R2/18701.
5. Cf. note 2.
6. Karl Schiller, *Arbeitsbeschaffung und Finanzordnung in Deutschland* (Berlin, 1936), pp. 59ff.
7. *RABl*, Teil II, 1934, p. 185.
8. Mitteilungen der Deputation für Handel, Schiffahrt und Gewerbe, Abt. II C, 3.1.1933, in Staatsarchiv Hamburg (StAH) Staatliche Pressestelle I-IV 7208.
9. Ursula Büttner, 'Hamburg in der Grossen Depression. Wirtschaftsentwicklung und Finanzpolitik 1928–1931' (Phil.Diss., Hamburg, 1979), p. 28.
10. *Hamburger Fremdenblatt* no. 32a, 1 Feb. 1933.
11. Mitteilungen des Handelsstatistischen Amtes vom 10.2.1939, in StAH Deputation für Handel, Schiffahrt und Gewerbe, Spezialakte XXI A 18 83.
12. These unemployment figures include only those unemployed persons who were in receipt of benefits from the labour office, from the welfare authorities or from other official bodies. Those who received no benefits are not included. The number of those seeking work, including those already in employment but seeking other jobs, progressively declined during the Depression, since more and more people gave up going to the job centres as a waste of time. These 'invisible' unemployed disappeared from the statistics altogether. See *SJDR*, 1934, p. 309; and Volks- und Berufszählung 16.6.1933, Nachtrag zum *SJH* 1933/34. See also Büttner, 'Hamburg', p. 349.
13. *Nordwestdeutsche Wohlfahrtsstatistik für das Rechnungsjahr 1932/33.* (Veröffentlichung der Vereinigten Nordwestdeutschen Wohlfahrtsämter, H.3), p. 4–5.

14. The Law on Labour Exchanges and Unemployment Insurance of 16 July 1927 (*RGBl* 1, p. 187), which was repeatedly amended and extended in subsequent years, created a threefold division of unemployment benefit provisions in Germany. Initially, an unemployed person was supported for 26 weeks by the unemployment insurance scheme of the *Reichsanstalt für Arbeitsvermittlung* und Arbeitslosenversicherung. Benefits were automatic only for the first 6 weeks. After this, the claimant had to prove eligibility (i.e. he or she needed the benefit in order to survive or to maintain a family). Proof of eligibility was also required for the second period of 26 weeks support, this time on the basis of 'crisis benefits'. After these 52 weeks had passed, payment of benefits by the *Reichsanstalt* ceased and the claimant was dependent on welfare from the local authority.

15. *SJH* 1933/34, pp. 154–5.

16. *Hamburgischer Correspondent* no. 418, 7 Sep. 1933.

17. Nachrichten des Lohnkontors, 28.10.1934, in StAH, Blohm & Voss, Kasten 608, S11.

18. The figures refer to the wage class 1 for workers without children in places where there was a special class and a local Class A. An earnings unit of 10 RM formed the basis of calculation for unemployment insurance in Wage Class 1. See *RGBl* I, 1932, pp. 305ff. For welfare benefits for married couples, see Mitteilungen des Deutschen Städtetages vom 25.10.1932 in StAH, Sozialbehörde I, Stat. 20.40; Rundschreiben no. 17, ibid., Arbeitsbehörde I, 132; Zusatzunterstützung für Arbeitslosen- und Krisenunterstützungsempfänger 1933–41, in ibid., Sozialbehörde I, AF 20.28 und AF 20.13.

19. A maximal rent element of 40 RM is included in the calculation: Richtsätze der Hansestadt Hamburg, Bd.1, 1933–42, in StAH, Sozialbehörde I, AF 20.13, Entwurf über die Bemessung der Unterstützung der Hilfsbedürftigen in der Arbeitslosenunterstützung und in der Krisenfürsorge, in ibid., Sozialbehörde I, Gen. IIIA I 1; *SJH* 1933/34, p. 10, Tab. 9.

20. StAH, Medizinalkollegium I K 27a, Bd.V.

21. StAH, Staatshaushaltsabrechnung über das Rechnungsjahr 1932, Ende März 1934, pp. I and II.

22. SJH 1933/34, p. 160; 1934/35, p. 172.

23. Staatsamt für auswärtige Angelegenheiten to Reichskommissar für Arbeitsbeschaffung, Berlin, 24.1.1933, in StAH, Senatskanzlei-Präsidialabteilung 1932 A 50.

24. Cf. note 22. The money for public utilities is included in these contributions, in so far as they had applied for support themselves.

25. Cf. StAH, Hamburgische Beleihungskasse für Hypotheken, Tätigkeits-berichte 1932/33, 1933/34.

26. StAH, Finanzdeputation IV, DV V B 6b IIA 15a.

27. Vertretung Hamburgs in Berlin to Staatsamt, 16 Mar. 1934 in StAH, Staatsamt 139.

28. The other institutions included the Deutsche Rentenbank-Kreditanstalt, the Deutsche Siedlungsbank and the Deutsche Bau- und Bodenbank AG.

29. Notiz der Baubehörde 11.3.1933, in StAH, Staatliche Pressestelle I–IV 3332, Bd.III.

30. *Aus Hamburgs Verwaltung und Wirtschaft 1934*, pp. 1–12, Tab. 5; Stand der Arbeitslosigkeit 1933–1937, in StAH, Staatsamt 90, Bd. 1–2.

31. StAH, Sozialbehörde I, Stat 20.13, Bd.II, Staatsamt 90, Bd. 1–2.

32. *SJDR* 1935, pp. 509–510.

33. Durchführungsbestimmung zur Wohlfahrtshilfeverordnung; der Deutschen Gemeindetag an seine Mitglieder vom 29.1.1934, in StAH Sozialbehörde I, Stat. 20.13, Bd. II.

34. *SJH* 1933/34, pp. 154–5; 1934/35, pp. 166–7; Monatliche Übersicht über die Arbeitsvermittlung in Hamburg 1934, in StAH, Arbeitsbehörde I, 35.

35. Handelskammer Hamburg an die hiesigen kaufmännischen und industriellen Vereinigungen, 23.10.1933 in StAH Blohm & Voss, Kasten 585, B/H la XI.

36. *Hamburger Nachrichten* 4 Aug. 1933. From 1932–3 the number of marriages in Hamburg increased from 10,700 to 13,800 (StAH, Staatliche Pressestelle, I–IV 7497 and 3335).

37. This also included relief work, undertaken as part of the work-creation programmes and the work that was financed by the Reichsanstalt from unemployment insurance funds.

38. Entwicklung und Lage des Hamburger Arbeitsmarktes, 8.2.1935, in StAH, Staatliche Pressestelle I–IV 7494; *SJDR* 1935, pp. 317 and 319; *SJH* 1933/34, pp. 156 and 166; 1934/35, pp. 180 and 186.

39. Finanzdeputation to Innere Verwaltung vom 23.9.1935, in StAH, Finanzdeputation DV V B 6b II A 24a. According to this there were 3,727 Hamburg workers employed on outside schemes.

40. Zahlen und Schicksale. Das Landesarbeitsamt Nordmark, 21.11.1935 in StAH, Staatliche Pressestelle I–IV 7494.

41. Aufzeichnungen der Fürsorgebehörde, 13.7.1937, in StAH, Sozialbehörde I, AW 40.16.

42. The measures were implemented on the basis of the *Gesetz zur Regelung des Arbeitseinsatzes*, 15 May 1934; cf. StAH, Staatliche Pressestelle I–IV 3329.

43. *Aus Hamburgs Verwaltung und Wirtschaft*, 1933, 1–12, Tab. 5; ibid., 1934, 1–12, Tab. 5; Monatliche Übersicht über die Arbeitsvermittlung, in StAH, Arbeitsbehörde I 35.

44. Cf. note 43.

45. The 'work comradeships' were set up at the beginning of 1934. They contained young people up to 25 years of age. The 'comradeships' served as preparation for labour service. The members were put up in camps, took lessons together, engaged in sporting activities and, like the members of the labour service, were mainly used in heavy manual labour on earthworks. 10,433 young people went through the 'comradeships' between 1 Jan. 1934 and 25 May 1935. See StAH, Staatsamt 54, Bd.1; Hamburg im Dritten Reich, Heft 6, Hamburgs Fürsorgewesen im Kampf gegen die Arbeitslosigkeit, Hamburg 1935.

46. Präsident des Landesarbeitsamtes Nordmark to the Regierender Bürgermeister vom 24.4. 1934, in StAH Staatsamt 90, Bd. 1; Fürsorgewesen to leitender Regierungsdirektor Dr Werdemann, Hamburgische Finanzverwaltung, vom 27.12.1933 in StAH, Finanzdeputation IV, DV V B 6b IIIB9a II, Teil.

47. Jahresbericht der Hilfsstelle Hafen 1935/36 unter besonderer Berücksichtigung der Entstehung der Dienststelle, in StAH Sozialbehörde I, VG 54.36.

48. *RGBl* I 1933, pp. 323–5. This did not apply to civil-engineering workers, according to Abschnitt I, Art. 1, Ziff, 7 of the Reinhardt Programme they obtained either their principal benefit from the labour office or the benefits payable by the welfare office, plus a token of 25 RM, to cover living expenses for four weeks at a time.

49. The figures give the rough average weekly wage of an unskilled worker. The wages are difficult to calculate because of the different calculations of additional elements in the different trades, for example piecework extras of between 0.60 and 0.80 RM in some trades and in a few cases also family extras of 0.01 marks per hour and person. See StAH, Sozialbehörde I, AF 20.13.

50. *SJH* 1933/34, p. 145, Tab. 9: Aus Hamburgs Verwaltung und Wirtschaft 1934, 1–12, Tab. 17; Jan. 1933, total cost of living for a family of five 153.61 RM — Dec. 1934 163.56 RM. In autumn 1934 the 'shopping basket' used as the basis of calculation since 1925 was enlarged. According to the new calculations the cost of living in Dec. 1934 was 178.07 RM.

51. Michael Wolffsohn, *Industrie und Handwerk im Konflikt mit staatlicher*

Wirtschaftspolitik? Studien zur Politik der Arbeitsbeschaffung in Deutschland 1930–1934 (Berlin, 1977), p. 50.

52. On 1 Oct. 1933 the basis of calculation for the Reich welfare assistance was altered. The local authorities did not have to contribute to the Reich 'crisis welfare' any more, but their contribution to the costs of welfare in general was fixed, so that the fall in welfare costs only benefited the Reich. The system was altered again in May 1934, reducing Reich welfare assistance once more. See Rechnungshof des Hamburger Staates vom 29.9.1933; Rundverfügung des Reichsministers der Finanzen vom 9.5.1934, in StAH, Finanzdeputation IV, V + O I D 11b IX B 2, 3.Akte.

53. StAH, Verwaltung für Wirtschaft, Technik und Arbeit I, 22, Bd. 1.

54. Gutachten des Reichssparkommissars, n.d., in StAH, Sozialbehörde I, AF 20.13, Bd. 1.

55. Reichsstatthalter in Hamburg to Reichsminister der Finanzen, 27.8.1934, in StAH, Verwaltung für Wirtschaft, Technik und Arbeit I, 22, Bd. 1.

56. In the Referendum on the Head of State of the German Reich, there were 20.4 per cent 'no' votes and 2.5 per cent spoilt ballots in Hamburg. A later examination of the results by the welfare office revealed that the percentage of 'no' votes was highest in the traditional working-class districts. See StAH, Sozialbehörde I, VG 30.70.

57. Henning Timpke, *Dokumente zur Gleichschaltung des Landes Hamburg 1933* (Frankfurt am Main, 1964), pp. 283 and 287.

58. 'Industrieförderung in Hamburg', Denkschrift der Handelskammer vom August 1934, in StAH, Deputation für Handel, Schiffahrt und Gewerbe II — Spezialakten — XXXIII A 1.

59. Vertretung Hamburgs in Berlin to Staatsamt, 11 Jan. 1934, in StAH, Staatsamt 94, Bd. 1: Aktenvermerk über eine Besprechung im Reichswirtschaftsministerium, n.d., in StAH, Staatsamt 96.

60. Vertretung Hamburgs in Berlin to Staatsamt, 29 May 1934, in StAH, Staatsamt 94, Bd. 1.

61. Hamburg's duty-free harbour was deprived of its function by an order of the Reich government that imports had to be accompanied by a customs receipt. This caused no difficulties for the industries that imported raw materials, but merchants used the duty-free harbour to store goods, to sell imported goods or for re-export. These businesses did not count as imports as far as the customs were concerned, so that importers could not get approval for foreign currency. See Vertretung Hamburgs in Berlin to Staatsamt, 24 Mar. 1934, in StAH, Staatsamt 93.

62. Bürgermeister Krogmann to Reichsminister der Finanzen, 26 Sep. 1933, in StAH, Senatskanzlei — Präsidialabteilung 1934 A 9/2.

63. Besprechung beim Führer am 2.11.1934, in StAH, Staatsamt 91. Also for the following account. The term 'discussion with the Leader' is the usual way of referring to this special session of the Reich cabinet in the Senate files.

64. The participation of Hamburg in the special programme amounted in 1935 to 4 million RM from the budget of the *Reichsanstalt*. In addition, Hamburg obtained about 6 million RM financed by the Society for Public Works and loans and grants from the Reichsanstalt, for harbour construction. In 1936 Hamburg's participation was 3 million RM and in 1937 2.5 million RM. In addition, the Senate won the approval of the Reich for raising loans on the free capital market, which made an additional 18 million RM for work creation in 1936. Cf. StAH, Finanzdeputation DV V B 6b III C 38, DV V B 6b III C39, DV V B 6b III C40.

65. StAH, Sozialbehörde I, AW 50.12.

66. The insurance principle had in fact already been breached by presidential decrees issued in the last phase of the Weimar Republic, but this law can certainly be seen as the final point of this process.

NOTES ON CONTRIBUTORS

Richard Bessel was born in Springfield, Massachusetts, in 1948 and studied at Antioch College from 1966 to 1970. From 1971 to 1973 he worked for *The Current Digest of the Soviet Press*, and from 1973 to 1977 he studied at St Peter's College, Oxford, taking his DPhil. in 1980. From 1977 to 1979 he was Parkes Library Fellow at the University of Southampton, and since 1979 he has been Lecturer in History at the Open University. He is co-editor (with E. J. Feuchtwanger) of *Social Change and Political Development in the Weimar Republic* (Croom Helm, 1981) and author of *Political Violence and the Rise of Nazism* (1984) and *Reconstituting German Society: Social Processes of Demobilisation after the First World War* (1986).

Richard J. Evans was born in Woodford, Essex in 1947 and studied Modern History at Jesus College, Oxford, graduating in 1969. From 1969 to 1972 he studied at St Antony's College, Oxford, holding a Hanseatic Scholarship from the FVS Foundation in Hamburg and Berlin from 1970 to 1972 and gaining his doctorate in 1972. From 1972 to 1976 he taught at Stirling University, and since 1976 he has been on the faculty of the University of East Anglia, Norwich, becoming Professor of European History in 1983. His latest book, *Death in Hamburg: Society and Politics in the Cholera Years 1830–1910*, is to be published in 1987.

Dick Geary was born in Leicester in 1945 and studied History at King's College, Cambridge, from 1964 to 1970, gaining his doctorate in 1971. From 1970 to 1973 he was a Research Fellow of Emmanuel College, Cambridge, and from 1973 Lecturer at the University of Lancaster, where he is currently Senior Lecturer in German Studies. He is the author of *European Labour Protest 1848–1939* (Croom Helm, 1981) and of numerous articles on Marxist theory, the German labour movement, and unemployment in the Weimar Republic. His latest book, *Revolution and the German Working Class 1848–1933*, is due to be published in 1987.

Elizabeth Harvey was born in Norwich in 1957 and studied History and German at Wadham College, Oxford, graduating in 1981. From 1981 to 1985 she studied at St Antony's College, Oxford, holding a Hanseatic Scholarship of the FVS Foundation in Hamburg from 1982 to 1984. Since 1985 she has been Lecturer in German at the University of Dundee. Her publications include a study of youth welfare and the Left in Berlin in the 1920s. She is currently completing her doctorate on state policies towards working-class youth in the Weimar Republic.

Heidrun Homburg was born in 1948 and studied History at the Universities of Rennes, Freiburg and Bochum. In 1974–5 she was a research assistant at the University of California, Berkeley. Her publications include *Industrie und Inflation*, a collection of documents on German industrialists from 1916 to 1923 edited jointly with Gerald D. Feldman, and a number of articles on the electrical industry during the inflation, the 'Taylor system' of management, and related subjects. She has been on the staff of the Department of History at the University of Bielefeld since 1976, and is working on problems of scientific management, labour market politics and business history in modern Germany.

Helgard Kramer was born in 1947 and studied Philosophy and Sociology at the Universities of Heidelberg, Berlin and Frankfurt. Since 1971 she has been a member of the Institute for Social Research in Frankfurt, where she currently holds the post of Research Sociologist. In 1983–4 she was a Fellow of the American Council of Learned Societies at the University of California, Berkeley. Her books include a collection of edited texts from the women's liberation movement in the USA and a study of women's work in factory and family. She is co-editor of *Feministische Studien* and co-author of *Grenzen der Frauenlohnarbeit*, a book on women's work since 1900, published in 1986.

Anthony McElligott was born in Slough in 1955 and studied at the Universities of Essex, Hamburg and Manchester. The author of various articles on the labour movement in Altona, he has also taught at the Universities of Hamburg and Salford. He is now Lecturer in History at Humberside Polytechnic, and is completing a PhD dissertation on Communal Politics and Conflict in Altona during the Weimar Republic.

Merith Niehuss was born in 1954 in Bielefeld and studied at the Universities of Regensburg and Munich, graduating with an MA in History in 1978, a Diploma in Sociology in 1979 and a PhD in 1982. Since 1983 she has taught History at the Institute for Modern History at the University of Munich. She is co-author (with Gerhard A. Ritter) of the *Wahlgeschichtliches Arbeitsbuch* (2 vols, 1980, 1986). Her book *Arbeiterschaft in Krieg und Inflation. Soziale Schichtung und Lage der Arbeiter in Augsburg und Linz 1910–1925* was published in 1985. She is currently preparing a study of the social history of the family in West Germany since 1945.

Detlev Peukert was born in 1950 and studied History, Philosophy and German at the Ruhr University, Bochum, graduating in 1975, gaining his doctorate in 1979 and his Dr habil. in 1984. He currently teaches History at the University of Essen. His books include *Die KPD im Widerstand* (Wuppertal, 1980) and *Grenzen der Sozialdisziplinierung. Aufstieg und Krise der deutschen Jugendfürsorge 1878–1932* (Cologne, 1986). His book *Inside Nazi Germany. A Social History of Daily Life, People's Opposition and Racism* also appeared in 1986. His current work includes studies of dictatorship and trade unionism in Latin America.

Eve Rosenhaft was born in New York City in 1951. She gained her BA from McGill University in 1969 and her doctorate from the University of Cambridge in 1979. During 1975–6 she was a fellow of the Institut für Europäische Geschichte, Mainz, and from 1978 to 1981 she held a research fellowship at King's College, Cambridge. She is the author of *Beating the Fascists? The German Communists and Political Violence 1929–1933* (1983). Her latest book, *The German Communist Party. A Social and Political History 1917–1936* is due to appear in 1987. Since 1981 she has been Lecturer in German at the University of Liverpool.

Birgit Wulff was born in 1952 and studied at the University of Hamburg, graduating in 1981. From 1981 to 1984 she was on the staff of Hamburg University as a *wissenschaftliche Mitarbeiterin*. Her publications include contributions to a dictionary of the Third Reich (1986) and an article in *Arbeiter in Hamburg* (1983), edited by Arno Herzig, Dieter Langewiesche and Arnold Sywottek. She completed a dissertation on work creation and unemployment in Hamburg 1933–9, in 1986.

INDEX

306